The Siberian Sturgeon (*Acipenser baerii,* Brandt, 1869) Volume 1 - Biology

Patrick Williot • Guy Nonnotte
Denise Vizziano-Cantonnet
Mikhail Chebanov
Editors

The Siberian Sturgeon (*Acipenser baerii*, Brandt, 1869) Volume 1 - Biology

Editors
Patrick Williot
Audenge
France

Guy Nonnotte
La Teste de Buch
France

Denise Vizziano-Cantonnet
Facultad de Ciencias
Universidad de la República Oriental
del Uruguay
Montevideo
Uruguay

Mikhail Chebanov
Kuban State University
Krasnodar, Krasnodar Territory
Russia

ISBN 978-3-319-61662-9   ISBN 978-3-319-61664-3   (eBook)
https://doi.org/10.1007/978-3-319-61664-3

Library of Congress Control Number: 2017959368

© Springer International Publishing AG, part of Springer Nature 2018
This work is subject to copyright. All rights are reserved by the Publisher, whether the whole or part of the material is concerned, specifically the rights of translation, reprinting, reuse of illustrations, recitation, broadcasting, reproduction on microfilms or in any other physical way, and transmission or information storage and retrieval, electronic adaptation, computer software, or by similar or dissimilar methodology now known or hereafter developed.
The use of general descriptive names, registered names, trademarks, service marks, etc. in this publication does not imply, even in the absence of a specific statement, that such names are exempt from the relevant protective laws and regulations and therefore free for general use.
The publisher, the authors and the editors are safe to assume that the advice and information in this book are believed to be true and accurate at the date of publication. Neither the publisher nor the authors or the editors give a warranty, express or implied, with respect to the material contained herein or for any errors or omissions that may have been made. The publisher remains neutral with regard to jurisdictional claims in published maps and institutional affiliations.

Printed on acid-free paper

This Springer imprint is published by the registered company Springer International Publishing AG part of Springer Nature
The registered company address is: Gewerbestrasse 11, 6330 Cham, Switzerland

Foreword

This book on the Siberian sturgeon (*Acipenser baerii*) presented by my friend Patrick Williot is both a body of scientific data for almost 40 years on this species and its breeding and exemplary history of the development of a new species in aquaculture. But how has this "saga" started and how has Patrick Williot been able to devote his entire life of being a scientist to this species?

The story was born in the region that remained the last sanctuary of the European sturgeon (*Acipenser sturio*), Aquitaine, and its rivers and estuaries. In the 1970s, it seemed that this species was doomed to disappear if no protective measures were taken. This then led the CEMAGREF (old Irstea) team, of which at this time I had the responsibility, to launch with the support of national and regional authorities, as well as professionals, a research program on the biology of this species. This program would probably not have achieved its current development if a fortuitous event had occurred. In 1975, I was called by the deputy director of the Aquaculture Department of CNEXO (ancestor of IFREMER) Pierre Rouzaud, who asked me if we were interested in hosting 300 young Siberian sturgeon (300 g each) that CNEXO was to receive from its Soviet counterpart, VNIRO, in exchange for sea bass fry from the Palavas station. My response was immediate and positive, while we had no program for this operation!

We did have some advantages: an experimental breeding facility with warm waters from a thermal power station in Ambés, a young and motivated team in aquaculture, and an early knowledge of this family of fish through the work on local species. The reception of these early fingerlings was made at the Roissy Airport (where I rode in a van), with an immediate transfer with a transport container to our station. Then it was the start of a first work program on the diet of this species, for which we had no data. Collective work between CEMAGREF, INRA, and the subsidiary factory of Grands Moulins de Paris specializing in fish feed allowed to move quickly on this issue and even publish the first scientific paper on the subject [1]

But everything remained to be known and learned to get the technical mastery of all the operations necessary to develop the breeding of this species. Indeed, until then, only the Soviet team had been working on it, but only on the reproduction and

---

[1] Barrucand M., Ferlin P., Lamarque P., Sabaut J.J., 1979. Alimentation artificielle de l'esturgeon *Acipenser baeri*. Proc. World Symposium on Finfish Nutrition and Fish feed Technology, Hambourg, 20-23 juin 1978, Vol. 1, 411-421.

larval rearing, aiming on restocking of fresh and estuarine waters. Jointly conducted missions by CEMAGREF (P. Ferlin) and INRA (P. Lamarque) in the 1970s in USSR and Iran showed that the larval rearing system (crucial phase for aquaculture) was absolutely not adapted to Western conditions. To feed the larvae and fingerlings of sturgeon, they first had to make a mashed cabbage (sort of "borsch") and then give it as food to small red worms (chironomids), which were recovered by hand in trays with light above (that they do not like!) and heating below (which attracts and concentrates them). On the other hand, in countries rich in native populations, obtaining eggs was done by sacrificing the females, while with our small population imported, we had to eventually reuse for other female pundits, developing a cesarean technique[2]. Many other and fundamental aspects for breeding also remained to be studied in terms of reproductive physiology, nutrition, diseases, etc.

This work started through an agreement between CEMAGREF and INRA, at the Donzacq station of INRA, but it was very soon necessary to create new facilities, such as the hatchery in Saint Seurin sur L'Isle, strengthen scientific laboratories, and train researchers and technicians, but above all find a leader to initiate and manage these actions. It is Patrick Williot who volunteered to take the lead in this ambitious research program that would be subject to his life of being a scientist.

His work had three consequences:

- An amount of scientific and technical knowledge on this species, subject of this book
- The development of a new aquaculture industry in France and abroad, creating jobs and producing high-quality products
- The ability to protect wild sturgeons, especially in Aquitaine, not only by controlling their capture but also by developing methods of breeding and larval rearing issued from the work of the team of Patrick

So I would like, as a prelude to this book, to pay tribute to the tenacious and efficient work of Patrick Williot and thank him for having contributed to the development of this sector and to the protection of these species that are so emblematic, the sturgeons. I also hope that this book will induce new vocations and develop new research in this area that has always fascinated me.

Philippe Ferlin
Former Chairman of OECD Committe of Fisheries
General Inspector of Agriculture and Fisheries (Hon.)
Boulogne-Billancourt, France

---

[2] The editor in chief of this two volumes (P. Williot) has been primarily trained in the former USSR in the 1970s. A publication (Charlon et Williot 1978) provided, among others, the different ways of rearing the live preys for sturgeon larvae (oligochete, dapnia, artemia) as well as basic surgery. In the early 1980s, ceasarian has been industrialized simultaneously in France and USA. Later on, obtaining of ovulated eggs without ceasarian has been developed in Russia (Podushka 1986, see also Chap. 27 in this issue).

Contents

Part I Biology and Ecology

1 Geographical Distribution, Ecological and Biological Characteristics of the Siberian Sturgeon Species 3
Georgii Igorevich Ruban

2 What Makes the Difference Between the Siberian Sturgeon and the Ponto-Caspian Sturgeon Species? 29
Georgii Igorevich Ruban

3 Anatomic Description 41
Mikhail Chebanov and Elena Galich

4 The Axial Skeleton of the Siberian Sturgeon. Development, Organization, Structure and New Insights on Mineralization and Ossification of Vertebral Elements. 53
Jean-Yves Sire and Amandine Leprévost

5 Evolution of Molecular Investigations on Sturgeon Sex Determination and Most Recent Developments in DNA Methylation with a Focus on the Siberian Sturgeon 71
Rémy Simide and Sandrine Gaillard

6 Sex Determination and Differentiation of the Siberian Sturgeon . . . 93
Denise Vizziano-Cantonnet, Santiago Di Landro, and André Lasalle

7 Analysis of Transposable Elements Expressed in the Gonads of the Siberian Sturgeon. 115
Frédéric Brunet, Alexia Roche, Domitille Chalopin, Magali Naville, Christophe Klopp, Denise Vizziano-Cantonnet, and Jean-Nicolas Volff

8 Early Ontogeny in the Siberian Sturgeon. 131
Enric Gisbert and Yoon Kwon Nam

9 Behaviour of Early Life Stages in the Siberian Sturgeon 159
Enric Gisbert and Mikhail Solovyev

viii   Contents

10 Olfaction and Gustation in *Acipenseridae*, with Special References to the Siberian Sturgeon 173
Alexander Kasumyan

11 Nutritional Requirements of the Siberian Sturgeon: An Updated Synthesis 207
Bahram Falahatkar

12 Swimming Characteristics of the Siberian Sturgeon 229
Ming Duan, Yi Qu, and Ping Zhuang

Part II   Biology and Physiology of Reproduction

13 Chemical Neuroanatomy of the Hypothalamo-Hypophyseal System in Sturgeons 249
Olivier Kah and Fátima Adrio

14 An Updated Version of Histological and Ultrastructural Studies of Oogenesis in the Siberian Sturgeon *Acipenser baerii* 279
Françoise Le Menn, Catherine Benneteau-Pelissero, and René Le Menn

15 Sperm and Spermatozoa Characteristics in the Siberian Sturgeon 307
Martin Pšenička and Andrzej Ciereszko

16 Gonadal Steroids: Synthesis, Plasmatic Levels and Biological Activities in Sturgeons 327
Denise Vizziano-Cantonnet

17 Steroid Profiles Throughout the Hormonal Stimulation in Females Siberian Sturgeon *Acipenser baerii* Brandt 351
Patrick Williot, Sylvain Comte, Françoise Le Menn, and Blandine Davail-Cuisset

Part III   Ecophysiology: Adaptation to Environment

18 Respiratory and Circulatory Responses to Hypoxia in the Sturgeon, *Acipenser baerii* 369
Guy Nonnotte, Patrick Williot, and Valérie Maxime

19 Effects of Exposure to Ammonia in Water: Determination of the Sublethal and Lethal Levels in Siberian Sturgeon, *Acipenser baerii* 391
Guy Nonnotte, Dominique Salin, and Patrick Williot

Contents

20 Consequences of High Levels of Ammonia Exposure on the Gills Epithelium and on the Haematological Characteristics of the Blood of the Siberian Sturgeon, *Acipenser baerii* 405
Guy Nonnotte, Dominique Salin, Patrick Williot, Karine Pichavant-Rafini, Michel Rafini, and Liliane Nonnotte

21 Acid-Base Balance and Ammonia Loading in the Siberian Sturgeon *Acipenser baerii*, **Exposed to High Concentrations of Ammonia** 425
Guy Nonnotte, Dominique Salin, Patrick Williot, Karine Pichavant-Rafini, Michel Rafini, and Liliane Nonnotte

22 Effects of Exposure to Ammonia on Plasma, Brain and Muscle Concentrations of Amino Acids and Adenyl Nucleotides in the Siberian Sturgeon, Acipenser baerii 435
Guy Nonnotte, Dominique Salin, Patrick Williot, Karine Pichavant-Rafini, Michel Rafini, and Liliane Nonnotte

23 The Importance of Water Quality in Siberian Sturgeon Farming: Nitrite Toxicity. 449
Enric Gisbert

Part IV Specific Methods

24 Cannulation in the Cultured Siberian Sturgeon, *Acipenser baerii* **Brandt** 465
Patrick Williot and Françoise Le Menn

25 Some Basic Methods in Respiratory Physiology Studies Applied in the Siberian Sturgeon. 475
Guy Nonnotte, Patrick Williot, Karine Pichavant-Rafini, Michel Rafini, and Liliane Nonnotte

Conclusions of the Volume 1: Recent Progress in Biology of *Acipenser baerii*, **an Exciting Subject for Further Academic Research and Development of Aquaculture** 491

About the Editors

Denise Vizziano Cantonnet Denise Vizziano Cantonnet is a professor of the Facultad de Ciencias at the Universidad de la República Oriental del Uruguay (UdelaR) and at the postgraduate program Programa de Desarrollo de las Ciencias Básicas (PEDECIBA). She is researcher of the Government System of National Researcher (ANII) and head of the Department of Animal Biology (UdelaR). She is doctor of the University of Rennes I (U-Rennes I, France) and made a postdoctoral stay at the INRA of Rennes (France). She worked as invited professor at U-Rennes I. She conducted original research on reproductive endocrinology of fish working in different species of commercial importance as trouts, silverside, and croakers and developed for 20 years a training course on reproductive physiology of fish for Latino American students promoting the development of this new area of research in Latin America. She was invited as professor at different universities of Latin America (Argentina, Brazil, Chile, Colombia, Peru) interested to develop endocrinology of reproduction of native species key for fish farming. She started to work in Siberian sturgeons when the producers contacted the university to reproduce the species in Uruguay. The first reproduction in captivity of the Southern Hemisphere has been done in Uruguay together with Dr. Patrick Williot by applying the methodology developed by this researcher in the CEMAGREF (France). She started the first worldwide investigations in the molecular basis of sex differentiation in sturgeons and built a complete gonadal transcriptome of Siberian sturgeons in order to search the master gene for sex determination and genes involved in sex differentiation. In cooperation with specialists on fish evolution (Dr. Jean Nicolas Volff and Dr. Frederic Brunet, University of Lyon), a first study on retrotransposons has been made using the gonadal transcriptome. A multi-tissue transcriptome is recently built for Siberian sturgeons in cooperation with Ing. Christophe Klopp (INRA, Toulouse), and the team of Dr. Vizziano is now searching for a genetic sex marker that will improve the fundamental research and will have an impact in sturgeon production.

Mikhail Chebanov Professor Mikhail Chebanov, Dr.Sc., is professor of the Department of Aquatic Bioresources and Aquaculture of Kuban State University and director of the State Regional Centre for Sturgeon Gene Pool Conservation "Kubanbioresursi," Ministry of Natural Resources (Krasnodar, Russia). He has developed a technology of the all-year-round reproduction of the different sturgeon species as well as the selection and breeding program and the formation of the largest sturgeon living gene bank. He has developed and widely implemented the method of ultrasonic diagnostics of sturgeon for optimization of broodstock management. In 2004, he was awarded the Prize of the Government of Russia in the field of science and technology for the "development and implementation of technology of control reproduction and commercial rearing of sturgeon." For many years Chebanov served as director of the South Branch of the Federal Center of Genetics and Selection for Aquaculture, consultant for FAO and convener of ad hoc meetings of Working Party on the Management of Sturgeons of the European Inland Fisheries Advisory Commission (EIFAC FAO), and deputy chair and member of the Sturgeon Specialist Group of the International Union for Conservation of Nature (IUCN). Professor Chebanov is cofounder and member of the board of directors of the World Sturgeon Conservation Society (www.wscs.info) and has chaired numerous workshops and conference sessions, including several international symposia on sturgeons (1993–2017). He is the author of more than 170 scientific papers, including 12 books.

Guy Nonnotte Doctor of the University of Strasbourg, Guy Nonnotte was appointed research officer at the CNRS in the Laboratory of Comparative Physiology of the Regulations at Strasbourg in 1974. He conducted original research especially on cutaneous respiration and skin ionic exchanges on many models of freshwater and seawater fish. In 1986, he joined the Laboratory of Neurobiology and Compared Physiology (CNRS and University of Bordeaux I, Arcachon) and pursued research on extracellular acid-base balance and cell volume regulation in fish exposed to environmental changes and pollutants with a specific interest for the model "Siberian sturgeon." He supervised a thesis on the toxicity of ammonia for the Siberian sturgeon in collaboration with Dr. Patrick Williot. He took also the opportunity to initiate an important collaboration concerning the respiration, the acid-base balance, and the physiological effects of environmental stress for the Siberian sturgeon with the Laboratory of Animal Physiology, Brest. In 1994, he was appointed professor of animal physiology (fish physiology) at the University of Brest (France) and managed the Laboratory of Cellular Biology and Physiology. He wrote numerous publications in international reviews and managed several thesis on fish physiology. He was also appointed for numerous evaluations and jury of thesis by INRA, IFREMER, INSERM, and CNRS. In 2002, he was designated vice president of the University of Brest, responsible of the research, and in 2007, emeritus professor of the university.

Patrick Williot Patrick spent the three quarters of his professional career primarily to take the challenge of the disappearing of the European sturgeon, *Acipenser sturio*.

To achieve this task, he mobilized all means of research (both applied and fundamental lines, biological model, national and international cooperation) and management (private-public partnership, transfer of know-how, installation of a new experimental facility, search for regular financial support). Patrick has developed his activity around the couple (aquaculture-conservation biology). Patrick stimulated the development of sturgeon farming primarily in France based on the biological model, the Siberian sturgeon, *Acipenser baerii*. He has been the kingpin of the First International Symposium on Sturgeons held in Bordeaux in 1989 where he launched the concept of the international association for the conservation of sturgeons which was installed (the WSCS; wscs.info) some year later. Later on, he edited the peer-reviewed proceedings (*Acipenser*), served as scientific committee member of further international symposia on sturgeons as well as the Sturgeon Specialist Group of the IUCN, and organized other symposia and workshops.

He succeeded in obtaining for the first time the controlled reproduction of farmed European sturgeon in 2007 which avoided the species from a complete disappearing and opened the door for a restoration. Patrick initiated and carried out the edition of a book on the biology and conservation of the European sturgeon (2011) and on the biology and farming of the Siberian sturgeon which is in process. Patrick published about 90 papers in peer-reviewed journals or books.

General Introduction to the Siberian Sturgeon Books with a Focus on Volume 1 Dedicated to the Biology of the Species

The Siberian sturgeon, *Acipenser baerii* Brandt, 1869, is a native species of the Siberian catchments of the Russian Federation. The wild populations of the species dramatically declined from the twentieth century onward. Overfishing, erection of dam, industrialization that resulted in pollutions, and poor management of fisheries are the most likely reasons for the deteriorations (Akimova and Ruban 1999; Ruban 2005; Williot et al. 2002).

In the mid-1970s and in the early 1980s, young specimen of Siberian sturgeon arrived in France as a biological model thanks to French-Soviet cooperation network. The origin of the arrival of the first batch in 1975 is described by Ferlin (2017) in the preface of the books, and the conditions for the development of farming of the species (including the arrival of the second batch in 1982) are described by Williot et al. (2017) in the introduction of volume 2 dedicated to the farming of the species. Initially, the objectives for the French part were to acquire experience on sturgeons and to set up methods useful for the restoration-conservation program of the critically endangered European sturgeon species, *Acipenser sturio* (Williot and Rouault 1982; Williot and Castelnaud 2011; Williot et al. 1997, 2004). Partly due to its freshwater status together with its plasticity, the species has been spread rapidly worldwide for production purposes. As a result, the species supported many investigations in different fields. Obviously, the peculiar phylogenetic position of the species as well as its availability thanks to reproduction improvements adds to its attractiveness.

The idea of the project of a book arose in the late 2000s as there was no book focused on the species with two complementary approaches, biology and farming, which in fact have strong links. The present project has the following objectives: (a) providing up-to-date data in the different fields of biology, (b) giving synthesis on some given issues, (c) valorizing unpublished data including gray literature and non-Latin literature, and (d) giving methods and/or methodologies.

At the time we proposed the project to Springer, the provisional table of contents was so expanded that the publisher requested the coeditors to envisage two volumes. We accepted in proposing to assign volume 1 to the biology of the species, while volume 2 would be focused on farming. The assignment of chapters to its volume has been easy for most of the contributions.

In so doing, volume 1, devoted to the biology of the species, has three main parts plus an additional one on methods. There are biology and ecology, biology and physiology of reproduction, and ecophysiology: i.e. adaptation to the environment. With regard to the first part, i.e., biology and ecology, beyond an extensive abstract of the ecology of wild populations of the species, a comparison between the species and the other sturgeon species present in Eurasia is provided. For those not familiar with the species, a brief anatomic overview is given. To tentatively answer a question from the farmers related to the recording of abnormalities in shape, a new specific investigation on axial skeleton has been conducted. Sex determination is a highly sensible and attractive subject due to the fact that most sturgeon farmers are producing caviar and then rearing of females is the priority. Two different updated approaches are given. A first investigation on the transferable elements of the gonad of sturgeon, namely, in the Siberian sturgeon, is provided. A precise description of the early ontogenesis through the embryogenesis and later the behavior of larvae is given. An updated synthesis on olfaction and gustation in Acipenseridae with a special reference on the Siberian sturgeon is then given. A synthesis on nutritional requirements of the species is presented. Swimming characteristics of the species are described for the first time. The second part of volume 1 deals with fields broadly related to the biology and the physiology of reproduction. A synthetic description of the chemical neuroanatomy of the changes occurring in the central nervous system is given. As the females are favored for most of the sturgeon farms, an updated description by microscopy of the oogenesis along the first oogenesis is provided. A complete description of the characteristics of both sperm and spermatozoa is provided. Two chapters related to steroid end this part. The first deals with gonadal steroidogenesis and describes the synthesis, the plasmatic levels, and the biological activities of steroids in both sexes. The second provide the profiles of the main steroids in the females during the induction of the spawning. The third part deals with ecophysiology, i.e., how the physiology changed facing the environment. Oxygen, ammonia, and nitrite are particularly considered with an extended development around ammonia. In the last part two methods are described, one is focused on how to install a cannulation and the other on the respiratory physiology. Much of the aforementioned issues (or part of their outcomes) will be further evoked in volume 2 because there are direct or indirect connections between the two volumes.

Audenge, France	P. Williot
La Teste de Buch, France	G. Nonnotte
Krasnodar, Russia	M. Chebanov

Acknowledgments Coeditors are grateful to the authors for their contributions especially because some of them accepted to rewrite and update previous studies or to publish new results, to Chantal Gardes (Irstea) who made easier the obtaining of some documents, and finally together to Karine Pichavant-Rafini (Brest University, France) and Michel Rafini (Brest University, France) for their decisive help in the editing of some chapters.

References

Akimova NV, Ruban GI (1999) A classification of reproductive disturbances in sturgeons (Acipenseridae) caused by an anthropognenic impact. J Ichthyol 36(1):65–80

Ferlin P (2017) Foreword. in: Williot P, Nonnotte G, Vizziano-Cantonet D, Chebanov M (eds), The Siberian sturgeon (*Acipenser baerii* Brandt), Volume 1 - Biology, Springer

Ruban GI (2005) The Siberian sturgeon *Acipenser baerii* Brandt. Species structure and Ecology. World Sturgeon Conservation Society – Special publication N°1, p 203. (ISBN 3-8334-4038-4)

Williot P and Castelnaud G (2011) Historic overview of the European sturgeon *Acipenser sturio* in France: surveys, regulations, reasons for the decline, conservation, and analysis. In: Williot P, Rochard E, Desse-Berset N, Kirschbaum F, Gessner J (eds) Biology and conservation of the European sturgeon *Acipenser sturio* L. 1758, The reunion of the European and Atlantic sturgeons, Springer, p 285–307

Williot P, Rouault T (1982) Compte rendu d'une première reproduction en France de l'esturgeon sibérien *Acipenser baeri*. Bulletin Français de Pisciculture 286:255–261

Williot P, Rochard E, Castelnaud G et al (1997) Biological characteristics of European Atlantic sturgeon, *Acipenser sturio*, as the basis for a restoration program in France. Environ Biol Fish 48:359–370

Williot P, Arlati G, Chebanov M et al. (2002) Status and management of Eurasian sturgeon: an overview. Intern Review Hydrobiol 87:483–506

Williot P, Rouault T, Rochard E et al (2004) French attempts to protect and restore *Acipenser sturio* in the Gironde: status and perspectives, the research point of view. In: Gessner J, Ritterhoff J (eds) Species differentiation and population identification in the sturgeons *Acipenser sturio* L. and *Acipenser oxyrhinchus*, Bundesamt für Naturschutz, 101:83–99

Williot P, Nonnotte G, Chebanov M (2017) Introduction to the Siberian sturgeon books with a focus on volume 2 dedicated to the farming of the species. In: Williot P, Nonnotte G, Chebanov M (eds), The Siberian sturgeon (*Acipenser baerii* Brandt), Volume 1 - Biology, Springer

Part I

Biology and Ecology

Geographical Distribution, Ecological and Biological Characteristics of the Siberian Sturgeon Species

Georgii Igorevich Ruban

Abstract

The chapter contains generalized results of long-term field investigations by author of almost all large wild populations of the sturgeon since 1982 as well as all published up today data on the subject. In this chapter data on (a) current and historical distributional range including maps of the range; (b) taxonomic status of the Siberian sturgeon and its species structure; (c) phylogenetic relations of the sturgeon; (d) diet composition; (e) growth of the sturgeon on different populations; and (f) spawning habitat, reproductive parameters, and peculiarities of gameto- and gonadogenesis were included.

Keywords

Geographic distribution • Species structure • Phylogenetic relations • Diet composition • Growth • Spawning habitat • Reproductive parameters • Age at maturity • Fecundity • Gonadogenesis • Gametogenesis

Introduction

As one of the most commercially valuable fish families, acipenserids have long attracted the attention of investigators. The Siberian sturgeon is no exception. However, in view of the sparse habitation of many regions in Siberia and difficult access to many of its habitats, this species was relatively poorly researched

---

G.I. Ruban
A.N. Severtsov Institute of Ecology and Evolution Russian Academy of Science, Leninsky Prospect 33, 119071 Moscow, Russia
e-mail: georgii-ruban@mail.ru

© Springer International Publishing AG, part of Springer Nature 2018
P. Williot et al. (eds.), The Siberian Sturgeon (*Acipenser baerii,* Brandt, 1869)
Volume 1 - Biology, https://doi.org/10.1007/978-3-319-61664-3_1

through the beginning of the 1960s, particularly in Eastern Siberia. More intensive research on the ecology was conducted in the Ob and Irtysh River basins. Since 1961, long-term investigations of the sturgeon in the lower reaches of the Lena were initiated by V.D. Lebedev, professor of ichthyology at the Moscow State University. These investigations, conducted by L.I. Sokolov, assistant professor of the *Department of Ichthyology at the Moscow State University*; V.S. Malyutin, researcher of Central Industrial Acclimatization Department of the Ministry of Fisheries *of the USSR (MFUSSR)*; and A.N. Akimova, researcher of the A.N. Severtsov Institute of Evolutionary Morphology and Animal Ecology AN SSSR (currently IPEE RAS), determined the principle biological characteristics of the Lena population which served as biological base for the subsequent including in aquaculture. The initiator of this trend of works was the chief fish breeder of the Konakovo warmwater farm of VNIIPRKh (*All-Union Research Institute of Pond Aquaculture*) I.I. Smol'yanov, who in 1973 received from the Lena River the first batch of eggs of the Lena sturgeon—61,000 items. The technology of cultivating the Lena sturgeon under conditions of a warmwater farm was developed jointly by specialists of VNIIPRKh (first of all, I.I. Smol'yanov), Moscow State University (L.I. Sokolov), IPEE RAS (N.V. Akimova), and CPAU (V.S. Malyutin). These researchers and their respective departments were responsible for the development of the technical aspects of caviar extraction, transportation, incubation, and rearing of sturgeon up to maturation in aquaculture and the creation of brood stocks at hatcheries.

Sturgeon spawners at Konakovo farm first matured in 1981 (Malyutin and Ruban 2009), and brood stock at Konakovo farm was formed. It for several years became the main supplier of stock material of the Lena sturgeon to other farms in Russia and abroad. This made it possible, with time, to produce caviar and sell stock material to other countries. In 1981, sturgeon fingerlings from Konakovo farm were sent also to Hungary (Ruban 1999). At the present time, the Lena sturgeon is the basis of sturgeon aquaculture in many countries; it is cultivated in Russia, Moldova, the Czech Republic, Hungary, Germany, France, Italy, Uruguay, and many other countries (Bronzi and Rosenthal 2013). The Lena sturgeon is currently one of the principle subjects of the commercial sturgeon fishery worldwide.

The catastrophic collapse of all sturgeon populations in Russia, including the Siberian sturgeon, brought about by excessive harvest and anthropogenic transformations of their habitats mandates the study of their populations (Williot et al. 2002; Ruban et al. 2015). The Siberian sturgeon is of particular concern since even in the 1930s it was harvested at rates comparable to contemporary catches of the anadromous species in the Caspian, Black, and Azov seas in the late 1990s (Khodorevskaya et al. 2007, 2009; Ruban 1999, 2005; Ruban et al. 2015).

The development of conservation strategies for the Siberian sturgeon must be based not only on estimation of its separate populations state but also on knowledge of various aspects of the species structure, the complex of adaptive features which determine the species' plasticity in diverse conditions of its large habitat range. The species structure, both the taxonomic and population aspects, the adaptive radiation of the species, the specific adaptive features as inferred from an establishment of its

1 Geographical Distribution, Ecological and Biological Characteristics

current range, and the condition of the populations whose habitats are undergoing anthropogenic transformations have not yet been satisfactorily studied.

At the beginning of our investigations in 1982, only three populations could be considered relatively well known: that of the Ob-Irtysh basin, the Lena basin, and Lake Baikal. Data on the remaining populations were fragmentary, making it impossible to work out integral notion on the species. In particular this was true for their phenotypic diversity, ontogeny, diet, reproduction, etc. The lack of satisfactory data on the extent of the habitat range made it impossible to obtain a complete idea of the taxonomic and population aspects of the species structure, the ecology (including growth, diet, reproduction, and migration patterns) of the separate populations, the current spatial distribution of the fish and its changes due to anthropogenic impact, the development and functioning of the reproductive system among different populations of the sturgeon under the impact of anthropogenic impact, and the complex of adaptive features of the Siberian sturgeon, which distinguish this particular species from other members of the sturgeon family.

Long-term investigations since 1982 of many other populations (Yenisei, Aldan, Indigirka, and Kolyma rivers) made it possible to close many gaps mentioned above by new data. These data were analyzed in many publications and generalized in the monograph (Ruban 2005).

The chapter contains concentrated data on (a) current and historical distributional range including maps of the range; (b) taxonomic status of the Siberian sturgeon and its species structure; (c) phylogenetic relations of the sturgeon; (d) diet composition; (e) growth of the sturgeon on different populations; and (f) spawning habitat, reproductive parameters, and peculiarities of gameto- and gonadogenesis.

1.1   Geographical Distribution

The range of the Siberian sturgeon is exceptionally broad, encompassing from latitudes of 72–74°N at the mouth of the Lena and Ob Bay to 48–49°N in the Chernyi (Black) Irtysh and Selenga rivers (Dryagin 1948a; Votinov et al. 1975) and longitudinally over 97° (Dryagin 1948a). The range of the Siberian sturgeon and its changes during the last 150 years are presented in Figs. 1.1 and 1.2.

Ob-Irtysh River basin. The Siberian sturgeon is spread very broad. Here, the northern boundary of its range is at 72°N in the Ob Bay (Yudanov 1935; Dryagin 1948b, 1949). In the Ob itself, the Siberian sturgeon is encountered along its entire length, at a distance of 3680 km from the Ob River delta to confluence of the Biya and Katun rivers where the Ob originates, to its delta inclusive. In the Katun River, it inhabits the area from the mouth to 50–70 km upstream (Dryagin 1949; Petkevich et al. 1950).

The Siberian sturgeon inhabits large tributaries of the Ob River (Dryagin 1948b, 1949). Siberian sturgeon range extends the entire length of the Irtysh River, from the mouth to the territory of China (Bogan 1938; Dryagin 1948b, 1949; Petkevich et al. 1950; Sedelnikov 1910; Votinov 1963; Votinov et al. 1975). The sturgeon also inhabits some large tributaries of Irtysh (Dryagin 1948b, 1949).

Fig. 1.1 Borders of the Siberian sturgeon range

Taz River basin. The Siberian sturgeon inhabits Taz River up to 300 km from the mouth of the river (Chupretov and Slepokurov 1979). They also occur throughout Taz Bay and in the near-mouth parts of Taz Bay tributaries. The sturgeon also inhabits the lower reaches of the Pur River up to 100 km from the mouth.

Yenisei River basin. The northern border of the Siberian sturgeon range in Yenisei River is about 72°N in the Yenisei Gulf. Before the construction of hydroelectric dams, the southern limit was concerned 3215 km upstream of the river mouth (Podlesnyi 1963). In the Yenisei, the construction of the Krasnoyarsk hydroelectric dam shortened the range by about 600 km.

In the major tributaries of the Yenisei—the Angara, Podkamennaya Tunguska, and Nizhnyaya (lower) Tunguska rivers—there are small resident populations of sturgeon (Podlesnyi 1955, 1958). In the latter two rivers, the upper limits of the range are not indicated. Siberian sturgeon has also been observed in the lower reaches of the tributaries of the Nizhnyaya (lower) Tunguska (Dryagin 1949).

Sturgeon also encountered in the Angara River tributaries. In Bratsk Reservoir, the sturgeon is distributed everywhere (Lukyanchikov 1967b).

In Lake Baikal, Siberian sturgeon habitats are mainly located near the mouth of mean tributaries. It is most abundant near the delta of the Selenga River as well as main Baikal bays. Sturgeon is less abundant in the northern part of Lake Baikal. The Baikal sturgeon mainly enters the major rivers of the lake basin where spawning sites are located. Along the Selenga it migrates up to 1000 km (Yegorov 1961; Sokolov and Shatunovskii 1983). Presumably, a resident form exists in the Selenga and its tributary (Dashi-Dorzhi 1955).

1 Geographical Distribution, Ecological and Biological Characteristics

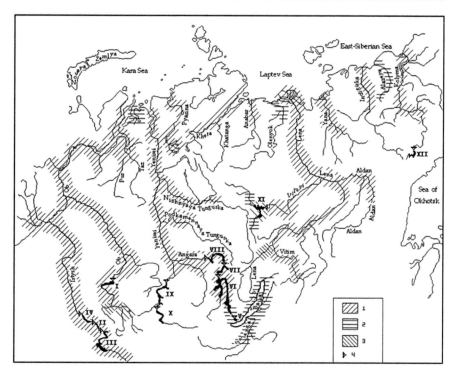

Fig. 1.2 Distribution of the Siberian sturgeon in various river systems. Legend: *1*, common; *2*, rare; *3*, extinct; *4*, hydroelectric dam's reservoir: *I* - Novosibirsk (1957–1959); *II* - Ust-Kamenogorsk (1950–1954); *III* - Bukhtarminsk (1960–1964); *IV* - Shulbinsk (1987); *V* - Irkutsk (1956); *VI* - Bratsk (1966); *VII* - Ust-Ilimsk (1974–1977); *VIII* - Boguchansk (1978); *IX* - Krasnoyarsk (1967–1970); *X* - Sayano-Shushenskaya (1978); *XI* - Vilyuysk (1965–1976); *XII* - Kolymsk (1981–1984)

Pyasina River basin. Sturgeon was of low abundance in the Pyasina River (Dryagin 1949). They have not been found in Pyasina Bay (Ostroumov 1937), though they are known to have inhabited some lakes of the Pyasina basin (Belykh 1940; Logashov 1940). In recent years, sturgeon has virtually disappeared from the Pyasina River (Savvaitova et al. 1994).

Khatanga River basin. In Khatanga basin sturgeon is distributed from the mouth of Kheta to Khatanga inlet inclusionary (Berg 1926, Lukyanchikov 1967a). In the Kheta, the left tributary of Khatanga, the sturgeon inhabits a range extending 460 km from the mouth. Occasionally, it can be found in the lakes in Khatanga floodlands upstream of the Kheta mouth where it enters with freshet during spring. The main habitat of the sturgeon in Khatanga basin is the middle and upper reaches of the Kheta, 350 km from its mouth (Berg 1926; Lukyanchikov 1967a).

Anabar River basin. The sturgeon is found in the lower and middle reaches of the Anabar River (Kirillov 1972).

Olenyok River basin. The Siberian sturgeon occurs mainly up to 1020 km from the mouth of the Olenyok (Kirillov 1972).

Lena River basin. The northern limit of the Siberian sturgeon in the Lena River is at 74°N in the river delta (Dryagin 1948a, 1949). Within the Lena, sturgeon can be found up to 3300 km upstream the river mouth (Borisov 1928; Dormidontov 1963; Dryagin 1933, 1949; Karantonis et al. 1956; Kirillov 1972). During the last 150 years, the southern extent of the range of the Lena sturgeon has been shortened by about 300 km (Ruban 2005).

The sturgeon inhabits a number of tributaries to the Lena including the Aldan and Vilyuy rivers (Dryagin 1949; Karantonis et al. 1956; Kirillov 1972). In the Aldan River, sturgeon inhabits the lower and middle reaches (Kirillov 1964). In the Vilyuy River, the sturgeon has been found in the lower and middle reaches of the river and its main tributaries (Kirillov 1972).

Yana River basin. Sturgeon inhabits the lower and middle reaches of the river from the delta to 67°N (Kirillov 1972); sturgeon abundance in the river is very low.

Indigirka River basin. In the Indigirka River, the sturgeon ranges from the mouth to 850 km (about 67°30′N) and is occasionally found in some upstreams (Kirillov 1955, 1972).

Alazeya River basin. In the Alazeya River, only isolated specimens occur. It has been encountered up to 68°N (Dryagin 1933).

Kolyma River basin. In the Kolyma River, sturgeon has been recorded from the delta up to 1500 km (Dryagin 1933, 1948a). According to historic data, separate individuals were met at 1665 km from river mouth (Koposov and Chekaldin 2009). Sturgeon has also been recorded in some large tributaries of the Kolyma.

1.2 Species Structure

Basing on results of the analysis of phenotypic diversity of the populations of Siberian sturgeon according to morphometric and meristic traits, it was shown that the species is monotypic (Ruban and Panaiotidi 1994; Ruban 2005). It is represented by isolated populations inhabiting separate river systems (Ob, Yenisei, Khatanga, Pyasina, Anabar, Olenyok, Lena, Yana, Indigirka, Alazeya, Kolyma, and now Pechora). The structure of the phenotypic diversity of the Siberian sturgeon is determined by parallel latitudinal clinal variability in morphometric and meristic traits among the populations of the major river basins (i.e., the Lena, Yenisei, and Ob) (Ruban 2005). In these river basins, this species exhibits a continuous row of populations (population continua).

The term "population continuum," introduced by Mayr (1969, 1970), was first applied to the Siberian sturgeon by Ruban (1998, 2005), since this species sturgeon presents a clinal variation in morphometric and meristic characters and different spawning times and sites within the same river (northern and southern reaches) that reveals that distinct populations of this species occur within the same river without significant mixture between them.

Within its range, the Siberian sturgeon has riverine and lacustro-reverine forms (Lake Baikal), some of which are resident within a local area, while others make extensive potamodromous migrations (Ob River) (Ruban 2005).

1 Geographical Distribution, Ecological and Biological Characteristics

An analysis of data on different ecological forms of Siberian sturgeon—riverine, lacustro-riverine, resident, and those that undergo extensive migrations—the location and time of their spawning, their capacity to adapt to various salinities, and data on clinal variability of morphological traits allows to conclude that the Siberian sturgeon cannot be considered as anadromous or even semi-anadromous. The entire life cycle of the Siberian sturgeon is bound to freshwater; its populations inhabiting the lower reaches of rivers do not leave the limits of fresh or weakly brackish waters more than other freshwater fish. Thus, the migrations of Siberian sturgeon, including the Ob and Baikal populations, whose migrations are extensive, can be considered potamodromous (Ruban 2005).

The principal differences between the annual migrations of the Ob sturgeon from the river into the Ob Bay in the fall and early winter and back upstream in the spring and the migrations of truly anadromous species that feed in saline marine waters and migrate to rivers only for spawning can be interpreted as follows (Ruban 2005). First, these annual migrations, which are undertaken by the Siberian sturgeon as well as several other freshwater fish including the sterlet (*Acipenser ruthenus*), orfe (*Leuciscus idus*), and others, are brought about by the annual winter hypoxia event that forces fish to move into the Ob Bay or to move upstream to remain above the hypoxic front. Second, the Siberian sturgeon in the Ob Bay remains within fresh or weakly saline waters. Finally, in contrast to the truly anadromous acipenserids, among which only fish preparing to spawn migrate, this migration of the Ob sturgeon includes fish of all ages and size classes and in reality is a feeding migration. Only for those fish preparing to spawn does this migration later become a spawning migration (Votinov 1963).

As it was mentioned above, the structure of the phenotypic diversity of the Siberian sturgeon is intricate and determined by parallel latitudinal clinal variability in morphometric and meristic traits among the populations of the major river basins (i.e., the Lena, Yenisei, and Ob) (Ruban 2005). Populations of the Siberian sturgeon inhabiting northern parts of large river basins are morphologically very similar. Morphological differences between populations from the Ob, Yenisei, Lena, Indigirka, and Kolyma rivers and Lake Baikal as well are below the subspecies level (Ruban 1999, 2005). The molecular-genetic investigations (Birstein et al. 2009) support these results.

The explanation of the fact of existence at the territory of Siberia of monotypic freshwater representative of sturgeons—the Siberian sturgeon inhabiting virtually every major river basin—is necessary to search in sturgeon phylogeny.

1.3   Phylogenetic Relations of the Siberian Sturgeon

The paleontological history of Acipenseriformes remains insufficiently studied. There is no single opinion on time and place of their origin. Some authors (Miller 1969) believe that the *Polyodon*, *Scaphirhynchus*, and *Acipenser* genus originated at the territory of modern North America corresponding in present time the Mississippi basin. Others (Yakovlev 1977) believe that freshwater Chondrosteidae, distributed at the beginning of the Jurassic era across the entire Palearctic and attaining moderate size and an archaic structure, gradually evolved in direction to modern Acipenseriformes.

According to the most recent data, Acipenseriformes emerged in the Jurassic epoch in the basin of the ancient Tethys Sea in modern Central Asia (Birstein and DeSalle 1998). The emergence of acipenserids as a result of the divergence of Acipenseridae and Polyodontidae is presumably due to a divergence from an extinct ancestor (*Peipiaostidae*) known from the late Jurassic sedimentary structures (Jin 1995, Jin et al. 1995; Grande and Bemis 1996) and also associated with the Jurassic period 200–135 million years ago (Birstein and DeSalle 1998).

The most ancient fossil records of acipenserids were filled in the date from the late Cretaceous, 95–65 million years (Nesov and Kaznyshkin 1983).

Currently, there is no consensus regarding the phylogenetic connection within the Acipenseridae family (Berg 1904, 1905; Findeis 1993, 1997; Mayden and Kuhajda 1996). Nor have the phylogenetic connections within the *Acipenser* genus been clarified, though during long time there were attempts to group species together (see the reviews of Dumeril 1870; Bemis et al. 1997). Such attempts, basing on biogeographical data (Artyukhin 2008) and molecular genetics (Birstein et al. 1997; Birstein and DeSalle 1998), were undertaken in the last years.

Molecular-genetic investigations (Birstein and DeSalle 1998) indicate that the closest relative to the Siberian sturgeon *Acipenser baerii* is the Persian sturgeon *A. persicus* and the Adriatic sturgeon *A. naccarii*. Recent investigations do not support the validity of the Persian sturgeon (Ruban et al. 2008, 2011). Less genetic similarity than between these species has been reported between the Siberian sturgeon and the Russian sturgeon (*A. gueldenstaedtii*) (Birstein and DeSalle 1998). Between the last two species, the morphological similarity was noted (Sokolov and Vasil'ev 1989).

Following the opinion by Birstein and DeSalle (1998), the ancestor of the Siberian sturgeon entered the Siberian rivers from the Ponto-Caspian basin during the middle of the Pleistocene through the system of periglacial lakes that existed during maximum glaciations (Berg 1928). The eastern and southeastern limits of the modern range of the Siberian sturgeon are determined by the extent of the last glaciation that occurred 18,000–20,000 years ago (Velichko et al. 1994), and from the late Pleistocene, glaciations of 7000 years ago to this day have not changed (Tsepkin 1995).

Thus, the monotypicity of the Siberian sturgeon is presumably a result of this being a phylogenetically young species of acipenserid, and its freshwater lifestyle is a result of the evolution of the species during isolation from marine basins in Ice Age.

1.4 Diet Composition

The Siberian sturgeon is a typical benthophagic fish, but foraging on fish is to some extent typical for all of the populations which have been studied. In general, a direct expected relationship was observed between the stomach contents and the composition of the benthos (Romanova 1948).

Ob-Irtysh basin. In Ob Bay the Siberian sturgeon feeds on the molluscs *Sphaerium corneum* and *Pisidium amnicum* and on the larvae of the tendipedids *Monodiamesa bathyphila*, *Procladius*, *Chironomus*, and *Paracladopelma* (Saldau 1948). In the lower Ob, the diet of sturgeon is dominated during summer by *Sphaerium* sp. molluscs but by Simuliidae larvae during winter (Saldau 1948). In middle reaches of the Ob River and in the Irtysh (Cheremkhovsk sands), the sturgeon feeds primarily on some 20 species of midge (Chironomidae) larvae, of which *Paracladopelma* sp. dominates in number and weight. During winter, the diet is largely composed of *Paracladopelma* and *Monodiamesa bathyphila*, whereas quantities of Simuliidae and Coleoptera larvae as well as *Sphaerium* molluscs are of no significance. During summer the quantity of *Paracladopelma* in the diet falls to 0.8%, while *Simulium* sp. larvae compose 85.6% of the stomach content by number and 37.2% by weight. This is due to an explosion in their populations in these warm months. During summer, the rate of molluscs and caddisworms also increases in the diet (Saldau 1948).

Sturgeon diet during hibernation contains exclusively midge larvae in young sturgeon (0.5–1.8 kg wet weight). The diet of larger sturgeon (7–10.4 kg, maturation stage II) was more variable with gammarids and caddisworms as main components. Sexually mature individuals (maturation stages III and IV) had empty stomachs (Votinov 1963).

Sturgeons of different sizes and ages in the Ob-Irtysh basin exhibit several differences in their diets. The juveniles (under 50 cm) feed primarily on larvae of midges, caddisworms, mayflies, and black flies (simuliids) and on several molluscs (*Spaerium*). Larger sturgeon feeds less on mayflies, consumes more molluscs, and occasionally feeds on eggs, larvae, and juveniles of other fish. Fish encountered in the diet include roach, dace, ruffe, minnows, burbot, lamprey, and pike (up to 25 cm) (Dryagin 1949).

Yenisei River basin. Similar to the Ob River basin, there is a difference in sturgeon diets in the Yenisei River and the bay. In the Yenisei Bay and Yenisei Inlet, the food mainly comprises the isopod *Mesidothea* (sea cockroaches) sometimes accounted for more than 90% of the diet, while midge larvae are almost absent (Isachenko 1912, Nikolskii 1971; Podlesnyi 1958). In the lower reaches of the river, the bulk of the diet is composed of isopods and amphipods (*Mesidothea*, *Pontoporeia*, and *Marenzelleria*) (Podlesnyi 1955). In the river, the sturgeon mostly fed on lamprey larvae, *Pisidium* molluscs, and on the larvae of Tabanidae, Chironomidae, and Ephemeridae (Isachenko 1912). In the Yenisei River stretch between the mouths of the Kureika River to the Golchikha settlement, the main food was midge larvae: *Polypedilum*, *Procladius*, *Trichocladium*, *Culicoides*, and others. Ephemerid larvae were the next most frequent item in the diet (Romanova 1948).

In the Angara River, the principal diet components were midge larvae, simuliids, and mayfly larvae. Gammarids were less important. Even rarer are molluscs and occasionally juvenile lamprey (up to 70 per stomach). Diet composition clearly reflects that habitat-related benthic fauna composition rocky beds contain more caddisworms, while silty beds contain more midge larvae (Dryagin 1949).

In Lake Baikal, the main foods of the sturgeon are gammarid shrimp, midge larvae, stonefly, and mayfly larvae, as well as Baikal bullhead (Cottocomephoridae). Molluscs, oligochaete worms, flatworms, caddisworms, fly, beetle, and gnat larvae were all commonly observed as well (Yegorov 1961). The Baikal sturgeon is marked by a distinct change in diet with age. Young sturgeon at the Selenga River near the mouth section feeds almost exclusively on gammarids, but not infrequent a young sturgeon's stomach will be filled with juvenile bullhead. Molluscs, worms, and flatworms are more rarely observed. The diet of yearling and juvenile fish was dominated by smaller forms of midge larvae and occasional gammarid young, which have more tender shells (Yegorov 1961). Older fish eat midge larvae and gammarids equally, and in the rivers, they consume mayfly and stonefly larvae, all in larger sizes. At age 3–4, sturgeons begin to eat young bullhead although midge larvae and gammarids still dominate the diet. By 5–6 years and older, gammarids become the principal component of diet, and young fish are more commonly found, primarily small bullhead but also, more rarely, some cyprinids. Adult fish feed primarily on bullhead but also on the young of other fish, including perch (Yegorov 1961). Fish wintering in Baikal (January–March) does not stop feeding (Yegorov 1961).

Lena River basin. It has been shown that the dependence of general diet patterns on the benthic fauna and section of the river is similar to that of other river basins. In the near-mouth parts of the rivers, the sturgeon feeds primarily on amphipods (*Pontoporeia*) and isopods (*Mesidothea*), and in other parts of the river, they feed primarily on midge larvae, including *Paracladopelma*, *Cryptochironomus*, and *Stictochironomus* (Pirozhnikov 1955; Sokolov 1966). The importance of the midge larvae declines as one advances up the river and the importance of the molluscs, caddisworms, mayfly, and stonefly larvae increases (Sokolov 1966). The proportion of various Chironomidae larvae is unequal and changes across different sections of the river, apparently reflecting a direct relationship to the composition of the benthos (Romanova 1948). In the Aldan River (the major right tributary of the Lena), sturgeon between 70 and 80 cm TL feeds mainly on the larvae of Chironomidae, Plecoptera, Trichoptera, Heleidae, and Ceratopoginidae, with smaller quantities of *Sphaerium* sp. and *Pisidium* molluscs. At a length of about 75 cm, the sturgeon begins to supplement its diet with fish. Larger, adult fish over 90 cm become entirely piscivorous (Sokolov et al. 1986).

The most common components in the diet of the Siberian sturgeon in the Indigirka River are midge larvae (*Paracladopelma camptolabis*, *Prodiamesa gr. bathyphila*, and *Procladius*) and larvae from Ceratopogonidae, Plecoptera, and Ephemeroptera (Ruban 2005).

In the diet of the Siberian sturgeon in the Kolyma River, the dominant foods are the larvae of Chironomidae and Heleidae. Copepoda, *Lynceus brachyurus*, and fish are frequently encountered in the guts of the Kolyma sturgeon. The maximal number of food components was observed in sturgeon from the lower reaches of the Kolyma River. The Kolyma sturgeon has a higher piscivory behavior than the Lena sturgeon (Sokolov and Novikov 1965).

Gathering up the data on the Siberian sturgeon's diet, it can be concluded that the sturgeon exhibits an extreme plasticity in terms of its diet choice. The composition of its diet varies significantly not only within the limits of species' range but also

1 Geographical Distribution, Ecological and Biological Characteristics

with time (e.g., age or size). As typical benthophagous species, some populations of the Siberian sturgeon occasionally switch opportunistically and completely to a piscivorous diet.

The changes in diet composition along the rivers of various basins point to several general trends. In estuaries of rivers flowing into the Arctic Ocean, marine benthos fauna dominates in the diet. In the majority of rivers such as the Yenisei, Lena, Indigirka, and Kolyma, sturgeons typically feed on amphipods and isopods. In the Ob Bay, specimens also feed primarily on molluscs. Further upstream, the spectrums of the diet widen, and food composition changes significantly. Here, the most important components are midge, caddisfly, mayfly, and stonefly larvae, as well as gammarids, molluscs, and other invertebrate species.

In all studied populations, changes toward large sizes and amounts of food organisms occurred with increasing age and size of sturgeon. In certain populations (e.g., Ob River), it observed a diversification of diet with age. In the majority of the populations (exception—Yenisei population), a partial transition to piscivory is observed at age 3–5 years. In some populations (e.g., Lake Baikal), older sturgeons feed primarily on fish.

A characteristic feature of the Siberian sturgeon is continuous feeding during winter over a large part of the range. Only in the Ob River, sexually mature fish stop feeding in winter. Probably the rich food resources in the Ob River basin allow the spawners to accumulate sufficient energy reserves for long-distance migration and spawning. The higher food available is indirectly indicated by the high growth rate compared to other populations. With few exceptions, sturgeons in the Lena River do not feed prior to spawning. In the Indigirka and Kolyma rivers, feeding proceeds during the spawning season.

Seasonal changes in the diet content were found in the majority of the populations of Siberian sturgeon studied. These data and differences between river basins and various river sections in diet composition support the existing opinion that sturgeon's diet composition reflects the benthic species composition in the foraging grounds (Romanova 1948; Podlesnyi 1955).

1.5 Growth of the Sturgeon in Different Populations

Siberian sturgeon can attain rather significant body sizes and mass. Sturgeons caught from the Ob River have been reported to weigh as much as 180–200 kg (Berg 1949; Dryagin 1949). However, the species is in its natural distributional range, one of the slowest growing acipenserids. This is due primarily to the ecological and environmental conditions of the habitat, in particular the availability of food and the low-temperature regimes of the water bodies it inhabits. Under artificial conditions with unlimited food availability and high temperatures, the growth rate of the Siberian sturgeon can significantly increase. In rearing experiments with the Lena River sturgeon in aquaria, it was demonstrated that its growth rate is comparable to that of the Russian sturgeon from the Volga River (Sokolov 1965a). An elevation of the temperature by 4–7 °C in the summer and by 9–10 °C in the winter in comparison to the water temperatures in the wild is accompanied by a 7–9 times increase in growth rates (Akimova et al. 1980).

In natural water bodies, growth in body length and weight is significantly different among populations and is directly related to the development of benthic food abundance in the respective water bodies as well as their thermal regimes (Ruban 2005). According to our results, the slowest rate of body length growth is exhibited by Siberian sturgeon populations from the lower reaches of the Lena and Indigirka rivers. Yenisei River sturgeons exhibit significantly higher growth rates. Sturgeons from the Baikal and the Kolyma River populations are characterized by higher growth rates than fish from the Lena and Indigirka rivers. Young specimens of Baikal and Kolyma sturgeon up to 11 years are of similar length, but after the twelfth year, the Baikal sturgeon surpasses the Kolyma sturgeon, and after the thirteenth year, it surpasses the Yenisei sturgeon. Among river populations of Siberian sturgeon, the fastest-growing population is the one from the Ob River, but the young, up to 8 years of age, are similar in size to the Yenisei population. At an age of 17, the Baikal sturgeon surpasses the Ob sturgeon as well. Growth rates for the Baikal sturgeon during benthophagy are similar to that of the Yenisei and Kolyma sturgeon but increase radically after transition to piscivorous diets. Thus, the ecological plasticity of the species allows it to attain high growth rates even under conditions of insufficient food abundance for the majority of populations. The Kolyma sturgeon is the fastest body length growing among those fish of the rivers of the Yakutia region. The Kolyma sturgeon (at the age of 17 years) surpasses the Yenisei sturgeon and, at age 21, the Ob sturgeon (Ruban 2005).

Differences in growth rate (length) of males and females are small expressed. As a rule, among young immature individuals, the females are of somewhat larger size. In older age classes, differences between the sexes diminish (Ruban 2005). But in separate populations, such as that of the lower Lena River, the males are of larger size (Koshelev et al. 1989).

Growth in weight of the Siberian sturgeon exhibits similar differences between populations and sexes.

For the Siberian sturgeon, as in other fish species, there is a higher mortality of faster-growing fish. As a result the maximal life span is found in slow-growing individuals. In populations inhabiting harsh conditions (Lena, Indigirka, and Kolyma), high longevity is with slow-growing females, which eventually attain also the largest sizes. This may be interpreted as adaptation to maintain a high reproductive capacity in areas of severe environmental conditions (e.g., low temperatures and food abundance) (Ruban 2005).

1.6 Spawning Habitat, Reproductive Parameters, and Peculiarities of Gameto- and Gonadogenesis

Spawning habitat. Siberian sturgeon spawns on gravelly sand or gravel substrates at a depth of 4–8 m in currents of 2–4 km/hr. (0.56–1.11 m/sec) (Podlesnyi 1955; Yegorov 1961; Votinov 1963; Sokolov and Malyutin 1977; Ruban and Akimova 1991, 1993).

1 Geographical Distribution, Ecological and Biological Characteristics

The spawning period in *Ob River* varies but is generally from late May to early June in waters between 12 °C and 18 °C (Votinov 1963, Votinov et al. 1975).

In *Yenisei River* spawning usually takes place between June and July at water temperatures of 16° to 21 °C (Podlesnyi 1955).

From *Lake Baikal* sturgeon enters the Selenga, Bagruzin, and Verkhnyaya (upper) Angara rivers to spawn. Migrations begin in April before the ice has completely receded and continues to the middle of June. The major migration takes place from the 5th–10th of May to the 5th–10th of June. Sturgeons spawn from mid-May to mid-June at water temperatures of 9 to 15 °C or higher (Yegorov 1961).

The spawning grounds of the *Lena* sturgeon extend a great length from the upper reaches (Nuya River mouth and possibly higher) to the lower reaches of the river (Kirillov 1972), possibly even to the upper ends of the Lena delta (Pirozhnikov 1955). Spawning time differs between the northern and southern parts of the Lena basin. In lower reaches (600 km from the Lena's mouth), sturgeon spawns from the middle of June to the middle of July (Akimova 1985a, b; Sokolov and Malyutin 1977). At the Lena River section adjacent to the mouth of the Olekma River, i.e., the middle reaches, spawning takes place between the end of May and early June (Koshelev et al. 1989).

Experiments have shown that egg development in the Lena sturgeon can occur in temperatures between 8 and 20 °C, while temperatures above 21 °C causes embryo mortality. The optimal incubation temperature is between 11.4 and 14.9 °C (Nikolskaya and Sytina 1974). Other authors (Reznichenko et al. 1979) identify the optimal temperature range for the Lena sturgeon egg incubation with 13–19 °C. From fish hatcheries it is known, however, that incubation temperatures of 23–25 °C produce quite viable offsprings from the lower Lena sturgeon (Sokolov and Malyutin 1977).

In the lower reaches of the *Indigirka River*, sturgeon's spawning ground located on the right bank between rkm 306 and 315 km. Appropriate spawning sites probably exist near the villages of Olenegorsk and Vorontsovo (rkm 327 and 363, respectively), where the right banks of the Indigirka are also rocky and high and the riverbed is gravelly (Ruban and Akimova 1991). Earlier, Kirillov (1955) assumed a spawning ground near the settlement of the Druzhina (rkm 714). In the lower Indigirka, sturgeon reproduces from July through the beginning of August in waters of 13°–16 °C, as evidenced by catches of males undergoing spermiation (stage V) and stage IV females in late July (Ruban and Akimova 1991).

In the *Kolyma River*, Siberian sturgeon spawning grounds are located on a section of the river adjacent to the mouth of the Ozhogina River, a left tributary, about 900 km from the mouth of the Kolyma (Ruban and Akimova 1993). However, many gravel bars in the Kolyma between the settlement of Zyryanka and the town of Srendnekolymsk (rkm 995 and 665, respectively) indicate that this section likely sustains other sturgeon spawning grounds.

Reproduction of sturgeon in the Kolyma takes place from the end of June to the end of July in water temperatures ranging from 16 °C to 21 °C, which is somewhat higher than in other rivers (Lena and Indigirka) (Ruban and Akimova 1993). Males already have mature sexual products in the autumn of the pre-spawning year and overwinter in this condition (Ruban and Akimova 1993).

Biometry at maturation. Size and age of Siberian sturgeon females and males at maturation and their reproductive periodicity in various bodies of water are rather different (Tables 1.1 and 1.2).

Spawners' life span and sex ratio. In separate populations of the Siberian sturgeon, as had been shown above, a significant difference in life span for males and females has been observed. Populations having a high food availability are consequently characterized by higher growth rates (Ob, Yenisei, Baikal), and the maximal age of males and females is similar. For sturgeons inhabiting the rivers of the Yakutia (Lena, Indigirka, and Kolyma rivers), females, with a lower growth rate, have a greater life span. In populations with the lowest growth rates (Lena, Indigirka) life expectancy of females is roughly twice that of males (Ruban 2005).

The ratio of males to females in spawning parts of all the examined populations is close to 1:1, with a slight predominance of males (Podlesnyi 1955;

Table 1.1 Size and age of Siberian sturgeon females at maturation and their reproductive periodicity for various bodies of water

Water body	Length (cm)	Body weight (kg)	Age (years)	Interspawning period	Source
Ob	103–110	9	9–12	4	Petkevich et al. (1950), Petkevich (1952), Dormidontov (1963)
	–	–	16	4	Votinov (1963)
	114–120	8.2	12	–	Dryagin (1947)
Irtysh	–	–	12	–	Menshikov (1936)
	–	–	17–18	–	Bogan (1938)
	–	15–17	16–17	–	Yereschenko (1970)
Yenisei	65–79	5–8	19–24	–	Podlesnyi (1955)
	85–90	4–6	–	–	Dryagin (1947, 1948b)
Lake Baikal	119–124	14	20–22	–	Yegorov (1961)
	152–167	19–38	26–34	–	Afanasieva (1977)
Lena	80	1.4–2.0	–	–	Dryagin (1947)
	70	1.5	16–20	–	Pirozhnikov (1955)
	–	4.5	15–18	–	Karantonis et al. (1956)
	70	1.5–2.0	11–13	3–5	Sokolov and Akimova (1976), Sokolov and Malyutin (1977), Akimova (1985a)
Vilyuy	97	2.7	18	–	Kirillov (1972)
Indigirka	70–75	1.2–2.0	12–14	4–5	Ruban and Akimova 1991
Kolyma	79	2.15	16	4–5	Ruban and Akimova 1993
Konakovo warmwater hatchery	109	5.0–6.8	7–9	1.5–2	Akimova (1985a, b)
Rostov region ponds	103	7.2	10		Berdichevskii et al. (1983)

1 Geographical Distribution, Ecological and Biological Characteristics   17

Table 1.2 Size and age of Siberian sturgeon males at maturation and their reproductive periodicity for various bodies of water

Water body	Length (cm)	Body weight (kg)	Age (years)	Interspawning period	Source
Ob	103–110	9	9–10	3	Petkevich et al. (1950), Petkevich (1952), Dormidontov (1963)
	–	–	8	3	Votinov (1963)
	100	5.4	10	–	Dryagin (1947)
Irtysh	–	–	11–13	–	Bogan (1938)
	–	–	8–9	–	Yereschenko (1970)
Yenisei	65–79	5–8	17–20	–	Podlesnyi (1955)
	75–80	2–3.2	–	–	Dryagin (1947, 1948b)
Lake Baikal	100	7	15	–	Yegorov (1961)
	115–154	9.6–24	17–29	–	Afanasieva (1977)
Lena	70–75	1.2–1.5	–	–	Dryagin (1947)
	60	1.0	15–18	–	Pirozhnikov (1955)
	60	1.5–2.0	9–10	3	Sokolov and Akimova (1976), Sokolov and Malyutin (1977), Akimova (1985a, b)
Indigirka	70–75	1.2–2.0	12–14	4–5	Ruban and Akimova 1991
Kolyma	79	2.15	16	4–5	Ruban and Akimova 1993
Konakovo warmwater hatchery	92–98	3.8–4.25	4+	1.5–2	Akimova (1985a, b)
Rostov region ponds	103	7.2	10		Berdichevskii et al. (1983)

Votinov et al. 1975; Sokolov and Malyutin 1977). However, in separate generations of the Ob sturgeon, this ratio occasionally shifts, and the predominance of males becomes significant to 2:1 (Votinov et al. 1975). According to our data, the sex ratios on the spawning grounds of the Indigirka and Kolyma are close to 1:1 (Ruban 2005).

Fecundity. The absolute fecundity of Siberian sturgeon varies over a wide range (Table 1.3), for the species as a whole as well as for separate populations. The maximum value of 1,459,000 eggs, attained by Ob sturgeon, is directly related to that being the largest of the sturgeon populations. Records exist of Ob females caught weighting 192 kg and yielding 64 kg of roe, equivalent to roughly 3–3.5 million eggs (Berg 1949; Votinov 1963).

The absolute fecundity of the Yenisei sturgeon is lower than that of the Ob sturgeon but is still relatively high. The absolute fecundity of the Baikal sturgeon is also rather high, and the relative fecundity varies within a range similar to that of the lower Lena sturgeon (Table 1.3). The diameter of the eggs ranges from 2.37 to 2.92 mm, and their weight ranges from 10.9 to 16.1 mg.

Table 1.3 Siberian sturgeon fecundity (a literature summary)

Water body	Absolute body length (cm)	Body weight (kg)	Age (years)	Absolute fecundity (×1000 eggs)	Relative fecundity (1000 eggs/kg)	Sources
Ob		7–69.2		79.0–1459.0	9.0–24.0	Votinov (1963)
Irtysh	137–151	14–31	20–26	146.0–420.0	7.0–17.0	Yegorov (1961)
Yenisei	127–173	7–50	11–32	83.2–607.6 101.0–286.0	6.0–21.0	Podlesnyi (1955) Yegorov (1961)
Lake Baikal	129–211	14–60	20–40	211.0–832.0	0.4–33.7	Yegorov (1961)
Lena	70–135	1.4–14.0 1.41–8.43	11–23	20.7–144.0 16.5–91.7	8.9–22.2 7.2–32.7	Sokolov (1965b) Sokolov and Ruban (1979)
Indigirka	115–141	10.05–21.0	33–58	105.56–245.34	15.0–17.0	Ruban and Akimova (1991)
Kolyma	105–137	5.8–17.1	18–42	65.6–227.84	9.34–24.49	Ruban and Akimova (1993)

The lowest absolute fecundity is seen among females in the lower Lena population and is probably related to the small size of the spawners. Their relative fecundity varies very widely. Minimum values of this character in the Ob, Yenisei, and lower Lena populations are close, while its maximum value in all studied populations is lower than in lower Lena population (Table 1.3).

Interrelationship between growth in weight and reproductive parameters. Considerable differences are revealed in the life span, growth rate, and parameters of reproduction between various populations of the Siberian sturgeon (Figs. 1.3 and 1.4; Tables 1.1, 1.2 and 1.3). The populations of sturgeon from the Ob River and Lake Baikal are characterized by the maximum rate of weight growth; those from the Lena and the Indigirka have the minimum rate. In the Kolyma and Yenisei rivers, the growth rate of sturgeon is intermediate (Fig. 1.3). The relationship of the body weight and the age is described by the exponential equation, the coefficient of determination (R^2) in all cases >0.9. According to the growth rate and life conditions, the populations of sturgeon may be conventionally subdivided into rapidly growing warm water (the Ob River), rapidly growing cold water (Lake Baikal), and slow-growing cold water (the Lena and Indigirka rivers).

The Siberian sturgeon of the *Ob River* population lives in the most favorable climatic and trophic conditions (Ruban 2005). In its main feeding grounds—the lower reaches of the Ob River and the Ob Gulf—the annual sum of heat varies within limits from 1500 to 2200 degree-days (Votinov 1963). The growth rate of the

1 Geographical Distribution, Ecological and Biological Characteristics

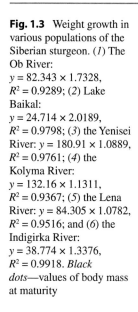
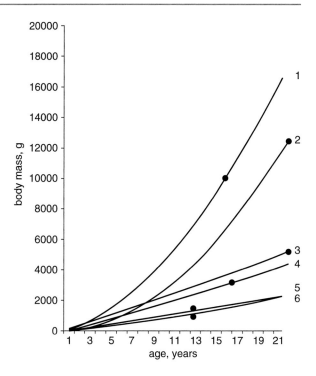

Fig. 1.3 Weight growth in various populations of the Siberian sturgeon. (*1*) The Ob River: $y = 82.343 \times 1.7328$, $R^2 = 0.9289$; (*2*) Lake Baikal: $y = 24.714 \times 2.0189$, $R^2 = 0.9798$; (*3*) the Yenisei River: $y = 180.91 \times 1.0889$, $R^2 = 0.9761$; (*4*) the Kolyma River: $y = 132.16 \times 1.1311$, $R^2 = 0.9367$; (*5*) the Lena River: $y = 84.305 \times 1.0782$, $R^2 = 0.9516$; and (*6*) the Indigirka River: $y = 38.774 \times 1.3376$, $R^2 = 0.9918$. *Black dots*—values of body mass at maturity

sturgeon of this population is higher than in other populations (Fig. 1.3). During the first 12 years of life, the weight growth rate of the Ob sturgeon is the highest. Then, it decreases, and in 4 years at the age of 16 years, the first females attain maturity (Votinov 1963). Retardation of the somatic growth seems to be related to the inability of benthos feeding to provide simultaneous requirements of growth and gonad development. The Ob River population is characterized by very high size and body mass at maturity and the minimal values for this species of the gonadosomatic index and individual relative fecundity. The life span of males and females is identical, and the age at maturity is low (Tables 1.1 and 1.2; Fig. 1.3).

The life strategy of this population living under conditions close to the optimum ones for this species is characterized by a high rate of growth and development, by identical life span of males and females, and by relatively low values of reproductive parameters, including the relative weight of gonads, which indicates the relatively low level of generative metabolism.

The *Lake Baikal* population of sturgeon should be attributed to cold-water populations. In spite of the fact that the annual sum of heat at the surface is relatively high, 1700 degree-days (Sokolov 1952), it should be taken into account that the sturgeon occurs in the lake at depths up to 200 m (Yegorov 1961), at a constant temperature of 3.1°C (Sokolov 1952), and the annual sum of heat is 1131.5 degree-days.

Prior to maturity the Lake Baikal sturgeon grows rapidly, just slightly less than the Ob River sturgeon (Fig. 1.3). However, its growth is irregular, and during the first 10 years of life, it grows slowly. The rate of its linear and weight growth is close to that in the river populations of sturgeon in Yakutia (Ruban 2005). In this period the Baikal sturgeon is a typical benthophage, but at the age of 10 years, it becomes a predator (Yegorov 1961). This is accompanied by an abrupt increase in the growth rate. The specimens of this population attain maturity at the maximum body weight and age for this species (Tables 1.1 and 1.2; Fig. 1.3). The size of the Baikal and Ob River sturgeon breeders as they spawn for the first time is similar. However, the Baikal sturgeon attains maturity 6 years later than the Ob sturgeon, at the age of 22 years due to retarded development caused by low temperature. These results agree with the conclusion made by Rass (1948) that an increase in the water temperature not exceeding the range possible for the species stimulates development more than growth.

The life strategy of the Baikal population of the Siberian sturgeon living at low water temperature but under conditions of a high food supply (beginning from the age of 10 years) is characterized by late maturation caused by retarded development due to low temperatures; by an identical life span of males and females; by low relative weight of gonads, indicating a relatively low level of generative metabolism; and by the maximal relative fecundity for this species (Fig. 1.4) attained by a decrease of the definitive weight of eggs.

The populations of the Siberian sturgeon from the *Lena and Indigirka* rivers live under the most unfavorable thermal and trophic conditions (Ruban 2005). The annual sum of heat in the lower reaches of the Lena and in its delta, the main feeding

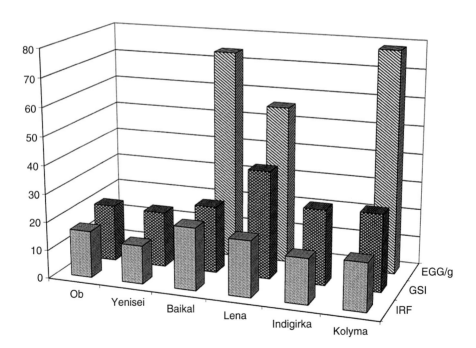

Fig. 1.4 Average values of reproductive characters in various populations of the Siberian sturgeon. *GSI*, gonadosomatic index of females at stage IV of maturity (%); *IRF*, individual relative fecundity (oocytes/g of body mass); *EGG/g*, number of definite eggs in 1 g

grounds of the sturgeon, varies from 1000 to 1200 degree-days; in the lower reaches of the Indigirka River, it is 1000–1100 degree-days (Ruban 2005). The sturgeon in these rivers is characterized by the minimal growth rate for this species (Fig. 1.3) and minimal age and the size of fish spawning for the first time (Tables 1.1 and 1.2). The life strategy of cold-water populations of the sturgeon from the Lena and Indigirka rivers is characterized by the following traits: the life span of males is twice as short as that in females; in these populations the ratio of somatic growth and generative metabolism shifts toward the latter (this is expressed in the maximal values of the gonadosomatic index for this species); the mean individual relative fecundity is close to the maximum for the species; and the mean definitive weight of eggs is maximal for the species (Fig. 1.4).

This difference of ratios of somatic growth and generative metabolism, relative fecundity, and egg weight in cold-water populations with different growth rates may be interpreted as adaptation targeted at higher survival of progeny in the populations with low growth rate living under the most unfavorable thermal and trophic conditions. It is known that larger eggs hatch larger larvae possessing enhanced viability (Meyen 1940).

The variation limits of the gonadosomatic index, weight of eggs, and relative fecundity are the widest in the slow-growing cold-water population of the sturgeon (Ruban 2005).

The parameters of growth and reproduction of the Siberian sturgeon in the *Yenisei* and *Kolyma* rivers are intermediate. This may be explained by the intermediate climatic and trophic conditions of these rivers. The basin of the Kolyma is located farther to the east of the Indigirka basin (with the cold pole of the Northern Hemisphere) and suffers the impact of the Pacific Ocean. The annual sum of heat in the lower reaches of the Kolyma River is 100 degree-days higher than in the lower reaches of the Indigirka River.

Gameto- and gonadogenesis. From juveniles through maturation and subsequent reproductive cycles after the first spawning follows the same general pattern in Siberian sturgeon as in all other acipenserids (Votinov 1963; Akimova 1978, 1985a, b). The specific difference in Siberian sturgeon relates to the timing within the reproductive cycle, determined both by age at maturity and the environmentally triggered periodicity. The unique feature of gametogenesis in Siberian sturgeon, which is related to the severity of the climate, the brevity of the arctic summer, and the exhaustion of spawners during their spawning migration, is the increased duration of reproductive cycles relative to other species (Votinov 1963).

The duration of separate stages of development of reproductive cells upon maturation of the Siberian sturgeon females has been more precisely established for the lower Lena River population. Juvenile females can be up to 3–7 years old and 35.5–56.5 cm in length before they show signs of maturation. Females whose oocytes are in the synaptic phase are 3–8 years old; their length and weight vary between 47 and 67 cm and from 320 to 1070 g. Beginning at age 5, females enter the cytoplasmatic growth phase. Females whose oocytes have entered the trophoplasmatic growth phase are no less than 7 years old. Upon attaining maturation, the whole vitellogenesis phase (stages III and IV) lasts no less than 4–5 years in females of the Lena sturgeon. The females of lower Lena sturgeon first spawn at 11–12 years (Akimova 1978, 1985a, b). However, the Lena population is also marked by a high variability in maturation rates and sizes of maturing females. Some 56.5 cm females can be juveniles but also at the same size can already have taken part in a spawning event (Koshelev et al. 1989). During repeated reproductive cycles, each phase of gonad development lasts at least a year (Akimova 1985a, b). As a whole, Lena sturgeon, as well as Kolyma and Indigirka sturgeon, is characterized by lower growth rates and earlier maturation at significantly smaller sizes than the sturgeon of the Ob, Yenisei, and Baikal populations. The longer interspawning period among Siberian sturgeon, as compared to other acipenserids, is related to the shorter growing season and lower accessibility to food in northern water bodies than in southern ones and, respectively, to longer periods of energetic resource accumulation which are necessary for the development of reproductive cells (Votinov 1963, Sokolov and Malyutin 1977, Akimova 1978, 1985a, b). The cultivation of Siberian sturgeon in warmwater environments is accompanied by a dramatic shortening of the time to maturation for males from 9–10 years to 3–4 years and for females from 11–12 years to 7–8 years, that is, by factors of 1.5–3. Interspawning intervals under these conditions also shorten from 4–5 years for females to 1.5–2 years, and males can reproduce annually (Berdichevskii et al. 1983, Akimova 1985a, b, Williot and Brun 1998).

References

Afanasieva VG (1977) O vosproizvodstve zapasov baikal'skogo osetra (On the artificial stock enhancement of Baikal sturgeon). Ryby I rybnoye khozyaistvo Vostochnoi Sibiri (Fish and and fish husbandry in eastern Siberia). Ulan-Ude 1:139–141

Akimova NV (1978) Gametogenez, funktsionirovaniye polovykh zhelez sibirskogo osetra (*Acipenser baeri* Brandt) r. Lena I ikh svyas' s obmenom veshchestv (gametogenesis, functioning of the sexual glands in Siberian sturgeon (*Acipenser baerii* Brandt) in the Lena River and their link with metabolites) Ekologomorfoligicheskie i ekologo fiziologicheskie issledovaniya razvitiya ryb (Ecomorphological and ecophysiological studies of fish development). Nauka, Moscow, pp 43–55

Akimova NV (1985a) Gametogenez i polovaya tsiklichnost' sibirskogo osetra v yestestvennykh i eksperimentalnykh usloviyakh (gametogenesis and reproductive cycles of the Siberian sturgeon under natural and experimental conditions). Osobennosti reproduktivnikh tsiklov u ryb v vodoemakh raznykh shirot USSR (characteristics of reproductive cycles in fish at different latitudes of the USSR). Nauka, Moscow, pp 111–122

Akimova NV (1985b) Gametogenez i razmnozheniye sibirskogo osetra (Gametogenesis and reproduction in Siberian sturgeon). PhD dissertation, Institute of Evolutionary Morphology and Animal Ecology, USSR Academy of Sciences, Moscow, p 243

Akimova NV, Sokolov LI, Smoljanov II, Malyutin VS (1980) Sravnitel'ny analiz rosta I gametogeneza sibirskogo osetra r. Leny v prirodnykh I experimental'nykh usloviyakh (comparative analyses of growth and gametogenesis of the Lena River Siberian sturgeon in natural and experimental conditions) Vnutrividovaya izmenchivost' v ontogeneze zhivotnykh (intraspecific variability in animals onthogenesis). Nauka, Moscow, p 1670176.

Artyukhin EN (2008) Osetrovye: ecokologiya, geograficheskoe rasprostraneniye i filogeniya (sturgeons: ecology, geographical distribution and phylogeny). St. Petersburg University Press, Moscow, p 137

1 Geographical Distribution, Ecological and Biological Characteristics

Barrucand M, Ferlin P, Lamarque P, Sabaut JJ (1979). Alimentation artificielle de l'esturgeon *Acipenser baeri*. In: Proc. world symposium on finfish nutrition and fish feed technology, Hambourg, 20–23 juin 1978, vol. 1, pp 411–421

Belykh FI (1940) Ozero Lama i yego rybokhozyaistvennoye ispol'zovanie (Lake Lama and its fishery use). Trudy Nauchno-Issledovatelskogo Instituta Polyarnogo Zemledeliya. Zhivotnovodstva i Promyslovogo Khozyaistva, Seriya Promyslovoe Khozyaistvo (Proseedings of Research Institute of Polar Agriculture, Animal husbandry and Fisheries. Fisheries series), vol 11, p 73–100

Bemis W, Findeis E, Grande L (1997) An overview of Acipenseriformes. Environ Biol Fish 48:25–71

Berdichevskii LS, Malyutin VS, Smolyanov II, Sokolov LI (1983) Itogi rybovodno-akklimatizatsionnykh rabor s sibirskim osetrom (the results of aquaculture and acclimatization researches of the Siberian sturgeon). Biologicheskiye osnovy osetrovodstva (Biological bases of sturgeon aquaculture). Nauka, Moscow, pp 259–269

Berg LS (1904) Zur systematic der Acipenseriden. Zool Anz XXI(22):665–667. (In German)

Berg LS (1905) Ryby Turkestana (Fish of Turkestan) Nauchnye Rezultaty Aralskkoi Ekspeditsii: Izvestiya Russkogo geogrtcheskogo obschestva (Scientific results of the Aral expedition: Transactions of Russian Geographical Society). St. Petersburg, vol 4, p 261

Berg LS (1926) Ryby basseina Khatangi (fishes of the Khatanga Basin). Materialy Komissii AN SSSR po Izucheniyu Yakutskoi ASSR (Materials by Committee of USSR Academy of Sciences on study of Yakutian Republic) 2:1–22

Berg LS (1928) On the evolution of the northern elements in the fauna of the Caspian Sea. Dokl Akad Nauk SSSR, Series A 7:107–112

Berg LS (1949) Ryby presnykh vod SSSR i sopredel'nykh stran (The freshwater fishes of the USSR and adjacent countries) Vols. 1–3 Akademia Nauk SSSR, Moscow & Leningrad, p 1382

Birstein VJ, Hanner R, DeSalle R (1997) Phylogeny of the Acipenseriformes: cytogenetic and molecular approaches. Environ Biol Fish 48:127–155

Birstein VJ, DeSalle R (1998) Molecular phylogeny of Acipenseridae. Mol Phylogenet Evol 5(1):141–155

Birstein VJ, DeSalle R, Doukakis P, Hanner R, Ruban G, Wong E (2009) Testing taxonomic boundaries and the limit of DNA barcoding in the Siberian sturgeon, *Acipenser baerii*. Mitochondrial DNA 20(5–6):110–118

Bogan FE (1938) K biologii sibirskogo osetra *(Acipenser baeri* Brandt) basseina reki Irtysha (contribution to the biology of the Siberian sturgeon (*Acipenser baeri* Brandt) of the Irtysh River basin). Uchenye Zapiski Permskogo Gosudarstvennogo Universiteta (Scientific Reports of the Perm State University) 3:145–163

Borisov PG (1928) Ryby reki Leny (fishes of the Lena River). Trudy Komissii ANSSSR po Izucheniyu Yakutskoi Respubliki (Materials by Committee of USSR Academy of Sciences on study of Yakutian Republic) 9:1–181

Bronzi P, Rosenthal HK (2013) Present and future sturgeon and caviar production and marketing: a global market overview. J Appl Ichthyol 30:1536–1546

Chupretov VM, Slepokurov VA (1979) O letnem radspredelenii sibirskogo osetra v Obskoi i Tazovskoi gubakh (On summer distribution of the Siberian sturgeon in the Ob and Taz bays). Osetrovoye khozyaistvo vo vnetrennikh vodoyemakh SSSR (Sturgeon Husbandry in Inland Water Bodies of the USSR), Abstracts of Papers at the 2nd All-Union Conference, Astrakhan, pp 270–271

Dashi-Dorzhi A (1955) Materialy po ichthyofaune verkhov'ev Selengi b Avura v predelakh Mongolii (Materials on the ichthyofauna of the upper Selenga and Amur rivers within the territory of Mongolia). Zoologicheskii Zhurnal (Zoological Journal) 34:570–577

Dormidontov AS (1963) Rybokhzyaistvennoye ispol'zovanie osetra r. Leny (fishery utilization of the Lena River sturgeon). Osetrovoye khozyaistvo v vodoyemakh SSSR (sturgeon Fisheries in the water Bodies of the USSR). Izdatelstvo Akademii Nauk SSSR, Moscow, pp 182–187

Dryagin PA (1933) Rybnye resursy Yakutii (fish resources of Yakutiya). Trudy Soveta po Izucheniyu Proisvoditelnykh Sil (Proceedings of the Committee on study of productive forces) 5:3–94

Dryagin PA (1947) Ulovy osetra v vodoyemakh Sibiri (sturgeon catches in the water bodies of Siberia). Rybnoe Khozyaistvo (Fish Husbandry) 1:34–38

Dryagin PA (1948a) O nekotorykh morfologicheskikh i biologicheskikh otlichiyakh osetra, obitayushchego v rekakh Yakutii ot sibirskogo osetra *Acipenser baerii* Brandt (on some morphological and biological distinctions between sturgeon occurring in the rivers of Yakutiya and the Siberian sturgeon, *Acipenser baerii* Brandt). Zoologicheskii Zhurnal (Zoological Journal) 27:525–534

Dryagin PA (1948b) Promyslovye ryby Ob'-Irtyshskogo basseina (commercial fishes of the Ob-Irtysh River Basin). Izvestiya Vsesouznogo Instituta Ozernogo i Rechnogo Rybnogo Khozyaistva (Bulletin of All Union Institute of Lake and River Fish Husbandry) 25(2):3–104

Dryagin PA (1949) Biologiya sibirskogo osetra, yego zapasy i ratsional'noye ispol'zovaniye (Biology of the Siberian sturgeon, its reserves and rational utilization). Izvestiya Vsesouznogo Instituta Ozernogo i Rechnogo Rybnogo Khozyaistva (Bulletin of All Union 0) 29:3–51

Dumeril AHA (1870) Histoire Naturelle des Poissons, ou Ichthyologie generale. Vol. 2, Paris, p 624

Ferlin P (2017) Foreword. In: Williot P, Nonnotte G, Vizziano-Cantonet D, Chebanov M (eds) The Siberian sturgeon (*Acipenser baerii* Brandt), vol 1—Biology, Springer

Findeis EK (1993) Osteology of the North American shovelnose sturgeon *Scaphirhynchus platorynchus Rafinesque* 1820, with comparisons to other Acipenseridae and Acipenseriformes. Ph. D. Thesis, Univ. Of Massachusets, Amherst, p 444

Findeis EK (1997) Osteology and phylogenetic relationships of recent sturgeons. Environ Biol Fish 48:73–126

Grande L, Bemis W (1996) Interrelationships of Acipenseriformes, with comments on "Chondrostei". In: Stiassny MLJ, Parenti LR, Jonson GD (eds) Interrelationsips of fishes. Academic Press, New York, pp 85–115

Jin F, (1995) Late Mesozoic acipenseriformes (*Osteichthyes: Actinopterigii*) in Central Asia and their biogeographical implications. Pages 15–22 *in* Sixth Simposium on Mesozoic Terrestrial Ecosystems and Biota, Short Papers, (A. Sun, and Y. Wang, Eds.). Cina Ocean Press, Beijing

Jin F, Tian Y, Deng S (1995) An early fossil sturgeon (*Acipenseriformes, Peipiaosteidae*) from Fengning of Hebei. China Vert Palasiatica 33:1–16. (in Chinese with English summary)

Isachenko VL (1912) Ryby Turukhanskogo kraya, vstrechayushchiesya v Yenisee i Yeniseiskom zalive (Fishes of Turukhanskii Krai Found in the Yenisei and Yenisei Gulf). Materialy po issledovaniyu reky Yeniseia v rybopromyslovom otnoshenii (Materials on study of the Yenisei River in relation of fishery). Krasnoyarsk, 4: p 111

Karantonis FE, Kirillov FN, Mukhomediyarov FB (1956) Ryby srednego techeniya reki Leny (fishes of the middle Lena reaches). Trudy Instituta Biologii Yakutskogo Filiala AN USSR. Proceedings of Biological Institute of Yakutian branch of USSR Academy of Science 2:3–144

Khodorevskaya RP, Ruban GI, Pavlov DS (2007) Povedeniye, migratsii, raspredeleniye i zapasy osetrovykh ryb Volgo-Kaspiyskogo basseina (behaviour, migrations, distribution, and stocks of sturgeons in the Volga–Caspian basin). KMK, Moscow, p 241

Khodorevskaya RP, Ruban GI, Pavlov DS (2009) Behaviour, migrations, distribution, and stocks of sturgeons in the Volga-Caspian Basin. World Sturgeon Conservation Society: Special Publication № 3, Germany. Norderstedt: Books on Demand GmbH, p 233

Kirillov FN (1955) Ryby reki Indigirki (fishes of the Indigirka River). Izvestiya VNII ozernogo i rech ryb Khozyayistva (Bulletin of All Union Institute of Lake and River Fish Husbandry) 35:141–167

Kirillov FN (1964) Vidovoi sostav ryb reki Aldana (species composition of fishes of the Aldan River). Vertebrate Animals of Yakutiya, Yakutsk, pp 73–82

Kirillov FN (1972) Ryby Yakutii (fishes of Yakutia). Nauka, Moscow, p 260

Koposov AE, Chekaldin YN (2009) Sibirskii osetr *Acipenser baerii* v srednem techenii reki Kolymy (the Siberian sturgeon *Acipenser baerii* in middle reaches of the Kolyma River). Voprosy rybolovstva 10(1–37):22–26

Koshelev BV, Ruban GI, Sokolov LI, Khalatyan OV, Akimova NV, Sokolova EL. (1989) Ecologocheskaya kharakteristika sibirskogo osetra *Acipenser baeri* Brandt basseina srednei i verkhnei Leny (ecological characteristics of Siberian sturgeon from the middle and upper Lena basin) Ekologiya, morfologiya i povedenie osetrovykh (ecology, morphology, and behavior of sturgeon). Nauka, Moscow, pp 16–33

Logashov FV (1940) Ozero Melkoye I ego rybokhozyaistvennoye ispol'zovaniye (Lake Melkoe and its utilization for fishery). Trudy Nauchno-Issledovatelskogo Instituta Polyarnogo Zemledeliya, Zhivotnovodstva i Promyslovogo Khozyaistva, Seriya Promyslovoe Khozyaistvo (Proseedings of Research Institute of Polar Agriculture, Animal husbandry and Fisheries Fisheries series) 11:7–71

Lukyanchikov FV (1967a) Ryby sistemy reki Khatangi (Fishes of the Khatanga River system). Trudy Krasnoyarskogo Otdeleniya SibNIIRKh (Proceedings of the Krasnoyrsk branch of the Siberian Research Institute of Fisheries) 9:11–93.

Lukyanchikov FV (1967b) Promyslovo-biologicheskaya kharakteristika i sostoyanie zapasov promyslovykh ryb Bratskogo vodokhranilishcha v pervye gody yego sushchestvovaniya (Commercial and biological characteristics and condition of commercial fish in the Bratskoye reservoir in the first years of its existence. Izvestiya Biologo-Geograficheskogo NII pri Irkutskom Gos. Univ. (Bulletion of Biological and Geographical research Institute of Irkutsk State University), XX, pp 262–286

Malyutin VS, Ruban GI (2009) On the history of fish husbandry of Siberian sturgeon *Acipenser baerii* from the Lena River for acclimatization and commercial cultivation. J Ichthyol 49(5):376–382

Mayden RL, Kuhajda BR (1996) Systematics, taxonomy, and conservation status of the endangered Alabama sturgeon, *Scaphirhynchus suttkusi* Williams and Clemmer (*Actinopterygii, Acipenseridae*). Copeia:241–275

Mayr E (1969) Principles of systematic zoology. McGraw-Hill, New York, p 428

Mayr E (1970) Populations, species, and evolution. The Belknap Press of Harvard University Press Cambridge, Massachusets, p 460

Meyen VA (1940) O prichinakh kolebaniya razmerov ikrinok kostistykh ryb (on the reasons behind the variability in egg size in bony fish). Doklady AN SSSR (Proceedings of Academy of Sciences of the USSR) 28(7):654–656

Menshikov MI (1936) K biologii sibirskogo osetra *Acipenser baeri* i sterlyadi r. Irtysha (on the biology of the Siberian sturgeon *Acipenser baeri* and the sterlet *Acipenser ruthenus* in the Irtysh River). Uchoniye Zapiski Permskovo Gosudarstvennogo Universirteta (Scientific Reports of the Perm State University) 2(1):41–65

Miller R (1969) Chetvertichnye presnovodnye ryby Severnoi Ameriki (freshwater fish of the Quaternary in North America). Chetvertichniy period v SSHA (the Quaternary period in the USA). MIR press, Moscow, pp 174–192

Nesov LA, Kaznyshkin MN (1983) Novye osetry mela I paleogena SSSR (new sturgeons from cretaceous and Paleogene of USSR). Sovremenyie problemy paleoikhtiologii (current Issues in Paleoichthyology). Nauka, Moscow, pp 68–76

Nikolskaya NG, Sytina LA (1974) Zona temperaturnykh adaptatsiy pri razvitii ikry osetra r. Leny (Zone of temperature adaptations during the Lena River sturgeon eggs development). Tezisy otchetnoi sessii Tsentralnogo Nauchno-Issledovatelskogo Instituta Osetrovogo Rybnogo Khozyaistva (Abstracts of report session of Central Research Institute of Sturgeon Fisheries), Astrakhan, p 108–109

Nikolskii GV (1971) Chastnaya Ikhtiologiya (Special ichthyology). Visshaya Shkola, Moscow, p 471

Ostroumov NA (1937) Ryby i rybnyi promysel r.Pysiny (Fishes and fishery of the Pyasina River). Trudy Polyarnoi Komissii AN USSR (Proceedings of the Polar Comission of USSR Academy of Science). Leningrad, 30, pp 1–115

Petkevich AN (1952) Biologiya i vosproizvodstvo osetra Srednei I Verkhnei Obi v svyazi s gidrostroitel'stvom (Biology and reproduction of sturgeon in the Middle and Upper Ob related to hydroconstruction). Trudy Tomskogo Gosudarstvennogo Universiteta (Proceedings of the Tomsk State University). 119, p 39–64

Petkevich AN, Bashmakov VN, AYa B (1950) Osetr srednei i verkhnei obi (sturgeons of the middle and upper reaches of the Ob River). Trudy Barabinskogo Otdeleniya Vsesoyuznogo Nauchno-Issledovatelskogo Instituta Ozernogo i Rechnogo Rybnogo Khozyaistva (Proceedings of the Barabinsk branch of State Institute of Lake and River Fish Husbandry) 4:3–54

Pirozhnikov PL (1955) Materialy po biologii promyslovykh ryb reki Leny (Materials on the Biology of commercial fishes of the Lena River). Izvestiya Gosudarstvennogo Nauchno-Issledovatelskogo Instituta Ozernogo i Rechnogo Rybnogo Khozyaistva (Bulletin of All Union Institute of Lake and River Fish Husbandry) 35:61–128

Podlesnyi AV (1955) Osetr (*Acipenser baerii stenorrhynchus* a. Nikolsky) reki Yeniseya (sturgeon (*Acipenser baerii stenorrhynchus* a. Nikolsky) of the Yenisey River). Voprosy Ikhtiologii (Issues in Ichtyology) 4:21–40

Podlesnyi AV (1958) Ryby Yeniseya, usloviya ikh obitaniya I ispol'zovvniya (Fishes of the Yenisey River, conditions of their life and utilization of them). Trudy Vsesoyuznogo Nauchno-Issledovatelskogo Instituta Ozernogo i Rechnogo Rybnogo Khozyaistva (Proceedings of All Union Institute of Lake and River Fish Husbandry) 44, p 97–178

Podlesnyi AV (1963) Sostoyaniye zapasov osetrovykh na Yenisee i puti ikh uvelicheniya (state of stock of sturgeons in the Yenisey River and ways to increase them). Osetrovoye khozyaistvo v vodoyemakh SSSR (sturgeon fishery in water Bodies of the USSR). Izdatelstvo Akademii Nauk USSR, Moscow, pp 200–205

Rass TS, (1948) O periodakh zhizni i zakonomernostyakh razvitiya i rosta ryb (On life periods and developmental and growth regularities in fish). Izvestiya Akademii Nauk SSSR, Seriya Biologicheskaya (Bulletin of USSR Academy of Sciences. Biology Series). 3, p 295–305

Reznichenko PN, Ziborova IN Malyutin VS (1979) Vliyaniye postoyannykh temperatur inkubatsii na razvitiye ikry sibirskogo osetra r.Lena (Effect of constant incubation temperatures on the development of eggs of the Siberian sturgeon in the Lena River). Ekologisheskaya fiziologiya i biokhimiya ryb. Astrakhan. II, p 162–164.

Romanova GP (1948) Pitaniye ryb v nizhnem Yenisee (Feeding of fishes in down streams of the Yenisei River). Trudy Sibirskogo otdeleniya Vsesoyuznogo nauchno-issledovatelskogo instituta ozernogo i rechnogo rybnogo khozyaistva (Proceedings of the Siberian branch of All Union Institute of Lake and River Fish Husbandry). Krasnoyarsk 2:151–203

Ruban GI (1998) On the species structure of the Siberian sturgeon *Acipenser baerii* Brandt (Acipenseridae). J Ichthyol 38(5):345–365

Ruban GI (1999) Sibirskii osetr *Acipenser baerii* Brandt (struktura vida i ekologiya) (the Siberian sturgeon *Acipenser baerii* Brandt (species structure and ecology)). GEOS, Moscow, p 235

Ruban GI (2005) The Siberian sturgeon *Acipenser baerii* Brandt. Species structure and ecology. World Sturgeon Conservation Society. Special Publication No 1, Norderstedt. Germany, p 203

Ruban GI, Akimova NV (1991) The ecological features of the Siberian sturgeon *Acipenser baerii* Brandt from the Indigirka River. J Ichthyol 31:118–129

Ruban GI, Akimova NV (1993) Osobennosti ecologii sibirskogo osetra *Acipenser baeri*, reki Kolymy (Ecological features of Siberian sturgeon, *Acipenser baeri*, from the Kolyma River). Voprosy Ikhtiologii (Issues in Ichtyology) 33(1):84–92

Ruban GI, Khodorevskaya RP, Koshelev VN (2015) O sostoyanii osetrovykh v Rossii (on the status of sturgeon in Russia). Astrakhanskii vestnik ecologicheskogo obrazovaniya 31:42–50

Ruban GI, Kholodova MV, Kalmykov VA, Sorokin PA (2008) Morphological and molecular-genetic study of the Persian sturgeon *Acipenser persicus* Borodin (Acipenseridae) taxonomic status. J Ichthyol 48(10):891–903

Ruban GI, Kholodova MV, Kalmykov VA, Sorokin PA (2011) A review of the taxonomic status of the Persian sturgeon (*Acipenser persicus*) Borodin. J Appl Ichthyol 27(2):470–477

Ruban GI, Panaiotidi AI (1994) Comparative morphological analysis of subspecies of the Siberian sturgeon, *Acipenser baerii stenorrhynchus* A. Nikolsky and *Acipenser baerii chatys* Drjagin (Acipenseriformes, Acipenseridae), in the Yenisey and Lena rivers. J Ichthyol 34:58–71

Saldau MP (1948) Pitaniye ryb Ob-Irtyshskogo basseina (diet of the fish of the Ob-Irtysh basin). Izvestiya Vsesouznogo Instituta Ozernogo i Rechnogo Rybnogo Khozyaistva (Bulletin of All Union Institute of Lake and River Fish Husbandry) 28:175–226

Savvaitova KA, Pichugin MI, Maksimov VA, Maksimov SV, Pavlov SD (1994) Izmeneniya sostava ikhtiofauny vodoyemov Norilo-Pyasinskoi vodnoi sistemy v usloviyakh intensvnogo antropogennogo vozdeistviya (Chagnes in the composition of the ichthyofauna of the water bodies of the Norilo-Pyasinsk water system under conditions of intensive anthropogenic influence). Voprosy Ikhtiologii (Issues in Ichtyology) 34(4):566–569

Sedelnikov AK (1910) Ozero Zaisan (Lake Zaisan). Zapiski Zapadno-Sibirskogo Otdela Imperatorskogo Russkogo Geograficheskogo Obshchestva (Bulletin of wetern Siberian department of Imperial Russian Geographic Society) 35:1–253

Sokolov AA (1952) Gidrografiya SSSR (hydrography of the USSR). Gidrometeoizdat, Leningrad, p 573

Sokolov VE, Shatunovskii MI (1983) Ryby Mongolskoi Narodnoi Respubliki (fish of the Peoples Republic of Mongolia). Nauka, Moscow, p 277

Sokolov LI (1965a) O roste sibirskogo osetra *Acipenser baeri* r. Leny (On growth of Siberian sturgeon *Acipenser baerii* Brandt of the Lena River). Vestnik MGU (Bulletin of Moscow State University). 1:3–12

Sokolov LI (1965b) Sozrevaniye i plodovitost' sibirskogo osetra *Acipenser baeri* r. Leny (maturation and fecundity of the Siberian sturgeon *Acipenser baeri* Brandt from the Lena River). Voprosy Ikhtiologii (Issues in Ichtyology) 5(1):70–81

Sokolov LI (1966) Pitaniye sibirskogo osetra *Acipenser baeri* r. Leny (diet of the Siberian sturgeon *Acipenser baerii* Brandt of the Lena River). Voprosy Ikhtiologii (Issues in Ichtyology) 6(3):550–560

Sokolov LI, Akimova NV (1976) K metodike opredeleniya vozrasta sibirskogo osetra *Acipenser baeri* r. Leny (a technique for determining the age of Siberian sturgeon of the Lena River). Voprosy Ikhtiologii (Issues in Ichtyology) 16(5):853–858

Sokolov LI, Koshelev BV, Khalatyan OV, Ruban GI, Akimova NV, Sokolova EL (1986) Ekologo-morfologicheskaya kharakteristika sibirskogo osetra *Acipenser baeri* reki Aldan (ecological and morphological characteristics of the Siberian sturgeon (*Acipenser baerii* Brandt) in the Aldan River). Voprosy Ikhtiologii (Issues in Ichtyology) 26(5):741–749

Sokolov LI, Malyutin VS (1977) Osobennosti struktury populyatsii i khrakteristiki proizvoditelei sibirskogo osetra r. Leny v raione nerestilishch (The structure of the population and spawners characteristics in the Siberian sturgeon from river Lena at spawning sites). Voprosy Ikhtiologii (Issues in Ichtyology) 17(2):327–246

Sokolov LI, Novikov AS (1965) Materialy po biologii sibirskogo osetra (*Acipenser baeri* Brandt) vodoyemov Yakutii (data on the biology of the Siberian sturgeon (*Acipenser baeri* Brandt) in Yakut waters). Nauchnye doklady vysshei shkoly Seriya Biologocheskikh nauk (High School scientific reports Biological series) 4:36–38

Sokolov LI, Ruban GI (1979) Rasnokachestvennost' samok sibirskogo osetra (*Acipenser baerii* Brandt) reki Leny i nekotorye pokazately ikh vosproizvoditelnoi sposobnosti (different properties among female Siberian sturgeon (*Acipenser baerii hatys* Drjagin) in the Lena River and some indicators of their reproductive capacity). Byul. MOIP. Otd. Biol. (Moscow Society of nature researchers Section f Biology) 84(6):67–73

Sokolov LI, Vasil'ev VP (1989) *Acipenser baeri* Brandt, 1869. The Freshwater Fishes of Europe. Vol. I/II General Introduction to Fishes Acipenseriformes. AULA-Verlag, Wiesbaden, pp 263–284

Tsepkin EA (1995) Izmeneniya promyslovoi fauny ryb kontinental'nakh vodoyemov Vostochnoi Yevropy I Severnoi Azii v chetvertichnom periode (Changes in the harvested fish fauna of continental water bodies in Eastern Europe and North Asia in the Quaternary period). Voprosy Ikhtiologii (Issues in Ichtyology) 35(1):3–18

Velichko AA, Kononov YM, Faustova MA (1994) Polednee oledeneniye zemli v pozdnem pleistotsene (the last glaciations of the Pleistocene). Priroda (Nature) 7:63–67

Votinov NP (1963) Biologicheskie osnovy iskusstvennogo vosproizvodstva obskogo osetra (Biological foundations of artificial reproduction of the Ob River sturgeon). Trudy Ob-Tazovskogo Otdeleniya Gosudarstvennogo Nauchno-Issledovatelskogo Instituta Ozernogo i Rechnogo Rybnogo Khozyaistva, Novaya Seria 3:5–102

Votinov NP, Zlokazov VN, Kasyanov VP, Setsko RI (1975) Sostoyanie zapasov osetra v vodoemakh Sibiri i meropriyatiya po ikh uvelicheniyu (status of sturgeon reserves in the rivers of Siberia and measures aimed to increase these reserves). Sredneuralskoe Knizhnoe Izdatelstvo, Sverdlovsk, p 94

Wei Q, Zou Y, Li P et al (2011) Sturgeon aquaculture in China: progress, strategies and prospects assessed on the basis of nation-wide surveys (2007–2009). J Appl Ichthyol 27:162–168

Williot P, Arlati G, Chebanov M, Gulyas T, Kasimov R, Kirschbaum F, Patriche N, Pavlovskaya L, Polyakova L, Pourkazemi M, Kim Y, Zuang P, Zholdasova I (2002) Status and management of Eurasian sturgeon: an overview. Internat Rev Hydrobiol 87(5–6):483–506

Williot P, Brun R (1998) Ovarian development and cycles in cultured Siberian sturgeon *Acipenser baeri*. Aquat Living Resour 11(2):111–118

Williot P, Arlati G, Chebanov M et al (2002) Status and management of Eurasian sturgeon: an overview. Intern Review Hydrobiol 87:483–506

Williot P, Nonnotte G, Chebanov M (2017) Introduction to the Siberian sturgeon books with a focus on volume 2 dedicated to the farming of the species. In: Williot P, Nonnotte G, Chebanov M (eds) The Siberian sturgeon (*Acipenser baerii* Brandt), vol 1—Biology, Springer

Yakovlev VN (1977) Filogenes osertoobraznykh (Phylogenesis of acipenserids). Ocherki filogenii i sistematika iskopayemikh ryb i bezchelyustnykh. Nauka, Moscow, pp 116–144

Yegorov AG (1961) Baikal'skii osetr *Acipenser baerii* stenorrhynchus natio baicalensis A. Nikolsky. Sistematika, biologiya, promysel, syr'yevaya baza I vosproizvodstvo zapasov (Baikal sturgeon, *Acipenser baerii* stenorrhynchus natio baicalensis A. Nikolsky (taxonomy, biology, fishery, and reproduction of reserves)). Ulan-Ude, p 121

Yereschenko VI (1970) Sostoyaniye stada Sibirskogo osetra v vodokhranilishchakh Verkhnego Irtysha i puti yego vosproizvodstva (condition of Siberian sturgeon populations in the reservoirs of the upper Irtysh and possibilities of its cultivation). Osetrovye SSSR i ikh vosproizvodstvo (Acipenserids of the USSR and their cultivation). Nauka, Moscow, pp 158–163. (in Russian)

Yudanov IG (1935) Perspectives on sturgeon fishing in the Ob Bay and Cape Novy Port. Trudy Ob-Tazovskogo Otdeleniya Gosudarstvennogo Nauchno-Issledovatelskogo Instituta Ozernogo i Rechnogo Rybnogo Khozyaistva, Novaya Seria 3:7–12

What Makes the Difference Between the Siberian Sturgeon and the Ponto-Caspian Sturgeon Species?

2

Georgii Igorevich Ruban

Abstract

Adaptive features of the Siberian sturgeon connected with species structure (adaptive radiation), population structure, ecology, physiology, reproduction and development in comparison with phylogenetically closely Ponto-Caspian sturgeon related species were analyzed. The adaptive radiation of freshwater Siberian sturgeon unlike anadromous Ponto-Caspian sturgeon was directed to formation of continuous rows of populations (so called "population continuums") over wide ranges in a single river basin allowing separate populations inhabiting different sections of the river to reproduce in the optimal time without performing energetically expensive migrations under conditions of the limited food supply. Unlike most of sturgeon species sex ratio in Siberian sturgeon is close to 1:1. In populations living under worse foraging conditions the life span of males is almost twice less than that of females. The Siberian sturgeon have to feed at the main part of the range through the winter at the water temperature close to 0 °C and in populations living in the most severe conditions (Lena and Indigirka rivers) it feeds through the spawning period. In these populations were found (1) significant increase in generative tissue synthesis level, (2) decrease of somatic growth efficiency, (3) maturation at earlier age and smaller size, (4) increase of relative fecundity and relative gonad weight of females, (5) increased duration of interspawning intervals of females. The efficiency of yolk utilization in the Siberian sturgeon is higher than in Russian sturgeon. Despite the smaller size of the eggs, this results in larger size of hatchling prelarva in Siberian sturgeon than in Russian sturgeon.

G.I. Ruban
A.N. Severtsov Institute of Ecology and Evolution Russian Academy of Science, Leninsky prospect 33, 119071 Moscow, Russia
e-mail: georgii-ruban@mail.ru

© Springer International Publishing AG, part of Springer Nature 2018
P. Williot et al. (eds.), *The Siberian Sturgeon (Acipenser baerii,* Brandt, 1869)
Volume 1 - Biology, https://doi.org/10.1007/978-3-319-61664-3_2

> **Keywords**
> Adaptations • Species structure • Adaptive radiation • Population structure
> Anadromous • Potamodromous • Seasonal races • Feeding patterns • Maturation
> Life span • Reproduction • Development • Metabolism

Introduction

As it was noted in the preceding chapter, the Siberian sturgeon phylogenetically is closely related with the group of anadromous Ponto-Caspian sturgeon species—the Adriatic sturgeon (*Acipenser naccarii*) and Russian sturgeon (*Acipenser gueldenstaedtii*). For Ponto-Caspian, anadromous species was a discovered system of adaptations determining tolerance to variable environmental conditions (Gerbilskii 1957, 1962, 1965, 1967). But for the Siberian sturgeon, such a system was not studied. That is why it is of most interest to compare adaptive features of anadromous Ponto-Caspian group of species with potamodromous Siberian sturgeon's features appeared, while their divergence and occupation of the most northwest part of sturgeon range in Eurasia.

2.1 Adaptive Features of the Siberian Sturgeon Connected with Species Structure (Adaptive Radiation)

Gerbilskii (1957, 1962, 1965, 1967) developed a theory on the biological progress of sturgeon based on the ideas of A. N. Severtsov (1925) on the biological progress on certain groups of animals. The main characteristics of such groups are high number, wide ranges, and adaptive radiation (=divergence). This theory was advanced on the basis of study on biological features of the genus *Acipenser* representatives. The main attention was given to anadromous species of the Ponto-Caspian basin. The freshwater species, such as Sterlet sturgeon *(Acipenser ruthenus)* and Siberian sturgeon, were not considered in this aspect. For the representatives of the genus *Acipenser*, Gerbilskii (1957, 1962, 1965, 1967) has investigated and described the system of adaptations providing the increasing of euryoky (ability to live under various environmental conditions) degree, intensification of reproduction, and reduction of offspring mortality.

The majority of adaptive features inherent in the whole genus *Acipenser* are also characteristic for the Siberian sturgeon living in freshwaters of Siberia. However, this species has a number of essential differences in this respect from the representatives of the genus inhabiting the Ponto-Caspian basin. These differences could not be reduced only to distinctions between anadromous and nonmigrating forms. We shall consider them using the systemization of adaptations proposed by Gerbilskii (1962, 1967).

The first group of adaptations was named by Gerbilskii (1962, 1967) as "many-sided ecological fitness" or "increasing of euryoky degree." It includes some interconnected features of species which first of all characterize adaptive radiation within

2 What Distinguishes the Siberian Sturgeon from Ponto-Caspian Sturgeon?

a genus, separate species, and populations: amphibionic and holobionic forms, biological groups, seasonal races, etc., spatial and temporal differentiation in utilization of foraging resources of water bodies, nature of a migration impulse, in time of spawning and location of spawning sites. Thus, this group of adaptations includes the features of species characterizing first of all their structure.

The analysis of phenotypic diversity and ecological traits of the Siberian sturgeon (Ruban 2005) demonstrated that it is a monotypic species. The absence of significant differences in morphological characters of the Siberian sturgeon from different river basins, especially among the northern populations, is probably connected with two causes. First, it is probably connected with features of colonization and isolation of separate populations. Acipenseridae exuviae in Siberia have been found dating to the Oligocene (37–22 million years) and the Pliocene (5–1.8 million years) (Shtylko 1934, Berg 1962, Gardiner 1984). The Glaciacion and huge freshwater near-glacial lakes at that time (Baulin 1970) could have had an effect on the range of the Siberian sturgeon (Baulin 1970). Most likely, the complete isolation of its populations occupying separate river basins has occurred rather recently. The last (Zyryanian) glaciation terminated about 10,000 years ago (Imbry and Imbry 1988). Second, the morphological similarity of the northern populations can also be convergential as a result of living under similar climatic conditions. This hypothesis is supported by the phenomenon of parallel latitudinal clinal variation in several morphological characters within large river basins (Ruban 1989a,b, 1992, 1998, Ruban and Panaiotidi 1994).

It can be assumed that the Siberian sturgeon as a freshwater species formed under the influence of multiple isolation events from the seas by glaciation during the quaternary period (Baulin 1970) as well as the lack of sufficiently productive zones in the Arctic Ocean. Consequently, adaptive radiation of Siberian sturgeon was directed not to the formation of distinct ecological forms (hiemal and vernal), typical for anadromous sturgeon (particularly of the Ponto-Caspian basin), but to formation of continuous spectrums (rows) of populations, so-called population continuums of resident forms over wide ranges in a single basin. Because of the lack of anadromous form migrating at various distances in the rivers, such structure allows this species to use efficiently both foraging and spawning grounds of extended river basins, using other mechanisms of spatial and temporal differentiation in spawning.

Under conditions of short hydrological summer in Siberian rivers, the existence of seasonal races of sturgeon, i.e., different groups of sturgeon that reproduce in the same spawning ground at different times, is impossible. The selection of habitat, including spawning habitat, occurs via the formation of population continua along the river, allowing different populations inhabiting different sections of the river to reproduce in the optimal time without performing energetically expensive migrations. It gives certain advantages under conditions of the limited food supply in northern rivers. This population structure is typical over the most part of the Siberian sturgeon's habitat. There are only two exceptions caused by specific ecological conditions of water bodies.

The population inhabiting the Ob-Irtysh basin has a more complex structure than the populations of other Siberian rivers. Besides the resident forms, there is also the so-called "anadromous" form performing potamodromous migrations. The principal differences of this form from anadromous sturgeons inhabiting Ponto-Caspian basin were described above. It is obvious that the existence of this form is determined by the unique feature of the Ob basin, the annual winter hypoxia, which compels all fish to abandon the lower reaches of the Ob and Irtysh rivers (Votinov 1963). In the absence of this event, the population structure of the Ob sturgeon would most likely be similar to that of the populations in other large river basins. Other authors (Malyutin 1980) share this opinion. In contrast to anadromous species of Ponto-Caspian basin, the Siberian sturgeons migration in the Ob River basin triggered by the oxygen saturation of water.

A separation between foraging and spawning grounds is characteristic for the Siberian sturgeon inhabiting Lake Baikal. On the one hand, food supply in the spawning rivers (Selenga, Bagruzin, and Upper Angara) is rather scarce. On the other, Lake Baikal does not provide appropriate spawning grounds. Thus, the sturgeon is compelled to perform extended potamodromous migrations from the feeding grounds of Lake Baikal to spawning grounds in the rivers.

2.2 Adaptive Features of the Siberian Sturgeon Connected with Population Structure and Ecology

The population structure, including multiple age cohorts, is typical of all sturgeon species and the Siberian sturgeon as well. This structure provides certain advantages under unstable conditions, since low numbers or even total collapse of a single generation does not have too great impact on the population as a whole (Nikolskii 1974). The features of the age and sexual structure of the Siberian sturgeon's populations such as lability of life span and ratio of males and females depending on environment conditions may be interpreted as adaptive. As was shown above, the life span of males and females is similar in populations with higher food supply and high growth rate. In populations living under worse foraging conditions, whose growth rate is much lower, the life span of males is almost twice less than that of females. For the majority of sturgeon species, it is typically the numerical prevalence of males at spawning grounds, but in populations of the Siberian sturgeon, this phenomenon is less expressed, and more often the sex ratio at spawning grounds is close to 1:1 (Ruban 2005). This allows the species to maintain high reproduction levels under low food supply conditions.

The first group of adaptations according to Gerbilskii (1962, 1967), providing high degree of euryoky, also includes feeding patterns. All acipenserids are to some extent omnivorous. However, in the Siberian sturgeon, this property is more expressed than in separate species of Caspian sturgeons. In overwhelming majority of cases, the Siberian sturgeon is nonselective benthophage. The diet composition of the sturgeon reflects the composition of benthos at foraging grounds (Romanova 1948). Moreover, only in the Siberian sturgeon, there are populations (Lake Baikal) that at certain age switch almost entirely to piscivory from benthic diet. This allows the Baikal sturgeon to attain even greater growth rates than the fastest growing riverine sturgeon (the Ob population) inhabiting the most favorable thermal conditions. A high level of omnivory is further supported by observations of Indigirka sturgeon foraging by drown rodents (Ruban 2005). Thus, intraspecific differentiation on the type of feeding in various populations of the Siberian sturgeon from primary benthophage to mainly carnivorous feeding reaches a level close to distinctions between representatives of *Acipenser* and *Huso* in the Caspian Sea basin.

As an adaptation, related to scarcity of Siberian sturgeon foraging source, the fact that most populations of the sturgeon feed at the main part of the range through the winter at the water temperature close to 0 °C and that populations in the more severe northeastern parts of the range feed through the spawning period as well can be considered. On the contrary, anadromous Ponto-Caspian sturgeon species stop feeding during long-term spawning migration and spawning time.

2.3 Adaptive Features of the Siberian Sturgeon Connected with Its Physiology, Reproduction, and Development

Following Gerbilskii (1967), two other groups of adaptations include the ecological features of sturgeons promoting intensification of their reproduction and reduction of offspring mortality. It is obvious that these groups of adaptations are interconnected and their division is rather conventional. It concerns to such features of sturgeons as wide limits of spawning temperatures and embryonic temperature. The analysis of our materials and data of other authors shows that the Siberian sturgeon does not only spawn at lower average temperatures than sturgeons of southern water bodies, but it can develop under wider limits of temperature. The threshold temperatures of embryonic development in the Siberian sturgeon are about 8–22 °C (Gisbert and Williot 2002), while the optimal temperature is 11.4–14.9 °C (Nikolskaya and Sytina 1974), whereas for stellate sturgeon from the Don River and Russian sturgeon from the Black Sea, the temperature ranges are 17–24 °C and 15–21 °C, respectively (Detlaf and Ginzburg 1954). The threshold temperatures of embryonic development are 8–25 °C for Siberian sturgeon, 12–25 °C for stellate sturgeon in the Volga River, 12–22 °C for stellate sturgeon migrating in the Don River in spring, 10–20 °C for Russian sturgeon migrating in the Don River in spring, and 14–23 °C for Russian sturgeon migrating to the Kura River in autumn (Detlaf and Ginzburg 1954).

Some physiological features of the Siberian sturgeon connected with maturation can be considered as adaptations promoting intensification of its reproduction. In general, acipenserids are characterized by late maturation. The Siberian sturgeon likewise matures at a relatively late age, though, as shown above, it displays considerably more plasticity in the relationship between growth and age to sexual maturity as a function of food supply. The size of maturing sturgeon females in high-growth rate populations (Ob, Baikal) is roughly 1.5 times greater than in the slow-growing populations of the Lena, Indigirka, and Kolyma. The age at sexual maturity for the fast-growing populations is likewise greater. These differences are highly significant. Females of the Baikal sturgeon, whose growth rate is high, spawn for the first time at the age which is almost two times more (20–22 years) than that in females of the sturgeon from the Lena (11–13 years). Observations from warmwater hatcheries cultivating Lena sturgeon have provided evidence of the relationship of size and age at sexual maturity to the availability of food. In an environment with elevated temperatures and abundant food, the Lena sturgeon, typically one of the slowest growing of the Siberian sturgeon in its natural habitat, exhibits a lowering of age at maturity by a factor of around 1.5 and an increase in length likewise by a factor of 1.5. The linear size of maturing females reared under such conditions attains the size of the first spawning females from the Ob population whose growth rate is high. Sexual maturation in the majority of fish species is connected with reaching a certain size characteristic of these species independently on age (Nikolskii 1974). However, at the boundaries of the habitat range, there are deviations in age and size at maturity from the given rule, expressed in the change of the relationship between generative metabolism and somatic growth (Shatunovskii 1986).

The significant differences in the level of generative metabolism were found among populations of the Siberian sturgeon with different growth rates. The greatest values of the gonadosomatic index (up to 66.8%) were observed in females from the Lena River, whose growth rate was the lowest. This parameter in females from the Ob River with high growth rate was much lower (up to 33%). The females from the Lena River at the repeated sexual cycles have the greatest values of relative fecundity and high content of fat in gonads at IV stage of maturity in comparison with other riverine populations (Akimova 1985a, b).

The same tendency can be found at interspecific level. The gonadosomatic index of mature females in Volga River Russian sturgeon living at more favorable conditions varies from 12.5 to 22.9% (Pavlov 1972) and is significantly lower than in the Siberian sturgeon.

Shatunovskii (1986) discovered some metabolic adaptations of fishes in the northern borders of the range. At these parts of the range, fish have low efficiency of somatic growth and elevated levels of energetic metabolism and generative tissue synthesis. In populations of Siberian sturgeon living in the most severe conditions (Lena and Indigirka rivers) and having the smallest growth rates, also a significant increase in generative metabolism (generative tissue synthesis) level and decrease of somatic growth efficiency were observed. It resulted in maturation at earlier age, smaller size, increase of relative fecundity, and relative gonad weight of females. The increase of generative metabolism level was accompanied by an increase in duration of interspawning intervals of females, which were not less than 5 years in this part of the range. These features can be interpreted as adaptive and directed to the maintenance of high numbers under conditions of low food supply.

According to the classification of Gerbilskii (1967), such features of Siberian sturgeon development as accelerated postembryonic development of its larvae in comparison with southern species of sturgeon and the absence of a pelagic feeding period can be attributed to adaptations ensuring the decrease of offspring mortality. Siberian sturgeons prelarva convert to a near-bottom lifestyle almost immediately after termination of yolk nutrition (Malyutin 1980).

2 What Distinguishes the Siberian Sturgeon from Ponto-Caspian Sturgeon?   35

By experimental data (Gisbert et al. 2002) at 4 days post hatching (14.8–15.6 mm TL) at water temperature 18 °C, there is a transition from pelagic to benthic behavior in free embryos. This period seems to mark the end of passive downstream migration in the river. At 5–6 days post hatching (17.5–18.3 mm TL), the fish's tactile bond with the ground is established, and free embryos become benthic swimmers that exhibit a positive rheotactic response and swim vigorously against the water current. Benthic-free embryos 7–8 days old (17.9–19.5 mm TL) aggregate in shoals to look for bottom-covered habitats and show a preference for dark bottoms (Gisbert et al. 1999). Similar behaviors have been reported in the pallid and shovelnose sturgeon (Kynard et al. 2002), while in the Russian sturgeon (*Acipenser gueldenstaedtii*), giant sturgeon (*Huso huso*), stellate sturgeon (*A. stellatus*), and sterlet (*A. ruthenus*), those changes in behavior take place after the transition to exogenous feeding, although those ontogenetic behavioral changes are not synchronous within each species (Khodorevskaya et al. 2009). Contrary to potamodromous species (Siberian sturgeon and sterlet), anadromous species (Russian sturgeon, giant sturgeon, and stellate sturgeon) possess the behavioral mechanisms described by Pavlov (1979) that neutralize rheotaxis and enable larvae and fingerlings to regulate and continue their downstream migration (Khodorevskaya et al. 2009).

At the juvenile stage, the Siberian sturgeon disperse into different sections of the tributary system of the river (side channels, inlets, oxbow lakes) (Votinov and Kas'yanov 1978) and look for bottom-covered habitats (Gisbert et al. 1999). This behavior differs from that of Ponto-Caspian anadromous sturgeon species (giant sturgeon, Russian sturgeon, and stellate sturgeon), whose juveniles continue to migrate downstream and forage in the riverbeds (Khodorevskaya et al. 2009). Juvenile and adult Siberian sturgeon migratory patterns depend on the ecological characteristics of each river basin. Because of the lack of anadromous forms, in all large Siberian rivers this species is represented by population continuums. In many cases, the foraging range also includes the spawning area (Ruban 2005). Ontogenetic changes in Siberian sturgeon behavior form a species-specific mechanism to maintain the population continuums without significant mixture of local populations within the river (Gisbert and Ruban 2003).

Accelerated postembryonic development of the Siberian sturgeon larvae in comparison with Ponto-Caspian species of sturgeon and the absence of a pelagic feeding period is based on general relationships between the rate and efficiency of yolk consumption, growth, and development of larvae depending on temperature during embryonic and postembryonic periods. A decrease in temperature during the period of embryonic development of fish results in hatching of larger prelarva with a greater amount of residual yolk and also greater protein increase of prelarvae during the period from hatching to the beginning of a mixed diet (Novikov 1991).

Siberian sturgeon developing at lower temperatures than the Ponto-Caspian sturgeons has the larger sizes of hatching prelarva than Russian sturgeon (10.5 mm and 9.4–9.6 mm, respectively) (Chusovitina 1963, Detlaf et al. 1981). The weight of mature eggs in Russian sturgeon varies from 15.1 to 26.0 mg (Trusov 1964, Barannikova 1970, Amirkhanov 1972). Siberian sturgeon ripe eggs are a little smaller—from 10.8 to 25 mg (Sokolov 1965b, Sokolov and Malyutin 1977, Akimova 1978). These data show that the efficiency of yolk utilization in the Siberian sturgeon developing at lower temperatures is higher than in Russian sturgeon. Despite the smaller size of the eggs, this results in larger size of hatchling prelarva in Siberian sturgeon than in Russian sturgeon. It is known that decrease in temperatures during the period from hatching to the stage of mixed feeding (simultaneous yolk and exogenous feeding) results in an increase in prelarvae protein accretion (Novikov 1991). The latter probably also explains the larger size of the Siberian sturgeon larvae at the stage of transition to active feeding in comparison with the Russian sturgeon larvae developing at higher temperature. Differences in the size of larvae between Siberian sturgeon and Russian sturgeon are about 12% at this stage of development. Most likely, the smaller amount of fat content in mid-intestine cells and in the spiral fold of the Siberian sturgeon prelarvae observed at the moment of their transition to mixed feeding in comparison with Russian sturgeon can be explained by more complete yolk utilization for protein growth in the Siberian sturgeon. Thus, the resistance to starvation of the Siberian sturgeon prelarvae is lower that of the Russian sturgeon. This difference is due to ecological dissimilarity of these species. Russian sturgeon larvae undertake an extended downstream migration, whereas the transition to external (exogenous) feeding in Siberian sturgeon larvae takes place near the spawning grounds (Bogdanova 1972).

Thus, it seems to be most probable that just the low temperature during the period of development of the Siberian sturgeon in comparison with Ponto-Caspian sturgeons (determining high efficiency and low rate of yolk utilization, high protein growth) also results in the large size of its larvae at the stage of transition to the active feeding and provides the opportunity of transition to near-bottom lifestyle missing the period of pelagic feeding.

These features of early ontogeny in Siberian sturgeon can be considered to be adaptations to conditions in northern water bodies with short vegetation periods and low planktonic biomass in the riverbeds where the development of Siberian sturgeon occurs.

In conclusion, the Siberian sturgeon exhibits a complex of specific adaptations determining the species' great ecological plasticity, allowing the Siberian sturgeon to inhabit a wide and diverse habitat range.

References

Akimova NV (1978) Gametogenez, funktsionirovaniye polovykh zhelez sibirskogo osetra (*Acipenser Baeri* Brandt) r. Lena I ikh svyas' s obmenom veshchestv (Gametogenesis, functioning of the sexual glands in Siberian sturgeon (*Acipenser Baeri* Brandt) in the Lena River and their link with metabolites) Ekologomorfoligicheskie i ekologo fiziologicheskie issledovaniya razvitiya ryb (Ecomorphological and ecophysiological studies of fish development). Nauka, Moscow, pp 43–55

Akimova NV (1985a) Gametogenez i polovaya tsiklichnost' sibirskogo osetra v yestestvennykh i eksperimentalnykh usloviyakh (Gametogenesis and reproductive cycles of the Siberian stur-

2 What Distinguishes the Siberian Sturgeon from Ponto-Caspian Sturgeon? 37

geon under natural and experimental conditions). Osobennosti reproduktivnikh tsiklov u ryb v vodoemakh raznykh shirot USSR (Characteristics of reproductive cycles in fish at different latitudes of the USSR). Nauka, Moscow, pp 111–122

Akimova NV (1985b) Gametogenez i razmnozheniye sibirskogo osetra (Gametogenesis and reproduction in Siberian sturgeon). PhD dissertation, Institute of Evolutionary Morphology and Animal Ecology, USSR Academy of Sciences, Moscow. p 243

Amirkhanov MI (1972) Sostoyaniye gonad osetra v period nerestovogo khoda v r. Tereke (the state of sturgeon gonads during the spawning migration in the Terek River). Trudy Tsentralnogo Nauchno-Issledovatelskogo Instituta Osetrovogo Rybnogo Khozyaistva 4:26–29

Barannikova IA (1970) Novye dannye o reaktsii populyatsii osetrovykh na narusheniye usloviy migratsiiy i razmnozheniya (new data on the reaction of sturgeon populations on disturbances of migration and reproduction conditions). Trudy Tsentralnogo Nauchno Issledovatelskogo Instituta Osetrovogo Rybnogo Khozyaistva 11:12–19

Baulin VV (1970) Istoriya podzemnogo oledeneniya Zapadnoi Sibiri v svyazi s transgressiei arkticheskogo basseyna (The history of the underground glaciacion at West Siberia in connection with the Arctic basin transgression). The Arctic Ocean and Its Seaboard During Cainozoe, Leningrad, pp 404–409

Berg LS (1962) Ryby basseyna Amura. Obshchaya biologiya, biogeoghrafiya i paleoichtyologiya (Fishes of the Amur River basin. General biology, biogeography and paleoichthyology) In: Nikolskii GV, Obrutchev DV (eds) L.S. Berg-selected transactions. 5, Moskva-Leningrad, p 320–360

Bogdanova LS (1972) Ekologitcheskaya plastichnost lichinok I molodi osetrovykh. Osetrovye I problemy osetrovogo khosyaistva. Pischevaya promyshlennost, Moscow, pp 244–250

Chusovitina LS (1963) Postemrional'noye razvitiye sibirskogo (*Acipenser baeri* Brandt) osetra (Postembryonal development of the Siberian sturgeon (*Acipenser baeri* Brandt)). Trudy Ob-Tazovskogo otdeleniya Gosudarstvennogo nauchno-issledovatelskogo instituta ozernogo i rechnogo rybnogo khozyaistva (GosNIIORKh) III:103–114

Detlaf TA, Ginzburg AS (1954) Zarodyshevoye razvitiye osetrovykh ryb (sevryugi, osetra i belugi) v svyazi s voprosami ikh razvedeniya (the embryonal development of sturgeons (Stellate sturgeon, Russian sturgeon, Giant sturgeon) in relation with problems of their breeding). Izdatelstvo Akademii Nauk SSSR, Moscow, p 204

Detlaf TA, Ginzburg AS, Shmalgauzen OI (1981) Razvitie osetrovykh ryb. Sozrevanie yaits, oplodotvorenie, razvitie zarodyshei i predlichinok (the development of sturgeons. Eggs maturation, fertilization, development of embryos and prelarvae). Nauka, Moscow, p 224

Gardiner BG (1984) Sturgeon as living fossils. Living fossils, New York, pp 148–152

Gerbilskii NL (1957) Puti razvitiya vnutrividovoi biologicheskoi differntsiatsii, tipy anadromnykh migrantov i vopros o migratsionnom impul'se u osetrovykh (the directions of development of intraspecific biological differentiation, types of anadromous migrants, and the problem of migration impulse in sturgeons). Uchenye zapiski Leningradskogo Gosudarstvennogo Universiteta Seria biologicheskikh nauk 228:11–32

Gerbilskii NL (1962) Teoriya biologicheskogo progressa osetrovykh i eyo primeneniye v praktike osetrovogo khozyaystva (the theory of sturgeon's biological progress, and its use in practice of sturgeon fishery). Uchenye zapiski Leningradskogo Gosudarstvennogo Universiteta Seriya biologicheskikh nauk 311:5–18

Gerbilskii NL (1965) Teoriya biologicheskogo progressa i eyo primeneniye v rybnom khozyaystve (The theory of biological progress, and its use in fishery). Theoretical basis of fishery. Nauka, Moskva, pp 77–84

Gerbilskii NL (1967) Izicheniye funktsional'nukh osnov vnutrividovoi evolyutsii v svyazi s problemoi chislennosti i areala v rybnom khozyaistve (the study of functional bases of intraspecific evolution in relation with problems of fish quantity and range in fishery). Vestnik Leningradskogo Gosudarstvennogo Universiteta 3:5–21

Gisbert E, Williot P, Castello-Orvay F (1999) Behavioral modifications in the early life stages of Siberian sturgeon (*Acipenser baeri*, Brandt). J Appl Ichthyol 15:237–242

Gisbert E, Williot P (2002) Duration of synchronous egg cleavage cycles at different temperatures in Siberian sturgeon (*Acipenser baerii*). J Appl Ichthyol 18:271–274

Gisbert E, Williot P, Berni P, Billard R (2002) Elevage intensif (Intensive rearing). Esturgeons et caviar, Editions TEC & DOC, Paris, pp 107–128

Gisbert E, Ruban G (2003) Ontogenetic behavior of Siberian sturgeon, *Acipenser baerii*: a synthesis between laboratory tests and field data. Environ Biol Fish 67:311–319

Imbry D, Imbry K (1988) Tainy lednikovykh epokh. Poltora veka v poiskakh razgadki (Mysteries of glacial epochs. One and a half century in search of a solution). Progress, Moscow, p 263

Khodorevskaya RP, Ruban GI, Pavlov DS (2009) Behaviour, migrations, distribution, and stocks of sturgeons in the Volga-Caspian Basin, World Sturgeon Conservation Society: Special Publication №. 3. Books on Demand GmbH, Norderstedt, Germany, p 233

Kynard B, Henyey E, Horgan M (2002) Ontogenetic behavior, migration and social behavior of pallid sturgeon. *Scaphirhynchus albus,* and shovelnose sturgeon, *S. platorynchus,* with notes on the adaptive significance of body color. Env Biol Fish 63:389–403

Malyutin VS (1980) Osobennosty ekologii lenskogo osetra i puti ego vosproizvodstva (Features of the Lena River sturgeon ecology and ways of its restocking). PhD Thesis, Moscow, p 159

Nikolskaya NG, Sytina LA (1974) Zona temperaturnykh adaptatsiy pri razvitii ikry osetra r. Lena (Zone of temperature adaptations during the Lena River sturgeon eggs development). Tezisy otchetnoi sessii Tsentralnogo Nauchno-Issledovatelskogo Instituta Osetrovogo Rybnogo Khozyaistva (VNIRO), Moscow, pp 108–109

Nikolskii GV (1974) Teoriya dinamiki stada ryb kak biologicheskaya osnova ratsionalnoi expluatatsii i vosproizvodstva rybnykh resursov (Theory of fish stock dynamics as a biological base of rational exploitation and restocking of fish resources). Pishchevaya promyshlennost, Moscow, p 447

Novikov GG (1991) Osobennosty rosta i energetiki razvitiya kostistykh ryb v rannem ontogeneze (Features of growth and developmental energetics of teleostean fishes during early onthogenesis). Abstract of doctors dissertation, Moscow, p 44

Pavlov AV (1972) Analiz nerestovoi populyatsii volzhskogo osetra v 1968 g. (Analysis of spawning population of the Russian sturgeon in the Volga River in 1968). Tr Tsentr Nauchno-Issled Inst Ozern Rechn Ryb Khoz 4:14–26

Pavlov DS (1979) Biologicheskie osnovy upravleniya povedeniem ryb v potoke vody (Biological principles of control over fish behaviour in water flow). Nauka, Moscow, p 319

Romanova GP (1948) Pitaniye ryb v nizhnem Yenisee (Feeding of fishes in down streams of the Yenisei River). Trudy Sibirskogo otdeleniya Vsesoyuznogo nauchno-issledovatelskogo instituta ozernogo i rechnogo rybnogo khozyaistva (Proceedings of the Siberian branch of All Union Institute of Lake and River Fish Husbandry). Krasnoyarsk 2:151–203

Ruban GI (1989a) Clinal variation of morphological characters in the Siberian sturgeon, *Acipenser baeri,* of the Lena River basin. J Ichthyol 29:48–55

Ruban GI (1989b) Morfologicheskaya izmenchivost' sibirskogo osetra basseina r. Leny (Morphological variation in the Siberian sturgeon of the Lena River basin). Morfologiya, ekologiya i povedenie osetrovikh (Morphology, ecology and behavior in acipenserids). Nauka, Moscow, pp 5–16

Ruban GI (1992) Plasticity of development in natural and experimental populations of siberian sturgeon *Acipenser baeri* Brandt. Acta Zool Fenn 191:43–46

Ruban GI (1998) On the species structure of the Siberian sturgeon *Acipenser baerii* Brandt (Acipenseridae). J Ichthyol 38(5):345–365

Ruban GI (2005) The Siberian sturgeon *Acipenser baerii* Brandt. Species structure and ecology, World Sturgeon Conservation Society. Special Publication No. 1. Books on Demand GmbH, Norderstedt, Germany, p 203

Ruban GI, Panaiotidi AI (1994) Comparative morphological analysis of subspecies of the Siberian sturgeon, *Acipenser baerii stenorrhynchus* A. Nikolsky and *Acipenser baerii chatys* Drjagin (Acipenseriformes, Acipenseridae), in the Yenisey and Lena rivers. J Ichthyol 34:58–71

Severtsov AN (1925) Morfologicheskie zakonomernosti evolyutsii (Morphological regularities of evolution). Moskow, p 132

Shatunovskii MI (1986) O nekotorykh metabolicheskikh adaptatsiyakh ryb na severnykh granitsakh arealov (On some metabolic adaptations' of fishes at the north borders of their range).

Tezisy doklada na Vsesoyuznom soveshchanii Organizmy, populatsii i soobshchestva v ekstremalnykh usloviyakh, Moscow, pp 143–144

Shtylko BA (1934) Neogenovaya Fauna presnovodnykh ryb Zapadnoi Sibiri (The Neogene fauna of freshwater fishes in West Siberia). Trudy Vsesoyuznogo geologorazvedochnogo obyedineniya 359:7–23

Sokolov LI (1965b) Sozrevanie i plodovitost' sibirskogo osetra *Acipenser baeri* Brandt r. Leny (Maturation and fecundity of the Siberian sturgeon *Acipenser baeri* Brandt from the Lena River). Voprosy Ikhtiologii (Issues in Ichtyology) 5(1):70–81

Sokolov LI, Malyutin VS (1977) Osobennosti struktury populyatsii i khrakteristiki proizvoditelei sibirskogo osetra r. Leny v raione nerestilishch (The structure of the population and spawners characteristics in the Siberian sturgeon from river Lena at spawning sites). Voprosy Ikhtiologii (Issues in Ichtyology) 17(2):327–246

Trusov VZ (1964) Nekotorye osobennosty sozrevaniya I shkala zrelosty polovykh zhelez osetra (Some peculiarities of maturation and the scale of sexual glands maturity in sturgeon). Trudy of All-Union Research Institute for Sea Fisheries and Oceanography (VNIRO) 56(3):69–78

Votinov NP (1963) Biologicheskie osnovy iskusstvennogo vosproizvodstva obskogo osetra (Biological foundations of artificial reproduction of the Ob River sturgeon). Trudy Ob-Tazovskogo Otdeleniya Gosudarstvennogo Nauchno-Issledovatelskogo Instituta Ozernogo i Rechnogo Rybnogo Khozyaistva, Novaya Seria 3:5–102

Votinov NP, Kas'yanov VP (1978) K voprosu o vosproizvodstve sibirskogo osetra v basseine r. Obi v usloviyakh zaregulirovannogo stoka (On the problem of of the Siberian sturgeon) *Acipenser baeri* Brandt reproduction in the Ob River as affected by hydraulic structure. Vopr Ikhtiol 18:25–35

Anatomic Description

3

Mikhail Chebanov and Elena Galich

Abstract

As known, sturgeon has a range of anatomical peculiarities; those are significantly different from the anatomy of other fish species. This is most strongly manifested in the structure of following internal organ systems: digestive, excretory, hemopoietic, and reproductive ones.

The current chapter provides a brief description of the anatomy of inner organs of Siberian sturgeon. The anatomic features, that distinguish Siberian sturgeon from other species of the family, are presented in more details.

In particular this pertains to the structure of the digestive system (length of the intestine), peculiarities in formation of tissues and organs during early ontogenesis (the dorsal embryonic process corresponds to the long lobe of the pancreas), and the pancreas that in terms of its histology and functions differs cardinally from other fish species, but is similar to one in mammals.

Keywords

Siberian sturgeon • Anatomy • Digestive system • Heart • Kidneys

---

M. Chebanov (✉) • E. Galich
State Regional Centre for Sturgeon Gene Pool Conservation "Kubanbioresursi",
275/1 Severnaya Street, Krasnodar 350020, Russian Federation

Department of Aquatic biological resources and Aquaculture, Kuban State University,
149 Stavropolskaya Street, Krasnodar 350040, Russian Federation
e-mail: MChebanov@gmail.com, sturg@land.ru

© Springer International Publishing AG, part of Springer Nature 2018
P. Williot et al. (eds.), The Siberian Sturgeon (*Acipenser baerii,* Brandt, 1869)
Volume 1 - Biology, https://doi.org/10.1007/978-3-319-61664-3_3

3.1 The Digestive System

Due to its type of anatomy, the digestive system in sturgeon is intermediate between bony and cartilaginous fish. Both the system of loops, formed by the intestine (typical for bony fish), and the preservation of the spiral valve in the midgut (peculiarity for cartilaginous fish) are peculiarities of the digestive tract, increasing the area of the suction surface of the gastrointestinal epithelium (Gurtovoy et al. 1976). Cataldi et al. (2002) reported the gut ontogenesis description of the *Acipenser naccarii*. The morphological and histochemical differences in the structure of the alimentary canal in feeding and runt (feed deprived) white sturgeon were studied by Domeneghini et al. (2002). Daprà et al. (2009) provided a detailed description of the gut of Siberian sturgeon. The digestive system in sturgeon comprises the digestive tube, liver, and pancreas (Fig. 3.1).

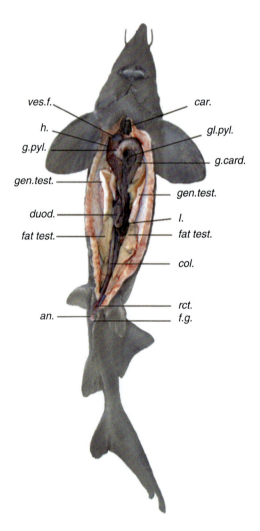

Fig. 3.1 Structure of inner organs of Siberian sturgeon. *Left*: *ves.f.*—gall bladder; *h.*—liver; *g.pyl.*—pyloric part of stomach; *duod.*—duodenum; *fat test,*—fat part of testes; *an.*—anal opening. *Right*: *car.*—heart; *gl.pyl.*—pyloric gland; *g.card*– cardial part of stomach; *gen.test*—generative part of testes; *l.*—spleen; *col.*—colon; *rct.*—rectum; *f.g.*—genital opening

3 Anatomic Description

The digestive tubes in male and female Siberian sturgeon appear similar. The digestive tube of different sturgeons has a similar shape (Cleveland and Hickman 1998; Cataldi et al. 2002; North et al. 2002) and differs primarily only in its volume and length (Randal and Buddington, 1985; Vajhi et al. 2013). Presumably, due to fish nutrition, the digestive tube is rather short (Hildebrand and Goslow 2001; Icardo and Colvee 2002; Vajhi et al. 2013).

The digestive tube is conventionally divided into four sections:

1. Headgut—from the mouth to the last pair of gill arches
2. Foregut—subdivided into the esophagus and stomach
3. Midgut—consisting of a small intestine and spiral intestine (the longest part of the digestive tube)
4. Hindgut (Schmalhauzen 1968)

For all sturgeon species, it is typical to have a retractable jaw apparatus in the form of the funnel mouth—a special device for setting the prey from the bottom. The lack of teeth is typical for adult sturgeon as distinct from the larvae.

A part of the headgut, namely, the throat (*pharynx*), that is behind the oral cavity is lined with multilayered epithelium; the gill slits are opened to the throat region. The muscles of the *pharynx* are cross striped. The upper and lower pharyngeal teeth are evident in the throat.

3.1.1　The Esophagus

The esophagus (*oesophagus*) is a continuation of the oral cavity, located under the spine (at a transverse section). The transversely striated muscles of the esophagus incorporate two layers: an inner that is longitudinal and outer that is circular. The mucosa of the esophagus comprises characteristic longitudinal folds from which epithelial outgrowths depart (Schmalhauzen 1968). These outgrowths contain a lot of mucus-secreting cells (Muskalu-Naji et al. 2006). Epithelial outgrowths in the form of leaf buds are arranged in longitudinal rows and, during the transition into the stomach, are gradually transformed into the longitudinal wavelike folds along all the cardinal part of the stomach as 6–7 rows.

3.1.2　The Stomach

The stomach (*gaster*) has a siphon form (U-shaped) and makes up to 40% of the digestive tube; a downward part (cardiac, adjacent to the esophagus) and an upward part (pyloric, separated from the midgut by a pyloric valve) are distinguished in the stomach.

The inner surface of the stomach, similar to that of the esophagus, has longitudinal folds, composed of collagen fibers and fibroblasts. The gastric epithelium is single-layered and cylindrical. The main function of the cylindrical cells is to secrete mucus. The muscular cover of the stomach is made of two layers of smooth muscles: the outer, longitudinal, and the internal, circular. The cardiac portion of the stomach is connected with the swimming bladder by a duct (*ductus pneumaticus*).

3.1.2.1 The Pyloric Part of the Stomach

The pyloric part of the stomach, surrounded by liver lobes, has a thicker muscle shell, especially in the pyloric valve, which switches into a muscular sphincter. One of the sphincter's prime functions is to adjust connection between the cavity of the stomach and the midgut cavity. It was supplemented for Persian sturgeon by Vajhi et al. (2013).

3.1.3 The Duodenum

The duodenum (*duodenum*), the front part of the anterior midgut, begins near the pyloric gland and is minimized in the form of a loop with a characteristic honeycomb relief of the mucous tunic.

Next to the sphincter of the pyloric stomach, the gallbladder, pancreas, and pyloric appendages flow into the intestine.

The histological structure of the midgut epithelium is one and the same in all its parts. The outer layers are represented by two strata of smooth muscles (a well-developed inner circular one and less developed outer longitudinal one). The inner single-layered columnar epithelium consists of three prime cellular elements: cylindrical cells, goblet, and ciliary ones.

3.1.4 The Spiral Gut

The spiral gut (*spiral colon*) is the posterior part of the midgut. It is formed by the tortuosity of the gut tube. The first appearance of the spiral valve was observed in lamprey (a spiral fold forms only half the turnover) (Suvorov 1940). Number of turns in sturgeon varies from 5 to 10 (Schmalgausen 1968). This is clearly discernable on the echogram of the Siberian sturgeon (see Chap. 49). Next to the valve that separates the spiral gut (*spiral colon*) from the intermediate one, a large lymph node is located. In addition to large folds, the small (secondary) ones are encountered on the inner surface of the gut; such a feature of the hindgut structure in sturgeon slows the passage of food and increases the suction surface of the gut (Suvorov 1940).

3.1.5 The Hindgut

The hindgut (*rectum*) is a very short part; its walls form longitudinal folds. The mucous membrane of the hindgut is composed of ciliated epithelium that contains many goblet cells. A circular layer of muscle in the muscular shell (with anal opening (*anus*) at its end) is well developed. Two small abdominal pores are located on either sides of the anal opening, while the genital opening is located behind it.

3 Anatomic Description

The entire digestive tract is surrounded from the outside by a serous membrane, consisting of a single-layered cylindrical epithelium, separated from the muscular shell by a layer of connective tissue (Schmalhauzen, 1968).

3.1.6 The Pyloric Gland

The pyloric gland (*glandula pilorica*) is made of pyloric appendages immersed in fat connective tissue and surrounded by a serous membrane, composing a single compact formation (Shmalhauzen 1968). The relief of the inner surface is cellular; the fanlike arranged cells are connected with each other by connective tissue and blood vessels (Suvorov 1940).

3.1.7 The Liver

The liver (*hepar*) is a large gland, shaped like a horseshoe and divided into two parts: right (larger) and left. The liver cells are polygonal with a central or eccentric spherical nucleus (Muskalu-Naji et al. 2006).

The liver consists of a network of glandular ducts and complex system of blood vessels—capillaries—through which whole blood from the vessels of the gastrointestinal tract enters the liver and takes off from the liver through the hepatic veins to the heart, forming a recirculation system of the liver (Shmalgauzen 1968). On the right lobe of the liver, there is the gallbladder (*vesica fellea*). The gallbladder is located at the cranio-ventral part of the right hepatic lobe and appears as a bubble-shaped sac. The protruded duct of the bladder enters the papilla near distal to the pyloric sphincter, at the same site as the orifice of the pyloric caecum in adult specimens (Vajhi et al. 2013).

On its ventral and front side, the gallbladder is covered with the small central lobe of the liver. The effluent from the gallbladder duct (*ductus choledochus*) encircles the pyloric part of the stomach, and the left at the bottom of the pyloric gland flows into the duodenum (Gurtovoy et al. 1976).

3.1.8 The Pancreas

The pancreas (*pancreas*) is located along the right side, partially entering the left side of the duodenum. Nicolas (1904) revealed that the structure named by him as "dorsal embryonic process" terminated in larvae of *Acipenser sturio* prior to the spiral valve, running dorsal of the visceral cavity.

According to Daprà et al. (2009), the dorsal embryonic process corresponds to the long lobe of the pancreas in Siberian sturgeon. The abovementioned researchers found that the long lobe showed a craniocaudal and ventrodorsal pathway. It was considerably different while being compared to the dorsal embryonic process that proved to be craniocaudal at the dorsal exposure. Moreover, the long lobe terminated in between the first two coils of the spiral valve, which was in contradiction with results obtained by Nicolas (1904) for the larvae. The presence of pancreas tissue in adult specimen of white sturgeon was reported by Gawlicka et al. (1996).

The latest studies of sturgeon pancreas have been conducted primarily in order to characterize hormone and digestive enzymes applying biochemical and proteomic approaches (Rusakov et al. 1998, Kim et al. 2000). The exocrine acini with zymogen granules in sturgeon pancreas were reported in a study of the digestive system development in the green sturgeon (*Acipenser medirostris*) larvae made by Gisbert and Doroshov (2003).

Gisbert et al. (1999) provided results of prior observations of sturgeon intestine tracts, liver, and pancreatic tissue reactivity, treated with different histological stains. However, the conducted studies were restricted to the ontogenesis of these organs, missing both the anatomical description and relations between the pancreas and other gut organs at successive stages of development.

Meanwhile, Daprà et al. (2009) focused on the pancreas as a part of detailed anatomical description of the gut in Siberian sturgeon. The pancreas in Siberian sturgeon is obvious as an organ upon dissection, separated from the liver and other abdominal organs, despite the fact that, to a considerable extent, it is associated with the gut wall and the liver by the serous membrane. The pancreatic duct ends in a papilla between the small intestine and the pyloric caecum.

The pancreas is clearly subdivided into three large lobes. The largest lobe originates from the pyloric caecum and passes laterally along the visceral cavity ending at the upper side of the spiral valve in between the first and second loop. This so-called pancreas long lobe comprises several small branches, linked with the spleen body, the right side spiral valve, and the dorsal ligament of spiral valve. Close to the spiral valve, the pancreas and the spleen are arranged in one organ, without any evidently separated connective tissue. The second lobe (left lobe) lies next to the small intestine with the ligament, which links it to the spleen. This lobe comes to its end near the S curve at the upper part of the small intestine. The third and smallest lobe (right lobe) continues along the right wall of the small intestine.

The walls of the glandular ducts and alveoli consist of two strata of cells: the glandular one and the layer of flat cells that line the alveoli of the cavity.

The exocrine cells are arranged by large eosinophilic zymogen granules and bunched together in structures similar to mammalian islets of Langerhans with inhomogeneous distribution. The morphometrical evaluation of the islet mean area provides the value 8.32×10^{-3} mm^2 (Daprà et al. 2009). The pancreas in Siberian sturgeon is a separate gland of similar histology and functionality as compared to the one in mammals, but of quite different anatomy as regards to other fish (Harder 1975).

3.1.9 The Swim Bladder

The swim bladder (*vesica pneumatika*) is linked with the esophagus and located on the dorsal side at the longitudinal section of the body, under the spine of the intestine—appearing as an oval-shaped hollow sack (Fig. 3.2). Interiorly, the

Fig. 3.2 Structure of inner organs of Siberian sturgeon. *Left*: *oes.*—oesophagus; *g.pyl.*—pyloric part of stomach; *d.pn.*—swim bladder duct; *g.card.*—cardial part of stomach; *ves.pn.*—swim bladder; *col.*—colon; *rct.*—rectum; *an.*—anal opening. *Right*: *h.*—liver; *ves.f.*—gall bladder; *gl.pyl.*—pyloric gland; *duod.*—duodenum; *pncr.*—pancreatic gland; *l.*—spleen; *s.v.*—spiral valve

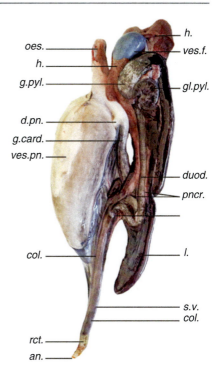

swim bladder is lined with a single layer of epithelium, followed by elastic and smooth muscle fibers. By an upper looser connective tissue cover, the swim bladder is tightly fused with the wall of the peritoneum. The excess gas from the swim bladder is removed through the air duct (*ductus pneumaticus*) (Suvorov 1940).

3.2 Blood Circulation and Blood-Making (Hemopoietic) Organs

3.2.1 The Heart

The heart (*cordis*) in sturgeons is located in the pericardial sac near the abdominal body cavity and from its ventral side is protected by a food belt. The pericardial sac—the pericardial cavity (*cavum pericardii*)—is a detached front part of the abdomen. The heart consists of four compartments (Gurtovoy et al. 1976): blood cone, ventricle, atrium, and venous sinus, possessing the ability of rhythmic pulsation.

3.2.1.1 The Arterial Cone

The arterial cone (*conus arteriosus*) is a short tube of pyramidal type, capable of independent reductions, following the contraction of the ventricle (Suvorov 1940). Cross-striped muscles constitute the basis of the walls. Inside the arterial cone are four semilunar valves (*valvula semilunaris*). The valves are endocardial folds; the presence of the valves determines the direction of blood flow in the cone. Furthermore, the blood cone proceeds to the arterial trunk, the initial part of the ascending or abdominal aorta (*aorta ventralis*).

3.2.1.2 The Ventricle

The ventricle (*ventriculus*) is a double-walled muscular sac; the inner wall is the endocarditis, while the outer is the epicardium, which has a trapezoidal shape. The outer surfaces of the ventricle and arterial cone are covered with lymphatic gland that appears as flat formation with numerous tubercles—follicular nodules (Ivanova 1983). The histological basis is the reticular syncytium and blood cells of all categories (Ivanova 1970). The fish reticular syncytium was identified in all organs, involved in the hematopoietic function. For sturgeons, it is typical to have multiple sites and organs of blood formation (kidney, spleen, vascular endothelium, intestine, gill apparatus, heart, lymphoid organ in the cranial compartment).

3.2.1.3 The Atrium

The atrium (*atrium*) is located behind the heart and appears as a kind of wrinkled sac with muscular walls (Gurtovoy et al. 1976) that is connected to the ventricle by the atrioventricular opening (*ostium atrioventricularis*). The atrium communicates with the rear compartment of the heart—the venous sinus (*sinus venosus*)—where venous blood flows from around the body of the fish.

3.2.2 The Spleen

The spleen (*lien*) is the largest lymphoid organ in which lymphocytes are produced, being as well the site where both a formation and destruction of red blood cells occurs (Shmalgauzen 1968). The spleen encircles the duodenum on the right and on the left and is attached with mesentery to the cardinal part of the stomach and then to the duodenum. The spleen has two lobes. Under the lower lobes of the liver, it forms a separate additional lobe. In Siberian sturgeon this extra lobe reaches a considerable size.

3.3 The Excretory System

3.3.1 The Kidneys

The kidneys (*ren*) in sturgeons are mesonephric, having the form of ribbon-shaped paired organs, merging into unpaired tail buds behind the swim bladder, reaching the anal opening, and are covered by the peritoneum lining. The glandular structure of the Siberian sturgeon kidneys is presented by cellular strands, embodying urinary tubules, ending with the Malpighian body of a spherical shape. The Malpighian body is a capsule with a ball of capillaries (*glomerulus*) encapsulated inside.

3 Anatomic Description

The presence of mesangial cells in glomerulae in sturgeons of the Caspian Sea was reported by Gambaryan (1984) and Charmi et al. (2009, 2010). The liquid part of urine is derived from the blood of sturgeon through glomerulae in the Malpighian ball, salt, and urea through the cells of the urinary tubules (Suvorov 1940).

According to Fange (1986), the head of this kidney in sturgeons is a versatile organ involved in the making of blood. This was supported lately by the results gained by Charmi et al. (2010).

Krayushkina et al. (1996) proved that the highest level of nephrons occurs mostly in caudal tissue of the kidney that is composed of nephrons dispersed through the organ blood-forming tissue.

The kidneys in sturgeon are a universal organ of hematopoiesis; the head part of the kidney fulfills the function of blood cell destruction and simultaneously is the site of all types of blood cell maturation (white blood cells and red blood cells). The kidneys are involved in the regulation of ion composition and the water-salt homeostasis and perform secretory functions (Natochin 1983). The kidneys are penetrated by blood vessels, forming a portal system of the kidneys (Gurtovoy et al. 1976).

3.3.2 The Ureters

The ureters (*ureter*) are primary renal ducts in sturgeon that also serve as sperm ducts (*vas deferens*). Starting at the front edge, from the outside the kidneys form a common duct by the renal tubules. The broad funnels are attached to this duct at the distal end of the swim bladder. Both in females and males, the mature sexual products are ejected through these funnels. Multiple nephridial channels are falling into the urethra. The urethra epithelium is a columnar epithelium.

3.4 Reproductive System

3.4.1 The Ovaries

The ovaries (*ovarian*) are paired female organs of irregular prolonged form, narrowed at the caudal ends. The ovaries are attached to the dorsal surface of the abdomen by mesenteries (*mesovarium*), lying on either side of the esophagus. They consist of ovicells of various generations. Each egg is separated by a thin layer of connective tissue.

3.4.2 The Oviducts

The oviducts (*oviductus*) are excretory channels lying on the outer side of the ovaries and appear as wide tubes opened into the body cavity via wide craters. The oviducts closer to the anus merge with each other and have a joint opening.

3.4.3 The Testes

The testes (*testis*) are a male paired organ of irregular elongated form, rounded at the ends. They are located on the side of the esophagus and attached to the dorsal surface of the body cavity by mesenteries (*mesorchium*). Unlike the ovaries, testes are covered with a thin connective tissue cover.

3.4.4 The Spermaducts

The spermaducts (*vas deferens*) are similar to female oviducts; at their external side, they have the oviduct funnels communicating with thin genital flow from both sides. From downward, the merging paired ducts are opened with a genital pore that is located behind the anal one.

Acknowledgments The authors wish to thank Dr. Karine Pichavant-Rafini (ORPHY laboratory, EA4324) and Michel Rafini (professor at the Language Dept.) of the Brest University (France) for their help in improving the quality of the English language.

References

Cataldi E, Albano C, Foglione C et al (2002) *Acipenser naccarii*: fine structure of the alimentary canal with references to its ontogenesis. J Appl Ichthyol 18:329–337

Charmi A, Bahmani M, Sajjadi MM et al (2009) Morpho-histological study of kidney in juvenile farmed beluga, *Huso huso*. Pak J Biol Sci 12:11–18

Charmi A, Parto P, Bahmani M et al (2010) Morphological and histological study of kidney in juvenile great sturgeon (*Huso huso*) and Persian sturgeon (*Acipenser persicus*). Am Eurasian J Agric Environ Sci 7(5):505–511

Cleveland P, Hickman JR (1998) Biology of animals. McGraw-Hill Companies. pp 591, 592, 603

Daprà F, Gai F, Palmegiano GB et al (2009) Siberian sturgeon (*Acipenser baerii*, Brandt JF 1869) pancreas. Int Aquat Res 3:15–43

Domeneghini C, Radaelli G, Bosi G et al (2002) Morphological and histochemical differences in the structure of the alimentary canal in feeding and runt (feed deprived) white sturgeons (*Acipenser transmontanus*). J Appl Ichthyol 18:341–346

Fange R (1986) Lymphoid organ in sturgeons (Acipenseridae). Vet Immunol Immunopathol 12:153–161

Gambaryan SP (1984) Micro dissection studies of the kidney of sturgeons (Acipenseridae) of the Caspian Sea basin. J Fish Biol 44:60–65

Gawlicka A, McLaughlin L, Hung SSO et al (1996) Limitations of carrageenan micro bound diets for feeding white sturgeon, *Acipenser transmontanus*, larvae. Aquac 141:245–265

Gisbert E, Saraquete MC, Williot P et al (1999) Histochemistry of the development of the digestive system of Siberian sturgeon during early ontogeny. J Fish Biol 55:596–616

Gisbert E, Doroshov SI (2003) Histology of the development digestive system and the effect of food deprivation in larval green sturgeon (*Acipenser medirostris*). Aquat Liv Res 16:77–89

Gurtovoy NN, Matveev BS, Dzerzhinsky FY (1976) Prakticheskaya zootomiya pozvonochnykh (practical zootomy of vertebrates). Vysshaya shkola (Higher School), Moscow, p 351 (in Russian)

Harder W (1975) In: Hans Richardz Publications ed. Anatomy of Fishes. Parts I and II. Schweizer-bartsche Verlagsbuchhandlung, Stuttgart, West Germany, E. Schweizerbart'sche Verlagsbuchhandlung, Pt.1:p 612, Pt.2:p 132

3 Anatomic Description

Hildebrand M, Goslow GE (2001) Analysis of vertebrate structure. Wiley, USA, p 145

Icardo JM, Colvee E (2002) Structure of the conus arteriosus of the sturgeon heart. Anat Rec 1:17–27

Ivanova NT (1983) Atlas kletok krovi ryb. Sravnitelnaya morfologiya I klassifikatsiya formennykh elementov krovi ryb (Atlas of blood cells in fish. Comparative morphology and taxonomy of formed blood elements in fish). Lyogkaya i pishchevaya promyshlennost, Moscow, p 184 (in Russian)

Ivanova NT (1970) Morfologiya krovi donskoy sevryugi pri fazovokontrastnoy mikroskopii (Blood cell morphology of Don stellate sturgeon at phase contrast microscopy). In: Materialy IV mezhvuzovskoy konferentsii fiziologov i morfologov pedinstititov (Materials of IV interacademic conference of physiologists and morphologists of pedagogical institutions), Yaroslavl. pp 148–149 (in Russian)

Kim BJ, Gadsbøll V, Whittaker J et al (2000) Gastroenteropancreatic hormones (insulin, glucagon, somatostatin, and multiple forms of PYY) from the Pallid sturgeon, *Scaphirhynchus albus* (Acipenseriformes). Gen Comp Endocr 120(3):353–363

Krayushkina LS, Panov AA, Gerasomov AA et al (1996) Changes in sodium, calcium and magnesium ion concentrations in sturgeon (*Huso huso*) urine and in kidney morphology. J Comp Physiol B 165:527–533

Muskalu-Naji K, Bergler G, Dumetresku G et al (2006) Histological aspects of the various organs of albino sterlet *Acipenser ruthenus*. Sturgeon aquaculture: achievements and prospects for development. In: Proceedings of the IV international conf. Astrakhan, 13–15 March 2006. VNIRO Publ, Moscow, pp 35–38

Natochin YV (1983) Evolyutsiya vodno-solevogo obmena i pochki (Evolution of water-salt metabolism and kidneys) Evolyutsionnaya fiziologiya (Evolutionary physiology). L. Nauka, Moscow, pp 371–426 (in Russian)

Nicolas A (1904) Reserches sur l'embryologie des reptiles. IV. La segmentation chez l'ovet. Arch Biol Paris 20:611–658

North JA, Farr RA, Vescei P (2002) A comparison of meristic and morphometric characters of green sturgeon *Acipencer medirostris*. J Appl Ichthyol 18(4–6):234–239

Randal K, Buddington J (1985) Digestive and feeding characteristics of the chondrosteans. In: North American sturgeon 2, pp 31–34

Rusakov Y, Moriyama S, Bondareva VM et al (1998) Isolation and characterisation of insulin in Russian sturgeon (*Acipenser gueldenstaedtii*). J Pept Res 51(6):395–400

Schmalhauzen OI (1968) Razvitiye pishchevaritelnoy sistemy osetrovykh (development of the digestive system in sturgeon). In Morfoekologicheskiye issledovaniya razvitiya ryb (morphological and ecological studies of fish development). Nauka, Moscow: pp 40–70 (in Russian)

Suvorov EK (1940) Osnovy obshchey ikhtiologii (Fundamentals of general ichthyology). Leningrad State University Publ, Leningrad, p 433 (in Russian)

Vajhi AR, Zehtabvar O, Masoudifard M et al (2013) Digestive system anatomy of the *Acipenser persicus*: new features. Iran J Fish Sci 12(4):939–946

The Axial Skeleton of the Siberian Sturgeon. Development, Organization, Structure and New Insights on Mineralization and Ossification of Vertebral Elements

4

Jean-Yves Sire and Amandine Leprévost

Abstract

Accurate data on axial skeleton development, structure and mineralization in Acipenseriformes are required not only to improve our knowledge on this functional region but also to know whether or not such information can help in understanding anomalies of the vertebral column in reared sturgeons. Recent data obtained using growth series of *Acipenser baerii* and various techniques such as X-ray microtomography, histological and ultrastructural descriptions, and solid-state NMR analyses have completed previous, fragmented information in Acipenseriformes, notably with regard to ossification and mineralization of the vertebrae. The axial skeleton is mostly cartilaginous, and each vertebra is composed of several elements organized around a persistent and unconstricted notochord. Typical perichondral bone is lately deposited around these elements during ontogeny and mineralizes. No resorption process of these mineralized structures was so far observed. A new mineralized tissue of unknown origin and function was discovered within the notochord. The nature of the mineral phase of all these structures is typical as they are mainly composed of amorphous calcium phosphate that progressively changes into hydroxyapatite. The few mineralized content of the vertebrae does not allow to explain the origin of the anomalies affecting the axial skeleton in farmed specimens. Future studies aiming to understand these pathologies should focus on (a) the mineral metabolism in order to identify variations of physiological markers with regard to environmental factors and (b) the structure and growth of the typical cartilage that constitutes the main component of the axial skeleton.

J.-Y. Sire (✉) • A. Leprévost
Sorbonne Universités, Université Pierre et Marie Curie-Paris 6, Institut de Biologie
Paris Seine, UMR 7138-Evolution Paris Seine, Equipe "Evolution et Développement du
Squelette", Paris, France
e-mail: jean-yves.sire@upmc.fr

© Springer International Publishing AG, part of Springer Nature 2018
P. Williot et al. (eds.), The Siberian Sturgeon (*Acipenser baerii*, Brandt, 1869)
Volume 1 - Biology, https://doi.org/10.1007/978-3-319-61664-3_4

Keywords

Acipenseriformes • Siberian sturgeon • Axial skeleton • Vertebrae • Structure
Development • Mineralization • Cartilage • Anomalies • Fish farming

Introduction

In fish farms, a significant percentage of Siberian sturgeons *Acipenser baerii*
(Brandt, 1869), mainly from 3 years old onwards, exhibit anomalies of their vertebral skeleton (lordosis, scoliosis). This pathology leads to death before being mature
enough to produce caviar, resulting in important financial loss for farmers. The factors that could be at the origin of these deformations are so far unknown, although
they are certainly related to inappropriate "environmental conditions" including
either external factors (temperature, water quality, etc.) or rearing factors (food,
pathologies, genetics, etc.), or both. Identifying causative factors is, therefore, difficult although we can believe that unfavourable environmental conditions could
result in physiological disturbances, which in turn weaken the organism and lead to
deformations of the axial skeleton.

Prior to trying to understand the causality of the axial skeleton deformities, it is
important to know as well as possible the biology of this important organ. The main
objective of this chapter is to summarize our current knowledge on the occurrence of
axial anomalies and on the biology of the vertebral column (development and growth,
organization and structural composition and mineralization) using mostly the Siberian
sturgeon as a model. Finally, we point out the fields that require further investigations
in order to understand the occurrence of axial anomalies in reared sturgeons.

4.1 Axial Anomalies in Sturgeons

Acipenser baerii was first obtained in 1975 from the river Lena (in the former
USSR) as a biological model to get experience in restoration of the European sturgeon (Williot and Rouault 1982; Williot et al. 1997) and then reared in intensive fish
farms. The first incidence of axial anomalies in reared sturgeons was reported by
Brun et al. (1991). These anomalies were observed in individuals aged 2–6 and
consisted mostly in scoliosis and in kyphosis to a lesser extent. In most cases, the
inflexion point took place either slightly in the posterior region of the body and/or
in the region near the anterior base of the dorsal fin (Fig. 4.1). The affected individuals showed difficulties in swimming, loss of balance and difficulties in finding food
and then died after a period of asthenia that could last several months. Similar disorders were currently observed in *A. baerii* in French sturgeon farms. In its last
report, the FAO (2013) indicates that this phenomenon is worldwide and not specific to a single farmed sturgeon species and that the origin of this pathology remains
unknown. However, axial anomalies are poorly documented in Acipenseriformes,

4 New Insights in the Axial Skeleton of the Siberian Sturgeon

Fig. 4.1 Severely deformed vertebral axis in a farmed specimen of *Acipenser baerii* (credit photo: A. Leprévost)

either in the wild (as the axial skeleton is weakly ossified, they may "disappear" rapidly after death) or in farming conditions (as they finally died, they were rarely recorded by the farmers). Also, it is not established, for instance, whether or not scoliosis, lordosis and kyphosis, which are externally noticeable, are the result of vertebrae anomalies. No incidence of axial anomalies in paddlefishes was reported in the literature, but this may be, in addition to the above reasons, because of the anecdotal nature of paddlefish aquaculture.

In Greek fish farms, juvenile *A. gueldenstaedtii* reared in open-flow systems were affected with scoliosis and lordosis. Histopathological studies were inconclusive for what concerned the aetiology of these symptoms, but these anomalies seemed to be related to horizontal transmission of nodavirus from infected sea bass (Athanassopoulou et al. 2004). Axial anomalies in sturgeons were also linked to pollution in the wild (Hu et al. 2009; Hou et al. 2011: *A. sinensis*), to thermal stress (Linares-Casenave et al. 2013: *A. medirostris*) or to environmental temperature (Dettlaff et al. 1993) resulting often in pathological processes (Chernyshov and Isuev 1980). However, to our knowledge these disorders affected mostly larval or juvenile stages and were expressed in young individuals. For instance, *A. baerii* larvae fed on a diet containing high levels of oxidized lipids showed accumulation of lipid peroxidation products in tissues and concomitant axial malformations (Fontagné et al. 2006).

Farmed sturgeons are nowadays still fed on commercial compound diets similar to those formulated for trout rearing (Williot et al. 2001; Williot 2009). Feed providers do not elaborate specific sturgeon food, as sturgeon production remains at a relatively low level. These diets may not meet the nutritional requirements of these species. Indeed, an experiment demonstrated that about 50% of a group of *A. sturio* fed on artificial diet showed axial anomalies, compared with only 6% in a group fed on a "natural" diet (mainly composed of shrimps) (P. Williot, pers. comm.).

Acipenseriformes are a hundred million years evolutionarily distant from teleost fishes, and their axial skeleton is mostly cartilaginous in contrast to teleost vertebrae that are composed of bone tissue. This large evolutionary distance and the important structural differences explain why our current knowledge of the origin of axial anomalies in teleost species cannot help in understanding and determining the causative factors involved in this pathology in sturgeons and why an accurate knowledge of the development, structure and mineralization of the axial skeleton is essential in order to understand any pathological change occurring during ontogeny. Now it clearly appears that the techniques generally used for the study of teleost vertebrae may not apply to sturgeon vertebrae (skeletal preparation, measures of bone mineralization rate and bone compactness) because of cartilage predominance.

4.2 Development and Growth of the Vertebral Axis

To date, a few studies only were devoted to the description of vertebral axis development in Acipenseriformes species (Arratia et al. 2001, *Acipenser ruthenus* and *Polyodon spatula*; Zhang et al. 2012, hybrid of *Huso huso* X *A. ruthenus*; Leprévost et al. 2017a, *A. baerii*).

In *A. baerii*, morphogenesis and differentiation of the vertebrae occur during prelarval stages, a period comprised between hatching [10.4–11.1 mm total length (TL)] and the first exogenous feeding [9–10 days post hatching (dph), 19.7–21.6 mm]; then vertebral elements form during larval stages, which are comprised between the first feeding and metamorphosis (19–20 dph, 28.0–32.0 mm) (Gisbert et al. 1998).

The development of the axial skeleton was recently studied using in toto clear and stain procedure and histological sections (Leprévost et al. 2017a). Cartilage anlages of basidorsals and basiventrals start to initiate at 9 dph, above and below the notochord, respectively. These elements extend in opposite directions, the basidorsals forming antero-posteriorly and the basiventrals developing postero-anteriorly (Fig. 4.2a, b), as previously reported in *A. brevirostrum* by Hilton et al. (2011). Such a typical development explains why basiventral anlages are not visible in sections of the abdominal region until 25 dph (Fig. 4.3a–c). At this development stage, each vertebra possesses paired basidorsals and paired basiventrals, while neural spines have started to develop

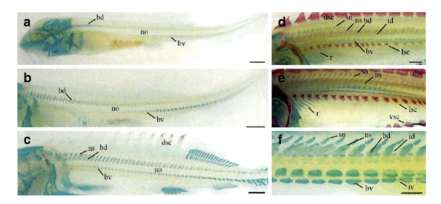

Fig. 4.2 Development of the vertebral elements in a growth series of *Acipenser baerii*. In toto cleared and stained specimens. Only the scutes are mineralized in this series (dorsal row first and then lateral and ventral rows). (**a**) 9 dph, 1.7 cm TL; (**b**) 15 dph, 2.0 cm TL; (**c**) 25 dph, 2.6 cm TL; (**d**) 31 dph, 5.6 cm TL; (**e**) 37 dph, 7.4 cm TL; (**f**) 43 dph, 8.1 cm TL. Modified after Leprévost et al. (2017a). Abbreviations: *bd*, basidorsal; *bv*, basiventral; *dsc*, dorsal scute; *id*, interdorsal; *iv*, interventral; *lsc*, lateral scute; *no*, notochord; *ns*, neural spine; *p*, parapophysis; *r*, rib; *sn*, supraneural; *vsc*, ventral scute. Scale bars = 1 mm

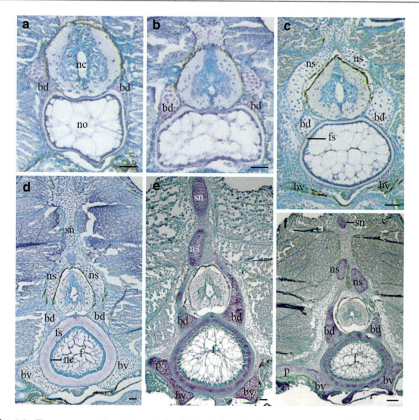

Fig. 4.3 Transverse sections through the abdominal region in a growth series of *Acipenser baerii* (toluidine *blue* staining) at the same developmental stages as in Fig. 4.2. (**a**) 1.7 cm; (**b**) 2.0 cm; (**c**) 2.8 cm; (**d**) 5.0 cm; (**e**) 6.5 cm; (**f**) 9.5 cm. Modified after Leprévost et al. (2017a). Abbreviations as in Fig. 4.2; *f*, funiculus; *fs*, fibrous sheet; *nc*, neural canal; *ne*, notochordal epithelium. Scale bars: $a–e = 100$ μm; $f = 200$ μm

antero-posteriorly through extension of the basidorsals (Figs. 4.2ca nd 4.3c). At 31 dph, interdorsals, interventrals and ribs have initiated, and they develop in the same direction as basidorsals and basiventrals, respectively (Fig. 4.2d). Supraneurals have also started to develop above the neural spines and in the antero-posterior direction. They form independently from the other vertebral elements (Figs. 4.2d and 4.3d) and then extend dorsally in an oblique direction (Fig. 4.2e, f).

In the larval stages of a hybrid of *Huso huso* X *A. ruthenus*, cartilage development begins by 7 dph (14.0 mm TL) around the notochord (Zhang et al. 2012). Chondroblasts accumulate in the dorsal and ventral regions of the notochord and deposit the cartilaginous matrix of basidorsals and basiventrals. Supraneurals develop above the neural arch by 44 dph (44.2 mm). All these cartilaginous elements are almost completely developed by 181 dph (179.0 mm). The ribs begin to form beneath the basiventrals in the anterior body by 122 dph (116.3 mm), but in the largest specimen (181 dph, 179 mm) studied by Zhang et al. (2012), rib development was still incomplete and neural spines had not developed yet.

The main changes occurring during late ontogeny are the progressive fusion of the vertebral bodies, a process that is not documented so far in the literature, and the slow ossification of some vertebral elements that was recently described (see below).

4.3 Organization and Structural Composition of the Axial Skeleton

Sturgeons vertebrae display three main characters that distinguish them from other actinopterygians: they are mainly cartilaginous with a reduced ossification of their elements, and they lack centra, which means that the notochord is the only support for the cartilaginous elements composing the so-called aspondylous vertebrae (Birstein 1993; Bemis et al. 1997; Findeis 1997; Zhang et al. 2012).

Until recently, and the description of the ontogeny and mineralization of the axial skeleton of *A. baerii* by Leprévost et al. (2017a, b), a few morphological studies of the skeleton in Acipenseriformes species were available in the literature. These gaps in our knowledge could be explained, a.o. by the relatively recent interest in sturgeon farming, by the difficulty to get specimens from the wild and by cartilaginous skeleton poorly preserved in archaeological remains and in fossil records. Several studies were devoted to comparative osteology and phylogenetic relationships using many extant and extinct species (Gadow and Abbott 1895; Grande and Bemis 1991; Bemis et al. 1997; Findeis 1997; Arratia et al. 2001). Hilton et al. (2011) described the skeletal anatomy of *A. brevirostrum* using dried specimens, cleared and double-stained specimens and scanning electron microscopy. The studies by Leprévost et al. (2017a) in *A. baerii* completed our knowledge not only in bringing comparative information using growth series but also in providing original description of the ossification and mineralization processes of the vertebral elements during ontogeny.

4.3.1 Organization of Vertebral Elements

In adult sturgeons, the vertebral bodies are fused, and the axial skeleton appears as a single, cartilaginous rod extending from below the head to the tail (Fig. 4.4). The limits between each vertebra are hardly visible, and the only presence of characteristic elements such as basidorsals, basiventrals and ribs allows to determine the vertebra number. The presence or absence of these elements allows to delimit the

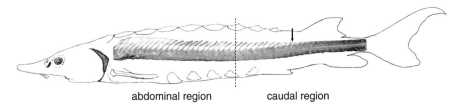

Fig. 4.4 Drawing of a sturgeon in lateral view including a photograph of the vertebral axis from head to tail; modified after Leprévost et al. (2017b). The dashed line delimits the abdominal and caudal regions. The arrow points to the end of the supraneural series before the dorsal fin

different regions of the vertebral column. In *A. brevirostrum*, Hilton et al. (2011) divided the vertebral axis into two regions, and, depending on the regions of the axial skeleton, the vertebrae display different morphologies. Anteriorly, the abdominal region is composed of vertebrae supporting ribs, neural spines and supraneurals (Fig. 4.5 and 4.6a). The supraneural series is interrupted slightly behind the dorsal fin. Posteriorly, in the caudal region, the ribs are absent, and the neural spines are shorter (Fig. 4.6b) and then disappear after the dorsal fin (Fig. 4.6c) so that the

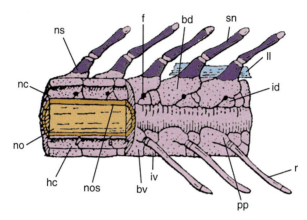

Fig. 4.5 Drawing of a segment including five vertebrae of the axial skeleton in the abdominal region of *Acipenser sturio* (modified from Gurtovoy et al. 1976). In adult, the vertebrae are fused, and each vertebra can only be recognized by its own elements: dorsally, the interdorsal (*id*) and the basidorsal (*bd*), to which is attached the neural spine (*ns*) prolonged with the supraneural (*sn*), and ventrally, the interventral (*iv*) and the basiventral (*bv*), to which is attached the rib (*r*) by means of the parapophysis (*pp*). These vertebral elements protect, *from top to bottom*, the neural canal (*nc*), the notochord (*no*) surrounded with the notochord sheath (*nos*) and the haemal canal (*hc*). Abbreviations: *f*, foramen; *ll*, longitudinal ligament

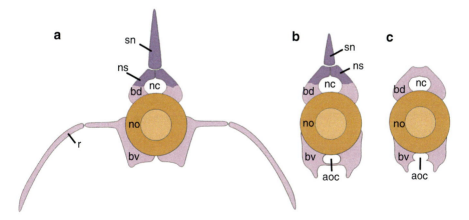

Fig. 4.6 Schematic drawings of cross sections of vertebrae in *Acipenser brevirostrum*. (**a**) Abdominal region; (**b**) caudal region anterior to dorsal fin; (**c**) caudal region posterior to dorsal fin (modified from Hilton et al. 2011). Abbreviations as in Fig. 4.5; *aoc*, aortic canal

caudal vertebrae look rounder. The five vertebrae of the anterior part of the vertebral column are considered to belong to the occipital region of the neurocranium. According to Bemis et al. (1997), this fusion is typical of Acipenseriformes.

Paired basidorsals form the neural arches in both the abdominal and caudal regions, and basiventrals form the haemal arches in the caudal region only. Neural arches are independent from each other along the vertebral column, except at the tip of the caudal region where they form a continuous series of large plates (Arratia et al. 2001; Hilton et al. 2011). In Acipenseriformes the neural spines are short and carried by neural arches (Fig. 4.5). They enclose the longitudinal ligament, which runs along the vertebral axis and connects vertebrae to each other (Gadow and Abbott 1895; Arratia et al. 2001). In Acipenserinae, the basiventrals anterior to the caudal fin never possess haemal spines (Findeis 1997).

In addition to paired ribs, the abdominal vertebrae also support a median supraneural (Fig. 4.5). In Acipenseridae, e.g. *A. brevirostrum* and *A. baerii*, the supraneural series is interrupted beneath the dorsal fin, contrary to Polyodontidae, which possess a continuous series of supraneurals. In sturgeons they are similar in size in most of the abdominal region and then become gradually smaller and disappear near the anterior base of the dorsal fin (Findeis 1997; Hilton et al. 2011; Leprévost et al. 2017a).

In the caudal region, Hilton et al. (2011) distinguished the preural region, located between the dorsal fin and the caudal fin, and the ural region that includes the caudal fin and its support. The border between the preural and the ural regions is the vertebra supporting the parhypural: this vertebra is the posteriormost preural vertebra.

The interpretation of the elements located above the neural arch, especially in the caudal region, gave rise to some discussion. Hilton et al. (2011) did not mention the presence of neural spines and include them in the term "basidorsals", considering that neural spines were the only dorsal growth of the basidorsals, while Arratia et al. (2001) make a distinction between these two elements. Zhang et al. (2012) wrongly termed the supraneural "neural spine". Recent descriptions of developmental stages of the axial skeleton in *A. baerii* allowed to close this debate (Leprévost et al. 2017a). Each vertebra is composed of four elements organized around the notochord: basidorsal and interdorsal, basiventral and interventral, and the neural spines form independently (Fig. 4.5). All these elements are initially cartilaginous, and fusion between adjacent elements seems to be common (Findeis 1997; Arratia et al. 2001; Hilton et al. 2011; Williot et al. 2011). Gadow and Abbott (1895) reported that interdorsals and interventrals can be pierced by motor roots of the spinal nerves and by sensory nerves.

4.3.2 Structure of the Vertebral Elements

The cartilaginous composition of the vertebral elements in Acipenseriformes is now largely admitted in the literature, although the real nature of the cartilage was not accurately described. Several authors (Grande and Bemis 1991; Arratia et al. 2001; Hilton et al. 2011) have reported the presence of "perichondral ossification" of the

4 New Insights in the Axial Skeleton of the Siberian Sturgeon

vertebral elements, but the presence of perichondral bone around cartilaginous elements (or of calcified cartilage) was not supported by histological data. Studying a transverse section of a "neurapophysis" of *Acipenser* sp. from a material dated 1876, Meunier and Herbin (2014) described the presence of primary bone: "a crown of periosteal bone tissue surrounding a large, circular surface resembling amorphous cartilage". This bony tissue contains osteocytes and presents Sharpey's fibres and concentric growth ridges.

Using a growth series of *A. baerii*, Leprévost et al. (2017a) provided new data on the structure and ultrastructure of the "perichondral ossification" of vertebral elements (neural spines and supraneurals). Histological observations of the demineralized vertebral elements revealed that bone matrix directly lines the cartilage, which typologically corresponds to perichondral bone (Fig. 4.7a). Transmission electron microscopy revealed that this matrix is entirely composed of randomly distributed bundles of collagen fibres, a feature that characterizes fibrous (woven-fibered) bone (Fig. 4.7b–d). A few osteocytes are located within the bone matrix, in which no vascular canals could be identified. Therefore, in sturgeon the mineralized matrix covering vertebral elements, supraneurals and neural spines is defined as perichondral, avascular, cellular, woven-fibered bone. An irregular layer of scattered osteoblasts lines the bone surface, roughly delimiting the interface between the unmineralized bone matrix, osteoid, and the surrounding mesenchyme, which is largely composed of layers of oriented fibre bundles of collagen (Fig. 4.7b). The osteoblasts facing the forming bone matrix exhibit many cytoplasmic extensions that surround patches or collagen bundles. These features strongly suggest that the osteoblasts are rather modelling the pre-existing collagen layers of the mesenchyme than synthesizing collagen bundles, a process which characterizes metaplastic ossification. Some cells, osteocytes, are entrapped into the forming bone matrix and become osteocytes (Fig. 4.7c). The osteoid tissue resulting from this process is first unmineralized; then is embedded within an electron-dense, thin, extrafibrillar matrix; and mineralizes, as a typical bone matrix.

The limit between perichondral bone and cartilage matrix is sharp and devoid of cells, and no remnants of pre-existing cells were present (Fig. 4.7a, d). These features indicate that the cartilage was deposited first by the chondroblasts located in the perichondrium, and then these cells stop depositing cartilage. They either disappear from the cartilage surface or they differentiate into osteoblasts (perichondrium to periosteum transition) and start to form bone tissue at the cartilage surface using the pre-existing collagen fibres of the surrounding mesenchyme and embedding them with proteins favouring mineralization, such as secretory calcium-binding proteins.

It is worth noting that in sturgeons, the cartilage matrix of the vertebral elements differs from various cartilage types by its dense matrix and its scarce and not hypertrophied chondrocytes looking like rounded osteocytes (Fig. 4.7a, d). However, precise descriptions of the ontogeny and growth of this typical cartilage are lacking in the literature, although this tissue is importantly developed in the vertebral axis.

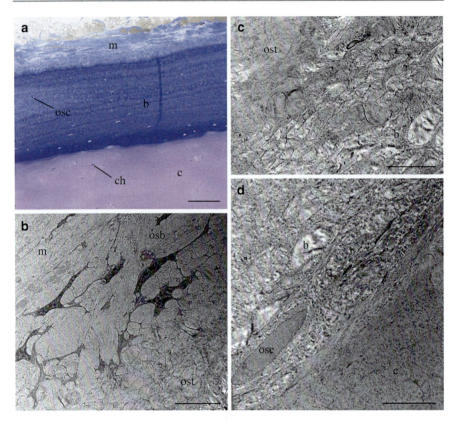

Fig. 4.7 Transverse sections of a supraneural in an adult *Acipenser baerii* (7 years old, 105 cm TL). (**a**) 2-µm-thick sections, toluidine *blue* staining. (**b–d**) TEM micrographs of 80 nm-thick sections. (**b**) Interface between the mesenchyme and the osteoid tissue; (**c**) transition area between osteoid and maturing bone; (**d**) interface between bone and cartilage. Modified after Leprévost et al. (2017a). Abbreviations: *b*, bone; *c*, cartilage; *ch*, chordocyte; *m*, mesenchyme; *osb*, osteoblast; *osc*, osteocyte; *ost*, osteoid. Scale bars, *a* = 100 µm; *b–d* = 10 µm

4.3.3 Development and Structure of the Notochord

In sturgeons the notochord is persistent throughout life and unconstricted and extends to the posterior end of the tail. In the absence of vertebral centra, the notochord is the only support for the cartilaginous elements composing the vertebrae (Schmitz 1998; Arratia et al. 2001; Hilton et al. 2011; Zhang et al. 2012; Leprévost et al. 2017a, b). The notochord grows uniformly and continuously throughout life (Gadow and Abbott 1895). Schmitz (1998) focused on the cellular ultrastructure of the notochord in young specimens of *A. brevirostrum* (6 to 28 cm total length) using light, transmission electron and scanning electron microscopy, and Leprévost et al. (2017a, b) described its development and histological structure using a large growth series of *A. baerii* (Fig. 4.3).

The notochord is surrounded by an acellular fibrous sheath and by an *elastica externa*. Two cell types have been identified in the core of the notochord: the first type, called notochordal epithelium or basal cells, forms the most peripheral cell layer and is located within the fibrous sheath. These cells are cuboidal in juvenile specimens and become columnar-shaped in adults (Fig. 4.3). Below this layer is the second cell type: the vacuolated cells, or chordocytes, which compose the largest portion of the medulla. A dense cytoplasmic network of intermediate filaments surrounds vacuoles, and chordocytes are connected by desmosomes. As the size of the notochord increases radially, the number of chordocyte layers also increases. The central region of the notochord is occupied by the *funiculus*, a flat condensation of the cytoskeletal network of vacuolated cells, perpendicular to the dorsoventral axis (Schmitz 1998; Leprévost et al. 2017b).

4.4 Mineralization of the Axial Skeleton

In toto alizarin red staining did not enable to visualize accurately the onset of mineralization of the vertebral elements, because from 10 months onwards, the muscles could not be cleared enough to see the skeleton. It is the reason why data on the mineralization of the axial skeleton of Acipenseriformes were scarce and incomplete until recently when Leprévost et al. (2017a, b) followed the mineralization process within the axial skeleton using X-ray microtomography and studied the mineral phase.

4.4.1 Mineralization Process of the Vertebral Elements

According to Hilton et al. (2011), all the elements composing the vertebrae, except the interdorsals and interventrals, can be mineralized in adults of extant Acipenseriformes, i.e. basidorsals, basiventrals, neural spines, supraneurals and ribs.

In adult specimens of *A. brevirostrum*, supraneurals are the first vertebral elements to mineralize during ontogeny (Hilton et al. 2011). In a large specimen of *Polyodon spathula* (160.0 cm TL), the only well-mineralized elements were supraneurals, posterior haemal spines and hypurals, which show a thin perichondral bone layer deposited at their surface. The basidorsals, basiventrals, interdorsals, interventrals and ribs are still cartilaginous. Interestingly, in specimens between 82.5 and 148.0 cm TL of the fossil polyodontid †*Crossopholis magnicaudatus* (early Eocene, ~50 million years), basidorsals and basiventrals are clearly mineralized in addition to supraneurals and haemal spines. This may suggest that mineralization occurs at a smaller size in †*Crossopholis magnicaudatus* than in extant species and that these elements can eventually mineralize in large, old individuals of extant polyodontids (Grande and Bemis 1991). Arratia et al. (2001) confirm that in some Acipenseriformes the mineralization of basidorsals and basiventrals, forming, respectively, the neural and haemal arches, develops late in ontogeny. In contrast, in adult *A. brevirostrum*, neural arches are mineralized (Hilton et al. 2011).

In *A. baerii* mineralization of the vertebral axis is limited to a few vertebral elements that become perichondrally mineralized at a relatively late ontogenetic stage (Leprévost et al. 2017a). Supraneurals are the first elements being mineralized at the age of 1 year (Fig. 4.8a), and then neural spines start to mineralize around 2 years (Fig. 4.8b). The mineralized matrix completely surrounds the neural spines in the

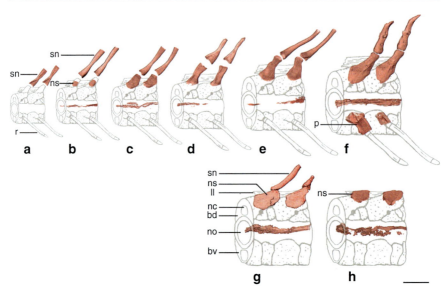

Fig. 4.8 3D modelization (lateral view) of two vertebrae of *Acipenser baerii* illustrating the development of mineralized elements during ontogeny using X-ray microtomography. Note the presence of a mineralized, elongated structure within the notochord from 2 years onwards. Anterior to the left. (**a–f**) Abdominal vertebrae from the same region. (**a**): 1 year old (y.o.), 57 cm TL; (**b**) 2 year old, 70 cm; (**c**) 3 year old, 80 cm; (**d**) 5 year old, 88 cm; (**e**) 7 year old, 108 cm; (**f**) 24 year old, 144 cm. (**g, h**) Same specimen as in (**f**). (**g**) Caudal vertebrae close to the region where the supraneural series is interrupted; (**h**) caudal vertebrae behind the posterior base of the dorsal fin. Modified after Leprévost et al. (2017a, b). Abbreviations: *bd*, basidorsal; *bv*, basiventral; *ll*, longitudinal ligament; *nc*, neural canal; *no*, notochord; *ns*, neural spine; *p*, parapophysis; *r*, rib; *sn*, supraneural. Scale bar = 2 cm

abdominal region of the specimen aged 3 years (Fig. 4.8c). Mineralization of neural spines and supraneurals follows the growth of these elements and then extends to basidorsals and parapophyses in older specimens (Fig. 4.8d–f). Along the vertebral axis, the mineralization is regionalized and progresses antero-posteriorly during development. In *A. baerii*, the mineralization of caudal vertebrae, anterior and posterior to the dorsal fin, was only detected in a 24-year-old specimen (Fig. 4.8g–h).

Bone thickness was measured in different regions of the supraneural on the X-ray microtomography sections: it was always higher in the medial region and increased in all regions with ageing (Leprévost et al. 2017a). Between 1 year and 24 years, bone thickness increases of about 22 μm per year in the proximal region, of 51 μm per year in the medial region and of 45 μm per year in the distal region. These findings indicate that the supraneurals extend in length by means of cartilage deposition at both extremities, allowing also extension of the diameter of the cartilage matrix. Once bone is deposited around the cartilage, the latter can no longer extend in diameter, and supraneural growth in diameter is then ensured by bone apposition as shown in the medial region.

To our best knowledge, no pattern of resorption of the perichondral bone surrounding the vertebral elements of Acipenseriformes was reported in the literature. This is strongly in contrast with what occurs in teleost vertebrae, in which bone remodelling is a permanent phenomenon. This process seems to be a compromise between the necessity to mobilize vertebral mineral ions in response to various physiological demands and the necessity to maintain vertebral strength against mechanical constraints, in optimizing the allocation of calcium and phosphorus (Deschamps et al. 2008, 2009a, b). In Acipenseriformes, the lack of bone remodelling in the vertebrae indicates that these mineralized elements are not involved in mineral homeostasis.

Surprisingly, X-ray microtomography images of the vertebrae in *A. baerii* revealed the unexpected presence of a mineralized, tubular-like tissue inside the notochord (Leprévost et al. 2017b). This feature was not previously described in the various species studied so far, but, at least, in *A. baerii*, it appears late in ontogeny, from 2 years old onwards.

4.4.2 Mineralization in the Notochord Funiculus

Leprévost et al. (2017b) studied the mineralization process in the notochord of a growth series of *A. baerii*, using X-ray microtomography (Fig. 4.8). Then, they described the histology and ultrastructure of the concerned tissues (Fig. 4.9) and monitored the mineralization rate during 1 year in two sturgeon populations reared at different temperature in French farms.

A tubular-like, mineralized structure located within the *funiculus* region of the notochord was identified in all specimens from 2 years old onwards (Fig. 4.8). In addition, this region was stained with alizarin red in the 7-year-old specimen, indicating the presence of calcium (Fig. 4.9a). This structure is roughly linear, continuous and irregularly shaped, parallel to the notochord axis and often located dorsally, appearing composed of juxtaposed, small amounts of mineralized matrix. From 2 to 7 years, the mineralized structure was only present in the abdominal region, but in older individuals this structure was well developed in the caudal region where the supraneural series is interrupted (Fig. 4.8g). These findings suggest a slow but regular, antero-posterior progression of the mineralization process. With ageing, the volume of the mineralized area increases, and from 3 to 24 years, notochord mineralization rate increased from 2.5% to 7.4%.

In histological sections in the abdominal region of *A. baerii*, the *funiculus* is already identified as flat condensations of cell materials in the centrodorsal region of the notochord in young specimens (Fig. 4.9a). From 3 years onwards, the *funiculus* appears as a compilation of cellular condensations containing a matrix that possesses metachromatic properties of toluidine blue revealing the presence of acidic components (Fig. 4.9c). By places, the *funiculus* region is linked dorsally to the fibrous sheath of the notochord, which also presents some metachromatic properties. In 7-year-old specimens, the mineralized region of the *funiculus* is easy to localize. Two µm-thick plastic sections reveal an accumulation of small patches (1–2 µm in diameter) of dark "granules" roughly distributed along the compressed membranes (Fig. 4.9d). Transmission electron microscopy (TEM) observations indicate by places some thin, electron-dense platelets (150 nm in length, less than

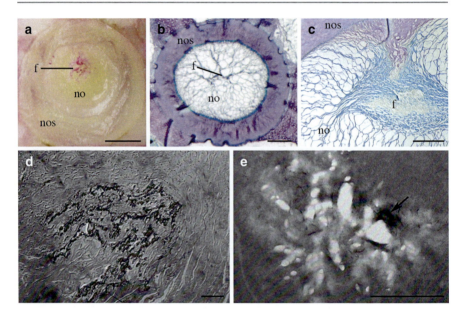

Fig. 4.9 (**a**) Photograph of a transverse slice of the notochord in the abdominal region of a 7-year-old *Acipenser baerii*. Alizarin red staining reveals the mineralized area in the *funiculus*. (**b, c**) Transverse, paraffin sections through the notochord in the abdominal region of a 60 dph, 12 cm TL (**b**) and a 3-year-old, 72 cm (**c**) specimens. Toluidine blue staining. Modified after Leprévost et al. (2016b). (**d**) Transverse, 2-μm-thick epon sections of the mineralized region in the *funiculus* of the notochord of *Acipenser baerii* (7-year-old, 105 cm TL); toluidine *blue* staining. Nomarski contrast. (**e**) Transmission electron micrograph of the mineralized region of the *funiculus* showing mineralized patches with artefactual, electron-lucent zones. The arrow points to thin, electron-dense hydroxyapatite crystals. Abbreviations: *f*, funiculus; *no*, notochord; *nos*, notochordal sheath. Scale bars, *a* = 1 cm; *b* = 200 μm; *c* = 500 μm; *d* = 50 μm; *e* = 1 μm

10 nm in width) located here and there in the granular region, and the features are similar to those reported for hydroxyapatite crystals (Fig. 4.9e).

The nature of the mineral phase of notochord was analysed in *A. baerii* using ^{31}P solid-state NMR and compared to bone mineral from mature ewe bone representative of mammalian bone hydroxyapatite (Leprévost et al. 2017a, b). The results revealed that the mineral phase in the notochord was slightly different from mature ewe bone in that it contained a large amount of amorphous calcium phosphates in addition to hydroxyapatite crystals as already demonstrated for bone mineral (Wang et al. 2013).

4.4.3 Seasonal Variations of Mineralization Rate in the Notochord

The presence of the mineralized region in the *funiculus* of the notochord in all studied individuals aged 2 years and more, along with the identification of an amorphous phase of calcium phosphates, prompted Leprévost et al. (2017b) to wonder whether or not this enigmatic structure could be a device allowing to store minerals that could be mobilized for the regulation of phosphocalcic metabolism (for instance, in a population that almost stops eating during winter). These authors measured the mineralization rate of the notochord during 1 year in two populations (2 to 3-year-old specimens): one farmed outdoor with seasonal variations of temperature that stops eating during winter and one farmed indoor at almost constant water temperature. No clear pattern of annual variation was revealed within each group. These results were inconsistent with the putative role of mineral store of the notochord, and the presence of these structures as well as their putative function is still an enigma.

4.5 Mechanical Constraints on the Axial Skeleton During Swimming

In sturgeon and paddlefish, the notochord provides the main axial support for the body (Zhang et al. 2012), and it resists to compression forces (Schmitz 1998; Liem et al. 2001). Ribs are part of the locomotor system too. They strengthen the myosepta and help in transferring the forces generated by muscular contraction to the vertebral axis. Sturgeons lack intervertebral joints (Long 1995), but possess a longitudinal ligament running along the vertebral column. Because of the asymmetrical tail, the vertebral column must also resist twisting or torsion.

Compared to many teleosts, Acipenseriformes do not swim rapidly and powerfully, and propulsion does not seem to be important for them (Long 1995; Liem et al. 2001). However, several sturgeon species are anadromous, and large specimens migrate into coastal environments and upstream rivers, which needs powerful swimming in some cases. Regionally, variable swimming forces apply along the vertebral column. In *A. transmontanus*, these variations were found to be correlated with variable notochordal mechanical properties but not with notochordal morphology (Long 1995). In *A. baerii*, prelarvae and larvae have an anguilliform swimming style, with flexure of large amplitude over a large part of their body. After metamorphosis, around 3 weeks after hatching, swimming changes to a subcarangiform type, in which only the caudal region of the body displays large amplitudes, while the rest of the body remains relatively rigid (Gisbert 1999). Such a subcarangiform swimming type applies the maximal constraints on median vertebrae, as in rainbow trout (Meunier and Ramzu 2006; Deschamps 2008). Therefore, sturgeon swimming type and the resulting mechanical constraints on vertebrae could have an effect on the emergence of axial anomalies.

4.6 Concluding Remarks

Recent studies of the Siberian sturgeon *A. baerii* have brought new, accurate information through ontogeny of the axial skeleton. In Acipenseriformes exhibiting intraspecific and ontogenetic variation, it would be interesting to examine numerous specimens from various species, including old individuals, in order to get a complete morphological assessment (Grande and Bemis 1991; Bemis et al. 1997).

Moreover, X-ray images of the axial skeleton mineralization of other Acipenseriformes species would help in comparative studies.

The axial skeleton is involved in body support and is subjected to strong mechanical constraints during swimming. Therefore, a poor vertebral mineralization will favour axial anomalies to appear in the region where the vertebrae support the strongest constraints. Identifying the region of the vertebral axis that is predominantly affected by anomalies would allow to (i) understand where the strongest constraints are applied during swimming and (ii) perform investigations on the vertebral structure in this particular region of healthy specimens in order to highlight adapted features resisting to these constraints, as previously demonstrated in the rainbow trout *Oncorhynchus mykiss*. The next step will be to study the effects of various rearing conditions, such as sustained exercise or rest, water temperature and quality or diet, on the emergence of axial anomalies.

In teleost fish, many factors were shown to be involved in the onset of skeletal anomalies; genetic factors, inappropriate nutrition and unfavourable abiotic conditions are the most possible causative factors (Boglione et al. 2013a, b). We cannot neglect any of these factors as being possibly involved in the vertebral pathology of sturgeons, but among them nutrition could be a key parameter. In teleosts, numerous nutrients (e.g. proteins and amino acids, lipids and fatty acids, water-soluble and fat-soluble vitamins, minerals) are responsible for the appearance of skeletal anomalies when their level and/or form of supply in the diet is inappropriate or unbalanced (for reviews, see Cahu et al. 2003; Lall and Lewis-Mc Crea 2007). Despite the negative consequences of skeletal anomalies for fish aquaculture, information about mineral nutrition is, however, limited although phosphorus and calcium are major structural components of mineralized tissues and are also important in osmoregulation, muscular activity and reproduction, among other physiological processes. From the few studies available in the literature (Leprévost et al. 2017a, b), it appears that the mineralization of vertebrae is so limited (along with the lack of bone resorption in these regions) that one can wonder whether it plays an important role in vertebral health, as shown in teleosts, and whether mineral defects could explain axial skeleton anomalies in some specimens within a population. A poor dietary mineral content would not have a direct consequence for the axial skeleton but rather indirect effects through physiological disturbances, which in turn disturb muscular activity and/or cartilage formation, and leads to deformations of the axial skeleton.

Given the importance of cartilage in the axial skeleton of Acipenseriformes, there is a crucial need of detailed studies of its composition, architecture and development, as previously performed for the bone tissue in the vertebrae of teleost fish (e.g. Deschamps et al. 2008, 2009a, b). Also, experimentations should be undertaken to understand the effects of various farming methods and environmental conditions on the biology of this particular skeletal tissue and on its putative role in vertebral health and "resistance" to anomalies.

Acknowledgments We thank Patrick Williot (*Sturgeon Consultant, Audenge, France*) and his colleagues (M. Chebanov and G. Nonnotte) who offer us the opportunity to write this review.

References

Arratia G, Schultze HP, Casciotta J (2001) Vertebral column and associated elements in dipnoans and comparison with other fishes: development and homology. J Morphol 250:101–172

Athanassopoulou F, Billinis C, Prapas T (2004) Important disease conditions of newly cultured species in intensive freshwater farms in Greece: first incidence of nodavirus infection in Acipenser sp. Dis Aquat Org 60:247–252

Bemis WE, Findeis EK, Grande L (1997) An overview of Acipenseriformes. Environ Biol Fish 48:25–71

Birstein VJ (1993) Sturgeons and paddlefishes: threatened fishes in need of conservation. Conserv Biol 7:773–787

Boglione C, Gavaia P, Koumoundouros G et al (2013a) Skeletal anomalies in reared European fish larvae and juveniles. Part 1: normal and anomalous skeletogenic processes. Rev Aquacult 5:S99–S120

Boglione C, Gisbert E, Gavaia P et al (2013b) Skeletal anomalies in reared European fish larvae and juveniles. Part 2: main typologies, occurrences and causative factors. Rev Aquacult 5:S121–S167

Brun R, Nougayrede P, Chene P et al (1991) Bilan sanitaire de 2 ans d'élevage d'*Acipenser baeri* en piscicultures intensives. In: Williot P (ed) Acipenser. CEMAGREF Publ., France, pp 429–437

Cahu C, Zambonino Infante JL, Takeuchi T (2003) Nutritional components affecting skeletal development in fish larvae. Aquaculture 227:245–258

Chernyshov VI, Isuev AP (1980) Etiology of the free-radical pathology in the sturgeons *Acipenser gueldenstaedtii* (Brandt), *Acipenser stellatus* (Pallas) and *Huso huso* (L.) during embryogenesis. Vopr Ikhtiol 20:334–344

Deschamps MH (2008) Etude histomorphométrique du squelette axial de la truite arc-en-ciel d'élevage, *Oncorhynchus mykiss*. PhD thesis, Université Pierre & Marie Curie, Paris, p 156. [http://sfi.mnhn.fr/sfi/8.theses/DeschampsThesis.pdf]

Deschamps MH, Girondot M, Labbé L et al (2009a) Changes in vertebral structure during growth of reared rainbow trout, *Oncorhynchus mykiss* (Walbaum): a new approach using modelling of vertebral bone profiles. J Fish Dis 32:233–246

Deschamps MH, Labbé L, Baloche S et al (2009b) Sustained exercise improves vertebral histomorphometry and modulates hormonal levels in rainbow trout. Aquaculture 296:337–346

Deschamps MH, Kacem A, Ventura R et al (2008) Assessment of "discreet" vertebral abnormalities, bone mineralization and bone compactness in farmed rainbow trout. Aquaculture 279:11–17

Dettlaff TA, Ginsburg AS, Schmalhausen OI (1993) Sturgeon fishes. Developmental biology and aquaculture. Springer-Verlag, Berlin, p 300

FAO (2013) Cultured aquatic species fact sheets, *Acipenser baerii* (Brandt, 1896). FAO Fisheries & Aquaculture Department

Findeis EK (1997) Osteology and phylogenetic interrelationships of sturgeons (Acipenseridae). Environ Biol Fish 48:73–126

Fontagné S, Bazin D, Breque J et al (2006) Effects of dietary oxidized lipid and vitamin a on the early development and antioxidant status of Siberian sturgeon (*Acipenser baeri*) larvae. Aquaculture 257:400–411

Gadow H, Abbott EC (1895) On the evolution of the vertebral column of fishes. Phil Trans R Soc Lond 186:163–221

Gisbert E (1999) Early development and allometric growth patterns in Siberian sturgeon and their ecological significance. J Fish Biol 54:852–862

Gisbert E, Williot P, Castello-Orvay F (1998) Morphological development of Siberian sturgeon (*Acipenser baeri*, Brandt) during prelarval and larval stages. Riv Ital Acquacolt 33:121–130

Grande L, Bemis WE (1991) Osteology and phylogenetic relationships of fossil and recent paddlefishes (Polyodontidae) with comments on the interrelationships of Acipenseriformes. J Vert Paleontol 11:1–121

Gurtovoy NN, Matveev BS, Dzerjinskiy FY (1976) Practical zootomy of vertebrates. Moscow, Vysshya Shkola. (in Russian)

Hilton EJ, Grande L, Bemis WE (2011) Skeletal anatomy of the shortnose sturgeon, *Acipenser brevirostrum* Lesueur, 1818, and the systematics of sturgeons (Acipenseriformes, Acipenseridae). Field Life Earth Sci 3:1–168

Hou JL, Zhuang P, Zhang LZ et al (2011) Morphological deformities and recovery, accumulation and elimination of lead in body tissues of Chinese sturgeon, *Acipenser sinensis*, early life stages: a laboratory study. J Appl Ichtyol 27:514–519

Hu J, Zhang Z, Wei Q et al (2009) Malformations of the endangered Chinese sturgeon, *Acipenser sinensis*, and its causal agent. Proc Natl Acad Sci (23):9339–9344

Lall SP, Lewis-Mc Crea LM (2007) Role of nutrients in skeletal metabolism and pathology in fish - an overview. Aquaculture 267:3–19

Leprévost A, Azais T, Trichet M et al (2017a) Vertebral development and ossification in the Siberian sturgeon (*Acipenser baerii*), with new insights on bone histology and ultrastructure in vertebral elements and in scutes. Anat Rec 300:437–449

Leprévost A, Azais T, Trichet M et al (2017b) Identification of a new mineralized tissue in the notochord of reared Siberian sturgeon (*Acipenser baerii*). J Morphol (in press)

Liem K, Bemis W, Walker WF et al (2001) Chapter 8: The postcranial skeleton: the axial skeleton. In: Lewis T (ed) Functional anatomy of the vertebrates. An evolutionary perspective. Emily Barrosse, New York

Linares-Casenave J, Werner I, Eenennaam JPV et al (2013) Temperature stress induces notochord abnormalities and heat shock proteins expression in larval green sturgeon (*Acipenser medirostris* Ayres, 1854). J Appl Ichtyol 29:958–967

Long JH (1995) Morphology, mechanics, and locomotion: the relation between the notochord and swimming motions in sturgeon. Environ Biol Fish 44:199–211

Meunier FJ, Herbin M (2014) La collection de préparations histologiques effectuées par Paul Gervais (1816-1879) sur le squelette des "poissons". Cybium 38(1):23–42

Meunier FJ, Ramzu MY (2006) La régionalisation morphofonctionnelle de l'axe vertébral chez les téléostéens en relation avec le mode de nage. C R Palevol 5:499–507

Schmitz RJ (1998) Comparative ultrastructure of the cellular components of the unconstricted notochord in the sturgeon and the lungfish. J Morphol 236:75–104

Wang Y, Von Euw S, Fernandes FM et al (2013) Water-mediated structuring of bone apatite. Nat Mater 12:1144–1153

Williot P (2009) L'élevage de l'esturgeon sibérien (*Acipenser baerii* Brandt) en France. Cah Agric 18:189–194

Williot P, Rouault T (1982) Compte rendu d'une première reproduction en France de l'esturgeon sibérien *Acipenser baerii*. Bull Fr Pisc 286:255–261

Williot P, Rochard E, Castelnaud G et al (1997) Biological characteristics of European Atlantic sturgeon, *Acipenser sturio*, as the basis for a restoration program in France. Environ Biol Fish 48:359–370

Williot P, Rochard E, Desse-Berset N et al (2011) Brief introduction to sturgeon with a special focus on the European sturgeon, *Acipenser sturio* L., 1758. Springer-Verlag, Berlin Heidelberg

Williot P, Sabeau L, Gessner J et al (2001) Sturgeon farming in Western Europe: recent developments and perspectives. Aquat Living Resour 14:367–374

Zhang X, Shimoda K, Ura K et al (2012) Developmental structure of the vertebral column, fins, scutes and scales in bester sturgeon, a hybrid of beluga *Huso huso* and sterlet *Acipenser ruthenus*. J Fish Biol 81:1985–2004

Evolution of Molecular Investigations on Sturgeon Sex Determination and Most Recent Developments in DNA Methylation with a Focus on the Siberian Sturgeon

5

Rémy Simide and Sandrine Gaillard

Abstract

Sturgeon aquaculture is largely based around females due to caviar production. In the absence of sexual dimorphism and differentiated gonads in juveniles, the gender sorting of sturgeon is carried out at about 2–3 years old depending on rearing conditions, which increases farming costs. Identification of a molecular sex determination mechanism or of a molecular sex marker could lead to earlier sex identification. For decades scientists have developed different methods and approaches to identify a way in which sturgeon can be sexed. In this chapter we gather together the different approaches employed: heterogametic sex chromosome identification, random identification of molecular polymorphisms, transcriptome sequencing, and targeting sequences of interest. We have included our own results from juvenile and adult Siberian sturgeon on the inter simple sequence repeat (ISSR) with the support of hierarchical cluster analysis and on the expression of genes known to be involved in sex differentiation, *Foxl2, So9, Igf1*, and *Fgf9*. To date, no sex marker has been identified following these methods. We also present the advantages of DNA methylation to assess gene expression regulation, which opens up new perspectives in sex determination and differentiation research in fish. The first investigation of DNA methylation of DMRT1 using MS-HRM technology in sturgeon will conclude this chapter.

R. Simide (✉)
Laboratoire Protee, Equipe de Biologie Moléculaire Marine, Université de Toulon. Campus de la Garde, CS 60584-83041 Toulon Cedex 9, France

Institut océanographique Paul Ricard. Ile des Embiez, 83140 Six Fours Les Plages, France
e-mail: remy.simide@institut-paul-ricard.org

S. Gaillard
Laboratoire Protee, Plateforme BioTechServices, Université de Toulon. Campus de la Garde, CS 60584-83041 Toulon Cedex 9, France
e-mail: Sandrine.gaillard@univ-tln.fr

Keywords

Fish • *Acipenser baerii* • Sexing • DNA methylation • MS-HRM (Methylation sensitive-high resolution melting)

Introduction

Sturgeon farming is dedicated to meat and caviar production (Williot et al. 2001; Bronzi and Rosenthal 2014). Sturgeons are large valuable fish with the production of mature females being costly for fish farmers. Indeed, female Siberian sturgeons (*Acipenser baerii*) need to be raised until 8 years or more before they can produce fully developed ovocytes, which are collected after slaughtering (Williot and Sabeau 1999). One way to reduce costs is to exclude males as soon as possible in order to raise only females until sexual maturity. In sturgeon there is no sexual dimorphism, and gonads are macroscopically indistinguishable between genders for years. Currently, sturgeons are sexed by measuring hormonal steroids or by observation of sexually differentiated gonads via a biopsy, laparoscopy, or using an ultrasound scanner (Williot 2002; Chebanov and Galich 2009; Webb et al. 2013). The ultrasound, which is the most current device used in the field, can't be used until about 2.5 years in Siberian sturgeon (see Chap. 49). Sturgeons have an equilibrated sex ratio between male and female regardless of the environmental breeding conditions (Keyvanshokooh and Gharaei 2010), and genome manipulation by gynogenesis induces an imbalanced sex ratio (Fopp-Bayat 2010; Havelka et al. 2011). That's why it is largely accepted that sturgeons have a genetic sex differentiation mechanism, meaning there is at least one genetic difference between males and females. In order to obtain a sex marker that is theoretically usable at any developmental stage, research should focus on the genome. Furthermore, before gonads are noticeably distinct between the sexes, some molecular pathways are already different in males and females (e.g., Piferrer and Guiguen 2008; Vizziano-Cantonnet et al. 2016; details in Chap. 16). Research on gene or protein expression is leading to advances in our fundamental knowledge on sex determination and differentiation mechanisms, with the aim of obtaining a sex marker via nonlethal sampling that can be used at a specific age but earlier than the current methods allow. The industrial advantage that could be obtained from a valid sturgeon sex marker has made this a dynamic scientific issue for decades. To our knowledge, Fontana and Colombo 1974 were the first to briefly address the question of sturgeon sex genetics in a pioneering description of the sturgeon karyotype. The first study specifically dedicated to this issue using a putative mammalian sex marker in sturgeon was achieved by Ferreiro et al. (1989). Since then the published scientific strategies used to obtain a molecular sex marker can be divided into four groups: (1) cytogenetic studies to identify heterogametic sex chromosome; (2) random search using random amplified polymorphic DNA (RAPD), amplified fragment length polymorphism (AFLP), or inter simple sequence repeat (ISSR); (3) transcriptome analysis; and (4) search or comparison of selected markers by analysis of the expression of genes involved in sexual differentiation or the targeting of the presence, structures, and copy

5 Evolution of Molecular Investigations on Sturgeon Sex Determination

number of genes linked to sex (e.g., Havelka et al. 2011; Vizziano-Cantonnet et al. 2016; Xiao et al. 2014; Yue et al. 2015). Despite international interest only one patent has been filed to date based on hormonal measurements of the TGF-β superfamily that enables the sexing of 16 month-old Siberian sturgeon (Wetzel et al. 2013; Wetzel and Reynolds 2013). If the current data fail to identify an early sex marker of juvenile sturgeon, it will nonetheless enable us to bring together the fundamental knowledge on genetic sex mechanisms. This is important because the handful of fish species whose genetic sex determination has been determined were species whose fundamental genetic knowledge, particularly that of the sex chromosome, were known. For a while, the only known mechanism of genetic sex determination in fish was the supplementary copy of the DMRT1 gene (named DMY or DMRT1Y) on the Y chromosome of Japanese medaka (*Oryzias latipes*). This was discovered simultaneously by two research teams following different but parallel approaches (Matsuda et al. 2002; Nanda et al. 2002). Apart from the closely related species, *Oryzias curvinotus*, this mechanism has to date never been observed in other fish species. The mechanism of sexual genetic determination in fish is probably highly species or phylogenetic group specific (Kobayashi et al. 2013). Recently, the AMHR2, AMH, GSDFY, SOX3, and SDY genes, probably involved in sex determination, were discovered in a range of fish species: pufferfish (*Takifugu rubripes*) (Kamiya et al. 2012), Patagonian pejerrey (*Odontesthes hatcheri*) (Hattori et al. 2012, 2013), Philippine medaka (*Oryzias luzonensis*) (Myosho et al. 2012), marine medaka (*Oryzias dancena*) (Takehana et al. 2014), and probably numerous salmonids including rainbow trout (*Oncorhynchus mykiss*) (Yano et al. 2012). However, a fundamental understanding of fish genetics may not be enough to identify the genes involved in sex determination. For example, zebrafish (*Danio rerio*) is a model species with several available transcriptomes and genomes (Howe et al. 2013; Li et al. 2004), but despite important collaborative efforts, to date, it has not been possible to determine any clearly identifiable genetic sex markers (Howe et al. 2013; Liew et al. 2012).

Recent studies on the mechanisms involved in the regulation of genome expression in fish have opened a new field of research in genetic sex determination. One of the most stable regulation mechanisms over time, after embryonic development, is DNA methylation (Cedar and Bergman 2009). This is the addition of a methyl group to the nitrogen base of cytosine giving methylcytosine (mC), which is generally followed by a guanine. This dinucleotide 5′-CG-3′ is called a CpG (Law and Jacobsen 2010). The DNA regions which are CpG-rich, called CpG islands, are often unexpectedly hypomethylated. CpG islands are generally present ±1 kb around the transcription start site (TSS) of genes and could act like cis-regulatory elements in promoter regions (Saxonov et al. 2006). A hypermethylated CpG island generally represses gene expression, while hypomethylation tends to increase gene expression (Lister et al. 2009). These features have been extensively studied in mammals and are also present in fish (Shao et al. 2014). The first study in fish to have made a link between the expression of a gene associated with sexual differentiation, methylation of the promoter, and sex determination was conducted in European seabass (*Dicentrarchus labrax*) on the aromatase (*Cyp19a1*) (Navarro-Martín et al. 2011). Recently, a comprehensive study in the half-smooth tongue sole (*Cynoglossus semilaevis*) has provided clues that suggest that regulation of DMRT1 gene expression by DNA methylation could be directly involved in sexual determination of this species (Graves 2014; Shao et al. 2014).

In this chapter we have gathered together studies on sturgeon that have either addressed to the identified molecular mechanisms involved in sexual determinism or focused on the acquirement of a sex marker. We also present our own results in Siberian sturgeon involving these different approaches. We have used ISSR to randomly scan DNA polymorphism of males and females with the aim of identifying a sex marker. We have presented *Foxl2*, *Sox9*, *Fgf9*, and *Igf1* expression data, which are genes involved in sex differentiation, in 3-month-old juveniles and 5-year-old males and females. Finally, we have presented the first results of DNA methylation in sturgeon from an analysis using methylation-sensitive high-resolution melting (MS-HRM) of the 5′ untranslated region (5′ UTR) of the DMRT1 gene.

5.1 Material and Methods

5.1.1 Study Fish and Sampling Procedures

To increase genetic variability for ISSR analysis, 230 Siberian sturgeons sexed by ultrasound were used (Table 5.1). In the first batch, sex was confirmed by direct observation of gonads, while in batches 2–9 we extrapolated that ultrasound technique gives the true sex of every fish. All fish came from the same hatchery (Ecloserie

Table 5.1 Details on fish sampled in the current study

Analysis	No. of samples	Raising conditions	No. of batch[d]	Age of fishes	Tissues
ISSR	88 males 142 females	Sources du Gapeau[a]	1	5 years	Pectoral fin (100% ethanol, −80 °C)
		Prunier Manufacture[b]	2 to 9	3 to 8 years	Pectoral fin (100% ethanol, −80 °C)
Gene expression	12 males 8 females	Sources du Gapeau[a]	10	5 years	Pectoral fin, gonads, and erythrocytes (RNAlater, −80 °C)
	24 juveniles	Ecloserie de Guyenne[c]	11	3 months	Pectoral fin and erythrocytes (RNAlater, −80 °C)
DNA methylation	15 males 15 females	Sources du Gapeau[a]	10	5 years	Pectoral fin and gonads (100% ethanol, −80 °C)
	20 juveniles	Ecloserie de Guyenne[c]	11	3 months	Pectoral fin and total gonads (100% ethanol, −80 °C)

[a]Stable temperature close to 16 °C. Males and females were raised in the same tank
[b]Seasonal temperature variation. Males and females were sorted between genders at 3 years old and then raised in different tanks
[c]Stable temperature close to 18 °C. Fish were raised in the same tank
[d]One batch was composed of males and/or females hatched from a single pool of genitors

5 Evolution of Molecular Investigations on Sturgeon Sex Determination

de Guyenne, France) and subsequently being raised in the Prunier Manufacture Company (France) or in the Sources du Gapeau fish farm (France). For gene expression and DNA methylation analysis, two other batches of Siberian sturgeons raised in the Ecloserie de Guyenne and in the Sources du Gapeau were used (Table 5.1).

5.1.2 ISSR Analysis

DNA was extracted using a column extraction kit (DNeasy Blood and Tissue, Qiagen, Germany) according to the manufacturer's manual with an additional step of RNAse A treatment. Purity and integrity of DNA were controlled by NanoDrop (Thermo Scientific, USA) measurement and standard electrophoresis on a 1% agarose gel. Sixty-one primers were randomly selected to amplify DNA regions between two microsatellites. Thirty-eight primers were used alone in two males and two females, while the 23 remaining primers were also crossed with one another in four males and four females for a total of 337 primer pairs tested. According to the results obtained, the number of tested fish increased until the maximum sample size. A PCR with a temperature gradient between 40 and 60 °C was used. Once the temperature was chosen, other PCR parameters can be adjusted by modification of the amplification program, DNA concentration or Mg^{2+} concentration. The most used reaction mix was 2.5 µL of buffer (including 2.5 mM of Mg^{2+}) (HotMaster *Taq*, 5Prime, Germany), 0.2 µL of *Taq* DNA polymerase (1 U final) (HotMaster *Taq*, 5Prime, Germany), 1 µL of dNTP (0.1 mM final), 16.3 µL of molecular grade water, 1.5 µL of each primer (6 µM final), and 2 µL of DNA (40 ng final). The most used PCR conditions were an initial denaturation step at 94 °C for 3 min followed by 35 amplification cycles at 94 °C for 1 min, the hybridization temperature for 45 s, and then 72 °C for 1 min which was followed by a final step at 72 °C for 15 min. A positive and a negative control were added to each template. All of the amplification products were compared by electrophoresis on an agarose gel (generally at 1.5%). Band size was determined for each primer combination in an agarose gel, and a comparison between tested individuals was scored in a presence/absence matrix.

5.1.3 Measurement of Gene Expression

Total RNA extraction was performed with a column extraction kit (NucleoSpin 8 RNA, Macherey-Nagel, Germany) following the manufacturer's recommendations with two supplementary steps. These were an initial purification step that involved mixing Extract-All (Eurobio, France) and chloroform with samples, followed by an rDNase treatment to remove any genomic DNA contamination. The absence of genomic DNA contamination and RNA integrity was controlled by a control PCR amplification of RNA and migration of total RNA by electrophoresis on a 1% agarose gel. Double-strand cDNA was synthesized from total RNA according to the manufacturer's manual (Omniscript, Qiagen, Germany). Real-time PCR quantification (qPCR) was performed using SYBR Green technology on a LightCycler 480 (Roche, Germany). All assays were carried out in duplicate, and a negative control and a calibrator were added to each run. Primers were designed from sequences available on GenBank (Table 5.2) from Siberian sturgeon for *β-actin*, *18S*, *B2m*, Foxl2, and *Igf1*. *Sox9* primers were taken from Berbejillo et al. (2012). *Fgf9* primers were designed from a conserved region of mammalian and amphibian sequences and then submitted to GenBank for Siberian sturgeon (no. KX589064). For each gene, specific amplifications were sequenced to confirm their identity. Expression of the four target genes Foxl2, *Igf1*, *Sox9*, and *Fgf9* was normalized to the mean expression of the three reference genes *β-actin*, *B2m*, and *18S*. The reaction mix consisted of 5 μL of master mix (Roche, Germany), 1 μL of molecular grade water, 1 μL of each primer (1 μM final), and 2 μL of cDNA. The qPCR conditions were an initial denaturation step at 95 °C for 10 min followed by 40 cycles at 95 °C for 10 s and then 20 s at the specific hybridization temperature (Tm) for each gene presented in Table 5.2 and 72 °C for 35 s, with a final denaturation step for 10 s and a final classic melting curve program. For each assay, the melting curve was analyzed to verify whether there was any by-product amplification.

5.1.4 MS-HRM Determination

The first step is to obtain a region upstream of the coding DNA sequence (CDS) of the DMRT1 gene. Based on the sequence of Siberian sturgeon available on GenBank (no. HQ110106.1), a genome-walking procedure was conducted using the GenomeWalker Universal kit (Clontech/Takara, USA). Using Dra1, EcoR5, Pvu2, or Stu1 as restriction enzymes, the libraries were completed from a pool of Siberian sturgeon DNA. These libraries were used in nested PCR with the two provided forward primers in the kit and the following reverse primers: DMRT1walking1-reverse (5′-CATGACTCGCTGTCGTTTTGCAA-3′) and DMRT1walking2-reverse (5′-GTTCCTGCAGCGAGAACACTT-3′). The reaction mix was 2.5 μL of LongRange PCR buffer (including 2.5 mM of Mg^{2+}) (LongRange PCR kit, Qiagen, Germany), 0.2 μL of *Taq* LongRange PCR enzyme (1 U final) (LongRange PCR kit), 1.25 μL of dNTP (2 mM final), 11.55 μL of molecular grade water, 5 μL of Q-solution (LongRange PCR kit), 1 μL of each primer (400 nM final), and 2.5 μL of the library. The PCR conditions were an initial denaturation step at 93 °C followed by 35 cycles at 93 °C for 15 s, 56 °C for 30 s, 68 °C for 2.5 min, and then a final step at 68 °C for 5 min. The resulting amplification products were cloned according to the manufacturer's protocol (Qiagen PCR Cloning kit, Qiagen, Germany) and then sequenced. The obtained sequence was submitted to GenBank (no. KX589063). The search for CpG islands was performed with CpG islands software (http://www.bioinformatics.org/sms2/cpg_islands.html) and CpGplot (EMBOSS; http://www.ebi.ac.uk/Tools/seqstats/emboss_cpgplot/).

In order to obtain a methylation standard range, 0–100% control methylation matrices were prepared. Cytosine methylation is lost during PCR amplification

5 Evolution of Molecular Investigations on Sturgeon Sex Determination 77

because there is no DNA methyltransferase in the PCR reaction mix (Laird 2010). The 0% control was acquired by LongRange PCR amplification of the entire DMRT1 region of interest from the DNA pool following a similar protocol to the one above. This was then purified using a commercial column kit (QIAquick PCR purification, Qiagen, Germany). The following pair of primers was used: DMRT1total-forward (5′-TGCATCAGTCAGTCCCTAATCTG-3′) and DMRT1total-reverse (5′-AGACCC GCACAATCCAGCAAC-3′). A portion of this matrix was then treated with CpG methyltransferase M.SssI according to the manufacturer's recommendations (Thermo Scientific, USA) and purified using a commercial column kit (MinElute Reaction Cleanup, Qiagen, Germany). Thus, all cytosines in a CpG context along the DMRT1 region of interest were methylated, which corresponds to the 100% control.

DNA was extracted as previously described for the ISSR analysis. Before MS-HRM runs, DNA and controls must be converted by bisulfite treatment, so that every unmethylated cytosine is converted to uracil while methylated cytosine will be left unchanged (Laird 2003). This step allows methylated and unmethylated cytosine to be discriminated following the MS-HRM run. This is because HRM enables to differentiate melting temperatures with the accuracy of a single nucleotide in short sequences. Bisulfite treatment was performed on 400 ng OF DNA using the EZ-96 DNA Methylation-Gold kit (Zymo Research, USA) according to the manufacturer's instructions. Converted DNA was then stored at −20 °C until further processing.

The DNA sequence targeted by the MS-HRM analysis is localized in the 5′ UTR of the DMRT1 gene. Primer selection is important in MS-HRM in order to avoid amplification biases. Primer design was conducted according to the instructions of Wojdacz et al. (2008). Briefly, primers should contain one or two CpG close to the 5′-end; the 3′-end should contain at least one converted thymine; primers should not have any hairpins or dimers and have melting temperatures that are close together. Information on primers and targeted sequences has been included in Table 5.3. MS-HRM was performed on a LightCycler 480. The PCR reaction mix was 5 μL of Master Mix HRM (Roche, Germany), 1 μL of Mg^{2+} (2.5 mM final), 2.6 μL of molecular grade water, 0.2 μL of each primer (200 nM final), and 1 μL of bisulfite-treated matrix. The PCR conditions were an initial denaturation step at 95 °C for 10 min followed by amplification cycles at 95 °C for 10 s, 56 °C for 15 s, and 72 °C for 10 s, then denaturation at 95 °C for 1 min, and an annealing step at 40 °C for 1 min. The HRM conditions were 25 fluorescence acquisitions every °C on a temperature ramp with a rate of 0.02 °C/s between 55 and 85 °C. All assays were carried out in triplicate. A negative control and standard range were added to each run. Repeated bisulfite treatment and MS-HRM templates were carried out on representative samples to validate the results. The methylation level of each sample was estimated by comparing its melting profile to the ones of the standard range derived from combinations of 100% and 0% controls using gene scanning software (Roche, Germany). The methylation levels were scored as no methylation level ($\leq 1\%$), very low methylation level (1–5%), low methylation level (5–10%), medium methylation level (10–50%), and high methylation level ($\geq 50\%$) (Fig. 5.1).

Table 5.2 Pair of primers used to analyze gene expression and their associated hybridization temperature (Tm), qPCR efficiency, and reference sequence on GenBank

Gene	Forward primer (5′–3′)	Reverse primer (5′–3′)	Tm (°C)	Efficiency (min–max %)	Reference sequence
Reference gene					
β-Actin	TATCCTGACCCTGAAGTACCC	CACGCAGCTCATTGTAGAAGG	55	96–100	JX027376.1
18S	TGAGAAACGGCTACCACATCC	GCCTCGAAAGAGTCCTGTATTG	55	97–99	AY904463.1
Beta-2-microglobulin (*B2m*)	GCTTCCACCCTCCCAACATC	GAGTAGTGCTCTCCCTCCTTGG	55	94–96	AJ133652.1
Target gene					
SRY-related HMG- box protein (*Sox9*)	AGCAGCAAAAACAAGCCTCA	AGCTCCGCGTTGTGAAGAT	55	96–100	EU241882.1
Fibroblast growth factor-9 (*Fgf9*)	TGATGGCTCCCTTGGGTGAAG	ATCCCCTTTAAATGATCCAA	55	91–94	KX589064
Forkhead box L2 (*FoxL2*)	CACAACCTGAGCTTGAACGA	TGCAAGTATTTTGGGGGAGA	60	97–100	JN182652.1
Insulin-like growth factor-1 (*Igf1*)	TTGTAGTTCTGGGATCCATGGG	ACATCACACAAGTGCCACTG	60	95–100	FJ428828.1

Table 5.3 Primer sequence and amplicon information for the MS-HRM assays of DMRT1. CpG site is underlined, and converted thymine is in bold

Forward primer (5′–3′)	Reverse primer (5′–3′)	Amplicon size (pb)	CpG between the primers
GGA<u>C</u>GGAGGAGG**T**ATTT**TT**	CCAAT**A**CCACAAA**C**TTTCTTTT**A**TTT**A**	126	6

5 Evolution of Molecular Investigations on Sturgeon Sex Determination

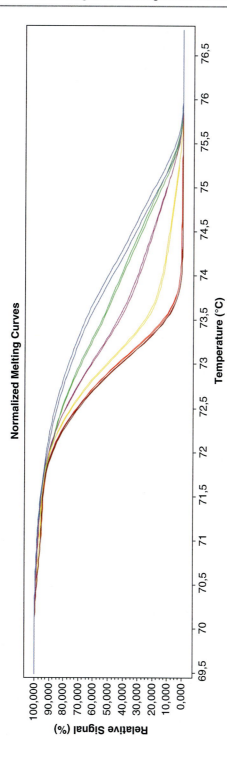

Fig. 5.1 Standard range of methylation to assess the sensitivity of MS-HRM assays. The y axis represents the relative fluorescence signal, and the x axis represents the melting temperature. The following control dilutions, represented in duplicate, were used: 100% methylated (*blue*), 50% methylated (*green*), 10% methylated (*pink*), 5% methylated (*yellow*), 1% methylated (*red*), and 0% methylated (*brown*)

5.1.5 Statistics

For gene expression measurements, the Mann-Whitney and Kruskal-Wallis tests were used because normality was not always achieved. To compare the presence or absence of an ISSR band between sexes and the presence of high or no methylation levels between sexes, we used a Khi2 test on contingency Tables. A hierarchical cluster analysis and a dendrogram produced by using the Jaccard similarity coefficient were performed from the presence/absence matrix of ISSR bands of interest using XLSTAT software (Addinsoft, France). Boxplots were performed with GraphPad Prism 6 (GraphPad Software, USA).

5.2 Results

5.2.1 ISSR

A set of 337 combinations of primers were used. Because of the presence of a smear or the absence of amplification in every assay, 27.6% were tested only once in a temperature gradient. Of the remaining combinations, a mean of 8.1 amplified bands (min, 2; max, 18) were observed per fish, ranged between 50 and 2000 bp. No sex-specific marker was identified in 230 fish. However, four pairs of primers were selected for further analysis because they had a band which was statistically different between males and females ($p < 0.05$). Hierarchical cluster analysis enables separation of fish according to several ISSR band combinations. A graphical representative dendrogram from the four ISSR amplicons from one batch of Siberian sturgeons is shown in Fig. 5.2. At first glance, this type of multivariate analysis seems to be a powerful means by which to discriminate sex despite a low Jaccard similarity coefficient. However, this result was not reproducible with the other batches.

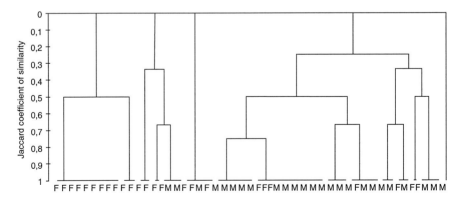

Fig. 5.2 Graphical representation of hierarchical cluster analysis based on ISSR bands founded on different proportions between the genders. In this dendrogram produced by using the Jaccard similarity coefficient, four ISSR bands were used from one batch of 25 males (M) and 25 females (F)

5.2.2 Gene Expression

No significant differences were observed between males, females, and juveniles for the three reference genes, *β-actin*, *18S*, and *B2m*. In differentiated gonads, *Foxl2* expression was sevenfold higher in ovaries compared to testes ($p = 0.0002$), while *Igf1* expression was threefold higher in testes than in ovaries ($p < 0.001$). The two other targeted genes, *Sox9* and *Fgf9*, showed similar expression between sexes (Fig. 5.3). In erythrocytes there was no expression of the four targeted genes in juveniles or 5-year-old Siberian sturgeon. In the pectoral fin samples, *Foxl2* was not expressed, while *Sox9* and *Fgf9* were expressed similarly between males, females, and juveniles. *Igf1* expression was comparable between males and females but was expressed more strongly in juveniles than in older fish ($p < 0.0001$) (Fig. 5.4). In juveniles, distinct expression groups that could be interpreted as a sexual pattern were never observed.

5.2.3 DNA Methylation

From the partial CDS (coding DNA sequence) of Siberian sturgeon available on GenBank (no. HQ110106.1), we have sequenced 517 bp upstream of the CDS in the 5′ UTR of the DMRT1 gene. This sequence corresponds to a CpG island. Among

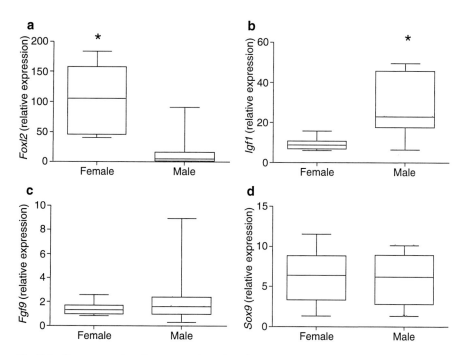

Fig. 5.3 Gene expressions in testes and ovaries of *Foxl2* (**a**), *Igf1* (**b**), *Fgf9* (**c**), and *Sox9* (**d**). Boxes indicate 75th percentile (*top line*), median (*middle line*), and 25th percentile (*bottom line*), and error bars indicate extreme values of data range. Superscript indicates a significant difference

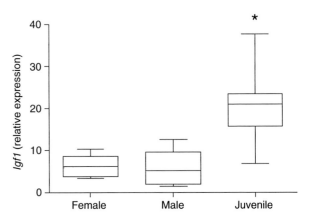

Fig. 5.4 Expression of *Igf1* in pectoral fin of males, females, and juveniles. Boxes indicate 75th percentile (*top line*), median (*middle line*), and 25th percentile (*bottom line*), and error bars indicate extreme values of data range. Superscript indicates a significant difference

the tested groups, we observed distinct methylation profiles from the six targeted CpG. In undifferentiated gonads, methylation levels were low (5–10%) except for one fish with medium levels (10–50%). The methylation level was high ($\geq 50\%$) in the ovaries of the 15 tested females and in the testes of ten males, while there was no methylation ($\leq 1\%$) for the five remaining males. Methylation patterns between male and female gonads were significantly different ($p < 0.014$). In pectoral fin samples, the methylation level was low in all males, females, and juveniles.

5.3 Discussion

5.3.1 Sex Chromosomes

To date, no studies have been able to identify heteromorphic sex chromosomes in sturgeon (Havelka et al. 2011; Keyvanshokooh and Gharaei 2010). This is a critical issue in the race to discover a gene involved in sex determination, because studies that have already described a gene linked to sex determination in fish species were based on targeted research of the Y sex chromosome (Hattori et al. 2012; Kamiya et al. 2012; Matsuda et al. 2002; Myosho et al. 2012; Takehana et al. 2014; Yano et al. 2012).

Sturgeons have a complex genome. The chromosome number is important and can be variable within a species (Havelka et al. 2011). For example, in Siberian sturgeon the chromosome number is $2n = 246 \pm 10$ (Fontana et al. 1997). The ploidy can also be labile between sturgeons of a single species including Siberian sturgeon (Zhou et al. 2011, 2013). These genome features suggest that sturgeon sex determination is maybe not limited to a single chromosome (Fontana et al. 2008).

5.3.2 Molecular Polymorphism

Study of molecular polymorphism is based on random screening of the genome, transcriptome, or proteome in order to identify at least one difference between males and females. This methodological approach is dedicated to the investigation of sex markers, but does not provide any fundamental knowledge of molecular sex determination mechanisms. RAPD is a PCR-based technique involving the design of arbitrary, short primers. AFLP uses restriction enzymes to digest the matrix followed by ligation of adaptors to the fragment extremities, which are then amplified. ISSR primers are designed to amplify sequences between microsatellites. These techniques have been used to identify sex markers in fish species (Penman and Piferrer 2008), such as Nile tilapia (*Oreochromis niloticus*) (Ezaz et al. 2004), rainbow trout (Felip et al. 2005), three-spined stickleback (*Gasterosteus aculeatus*) (Griffiths et al. 2000), or turbot (*Scophthalmus maximus*) (Vale et al. 2014). RAPD, AFLP, and ISSR were performed in Siberian sturgeon, Adriatic sturgeon (*A. naccarii*), Russian sturgeon (*A. gueldenstaedtii*), and sterlet (*A. ruthenus*), but failed to identify a genomic sex marker (Wuertz et al. 2006). RAPD was tested on white sturgeon (*A. transmontanus*) and lake sturgeon (*A. fulvescens*) with the same result (McCormick et al. 2008; Van Eenennaam 1997). AFLP was also used on beluga (*Huso huso*), Amur sturgeon (*A. schrenckii*), and Persian sturgeon (*A. persicus*) without success (Khodaparast et al. 2014; Xiao et al. 2014; Yarmohammadi et al. 2011a, b). To increase the chance of isolating sex markers, a suppression subtractive hybridization or a representational difference analysis was used to enrich the matrices in sex-specific sequences but without positive results (McCormick et al. 2008; Van Eenennaam 1997). ISSR seems to be an interesting technique because genes involved in sex determination are probably framed by repeated sequences (like microsatellites) (Piferrer and Guiguen 2008). Indeed, in zebrafish, one potential DNA region linked to sex is located on a chromosome rich in repeated sequences (Howe et al. 2013). In the present study, a large sampling effort has been performed to take into account genetic variability, using the most important combinations of ISSR primers. Of 337 primer combinations, no DNA sex marker was identified in Siberian sturgeon. The size and complexity of the sturgeon genome can contribute to the difficulty in isolating sex markers (Khodaparast et al. 2014; Xiao et al. 2014). Another explanation is that genetic sex determination systems could be much more complex than one sexual chromosome and one master sex gene (Bachtrog et al. 2014; Penman and Piferrer 2008), for example, the platy (*Xiphophorus maculatus*) which have a WXY system (Schultheis et al. 2009) or the platypus (*Ornithorhynchus anatinus*) which have a spectacular $X_1Y_1X_2Y_2X_3Y_3X_4Y_4X_5Y_5$ male sex chromosome (Rens et al. 2004) that involves a combination of sequences that determine sex. Therefore, taking into consideration these examples, the search for a single sex marker may not be the best option. Indeed, we have shown that with four pairs of primers, there was a band with an unbalanced sex ratio. With one batch of fish, we were able to separate a large proportion of Siberian sturgeon according to their gender (Fig. 5.2). However, this result seems highly dependent on the genitor pool, because we were unable to reproduce this with the other batches.

Random research has not only been performed on the genome. AFLP has been carried out on the transcriptome of Persian sturgeon testes and ovaries (Yarmohammadi et al. 2011a, b). In Siberian sturgeon, differential display, a technique based on random amplification of mRNA sequences, was tested on the transcriptome of the brain and gonads (Gaillard 2006). A proteome comparison from gonads of Persian sturgeon has also been conducted (Keyvanshokooh et al. 2009). These studies failed to detect true sex markers transposable to nonlethal sampling in juveniles. Recent advances in molecular biology techniques are enabling us to directly compare total transcriptome or proteome of individuals. Therefore, these previous methods should be abandoned in order to focus on next-generation sequencing (NGS) analysis.

5.3.3 Transcriptome Sequencing

Thanks to NGS, the transcriptome from the gonads of four sturgeon species has now been published. These databases can be used to identify genes, SNPs, and gene expression levels. This has been carried out with shovelnose sturgeon (*Scaphirhynchus platorynchus*) from 12 mature fish (Amberg et al. 2013), Chinese sturgeon (*A. sinensis*) from one male and one female (Yue et al. 2015), Adriatic sturgeon from one male and one female during early gonad differentiation (Vidotto et al. 2013), and lake sturgeon from two males, two females, and one juvenile with undifferentiated gonads (Hale et al. 2009, 2010). By comparing the transcriptome of the sexes, powerful information about mechanisms involved in sex determination and differentiation could be provided. However, used alone this technique is not sufficient for the detection of a sex marker because lethal sampling is used and it is rarely of any value for juveniles. Moreover, the first transcriptome of the zebrafish was published in 2004 by Li et al., and despite the multiplication of available genomes and transcriptomes for this species, no mechanism linked to sex determination has clearly been identified to date (Howe et al. 2013). The situation in zebrafish clearly shows the limit of this methodological approach to sexing fish species (Piferrer et al. 2012).

5.3.4 Targeted Sequences

Another approach consists of targeting specific sequences identified during experimental research, in a similar way to NGS, or based on knowledge acquired in other species (Piferrer et al. 2012; Piferrer and Guiguen 2008). After transcriptome sequencing of lake sturgeon, two selected genes, *Dmrt1* and *Tra-1*, found to be more abundant in one gender were analyzed using qPCR in four fish of each sex (Hale et al. 2010). In shovelnose sturgeon, four genes, *Dmrt1*, *Dkk1*, *Foxl2*, and *Dact1*, were targeted after NGS by qPCR assays in the 12 fish already used in the NGS (Amberg et al. 2013). Information from fish species with better understood molecular sex-linked mechanisms were thoroughly reviewed (Piferrer and Guiguen 2008). This data was used to target 25 genes via expression quantification of five, 9-month-old Russian sturgeon (Hagihara et al. 2014), and a total of 11 genes in adult and juvenile Siberian sturgeon (Berbejillo et al. 2012, 2013; Vizziano-Cantonnet et al. 2016). In sturgeon, gene expression seems highly variable between studies even for a single species. For example, *Dmrt1* was found in testes and ovaries of mature shovelnose sturgeon (Amberg et al. 2010) or only in testes (Amberg et al. 2013). A potential explanation for this was perhaps due to misinterpretation of gene expression because of variable expression levels of the reference genes (Amberg et al. 2010). Unlike the high expression variability of *β-actin* and *18S* in shovelnose sturgeon, we found no difference in expression levels between males, females, and juveniles in Siberian sturgeon for the three selected reference genes, *β-actin*, *18S*, and *B2m*. We found that *Foxl2* expression was greater in Siberian sturgeon ovaries (Fig. 5.3a), which confirms previous results (Vizziano-Cantonnet et al. 2016). This has also been demonstrated in Russian sturgeon (Hagihara et al. 2014), shovelnose sturgeon (Amberg et al. 2010, 2013), and Chinese sturgeon (Yue et al. 2015). *Igf1* was overexpressed in testes compared to ovaries (Fig. 5.3b), as observed in other batches of Siberian sturgeon (Berbejillo et al. 2011, 2012) and in Chinese sturgeon (Yue et al. 2015). We saw no difference in *Fgf9* expression between genders, even though *Fgf9* is known to play a role in male cellular proliferation by regulation of *Sox9* in vertebrate gonads (Brennan and Capel 2004; Piferrer and Guiguen 2008). No difference was also observed in *Sox9* expression. This gene expression seems highly variable between development states in Siberian sturgeon (Vizziano-Cantonnet et al. 2016). Indeed, mechanisms of sex determination and differentiation are established gradually, with the stage of gonadal maturation, potentially, highly influencing gene expression levels (Piferrer et al. 2012).

A usable aquaculture sex marker is required, but very little gene expression data is available from the nonlethal sampling of juvenile sturgeon. Nonlethal sampling has been achieved from the fins of European sturgeon (*A. sturio*) (Hett and Ludwig 2005) and from the muscle of Siberian sturgeon (Berbejillo et al. 2012, 2013). An assessment of gene expression in 3-month-old Siberian sturgeon has recently been performed but only with lethal sampling (Vizziano-Cantonnet et al. 2016). Genes linked to sex determination and differentiation would appear to be more greatly expressed in gonads than in somatic cells. However, somatic cell lines are not devoid of interest because they can be directly involved in sexual differentiation in chicken (Zhao et al. 2010). In this study, in the pectoral fins of adults and juveniles, only *Foxl2* was unexpressed, with no gender difference for *Igf1*, *So9*, and *Fgf9*. In nucleated erythrocytes we did not detect expression of these four genes (but we did for the three reference genes). To date, gene expression studies have provided us with information on sex differentiation mechanisms, but have not been able to lead us to a useful sex marker in sturgeon.

After years, sturgeon can be sexed by measurements of steroid hormones, testosterone, 11-ketotestosterone, and 17β-estradiol (Williot and Sabeau 1999; Ceapa et al. 2002; Webb and Doroshov 2011). To our knowledge only one study has targeted proteins that may be of interest for the sexing of young juveniles. The only available information on the measurements of hormones in juveniles is a patent which enables a 16-month-old (bw ≈ 1.1 kg) Siberian sturgeon to be sexed (Wetzel et al. 2013; Wetzel and Reynolds 2013). Although this technique cannot be used to detect sex markers in sturgeon, a few months old, it is nonetheless surprising that no further studies have been published on the role of sturgeon hormones during gonad differentiation. Potential targeted hormones in fish were deeply reviewed by Devlin and Nagahama (2002).

Studies on targeted genomic sequences have the advantage, in theory, of being applicable to all types of tissue and at any age. This approach was used by Ferreiro et al. (1989) in their pioneering study entirely dedicated to the identification of sex markers in sturgeon. In this study, the zinc finger Y-chromosomal protein (ZFY), considered at the time to be a putative sex-determining gene in mammals, was not detected in white sturgeon erythrocytes. No gender difference was detected in the sequence and copy number of SOX9 in fin tissue from European sturgeon (Hett et al. 2005) or in the presence/absence of seven SOX family genes in lake sturgeon (McCormick et al. 2008). Sequence targeting is a methodology which, for the moment, has failed to detect a sex marker in juvenile sturgeon.

5.3.5 DNA Methylation

DNA methylation of CpG islands in promoter regions is one of the most studied areas in genome expression regulation research, including that of sex determination. In European seabass gonads, a link between *Cyp19a1* expression, the regulation of *Cyp19a1* expression by DNA methylation, and sex determination has been studied (Navarro-Martín et al. 2011). In males, a high-CpG methylation-level upstream of the CDS was associated and probably responsible for the low level of *Cyp19a1* expression, with the opposite trend observed in females (Navarro-Martín et al. 2011). The influence of DNA methylation on this gene in gonads was also supposed in swamp eel (*Monopterus albus*) (Zhang et al. 2013), European eel (*Anguilla anguilla*) (Pierron et al. 2014), Japanese flounder (*Paralichthys olivaceus*) (Si et al. 2015; Wen et al. 2014), and probably also in Japanese medaka (Contractor et al. 2004). DNA methylation not only has an impact on *Cyp19a1* but also on many genes involved in sex determination and differentiation in the half-smooth tongue sole (Shao et al. 2014). The study by Shao et al. (2014) provided clues that sex determination in embryos was induced by the regulation of DNA methylation of the DMRT1 gene (Graves 2014). Studying DNA methylation of genes involved in sex determination and differentiation seems to be a promising methodological approach to acquiring evidence of a sex determination mechanism in sturgeon. In gonads we found three distinct patterns of the DNA methylation level of CpG in the 5′ UTR of the DMRT1 gene between males, females, and juveniles. In undifferentiated gonads the DNA methylation level is low and increases in ovaries and some of the testes, whereas in a third of the tested males, DNA methylation seemed to be null. Unfortunately we were not able to compare the DMRT1 methylation level to *Dmrt1* expression because we found one or two distinctive products in qPCR assays, regardless of gender. Similar results have already been discussed by Amberg et al. (2013). It is thought that high methylation levels repress gene expression in fish (Navarro-Martín et al. 2011; Shao et al. 2014) and high *Dmrt1* expression is generally observed in male gonads (Piferrer and Guiguen 2008). If this is the case, the possible absence of DNA methylation of DMRT1 in several males could be consistent with what is expected. In our study we found only one pattern of DNA methylation levels in 3-month-old juveniles, except for one fish, with no difference observed in pectoral fin samples. Recently, Vizziano-Cantonnet et al. (2016) showed that *Dmrt1* was induced only at 4 months old and after *Amh* and *Sox9* expression peaked in undifferentiated gonads of Siberian sturgeon. Other genes, including *Amh* and *Sox9*, should be assessed before reaching any conclusion. We have shown that MS-HRM technology has been able to differentiate patterns of DNA methylation levels. Further studies could provide relevant information on the sex determination mechanism in sturgeon.

Conclusion

Despite decades of intensive investigation and the diversity of molecular methodological approaches that have been used, no sex marker has been identified in sturgeon to date. Consequently, no sex determination mechanism has yet been revealed. The increasing knowledge accumulated on molecular sex differentiation, particularly thanks to NGS and qPCR analyses, could help to focus research efforts on more relevant sequences. In addition, the recent sex determination field of investigation that has been made possible by DNA methylation and which has been discussed in this chapter could provide new information in the near future. To assess DNA methylation, targeted genes such as *Amh*, *So9*, and *Dmrt1* could be selected, or DNA methylome analysis could be performed to access a comprehensive dataset of sturgeon gene regulation.

Acknowledgments We thank the Ecloserie de Guyenne, the Prunier Manufacture Company, and the Sources du Gapeau for generously providing sturgeon and O. Brunel and P. Benoit from the Sturgeon SCEA Company for their help. We also thank the Sturgeon SCEA Company and the region PACA who funded this study; L. Jaffrelot, E. Macarry, R. Ciarlo, and M. Lechable for the help with laboratory analyses; and A. Smith for the English corrections.

References

Amberg JJ, Goforth R, Stefanavage T, Sepúlveda MS (2010) Sexually dimorphic gene expression in the gonad and liver of shovelnose sturgeon (*Scaphirhynchus platorynchus*). Fish Physiol Biochem 36:923–932

Amberg JJ, Goforth RR, Sepúlveda MS (2013) Antagonists to the Wnt Cascade Exhibit Sex-Specific Expression in Gonads of Sexually Mature Shovelnose Sturgeon. Sex Dev 7:308–315

Bachtrog D, Mank JE, Peichel CL, Kirkpatrick M, Otto SP, Ashman TL, Hahn MW, Kitano J, Mayrose I, Ming R, Perrin N, Ross L, Valenzuela N, Vamosi JC (2014) The Tree of Sex Consortium. Sex Determination: Why So Many Ways of Doing It? PLoS Biol 12:e1001899. doi:10.1371/journal.pbio.1001899

Berbejillo J, Martinez-Bengochea A, Bedó G, Vizziano-Cantonnet D (2011) Molecular Characterization of Testis Differentiation in the Siberian Sturgeon, *Acipenser baerii*. Indian J Sci Technol 4:71–72

Berbejillo J, Martinez-Bengochea A, Bedo G, Brunet F, Volff JN, Vizziano-Cantonnet D (2012) Expression and phylogeny of candidate genes for sex differentiation in a primitive fish species, the Siberian sturgeon, *Acipenser baerii*. Mol Reprod Dev 79:504–516

Berbejillo J, Martinez-Bengochea A, Bedó G, Vizziano-Cantonnet D (2013) Expression of dmrt1 and sox9 during gonadal development in the Siberian sturgeon (*Acipenser baerii*). Fish Physiol Biochem 39:91–94

Brennan J, Capel B (2004) One tissue, two fates: molecular genetic events that underlie testis versus ovary development. Nat Rev Genet 5:509–521

Bronzi P, Rosenthal H (2014) Present and future sturgeon and caviar production and marketing: a global market overview. J Appl Ichthyol 30:1536–1546

Ceapa C, Williot P, Le Menn F, Davail-Cuisset B (2002) Plasma sex steroids and vitellogenin levels in stellate sturgeon (*Acipenser stellatus* Pallas) during spawning migration in the Danube River. J Appl Ichthyol 18:391–396

Cedar H, Bergman Y (2009) Linking DNA methylation and histone modification: patterns and paradigms. Nat Rev Genet 10:295–304

Chebanov M, Galich E (2009) Ultarsound diagnostics for sturgeon broodstock management. South Branch Federal Center of Selection and Genetics for Aquaculture, Krasnodar, Russian Federation, 116

Contractor RG, Foran CM, Li S, Willett KL (2004) Evidence of gender-and tissue-specific promoter methylation and the potential for ethinylestradiol-induced changes in Japanese medaka (*Oryzias latipes*) estrogen receptor and aromatase genes. J Toxicol Environm Health A 67(1):1–22

Devlin RH, Nagahama Y (2002) Sex determination and sex differentiation in fish: an overview of genetic, physiological, and environmental influences. Aquaculture 208:191–364

Ezaz MT, Harvey SC, Boonphakdee C, Teale AJ, McAndrew BJ, Penman DJ (2004) Isolation and physical mapping of sex-linked AFLP markers in Nile tilapia (*Oreochromis niloticus* L.) Mar Biotechnol 6:435–445

Felip A, Young WP, Wheeler PA, Thorgaard GH (2005) An AFLP-based approach for the identification of sex-linked markers in rainbow trout (*Oncorhynchus mykiss*). Aquaculture 247:35–43

Ferreiro C, Medrano JF, Gall GA (1989) Genome analysis of rainbow trout and sturgeon with restriction enzymes and hybridization with a ZFY gene derived probe to identify sex. Aquaculture 81:245–251

Fontana F, Colombo G (1974) The chromosomes of Italian sturgeons. Experientia 30:739–742

Fontana F, Rossi R, Lanfredi M, Arlati G, Bronzi P (1997) Cytogenetic characterization of cell lines from three sturgeon species. Caryologia 50(1):91–95

Fontana F, Congiu L, Mudrak VA, Quattro JM, Smith TIJ, Ware K, Doroshov SI (2008) Evidence of hexaploid karyotype in shortnose sturgeon. Genome 51:113–119

Fopp-Bayat D (2010) Meiotic gynogenesis revealed not homogametic female sex determination system in Siberian sturgeon (*Acipenser baeri* Brandt). Aquaculture 305:174–177

Gaillard S (2006) Détermination et différentiation sexuelles chez les poissons "le sexe des esturgeons." PhD thesis. University of Toulon

Graves JAM (2014) The epigenetic sole of sex and dosage compensation. Nature Genetics 46(3):215–217

Griffiths R, Orr KL, Adam A, Barber I (2000) DNA sex identification in the three-spined stickleback. J Fish Biol 57:1331–1334

Hagihara S, Yamashita R, Yamamoto S, Ishihara M, Abe T, Ijiri S, Adachi S (2014) Identification of genes involved in gonadal sex differentiation and the dimorphic expression pattern in undifferentiated gonads of Russian sturgeon *Acipenser gueldenstaedtii* Brandt & Ratzeburg, 1833. J Appl Ichthyol 30:1557–1564

Hale M, McCormick C, Jackson J, DeWoody JA (2009) Next-generation pyrosequencing of gonad transcriptomes in the polyploid lake sturgeon (*Acipenser fulvescens*): the relative merits of normalization and rarefaction in gene discovery. BMC Genomics 10:203

Hale MC, Jackson JR, DeWoody JA (2010) Discovery and evaluation of candidate sex-determining genes and xenobiotics in the gonads of lake sturgeon (*Acipenser fulvescens*). Genetica 138:745–756

Hattori RS, Murai Y, Oura M, Masuda S, Majhi SK, Sakamoto T, Fernandino JI, Somoza GM, Yokota M, Strüssmann CA (2012) A Y-linked anti-Müllerian hormone duplication takes over a critical role in sex determination. Proc Natl Acad Sci 109:2955–2959

5 Evolution of Molecular Investigations on Sturgeon Sex Determination

Hattori RS, Strüssmann CA, Fernandino JI, Somoza GM (2013) Genotypic sex determination in teleosts: Insights from the testis-determining amhy gene. Gen Comp Endocrinol 192:55–59

Havelka M, Kaspar V, Hulak M, Flajshans M (2011) Sturgeon genetics and cytogenetics: a review related to ploidy levels and interspecific hybridization. Folia Zool 60:93–103

Hett AK, Ludwig A (2005) SRY-related (Sox) genes in the genome of European Atlantic sturgeon (*Acipenser sturio*). Genome 48:181–186

Hett AK, Pitra C, Jenneckens I, Ludwig A (2005) Characterization of sox9 in European Atlantic sturgeon (*Acipenser sturio*). J Hered 96:150–154

Howe K, Clark MD, Torroja CF, Torrance J et al (2013) The zebrafish reference genome sequence and its relationship to the human genome. Nature 496:498–503

Kamiya T, Kai W, Tasumi S, Oka A, Matsunaga T, Mizuno N, Fujita M, Suetake H, Suzuki S, Hosoya S, Tohari S, Brenner S, Miyadai T, Venkatesh B, Suzuki Y, Kikuchi K (2012) A Trans-Species Missense SNP in Amhr2 Is Associated with Sex Determination in the Tiger Pufferfish, *Takifugu rubripes* (Fugu). PLoS Genet 8:e1002798

Keyvanshokooh S, Kalbassi MR, Hosseinkhani S, Vaziri B (2009) Comparative proteomics analysis of male and female Persian sturgeon (*Acipenser persicus*) gonads. Anim Reprod Sci 111:361–368

Keyvanshokooh S, Gharaei A (2010) A review of sex determination and searches for sex-specific markers in sturgeon. Aquac Res 41:e1–e7

Khodaparast M, Keyvanshokooh S, Pourkazemi M, Hosseini SJ, Zolgharnein H (2014) Searching the genome of beluga (*Huso huso*) for sex markers based on targeted Bulked Segregant Analysis (BSA). CJES 12:185–195

Kobayashi Y, Nagahama Y, Nakamura M (2013) Diversity and plasticity of sex determination and differentiation in fishes. Sex Dev 7:115–125

Laird PW (2003) The power and the promise of DNA methylation markers. Nat Rev Cancer 3:253–266

Laird PW (2010) Principles and challenges of genome-wide DNA methylation analysis. Nat Rev Genet 11:191–203

Law JA, Jacobsen SE (2010) Establishing, maintaining and modifying DNA methylation patterns in plants and animals. Nat Rev Genet 11:204–220

Li Y, Chia JM, Bartfai R, Christoffels A, Yue GH, Ding K, Ho MY, Hill JA, Stupka E, Orban L (2004) Comparative analysis of the testis and ovary transcriptomes in zebrafish by combining experimental and computational tools. Comp Funct Genomics 5:403–418

Liew WC, Bartfai R, Lim Z, Sreenivasan R, Siegfried KR, Orban L (2012) Polygenic Sex Determination System in Zebrafish. PLoS One 7(4):e34397

Lister R, Pelizzola M, Dowen RH, Hawkins RD, Hon G, Tonti-Filippini J, Nery JR, Lee L, Ye Z, Ngo QM et al (2009) Human DNA methylomes at base resolution show widespread epigenomic differences. Nature 462:315–322

Matsuda M, Nagahama Y, Shinomiya A, Sato T, Matsuda C, Kobayashi T, Morrey CE, Shibata N, Asakawa S, Shimizu N et al (2002) DMY is a Y-specific DM-domain gene required for male development in the medaka fish. Nature 417:559–563

McCormick CR, Bos DH, DeWoody JA (2008) Multiple molecular approaches yield no evidence for sex-determining genes in lake sturgeon (*Acipenser fulvescens*). J Appl Ichthyol 24:643–645

Myosho T, Otake H, Masuyama H, Matsuda M, Kuroki Y, Fujiyama A, Naruse K, Hamaguchi S, Sakaizumi M (2012) Tracing the Emergence of a Novel Sex-Determining Gene in Medaka, *Oryzias luzonensis*. Genetics 191:163–170

Nanda I, Kondo M, Hornung U, Asakawa S, Winkler C, Shimizu A, Shan Z, Haaf T, Shimizu N, Shima A, Schmid M, Schartl M (2002) A duplicated copy of DMRT1 in the sex-determining region of the Y chromosome of the medaka, *Oryzias latipes*. Proc Natl Acad Sci 99:11778–11783

Navarro-Martín L, Viñas J, Ribas L, Díaz N, Gutiérrez A, Di Croce L, Piferrer F (2011) DNA methylation of the gonadal aromatase (cyp19a) promoter is involved in temperature-dependent sex ratio shifts in the European sea bass. PLoS Genet 7:e1002447

Penman DJ, Piferrer F (2008) Fish gonadogenesis. Part I: genetic and environmental mechanisms of sex determination. Rev Fish Sci 16:16–34

Pierron F, Bureau du Colombier S, Moffett A, Caron A, Peluhet L, Daffe G, Lambert P, Elie P, Labadie P, Budzinski H, Dufour S, Couture P, Baudrimont M (2014) Abnormal Ovarian DNA Methylation Programming during Gonad Maturation in Wild Contaminated Fish. Environ Sci Technol 48:11688–11695

Piferrer F, Guiguen Y (2008) Fish gonadogenesis. Part II: molecular biology and genomics of sex differentiation. Rev Fish Sci 16:35–55

Piferrer F, Ribas L, Díaz N (2012) Genomic approaches to study genetic and environmental influences on fish sex determination and differentiation. Mar Biotechnol 14:591–604

Rens W, Grützner F, O'Brien PC, Fairclough H, Graves JA, Ferguson-Smith MA (2004) Resolution and evolution of the duck-billed platypus karyotype with an X1Y1X2Y2X3Y3X4Y4X5Y5 male sex chromosome constitution. Proc Natl Acad Sci U S A 101(46):16257–16261

Saxonov S, Berg P, Brutlag DL (2006) A genome-wide analysis of CpG dinucleotides in the human genome distinguishes two distinct classes of promoters. Proc Natl Acad Sci U S A 103:1412–1417

Schultheis C, Böhne A, Schartl M, Volff JN, Galiana-Arnoux D (2009) Sex determination diversity and sex chromosome evolution in poeciliid fish. Sex Dev 3:68–77

Shao C, Li Q, Chen S, Zhang P, Lian J, Hu Q, Sun B, Jin L, Liu S, Wang Z et al (2014) Epigenetic modification and inheritance in sexual reversal of fish. Genome Res 24:604–615

Si Y, Ding Y, He F, Wen H, Li J, Zhao J, Huang Z (2015) DNA methylation level of cyp19a1a and Foxl2 gene related to their expression patterns and reproduction traits during ovary development stages of Japanese flounder (*Paralichthys olivaceus*). Gene 575:321–330

Takehana Y, Matsuda M, Myosho T, Suster ML, Kawakami K, Shin T, Kohara Y et al (2014) Co-option of Sox3 as the male-determining factor on the Y chromosome in the fish *Oryzias dancena*. Nat Com 5:4157

Vale L, Dieguez R, Sánchez L, Martínez P, Viñas A (2014) A sex-associated sequence identified by RAPD screening in gynogenetic individuals of turbot (*Scophthalmus maximus*). Mol Biol Rep 41:1501–1509

Van Eenennaam V (1997) Genetic Analysis of the Sex Determination Mechanism of White Sturgeon (*Acipenser transmontamus* Richardson). Chapter 3: Experimental approaches Used in an Attempt to isolate molecular genetic marker for the identification of sexe on white sturgeon—PhD thesis. University of California, Davis

Vidotto M, Grapputo A, Boscari E, Barbisan F, Coppe A, Grandi G, Kumar A, Congiu L (2013) Transcriptome sequencing and de novo annotation of the critically endangered Adriatic sturgeon. BMC Genomics 14:407

Vizziano-Cantonnet D, Di Landro S, Lasalle A, Martínez A, Mazzoni T, Quagio-Grassiotto I (2016) Identification of the molecular sex-differentiation period in the Siberian sturgeon. Mol Reprod Dev 83:19–36

Webb MAH, Doroshov SI (2011) Importance of environmental endocrinology in fisheries management and aquaculture of sturgeons. Gen Compar Endocrinol 170:313–321

Webb M, Van Eenennaam J, Chapman FA, Vasquez D, Hammond G (2013) Techniques to determine sex and stage of maturity in sturgeons and paddlefish: a brief overview. In: workshop of the 7th international symposium on sturgeons, Nanaimo, BC, Canada, 21–25 July 2013

Wen AY, You F, Sun P, Li J, Xu DD, Wu ZH, Ma DY, Zhang PJ (2014) CpG methylation of dmrt1 and cyp19a promoters in relation to their sexual dimorphic expression in the Japanese flounder *Paralichthys olivaceus*. J Fish Biol 84:193–205

Wetzel DL, Reynolds JE (2013) Advances in application of LP9 analyses for determination of sex in young sturgeon. In: poster presentation at the 7th international symposium on sturgeons, Nanaimo, BC, Canada, 21–25 July 2013

Wetzel DL, Reynolds JE, Roudebush WE (2013) Fish sexual characteristic determination using peptide hormones. Published patent, EP2646833 A2 https://www.google.com/patents/EP2646833A2?cl=en

Williot P, Sabeau L (1999) Elevage d'esturgeons et production de caviar: exemple de l'esturgeon sibérien (*Acipenser baeri*) en France. CR Acad Agric 85(8):71–83

5 Evolution of Molecular Investigations on Sturgeon Sex Determination

Williot P, Sabeau L, Gessner J, Arlati G, Bronzi P, Gulyas T, Berni P (2001) Sturgeon farming in Western Europe: recent developments and perspectives. Aquat Living Ressour 14:367–374

Williot P (2002) Reproduction des esturgeons. In: Billard R (ed) Esturgeons et caviar. Tech et Doc, Lavoisier, pp 63–90

Wojdacz TK, Dobrovic A, Hansen LL (2008) Methylation-sensitive high-resolution melting. Nat Protoc 3:1903–1908

Wuertz S, Gaillard S, Barbisan F, Carle S, Congiu L, Forlani A, Aubert J, Kirschbaum F, Tosi E, Zane L et al (2006) Extensive screening of sturgeon genomes by random screening techniques revealed no sex-specific marker. Aquaculture 258:685–688

Xiao TQ, Lu CY, Li C, Cheng L, Cao DC, Sun XW (2014) An AFLP-based approach for the identification of sex-linked markers in Amur sturgeon *Acipenser schrenckii* Brandt, 1869. J Appl Ichthyol 30:1282–1285

Yano A, Nicol B, Jouanno E, Quillet E, Fostier A, Guyomard R, Guiguen Y (2012) The sexually dimorphic on the Y-chromosome gene (sdY) is a conserved male-specific Y-chromosome sequence in many salmonids. Evol Appl 6:486–496

Yarmohammadi M, Pourkazemi M, Chakmehdouz F, Kazemi R (2011a) Comparative study of male and female gonads in Persian sturgeon (*Acipenser persicus*) employing DNA-AFLP and CDNA-AFLP analysis. J Appl Ichthyol 27:510–513

Yarmohammadi M, Pourkazemi M, Ghasemi A, Hassanzadeh M, Chakmehdouz F (2011b) AFLP reveals no sex-specific markers in Persian sturgeon (*Acipenser persicus*) or beluga sturgeon (*Huso huso*) from the southern Caspian Sea, Iran. Prog Biol Sci 1:55–114

Yue H, Li C, Du H, Zhang S, Wei Q (2015) Sequencing and De Novo Assembly of the Gonadal Transcriptome of the Endangered Chinese Sturgeon (*Acipenser sinensis*). PLoS One 10:e0127332

Zhang Y, Zhang S, Liu Z, Zhang L, Zhang W (2013) Epigenetic Modifications During Sex Change Repress Gonadotropin Stimulation of Cyp19a1a in a Teleost Ricefield Eel (*Monopterus albus*). Endocrinology 154:2881–2890

Zhao D, McBride D, Nandi S, McQueen HA, McGrew MJ, Hocking PM, Lewis PD, Sang HM, Clinton M (2010) Somatic sex identity is cell autonomous in the chicken. Nature 464:237–242

Zhou H, Fujimoto T, Adachi S, Yamaha E, Arai K (2011) Genome size variation estimated by flow cytometry in *Acipenser mikadoi*, *Huso dauricus* in relation to other species of Acipenseriformes. J Appl Ichthyol 27:484–491

Zhou H, Fujimoto T, Adachi S, Abe S, Yamaha E, Arai K (2013) Molecular cytogenetic study on the ploidy status in *Acipenser mikadoi*. J Appl Ichthyol 29:51–55

Sex Determination and Differentiation of the Siberian Sturgeon

6

Denise Vizziano-Cantonnet, Santiago Di Landro, and André Lasalle

Abstract

This is a synthesis of the published information on sex determination and differentiation in sturgeons with special emphasis in Siberian sturgeon *Acipenser baerii*. The sex-determination system has been poorly studied in sturgeons and paddlefishes. Among 27 species of Acipenseriformes, the WZ-female model has been proposed in five species and one hybrid based on gynogenetic studies. The sex-determining gene remains unknown, and there are no studies on the influence from autosomal genetic factors or environmental cues. The lack of genetic sex marker is limiting the studies on sex determination and differentiation, as well as the application in the early sexing for culture purposes.

The gonad sex differentiation show similar histological features in different sturgeons studied and occur at juvenile stage (4–9 months of age). Siberian sturgeon reach the sex differentiation around the 8 months of age when reared in a range of temperature of 14–19 °C, coincident with the upper range of spawning temperature of species at wild. However, the potential effect of temperature on sex determination has not been studied in sturgeons. The molecular sex differentiation period precedes the sex differentiation and occurs between 3 and 6 months old in Siberian sturgeons. The factors involved in male and female pathways are currently studied at molecular level. Gonad transcriptomes are emerging in sturgeons stimulating the knowledge of factors that must direct a cascade of gene regulatory controls that provide sex-specific phenotypes. There is a lack of functional studies in sturgeons.

Keywords

Sturgeons • Sex determination • Sex differentiation • Gonad histology • Gene expression • *Acipenser baerii*

D. Vizziano-Cantonnet (✉) • S. Di Landro • A. Lasalle
Facultad de Ciencias, Laboratorio de Fisiología de la Reproducción y Ecología de Peces, Iguá 4225, Montevideo 11400, Uruguay
e-mail: vizziano@gmail.com

© Springer International Publishing AG, part of Springer Nature 2018
P. Williot et al. (eds.), The Siberian Sturgeon (*Acipenser baerii*, Brandt, 1869)
Volume 1 - Biology, https://doi.org/10.1007/978-3-319-61664-3_6

Introduction

Acipenseriformes is an order of fishes in which are included sturgeons and paddlefishes. Some of their morphological characteristics as cartilaginous skeletons, heterocercal tail, scaleless skin, and the presence of the notochord in adults are unique of this order and make them different from all other bony fish (Birstein 1993; LeBreton et al. 2005). They are found only in the Northern Hemisphere more specifically in Asia, Europe, and North America (Magnin 1959; Bemis and Kynard 1997; Sokolov and Vasili'ev 1989) but were introduced in the Southern Hemisphere, particularly in Uruguay for their culture.

Sturgeons and paddlefishes are represented by two families, Acipenseridae and Polyodontidae, respectively. Twenty-five species of living Acipenseriformes are grouped in four genera (*Acipenser, Huso, Scaphirhynchus, and Pseudoscaphirhynchus*) included in the Acipenseridae family, while the Polyodontidae family is represented only by two extant species (Bemis et al. 1997). All those species have critical problems of conservation (Sokolov and Vasili'ev 1989; Birstein 1993) because of overfishing, incidental fishing, river pollution, dam construction, environmental disruptions, and the poor organization system of fisheries management, among others (Bacalbaşa-Dobrovici 1997; Bemis et al. 1997; Sokolov and Vasili'ev 1989; Wei et al. 1997).

The sturgeons and paddlefish are target species for caviar production and are considered one of the most valuable commercial fishes. The studies of sex determination and differentiation are key points for the understanding of their biology and have an application both in the biology of conservation and in fish production. Nevertheless, there are very scarce data on these topics as it was summarized in Table 6.1.

6.1 Sex Determination Mechanisms

The mechanism of sex determination has been extensively reviewed in fish (Penman and Piferrer 2008) and shows that gonochoric species has two main types of sex determination: a genetic sex determination (GSD) that occurs during conception and an environmental sex determination (ESD) determined after fertilization. Penman and Piferrer (2008) also summarized the types of sex-determining mechanisms in fish that are very variable. In fact the inheritance of sex is based on three main factors: (a) the presence of major genetic sex factors that determines a monofactorial system, (b) the presence of minor factors determining a polyfactorial system, and (c) the environmental differences that determine an environmental sex-determination system. A combination of these different factors can be observed in fish showing their variable systems of sex determination.

The mechanism of sex determination in basal and gonochoric species as sturgeons is not completely elucidated (Flynn et al. 2006). The presence of sexual chromosomes has not been demonstrated by cytological studies in sturgeons (Fontana and Colombo 1974; Van Eenennaam 1997) probably because these basal fish exhibit a relative primitive sex chromosome evolution (Volff 2005). In the absence of cytological demonstrable chromosomes, genetic approaches have been used to show whether the

Table 6.1 List of extant Acipenseriformes, their distribution area, status of conservation, sex-determination system, and sex differentiation time

	Species name	Distribution area[a]	IUCN	Sex-determination system	References	Sex differentiation time	References
Asian species	*Acipenser baerii*	Rivers of north coast of Russia (Ob, Aldan, and Lena rivers)	VU	ZZ-ZW[b]	Fopp-Bayat (2010)	8 months 4 months	Vizziano-Cantonnet et al. (2016) Rzepkowska and Ostaszewska (2014)
	Acipenser dabryanus	Yangtze River and Tungting Lake	CR	–	–	–	–
	Acipenser mikadoi	Siberia Pacific coast	CR	–	–	–	–
	Acipenser schrenckii	Amur River (Russia, China)	CR	–	–	–	–
	Acipenser sinensis	Yangtze and Pearl rivers	CR	–	–	9 months	Chen et al. (2006)
	Pseudoscaphirhynchus fedtschenkoi	Syr-Darya basin	CR	–	–	–	–
	Pseudoscaphirhynchus hermanni	Amu-Darya basin	CR	–	–	–	–
	Pseudoscaphirhynchus kaufmanni	Amu-Darya basin	CR	–	–	–	–
	Huso dauricus	Amur River drainage	CR	–	–	–	–
	Psephurus gladius	Yangtze River	CR	–	–	–	–

(continued)

Table 6.1 (continued)

Species name	Distribution area[a]	IUCN	Sex-determination system	References	Sex differentiation time	References
European species						
Acipenser gueldenstaedtii	Ponto-Caspian distribution	EN	–	–	–	–
Acipenser stellatus	Black, Caspian, and Azov seas	CR	–	–	–	–
Acipenser ruthenus	Rivers emptying into the Azov, Black, Caspian, Baltic, and Arctic seas and in the Amur River	VU	–	–	8 months	Wrobel et al. (2002)
Acipenser nudiventris	Danube River, Aral and Caspian seas	CR	ZZ-ZW[b]	Saber and Hallajian (2014)	–	–
Acipenser persicus	Caspian Sea and the Volga, Kura, and Ural rivers	CR	–	–	–	–
Acipenser sturio	Gironde the Atlantic and the North Sea	CR	–	–	–	–
Acipenser naccarii	Adriatic Sea	CR	–	–	6 months	Grandi et al. (2007)
Huso huso	Black, Caspian, Azov, and Mediterranean seas and Po and Danube rivers	CR	–	–	–	–

6 Sex Determination and Differentiation of the Siberian Sturgeon

	Species	Distribution	IUCN status	Sex-determination system	Reference	Differentiation	Reference
North American species	*Acipenser medirostris*	North Pacific, North American rivers, from Southern California to Alaska	NT	–	–	–	–
	Acipenser transmontanus	California, Oregon, Washington, and the province of British Columbia	LC	ZZ-ZW[b]	Van Eenennaam et al. (1999)	Before 18 months	Doroshov et al. (1997)
	Acipenser fulvescens	Mississippi River, the Great Lakes, and the Hudson Bay	LC	–	–	–	–
	Acipenser brevirostrum	Rivers by the East Atlantic coast	VU	ZZ-ZW[b]	Flynn et al. (2006)	6 months	Flynn and Benfey (2007)
	Acipenser oxyrinchus	Gulf of Mexico to Labrador	NT	–	–	–	–
	Scaphirhynchus platorynchus	Mississippi River	VU	–	–	–	–
	Scaphirhynchus albus	Mississippi, Missouri, and Yellowstone rivers	EN	–	–	–	–
	Scaphirhynchus suttkusi	"Mobile Basin" of the Mississippi and Alabama rivers	CR	–	–	–	–
	Psephurus spathula	Mississippian Basin and contiguous rivers	VU	ZZ-ZW[b]	Shelton & Mims (2012)	–	–
	Huso huso female x Acipenser ruthenus male	Sturgeon farming in Russia, Germany, Hungary, and Japan	/	ZZ-ZW[b]	Omoto et al. (2005)	6 months	Omoto et al. (2001)

[a]Data mostly taken from Billard and Lecointre (2000), Bemis et al. (1997), and IUCN (2015)

[b]Sex-determination system suggested by sex ratios of meiotic gynogenetic progeny; "–" no information found; "/" does not correspond

VU vulnerable, *CR* Critically endangered, *EN* endangered, *NT* near threatened, *LC* least concerned

sex determination can be explained by sexual chromosomes. The gynogenetic studies support the contention of a female heterogametic genetic sex-determining system (ZZ/ZW) in many sturgeons (*Acipenser transmontanus*, Van Eenennaam et al. 1999; bester, *Huso huso* x *Acipenser ruthenus*, Omoto et al. 2005; *Acipenser brevirostrum*, Flynn et al. 2006; *Acipenser baerii*, Fopp-Bayat 2010; *Acipenser nudiventris*, Saber and Hallajian 2014) (Table 6.1). Recently it has been shown that *Polyodon spathula*, another fish of the order of Acipenseriformes, can have a ZZ/ZW sex-determination system (Shelton and Mims 2012). Collectively, these data suggest that sex determination among Acipenseriformes conforms to the WZ-female model (Shelton and Mims 2012). The emerging pattern of sex determination in Acipenseriformes is important for the long-term genetic management of cultured species of sturgeons and paddlefish. However, the number of species studied is scarce to consider that this is the common system of sex determination in this fish order.

For culture purposes the obtention of monosex populations is of great interest in particular a cohort of monosex genetic females producing caviar. In species with female heterogamety (ZZ male; ZW female) as proposed for Siberian sturgeon (Fopp-Bayat 2010), gynogenesis may produce ZZ males, WW "super" females, and/or ZW females (Van Eenennaam et al. 1999). These authors mentioned that the viability of "super females" WW after gynogenic experiments could not be identified in their experimental conditions (Van Eenennaam et al. 1999).

Another way for the production of super females (WW) is to obtain the sex inversion using hormones. More precisely the masculinization of females (ZW) is needed in order to obtain phenotypic and functional ZW neo-males to cross with normal females (ZW) to obtain super females (WW). However, very few efforts have been made to induce sex inversion by hormonal treatments in sturgeons, and most of them have been focused on feminization (Falahatkar et al. 2014; Flynn and Benfey 2007; Omoto et al. 2002). Only one report has been found in the literature on the masculinization of the hybrid bester (*Huso huso* female x *Acipenser ruthenus* male) using 17α-methyltestosterone (Omoto et al. 2002), but this protocol has not been used in females of non-hybrid sturgeons, reared until adult and crossed to normal females to confirm the ZZ/ZW genetic system and to produce monosex genetic females.

The lack of sex chromosome-specific markers—searched for many years in sturgeons (Keyvanshokooh and Gharaei 2010; Khodaparast et al. 2014; Wuertz et al. 2006)—hinders the identification of the presence of super females (WW). Since the survival of WW super females has been observed in at least one species with ZZ/ZW genetic system (tilapia *Oreochromis aureus*) (Guerrero III and Guerrero 1975; Mair et al. 1991), there are no reasons to discard the possibility of survival of WW in sturgeons.

Finally, whether sturgeons use a monofactorial ZZ/ZW genetic system or has an influence from autosomal genetic factors or environmental influences remains to be elucidated. The sex ratio in wild (*Acipenser trasnmontanus*, Chapman et al. 1996) and in captivity (*A. baerii*, Williot and Brun 1998; *Acipenser ruthenus*, Williot et al. 2005) remains stable at 50:50 sustaining the idea of a monofactorial sex-determination system. However, studies on temperature effects on sex ratio or the sex ratio of progeny from masculinized females crossed to regular females have not been made in order to better understand the sex determination mechanism present in sturgeons.

6.2 Master Gene Driving Sex Determination

In fish the master gene driving sex determination (SDG) is not well conserved and has been discovered in species with male heterogamety (XX/XY genetic system) in which the SDG is linked to the Y chromosome and acts as male-dominant gene. In teleost fish SDG are transcription factors as *dmY* (Matsuda et al. 2002) and *sox3* (Takehana et al. 2014), members of the transforming growth factors-beta (TGF-β) family as *gsdf-Y* (Myosho et al. 2012) and *amh-Y* (Hattori et al. 2012; Yamamoto et al. 2014), or factors related to immune system as the *sdY* (Yano et al. 2012). There is no data on the master gene that control the sex in fish with ZZ/ZW sex-determination system, as it is the case of sturgeons. In other vertebrates with ZZ/ZW, two different mechanisms of sex determination have been described. The first one is a double dosage of the transcription factor *dmrt1* in Z that masculinizes some birds (Smith et al. 2009); and the second is the presence of a female-dominant gene associated with W chromosome in some frogs (*dm-W* a paralog coming from a duplication of the gene *dmrt1*) (Yoshimoto et al. 2008, 2010). This opens the question about the presence of a female-dominant master gene linked to the W chromosome or the presence of a double dosage of a gene in males (ZZ). In the Siberian sturgeons the autosomal *dmrt1* is expressed in gonads (Berbejillo et al. 2012, 2013), but it is not the factor that triggers gonad differentiation, since it is overexpressed 1 month after genes involved in male (*sox9*, *amh*) and female (*foxl2*, *cyp19a1a*) gonad differentiation (Vizziano-Cantonnet et al. 2016). However whether a similar W-linked *dmrt1* paralog described in Amphibia is active in sturgeons remains to be elucidated.

The sex-determining gene and sex genetic markers have been searched for many years in sturgeons at genetic level without success (recent review in Keyvanshokooh and Gharaei 2010; Yarmohammadi et al. 2011). The lack of genotypic markers linked to the sex prevented a genetic mapping or a sex genetic identification helping to discover the SDG as in other fishes. An alternative to genetic methodologies for SDG identification is the analysis of gonad transcriptome at undifferentiated stages in which the SDG is potentially upregulated. In teleost fish in which SDG was discovered, this gene has been found to be upregulated during the molecular sex differentiation period (Hattori et al. 2012; Matsuda et al. 2002; Myosho et al. 2012; Vizziano et al. 2007; Yamamoto et al. 2014; Yano et al. 2012). In the Siberian sturgeon the molecular sex differentiation period has been identified (Vizziano-Cantonnet et al. 2016), and the analysis of gonad transcriptome at this stage is under current investigation. Another recent methodology that could help to discover a genetic sex marker or the SDG is the RAD-seq that compares the genomic ADN of male and females to discover differences between sexes as it was down in different animals from invertebrates (copepod, Carmichael et al. 2013) to vertebrates (lizard, Gamble and Zarkower 2014).

The period of molecular sex differentiation goes before the time of sex differentiation recognized at histological level; thus the gonad morphology at early stages of development and the first signs of gonad sex differentiation are key points for the Siberian sturgeon biology. These topics are developed in Sects. 6.4 and 6.5.

6.3 Morphology of Gonads During Sex Differentiation and Early Gametogenesis

The sex differentiation time can be considered as the moment in which the first morphological signs can be recognized at histological level between male and female gonads. The morphological criteria used to distinguish the sex of the gonads have been reported and reviewed in many species of teleost fish (see reviews of Bruslé and Bruslé (1983); Devlin and Nagahama (2002); Nakamura et al. (1998)), and the first signs of gonad sex differentiation vary among species. In some species the first signs are observed in somatic tissues having special features or arrangements, and in other cases there is a clear and significant increase in germ cell numbers.

In contrast to the amount of information available regarding teleost fish, scarce investigation has been dedicated to sex differentiation in basal fish such as sturgeons. Early gonad morphology has been described for the Adriatic sturgeon (*Acipenser naccarii*) (Grandi and Chicca 2008; Grandi et al. 2007), the Chinese sturgeon (*Acipenser sinensis*) (Chen et al. 2006), the shortnose sturgeon (*Acipenser brevirostrum*) (Flynn and Benfey 2007), and more recently, the Siberian sturgeon (*Acipenser baerii*) (Rzepkowska and Ostaszewska 2014). The criteria used to recognize a presumptive female Adriatic sturgeon or Siberian sturgeon are the presence of invaginations at the gonadal surface (Grandi and Chicca 2008; Grandi et al. 2007; Rzepkowska and Ostaszewska 2014); however, the presence of mitotic or meiotic germ cell clusters is the unequivocal sign of ovarian development in the Siberian sturgeon (Vizziano-Cantonnet et al. 2016). On the other hand, fish that have gonads with a homogeneous distribution of non-meiotic germ cells are considered presumptive males of Siberian sturgeons (Vizziano-Cantonnet et al. 2016).

At histological level the undifferentiated gonad of *Acipenser baerii* shows the primordial germ cells restricted to the gonadal periphery among the epithelial cells or located inside the gonad, surrounded by somatic cells (Rzepkowska and Ostaszewska 2014; Vizziano-Cantonnet et al. 2016). As differentiation progresses, two distinct morphological patterns of primordial germ cell distribution were established in Siberian sturgeons: (a) the germ cells organized in clusters that indicates the presence of a female (Fig. 6.1a, b) and (b) the germ cells homogeneously distributed that indicates the presence of a male (Fig. 6.2a) (Vizziano-Cantonnet et al. 2016). The immunohistology using anti-cyp17 elevated against zebra fish protein showed a positive reaction in the cytoplasm of meiotic cells in clusters of the recently differentiated ovaries (Fig. 6.1b, d), while non-meiotic cells present in putative males have no reaction for this antiserum (Fig. 6.2b).

In late pachytene oocytes, the cysts are individualized by follicle cells, forming the ovarian follicles. In the ovarian follicles within the gonadal tissue, the now diplotene oocytes enter into the primary growth period. The ovary develops, and the ovigerous lamellae become prominent and develop on the border an active germinal epithelium, made up of somatic and proliferating germ cells. Some primary growth oocytes emerge among the less-developed germ cells (Fig. 6.3a).

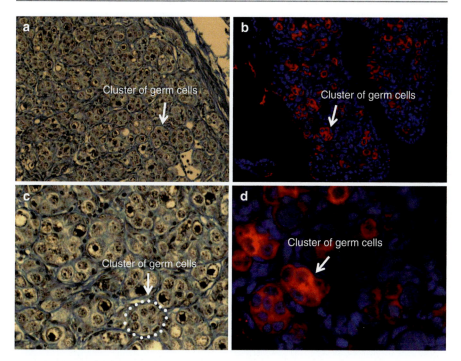

Fig. 6.1 Early differentiated ovary of *A. baerii*. The germ cells at different stages of meiotic prophase are surrounded by somatic cells and form cell nests or "clusters" (**a**, **c**). The cytoplasm of these germ cells in clusters is positive for the enzyme cyp17a as it is showed in *orange* in (**b**, **d**). The nucleus is colored in *blue*

In the putative male, the primordial germ cells differentiate into spermatogonia, proliferate, and increase in number, forming sinuous and cord-like structures that occupy the entire gonadal tissue. At this time, each spermatogonium is individualized by somatic cells, now differentiated into pre-Sertoli cells and form cysts. The interstitial tissue is organized among the cysts, consisting of fibroblasts, blood vessels, and amorphous mesenchymal cells. The testicular duct is formed in the dorsal gonad region (Fig. 6.3b).

The meiotic germ cells express the enzyme 17-hydroxylase (17OH or cyp17) that mediates two main steps of steroid synthesis: the conversion of progesterone into the precursor 17-hydroxy progesterone and the formation of androgens from progestogens. In teleost fish it has been proposed that meiosis is triggered by progestins as 17,20P (Miura et al. 2007), and the 17,20P is produced by spermatogonias that converts 17P into 17,20P by an activated 20βHSD (Vizziano et al. 1996). The present data obtained in sturgeons comfort the idea of an autocrine regulation of germ cells by self-produced steroid produced. The absence of reactivity in male germ cells that remains at spermatogonial stage supports this idea.

The correct identification of the sex of gonads is essential to understand the process of sex determination and differentiation of sturgeons. The period of sex differentiation in the Siberian sturgeon reared in Uruguay has been identified based in

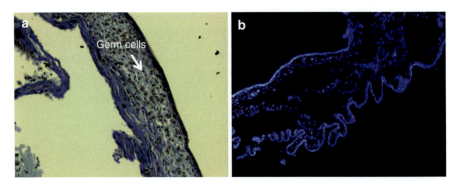

Fig. 6.2 Early differentiated testis of *A. baerii*. (**a**) Germ cells are located in the central region of the gonads and form cord-like structures delimited by somatic cells. (**b**) The cytoplasm of germ cell is not reactive for the cyp17a enzyme as it can be observed by the absence of *orange* color. The nucleus is colored in *blue*

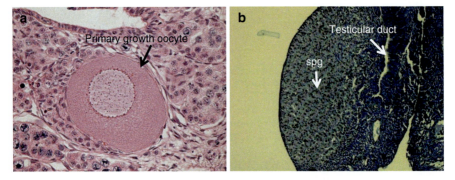

Fig. 6.3 Ovary (**a**) and testis (**b**) of *A. baerii*. A primary growth oocyte can be observed among less-differentiated germ cells in early developing ovaries (**a**). In testis a testicular duct is formed and the spermatogonias proliferate (**b**)

regular samplings of gonads during 3–27 months of age, and results were reported in the next section.

Gonad morphology is also a key point to consider when sex determination model needs to be discovered. In the Acipenseriformes as *Polyodon spathula*, it was concluded that the sex determination genetic model was XX/XY because of a wrong classification of the gonads of a cohort of gynogenote progeny (coming from gynogenesis) (Mims et al. 1997; Shelton and Mims 2012). The authors examined at 70 weeks-of-age fish (around 17 months old of age) and identified them as females, largely based on the developing lamellae in the anatomically differentiating gonads. Ten years later, the authors reinvestigate these fish and observed the presence of a significant number of males. Therefore, they revised their original hypothesis of female homogamety and corrected their earlier erroneous report (Shelton and Mims 2012). They conclude that *Polyodon* as sturgeons has a ZZ/ZW sex genetic-determining system.

6.4   Sex Differentiation Period

The sex differentiation is considered here as the time in which the fish has a gonad that can be recognized as a future testis or a future ovary by histological features.

The fish exhibit a great variability in the period of sex differentiation that can occur: (1) very early previous or around the hatching time (i.e., in the guppy *Poecilia reticulata*, the medaka *Oryzias latipes*), (2) during larval development (i.e., in the trout *Oncorhynchus mykiss* and tilapia, *Tilapia zillii*), or (3) very late during juvenile period between 1 and 2 years in the sea bass (*Dicentrarchus labrax*) or between 1.5 and 6 years in the eel (*Anguilla anguilla*) (Bruslé and Bruslé 1983). The sturgeons have a juvenile sex differentiation, and the age at sex differentiation varied from 4 to 9 months of age for the various sturgeons species analyzed (Chen et al. 2006; Grandi et al. 2007; Flynn and Benfey 2007; Grandi and Chicca 2008; Rzepkowska and Ostaszewska 2014; Vizziano-Cantonnet et al. 2016). It is not clear if the age at sex differentiation is species specific or depends on rearing conditions. For the Siberian sturgeon, when the presence of invaginations was the criterion used to define sex differentiation, age at sex differentiation was 4 months (Rzepkowska and Ostaszewska 2014). However, this distinction does not provide conclusive evidence of morphological sex differentiation in sturgeons (Hagihara et al. 2014; Vizziano-Cantonnet et al. 2016). The presence of mitotic or meiotic germ cell clusters was considered as a clear sign of ovarian development and the homogeneous distribution of non-meiotic germ cells as a sign of presumptive males. Using this criteria Siberian sturgeons are differentiated at 8 months of age in the Uruguayan rearing conditions (Figs. 6.1 and 6.2). It is interesting to note that the temperatures in Uruguay were warmer (from 12.5 °C in June to 26.5 °C in January, Table 6.2) than those of the original rivers in which the Siberian sturgeon lives as native species in the Northern Hemisphere (1–19 °C, Ruban, Chap. 1, this book). The lack of information on the time of sex differentiation at wild prevents a comparison between the wild and culture conditions. However the huge plasticity in the age of first maturity and in growth showed by Siberian sturgeons in the different rivers and lakes of Siberia (Ruban, Chap. 1, this book) suggests that the kinetics of sex differentiation can also be regulated by environmental factors.

At Uruguayan rearing conditions, the gonads remain undifferentiated until 6 months of age. We have no data on the 7 months of age, but we observed signs of sex differentiation at 8 months of age (Vizziano-Cantonnet et al. 2016). The kinetics of gonad development of Siberian sturgeons in Uruguay is shown in the Fig. 6.4a, and the temperature changes during the period of gonad development are shown in Fig. 6.4b. In the case of this cohort, the fecundation has been made in April in the Northern Hemisphere and transported and reared in Uruguay (Southern Hemisphere). In light gray, we are showing the period for which fish remained sexually undifferentiated and in dark gray the period of early gametogenesis (Vizziano-Cantonnet et al. 2016). The first 6 months of development occurred when the temperature ranged from 14 °C to 19 °C for this cohort (Fig. 6.4, Table 6.2), corresponding to the upper natural range of temperature at the spawning sites of the species (9 to 18 °C, Sokolov and Vasili'ev 1989). However, other batches of fertilized eggs can be received in February and March, when the temperatures are higher in Uruguay

Table 6.2 Annual average of temperature at the cages of sturgeons at the Rio Negro River

	2013												2014						
	Jan	Feb	Mar	Apr	May	Jun	Jul	Aug	Sep	Oct	Nov	Dec	Jan	Feb	Mar	Apr	May	Jun	Jul
T (°C)	26,48	24,91	23,37	23,37	17,57	15,57	14,45	14,02	15,24	19,03	22,65	25,74	26,20	25,03	23,3	20,05	16,14	12,5	13,18
s.d. (±)	0,73	3,00	0,81	0,81	1,14	0,80	1,72	0,68	1,99	2,10	1,00	0,94	0,65	0,99	1,04	1,95	1,88	0,5	0,95
n	25	19	31	31	31	30	31	31	30	31	30	31	31	28	31	30	31	30	31

T (°C) = temperature; n = number of water temperature measures taken each month, s.d. (±) = standard deviation

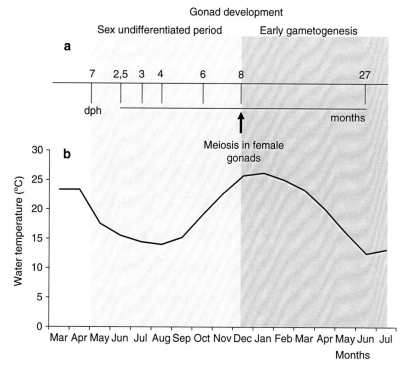

Fig. 6.4 Sex differentiation period and early gametogenesis in the Siberian sturgeon (**a**). dph = days post-hatching. The early meiosis in gonads with germ cells in clusters is one of the criteria to recognize the presence of a female gonad. Changes in temperature during the rearing conditions of embryos imported from the Northern Hemisphere are shown in (**b**)

(from 24 °C in February to 23 °C in April, Table 6.2). The sex ratio of Siberian sturgeons cultured in Uruguay has not been studied until now to understand whether high temperatures can affect the sexual development and sex ratio of the species. There are no comprehensive published data on the effect of temperature on sex differentiation of the Siberian sturgeon.

6.5 Period of Molecular Sex Differentiation

In fish, the period preceding gonad sex differentiation is characterized by early expression of sexually dimorphic genes involved in male or female pathways, controlled partly via molecular mechanisms; this stage is referred to as the "molecular sex differentiation period" (Vizziano et al. 2007). During the undifferentiated period, gonads show few histological changes and contain basically germ cells and somatic cells undergoing mitosis. Contrasting with that, they are very active at molecular level, and the pattern of male and female development can be recognized in particular when large-scale gene expression methodologies are applied to gonads of monosex genetic populations (Baron et al. 2005; Ijiri et al. 2008; Vizziano et al. 2007). These works combined with the disruption of gonads to change the fish sex, provided an idea of the group of genes repressed or activated during the process of molecular sex differentiation that precede the gonad sex differentiation in trout (Raghuveer et al. 2005; Vizziano-Cantonnet et al. 2008; Vizziano et al. 2008). Several candidate genes were selected from large-scale gene expression studies, and their sex dimorphic expression was validated using monosex genetic populations (Baron et al. 2005; Cavileer et al. 2009; Guiguen et al. 1999; Lareyre et al. 2008; Vizziano et al. 2007; Yano et al. 2011b) or sex markers (Ijiri et al. 2008; Piferrer and Guiguen 2008). Recently, the period of molecular sex differentiation around 3–4 months for the Siberian sturgeon has been described (Vizziano-Cantonnet et al. 2016).

In teleost fish with an XX/XY sex-determining system, which is the type of fish in which the master sex-determining gene was discovered (*O. latipes*, Matsuda et al. 2002; *O. hatcheri*, Hattori et al. 2012; *O. luzonensis*, Myosho et al. 2012; *O. mykiss*, Yano et al. 2012; *O. dancena*, Takehana et al. 2014; *O. bonariensis*, Yamamoto et al. 2014), the male pathway is characterized by early dimorphic expression of transcription factors such as *dmrt1*, *sox9*, *nr0b1*, *dax1, and tbx1* (Baron et al. 2005; Marchand et al. 2000; Vizziano et al. 2007; Yano et al. 2011a) and type beta-transforming growth factors such as *amh* and *gsdf2* (Baron et al. 2005; Hattori et al. 2012; Lareyre et al. 2008; Myosho et al. 2012; Shibata et al. 2010; Vizziano et al. 2007; Yamamoto et al. 2014). Androgens appear to be involved in molecular mechanisms of sex differentiation in some teleosts (trout, Vizziano et al. 2007, 2008; pejerrey, Hattori et al. 2009, Blasco et al. 2013) but not in other fish (tilapia, Nakamura et al. 1998, Ijiri et al. 2008). In females, estrogens (mediated by *cyp19a1a*), and other factors as *foxl2* and *follistatin (fst)*, are proposed as essential for ovarian differentiation (Guiguen et al. 2010; Piferrer and Guiguen 2008). Disruption of aromatase using anti-aromatase substances inhibits not only aromatase expression in all-female populations but also inhibits *foxl2a* expression during the ovary-to-testis transdifferentiation process (Vizziano et al. 2008), supporting the idea that *foxl2a* plays a key role in the early stages of ovarian development.

The gene candidates for regulation of sex differentiation in the Siberian sturgeon were *amh* and *sox9* for the masculine pathway and *cyp19a1* and *foxl2a* for the feminine pathway (Vizziano-Cantonnet et al. 2016). The expressions of *cyp19a1* and *foxl2a* were reinvestigated in the Siberian sturgeon in a larger period of development from 3 to 6 months of age following the methodology described previously (Vizziano-Cantonnet et al. 2016). We confirmed (a) the period of molecular sex differentiation is extended at least into 6 months of age (Fig. 6.5); (b) the presence of a group of fish with elevated *cyp19a1a* and *foxl2a* levels, presumably future females; and a second group with low *cyp19a1a* and *foxl2a* levels presumably indicating a male development (Fig. 6.5).

It is interesting to note that the percentage of fish with higher levels of *cyp19a1* and *foxl2* increases with the age of fish, being 31% at 3 months and 44–60% for 5–6 months (Table 6.3, Fig. 6.5) indicating that possibly the molecular sex differentiation occurs progressively during the period preceding the sex differentiation. This supports the idea of coexistence of fish engaged in the molecular sex differentiation

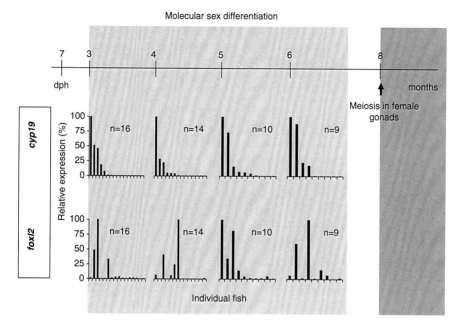

Fig. 6.5 Gonadal relative expression of *cyp19* and *foxl2* studied during the molecular sex differentiation period of Siberian sturgeon from 3 to 6 months of age. Relative quantification was carried out by normalizing the values to 18S ribosomal RNA gene abundance. Relative expression was calculated as a percentage of the highest expression level recorded for each gene tested. Dph = days post-hatching. The number of fish studied for each month is signaled as *n*

Table 6.3 Percentage of fish with higher levels of aromatase (*cyp19a1*) and *foxl2* during the molecular sex differentiation period (3–6 months old fish)

	3 M	4 M	5 M	6 M
n	16	14	10	9
cyp19a1	31%	43%	60%	44%
foxl2	31%	36%	60%	55%

M = months, *n* = number of fish studied

with fish undifferentiated at molecular level with bipotential gonads. The bipotential stage is the period in which the gonad is not sex differentiated at molecular level (mammals, Brennan and Capel 2004). This stage has not been studied in fish and sturgeons seem to be a good model for study this period.

Databases of gonadal transcriptome of sturgeon were published (*Acipenser fulvescens*, Hale et al. 2010; *Acipenser naccarii*, Vidotto et al. 2013; *Acipenser sinensis*, Yue et al. 2015), but these studies were unable in identifying the sex-determining gene or the pattern of sex differentiation at molecular level because they were based in one ovary compared to one testis. In order to better understand the process of molecular sex differentiation, fish at undifferentiated and differentiated stages are needed.

The knowledge on sex differentiation opens the possibility to study this period using multigenic approaches (i.e., using transcriptomics) in a key period in which the sex-determinant gene is known to be overexpressed in vertebrates to repress or activate the male or female pathways.

Conclusions

Sturgeons are basal and menaced fish very valuable for caviar production for which there is scarce information on sex determination and differentiation. The emerging pattern of sex determination is the WZ-female model (including Siberian sturgeon), but the sex-determining gene remains to be elucidated. The sex differentiation occurs at juvenile stage (4–9 months old) for different sturgeons studied. The Siberian sturgeon is sex differentiated at 8 months of age in Uruguayan rearing conditions (12–26 °C) and considering the clusters of meiotic germ cells in females as reference. The molecular sex differentiation occurs between 3 and 6 months of age, when *foxl2* and *aromatase* are taken into account. But the molecular pathway preceding the building of a testis or an ovary is largely unknown for sturgeons. The basic data to advance in this topic are emerging in the literature. The knowledge of sex determination and differentiation is less developed in sturgeons when compared to teleost fish, and an effort needs to be made to characterize at molecular level the changes that conduce to a male or a female gonad. There are no functional studies in sturgeons using the gene knockout by mutagenesis or the gain of function by transgenesis. These methodologies need to be developed for these basal non-teleost and non-model fish of commercial interest.

Acknowledgments Many thanks are due to the enterprise Estuario del Plata (Uruguay) for their cooperation with Universidad de la República (UdelaR). Dr. Vet. Andrés Ryncowski gave us the temperature records. We are grateful with Dr. François Brion (INERIS, Paris, France) for the donation of the anti-cyp17 elevated against zebra fish protein.

Glossary

amh	Anti-Müllerian hormone
amh-Y	Y-linked amh
ar	Androgen receptor
cyp17a1	Cytochrome P450, family 17, subfamily a, polypeptide 1
cyp19a1	Cytochrome P450, family 19, subfamily a, polypeptide 1a
dmrt1	Doublesex and Mab-3-related transcription factor 1
dm-W	W-linked dmrt1 paralog
dmY	Y-specific DM-domain gene required for male development
ESD	Environmental sex determination
figα	Factor in germ line alpha
foxl2a	Forkhead box L2 a

fst	Follistatin
GSD	Genetic sex determination
gsdf-Y	Gonadal soma-derived growth factor on the Y
gsdf2	Gonadal soma-derived growth factor
GSI	Gonadosomatic index
hsd17b1	Hydroxysteroid (17-beta) dehydrogenase 1
nrOb1	Nuclear receptor subfamily 0, group b, member 1
SDG	Sex-determining gene
sdY	Sexually dimorphic on the chromosome Y
sox9	Sex-determining region Y-box containing gene 9
sox9a1	Sex-determining region Y-box containing gene 9
sox9a2	Sex-determining region Y-box containing gene 9
star	Steroidogenic acute regulatory protein
tbx1	T-box transcription factor gene family
17,20P	17,20ß-dihydroxyprogesterone
17P	17-hydroxyprogesterone
20βHSD	20-hydroxysteroid dehydrogenase

References

Bacalbaşa-Dobrovici N (1997) Endangered migratory sturgeons of the lower Danube River and its delta. Env Biol Fish 48(1–4):201–207

Baron D, Houlgatte R, Fostier A, Guiguen Y (2005) Large-scale temporal gene expression profiling during gonadal differentiation and early gametogenesis in rainbow trout. Biol Reprod 73(5):959–966

Bemis WE, Kynard B (1997) Sturgeon rivers: an introduction to acipenseriform biogeography and life history. Env Biol Fish 48(1–4):167–183

Bemis WE, Findeis EK, Grande L (1997) An overview of Acipenseriformes. In: Birstein VJ, Waldman JR, Bemis WE (eds) Sturgeon biodiversity and conservation, 1st edn. Kluwer Academic/Plenum Publishers, New York, pp 25–71

Berbejillo J, Martinez-Bengochea A, Bedo G, Brunet F, Volff JN, Vizziano-Cantonnet D (2012) Expression and phylogeny of candidate genes for sex differentiation in a primitive fish species, the Siberian sturgeon, *Acipenser baerii*. Mol Reprod Dev 79(8):504–516

Berbejillo J, Martinez-Bengochea A, Bedo G, Vizziano-Cantonnet D (2013) Expression of *dmrt1* and *sox9* during gonadal development in the Siberian sturgeon (*Acipenser baerii*). Fish Physiol Biochem 39(1):91–94

Billard R, Lecointre G (2000) Biology and conservation of sturgeon and paddlefish. Rev Fish Biol Fisher 10(4):355–392

Birstein VJ (1993) Sturgeons and paddlefishes: threatened fishes in need of conservation. Conserv Biol 7(4):773–787

Blasco M, Somoza GM, Vizziano-Cantonnet D (2013) Presence of 11-ketotestosterone in pre-differentiated male gonads of *Odontesthes bonariensis*. Fish Physiol Biochem 39(1):71–74

Brennan J, Capel B (2004) One tissue, two fates: molecular genetic events that underlie testis versus ovary development. Nat Rev Genet 5(7):509–521

Bruslé J, Bruslé S (1983) La gonadogenèse des Poissons. Reprod Nutr Dévelop 23(3):453–449

Carmichael SN, Bekaert M, Taggart JB, Christie HRL, Bassett DI, Bron JE, Skuce PJ, Gharbi K, Skern-Mauritzen R, Sturm A (2013) Identification of a sex-linked SNP marker in the Salmon louse (*Lepeophtheirus salmonis*) using RAD sequencing. PLoS One 8(10):e77832

Cavileer T, Hunter S, Okutsu T, Yoshizaki G, Nagler JJ (2009) Identification of novel genes associated with molecular sex differentiation in the embryonic gonads of rainbow trout (*Oncorhynchus mykiss*). Sex Dev 3(4):214–224

Chapman F, Van Eenennaam J, Doroshov S (1996) The reproductive condition of white sturgeon, *Acipenser transmontanus*, in San Francisco Bay, California. FB 94(4):628–634

Chen X, Wei Q, Yang D, Zhu Y (2006) Observations on the formation and development of the primary germinal tissue of cultured Chinese sturgeon, *Acipenser sinensis*. J Appl Ichthyol 22:358–360

Devlin RH, Nagahama Y (2002) Sex determination and sex differentiation in fish: an overview of genetic, physiological, and environmental influences. Aquaculture 208(3):191–364

Doroshov SI, Moberg GP, Van Eenennaam JP (1997) Observations on the reproductive cycle of cultures white sturgeon, *Acipenser transmontanus*. Environ Biol Fish 48(1–4):265–278

Falahatkar B, Poursaeid S, Meknatkhah B, Khara H, Efatpanah I (2014) Long-term effects of intraperitoneal injection of estradiol-17beta on the growth and physiology of juvenile stellate sturgeon *Acipenser stellatus*. Fish Physiol Biochem 40(2):365–373

Flynn S, Benfey T (2007) Sex differentiation and aspects of gametogenesis in shortnose sturgeon *Acipenser brevirostrum* Lesueur. J Fish Biol 70(4):1027–1044

Flynn S, Matsuoka M, Reith M, Martin-Robichaud D, Benfey T (2006) Gynogenesis and sex determination in shortnose sturgeon, *Acipenser brevirostrum* Lesuere. Aquaculture 253(1):721–727

Fontana F, Colombo G (1974) The chromosomes of Italian sturgeons. Experientia 30(7):739–742

Fopp-Bayat D (2010) Meiotic gynogenesis revealed not homogametic female sex determination system in Siberian sturgeon (*Acipenser baeri* Brandt). Aquaculture 305(1–4):174–177

Gamble T, Zarkower D (2014) Identification of sex-specific molecular markers using restriction site-associated DNA sequencing. Mol Ecol Resour 14(5):902–913

Grandi G, Chicca M (2008) Histological and ultrastructural investigation of early gonad development and sex differentiation in Adriatic sturgeon (*Acipenser naccarii*, Acipenseriformes, Chondrostei). J Morphol 269(10):1238–1262

Grandi G, Giovannini S, Chicca M (2007) Gonadogenesis in early developmental stages of *Acipenser naccarii* and influence of estrogen immersion on feminization. J Appl Ichthyol 23(1):3–8

Guerrero RD III (1975) Use of androgens for the production of all-male *Tilapia aurea* (Steindachner). Trans Am Fish Soc 104(2):342348

Guiguen Y, Baroiller JF, Ricordel MJ, Iseki K, McMeel O, Martin S, Fostier A (1999) Involvement of estrogens in the process of sex differentiation in two fish species: the rainbow trout (*Oncorhynchus mykiss*) and a tilapia (*Oreochromis niloticus*). Mol Reprod Dev 54(2):154–162

Guiguen Y, Fostier A, Piferrer F, Chang C-F (2010) Ovarian aromatase and estrogens: a pivotal role for gonadal sex differentiation and sex change in fish. Gen Comp Endocrinol 165(3):352–366

Hagihara S, Yamashita R, Yamamoto S, Ishihara M, Abe T, Ijiri S, Adachi S (2014) Identification of genes involved in gonadal sex differentiation and the dimorphic expression pattern in undifferentiated gonads of Russian sturgeon *Acipenser gueldenstaedtii* Brandt & Ratzeburg, 1833. J Appl Ichthyol 30(6):1557–1564

Hale MC, Jackson JR, Dewoody JA (2010) Discovery and evaluation of candidate sex-determining genes and xenobiotics in the gonads of lake sturgeon (*Acipenser fulvescens*). Genetica 138(7):745–756

Hattori RS, Fernandino JI, Kishii A, Kimura H, Kinno T, Oura M, Somoza GM, Yokota M, Strüssmann CA, Watanabe S (2009) Cortisol-induced masculinization: does thermal stress affect gonadal fate in pejerrey, a teleost fish with temperature-dependent sex determination? PLoS One 4(8):e6548

Hattori RS, Murai Y, Oura M, Masuda S, Majhi SK, Sakamoto T, Fernandino JI, Somoza GM, Yokota M, Strussmann CA (2012) A Y-linked anti-Müllerian hormone duplication takes over a critical role in sex determination. Proc Natl Acad Sci U S A 109(8):2955–2959

Ijiri S, Kaneko H, Kobayashi T, Wang D-S, Sakai F, Paul-Prasanth B, Nakamura M, Nagahama Y (2008) Sexual dimorphic expression of genes in gonads during early differentiation of a teleost fish, the Nile tilapia *Oreochromis niloticus*. Biol Reprod 78(2):333–341

IUCN (2015) The IUCN Red List of Threatened Species. Version 2015-4. www.iucnredlist.org

6 Sex Determination and Differentiation of the Siberian Sturgeon

Keyvanshokooh S, Gharaei A (2010) A review of sex determination and searches for sex-specific markers in sturgeon. Aquac Res 41(9):e1–e7

Khodaparast M, Keyvanshokooh S, Pourkazemi M, Hosscini S, Zolgharnein H (2014) Searching the genome of beluga (*Huso huso*) for sex markers based on targeted bulked Segregant analysis (BSA). Caspian J Env Sci 12(2):185–195

Lareyre J-J, Ricordel M-J, Mahé S, Goupil A-S, Vizziano D, Bobe J, Guiguen Y, Le Gac F (2008) Two new TGF beta members are restricted to the gonad and differentially expressed during sex differentiation and gametogenesis in trout. Cybium 32(suppl):202

LeBreton GT, Beamish FWH, McKinley SR (eds) (2005) Sturgeons and paddlefish of North America. Springer Science & Business Media, INC, New York

Magnin E (1959) Répartition actuelle des acipenseridés. Revue des Travaux de l'Institut Des Pêches Maritimes 23(3):277–285

Mair G, Scott A, Penman D, Beardmore J, Skibinski D (1991) Sex determination in the genus Oreochromis: 1. Sex reversal, gynogenesis and triploidy in *O. niloticus* (L.) Theor Appl Genet 82(2):144–152

Marchand O, Govoroun M, D'Cotta H, McMeel O, Lareyre JJ, Bernot A, Laudet V, Guiguen Y (2000) DMRT1 expression during gonadal differentiation and spermatogenesis in the rainbow trout, *Oncorhynchus mykiss*. Biochim Biophys Acta 1493(1–2):180–187

Matsuda M, Nagahama Y, Shinomiya A, Sato T, Matsuda C, Kobayashi T, Morrey CE, Shibata N, Asakawa S, Shimizu N (2002) DMY is a Y-specific DM-domain gene required for male development in the medaka fish. Nature 417(6888):559–563

Mims SD, Shelton WL, Linhart O, Wang C (1997) Induced meiotic gynogenesis of paddlefish *Polyodon spathula*. JWAS 28(4):334–343

Miura C, Higashino T, Miura T (2007) A progestin and an estrogen regulate early stages of oogenesis in fish. Biol Reprod 77(5):822–828

Myosho T, Otake H, Masuyama H, Matsuda M, Kuroki Y, Fujiyama A, Naruse K, Hamaguchi S, Sakaizumi M (2012) Tracing the emergence of a novel sex-determining gene in medaka, *Oryzias luzonensis*. Genetics 191(1):163–170

Nakamura M, Kobayashi T, Chang XT, Nagahama Y (1998) Gonadal sex differentiation in teleost fish. J Exp Zool 281(5):362–372

Omoto N, Maebayashi M, Mitsuhashi E, Yoshitomi K, Adachi S, Yamauchi K (2001) Histological observations of gonadal sex differentiation in the F2 hybrid sturgeon, the bester. Fisheries Sci 67(6):1104–1110

Omoto N, Maebayashi M, Mitsuhashi E, Yoshitomi K, Adachi S, Yamauchi K (2002) Effects of estradiol-17β and 17α-methyltestosterone on gonadal sex differentiation in the F2 hybrid sturgeon, the bester. Fish Sci 68(5):1047–1054

Omoto N, Maebayashi M, Adachi S, Arai K, Yamauchi K (2005) Sex ratios of triploids and gynogenetic diploids induced in the hybrid sturgeon, the bester (*Huso huso* female× *Acipenser ruthenus* male). Aquaculture 245(1):39–47

Penman DJ, Piferrer F (2008) Fish gonadogenesis. Part I: genetic and environmental mechanisms of sex determination. Rev Fish Sci 16(S1):16–34

Piferrer F, Guiguen Y (2008) Fish gonadogenesis. Part II: molecular biology and genomics of sex differentiation. Rev Fish Sci 16(S1):35–55

Raghuveer K, Garhwal R, Wang DS, Bogerd J, Kirubagaran R, Rasheeda MK, Sreenivasulu G, Bhattachrya N, Tarangini S, Nagahama Y, Senthilkumaran B (2005) Effect of methyl testosterone- and ethinyl estradiol-induced sex differentiation on catfish, *Clarias gariepinus*: expression profiles of DMRT1, cytochrome P450aromatases and 3 beta-hydroxysteroid dehydrogenase. Fish Physiol Biochem 31(2–3):143–147

Rzepkowska M, Ostaszewska T (2014) Proliferating cell nuclear antigen and vasa protein expression during gonadal development and sexual differentiation in cultured Siberian (*Acipenser baerii* Brandt, 1869) and Russian (*Acipenser gueldenstaedtii* Brandt & Ratzeburg, 1833) sturgeon. Rev Aquaculture 6(2):75–88

Saber MH, Hallajian A (2014) Study of sex determination system in ship sturgeon, *Acipenser nudiventris* using meiotic gynogenesis. Aquacult Int 22(1):273–279

Shelton WL, Mims SD (2012) Evidence for female heterogametic sex determination in paddlefish *Polyodon spathula* based on gynogenesis. Aquaculture 356:116–118

Shibata Y, Paul-Prasanth B, Suzuki A, Usami T, Nakamoto M, Matsuda M, Nagahama Y (2010) Expression of gonadal soma derived factor (GSDF) is spatially and temporally correlated with early testicular differentiation in medaka. Gene Expr Patterns 10(6):283–289

Smith CA, Roeszler KN, Ohnesorg T, Cummins DM, Farlie PG, Doran TJ, Sinclair AH (2009) The avian Z-linked gene DMRT1 is required for male sex determination in the chicken. Nature 461(7261):267–271

Sokolov L, Vasili'ev V (1989) *Acipenser baerii* Brandt, 1869. In: Holcik J (ed) The freshwater fish of Europe, Vol I, Part II, general introduction to fishes-Acipenseriformes. AULA-Verlag, Wiesbaden, pp 263–284

Takehana Y, Matsuda M, Myosho T, Suster ML, Kawakami K, Shin IT, Kohara Y, Kuroki Y, Toyoda A, Fujiyama A, Hamaguchi S, Sakaizumi M, Naruse K (2014) Co-option of Sox3 as the male-determining factor on the Y chromosome in the fish *Oryzias dancena*. Nat Commun 5:4157

Van Eenennaam AL (1997) Genetic Analysis of the Sex Determination Mechanism of White Sturgeon (*Acipenser transmontanus* Richardson). Research Theses and Dissertations, California Sea Grant College Program UC San Diego. Available via https://escholarship.org/uc/item/9x3272ww. Accessed 1 Jan 1997

Van Eenennaam A, Van Eenennaam J, Medrano J, Doroshov S (1999) Brief communication. Evidence of female heterogametic genetic sex determination in white sturgeon. J Hered 90(1):231–233

Vidotto M, Grapputo A, Boscari E, Barbisan F, Coppe A, Grandi G, Kumar A, Congiu L (2013) Transcriptome sequencing and de novo annotation of the critically endangered Adriatic sturgeon. BMC Genomics 14:407

Vizziano D, Fostier A, Le Gac F, Loir M (1996) 20 beta-hydroxysteroid dehydrogenase activity in nonflagellated germ cells of rainbow trout testis. Biol Reprod 54(1):1–7

Vizziano D, Randuineau G, Baron D, Cauty C, Guiguen Y (2007) Characterization of early molecular sex differentiation in rainbow trout, *Oncorhynchus mykiss*. Dev Dyn 236(8):2198–2206

Vizziano D, Baron D, Randuineau G, Mahe S, Cauty C, Guiguen Y (2008) Rainbow trout gonadal masculinization induced by inhibition of estrogen synthesis is more physiological than masculinization induced by androgen supplementation. Biol Reprod 78(5):939–946

Vizziano-Cantonnet D, Baron D, Mahe S, Cauty C, Fostier A, Guiguen Y (2008) Estrogen treatment up-regulates female genes but does not suppress all early testicular markers during rainbow trout male-to-female gonadal transdifferentiation. J Mol Endocrinol 41(5):277–288

Vizziano-Cantonnet D, Di Landro S, Lasalle A, Martínez A, Mazzoni TS, Quagio-Grassiotto I (2016) Identification of the molecular sex-differentiation period in the Siberian sturgeon. Mol Reprod Dev 83(1):19–36

Volff J (2005) Genome evolution and biodiversity in teleost fish. Heredity 94(3):280–294

Wei Q, Fe K, Zhang J, Zhuang P, Luo J, Zhou R, Yang W (1997) Biology, fisheries, and conservation of sturgeons and paddlefish in China. In: sturgeon biodiversity and conservation. Environ Biol Fish 17:241–255

Williot P, Brun R (1998) Ovarian development and cycles in cultured Siberian sturgeon, *Acipenser baeri*. Aquat Living Resour 11(02):111–118

Williot P, Brun R, Rouault T, Pelard M, Mercier D, Ludwig A (2005) Artificial spawning in cultured sterlet sturgeon, *Acipenser ruthenus* L., with special emphasis on hermaphrodites. Aquaculture 246(1):263–273

Wrobel K-H, Hees I, Schimmel M, Stauber E (2002) The genus Acipenser as a model system for vertebrate urogenital development: nephrostomial tubules and their significance for the origin of the gonad. Anat Embryol 205(1):67–80

Wuertz S, Gaillard S, Barbisan F, Carle S, Congiu L, Forlani A, Aubert J, Kirschbaum F, Tosi E, Zane L, Grillasca J-P (2006) Extensive screening of sturgeon genomes by random screening techniques revealed no sex-specific marker. Aquaculture 258(1):685–688

Yamamoto Y, Zhang Y, Sarida M, Hattori RS, Strussmann CA (2014) Coexistence of genotypic and temperature-dependent sex determination in pejerrey *Odontesthes bonariensis*. PLoS One 9(7):e102574

Yano A, Nicol B, Guerin A, Guiguen Y (2011a) The duplicated rainbow trout (*Oncorhynchus mykiss*) T-box transcription factors 1, tbx1a and tbx1b, are up-regulated during testicular development. Mol Reprod Dev 78(3):172–180

Yano A, Nicol B, Valdivia K, Juanchich A, Desvignes T, Caulier M, Zadeh AV, Guerin A, Jouanno E, Nguyen T (2011b) Sex in salmonids: from gonadal differentiation to genetic sex determination. Indian J Sci Technol 4(S8):60–61

Yano A, Guyomard R, Nicol B, Jouanno E, Quillet E, Klopp C, Cabau C, Bouchez O, Fostier A, Guiguen Y (2012) An immune-related gene evolved into the master sex-determining gene in rainbow trout, *Oncorhynchus mykiss*. Curr Biol 22(15):1423–1428

Yarmohammadi M, Pourkazemi M, Chakmehdouz F, Kazemi R (2011) Comparative study of male and female gonads in Persian sturgeon (*Acipenser persicus*) employing DNA-AFLP and CDNA-AFLP analysis. J Appl Ichthyol 27(2):510–513

Yoshimoto S, Okada E, Umemoto H, Tamura K, Uno Y, Nishida-Umehara C, Matsuda Y, Takamatsu N, Shiba T, Ito M (2008) A W-linked DM-domain gene, DM-W, participates in primary ovary development in *Xenopus laevis*. Proc Natl Acad Sci U S A 105(7):2469–2474

Yoshimoto S, Ikeda N, Izutsu Y, Shiba T, Takamatsu N, Ito M (2010) Opposite roles of DMRT1 and its W-linked paralogue, DM-W, in sexual dimorphism of *Xenopus laevis*: implications of a ZZ/ZW-type sex-determining system. Development 137(15):2519–2526

Yue H, Li C, Du H, Zhang S, Wei Q (2015) Sequencing and de novo assembly of the gonadal transcriptome of the endangered Chinese sturgeon (*Acipenser sinensis*). PLoS One 10(6):e0127332

Analysis of Transposable Elements Expressed in the Gonads of the Siberian Sturgeon

7

Frédéric Brunet, Alexia Roche, Domitille Chalopin, Magali Naville, Christophe Klopp, Denise Vizziano-Cantonnet, and Jean-Nicolas Volff

---

Abstract

Transposable elements (TEs) are mobile and repeated sequences that are major factors of diversity and evolution in genomes. We report here through the analysis of gonad transcriptomes of the Siberian sturgeon *Acipenser baerii*, a non-teleost ray-finned fish, that sturgeon genomes contain many families of TEs, which are expressed in gonads and might be involved in the evolution of this divergent fish lineage. The high diversity of TEs observed in sturgeons, which is also found in teleost fish, coelacanth, and amphibians but not in birds and mammals, strongly supports that many TE families were present in ancestral vertebrate genomes. Two types of transposable elements potentially differing in their

---

F. Brunet (✉) • A. Roche • M. Naville • J.-N. Volff • D. Chalopin
Institut de Genomique Fonctionnelle de Lyon, Univ Lyon, CNRS UMR 5242, Ecole Normale Superieure de Lyon, Universite Claude Bernard Lyon 1, 46, allée d'Italie, 69364 Lyon, France
e-mail: frederic.brunet@ens-lyon.fr; fgb27@psu.edu

C. Klopp
Genotoul BioInformatic, INRA, UR 875 Unité de Mathématique et Informatique Appliquées, Bioinformatics plateforme Toulouse Midi-Pyrenees, Auzeville, CS 52627, Castanet-Tolosan 31326, France

Plate-forme SIGENAE, INRA, GenPhyse, Auzeville, CS 52627, 24 chemin de Borde-Rouge, Castanet-Tolosan 31326, France

D. Vizziano-Cantonnet
Facultad de Ciencias, Laboratorio de Fisiología de la Reproducción y Ecología de Peces, Iguá 4225, Montevideo 11400, Uruguay

© Springer International Publishing AG, part of Springer Nature 2018
P. Williot et al. (eds.), The Siberian Sturgeon (*Acipenser baerii*, Brandt, 1869)
Volume 1 - Biology, https://doi.org/10.1007/978-3-319-61664-3_7

evolutionary dynamics have been further characterized: DIRS-like retrotransposons, with a single lineage mainly transmitted vertically, and Tc1/mariner DNA transposons, with multiple lineages and the possible involvement of horizontal transfer. This first global analysis is a new step toward the understanding of TE evolution and evolutionary impact in non-teleost ray-finned fish and will help to annotate the upcoming sequences of the large sturgeon genomes.

Keywords

Transposable elements • Retrotransposons • DNA transposons • Whole genome duplications • Actinopterygii • Teleostei • Sturgeons

Introduction

With ca. 28,000 species, ray-finned fish (Actinopterygii) represent more than half of vertebrate species (Nelson 2006). This species richness is associated with a wide range of biological diversity affecting many facets of the life of fish, including development, physiology, reproduction, behavior, and ecology. It has been proposed that ray-finned fish genomes, and especially teleost fish genomes, might have some peculiar properties making them highly evolvable (Volff 2005; Ravi and Venkatesh 2008). Particularly, a teleost-specific whole genome duplication (TWGD) has taken place at the base of the teleost fish lineage (Meyer and Schartl 1999; Taylor et al. 2003; Jaillon et al. 2004). Such WGDs are assumed to provide a genomic framework enabling the massive differential evolution of duplicated genes, leading, for example, to the emergence of new functions (Ohno 1970). Lineage-specific evolution of duplicates might promote speciation. The age of the TWGD has been estimated to lie between 225 and 316 million years (Hurley et al. 2007; Santini et al. 2009). Additional events of WGDs have been detected in some teleost lineages including salmonids (Berthelot et al. 2014, Lien et al. 2016), as well as in more divergent non-teleost ray-finned fishes such as sturgeons (Ludwig et al. 2001).

Repeated sequences constitute a ubiquitous component of eukaryotic genomes. The prevalence of these sequences is highly variable in terms of copy number and type of sequences. It was recently estimated that repetitive or repeat-derived sequences compose more than two third of the human genome (de Koning et al. 2011). Repeated sequences are mainly grouped in two classes. Tandem repeats correspond to multiple adjacent repetitions of a DNA motif; they are often found at centromeres and telomeres. Interspersed repeats correspond to the dispersion of a DNA sequence throughout the genome. They mainly come from transposable elements (TEs), typically in the 100–10,000 bp size range. Transposable elements (TEs) are a component of the ubiquitous interspersed mobile and repetitive sequences present in the genomes of prokaryotes and eukaryotes including vertebrates (Deininger et al. 2003; Kazazian 2004; Biémont and Vieira 2006; Böhne et al. 2008; Chalopin et al. 2015; Warren et al. 2015). In vertebrates, they constitute from ~6% (*Tetraodon*, a pufferfish with compact genome) to more than half of the genome in zebra-fish and opossum, for example (Chalopin et al. 2015). Teleost fish genomes present a high diversity of transposable elements (TEs), with many more active families of TEs than mammals and birds (Volff et al. 2003; Volff 2005; Chalopin et al. 2015). TEs are typically sorted in two major classes based on their mechanism of transposition (Fig. 7.1). Class I TEs, also called retroelements, generate a new copy of themselves through the reverse transcription of an RNA intermediate. The new cDNA copy is integrated at random somewhere else into the genome without excision of the original copy. Retrotransposition is therefore a mechanism of duplication, which directly increases the copy number of a TE. In contrast, class II DNA transposons generally use a cut-and-paste mechanism to transpose, with excision of the copy from its original position and reinsertion into a new genomic location. This mechanism does not require reverse transcription.

TE classes are further subdivided into orders, superfamilies, families, and subfamilies (Fig. 7.1) according to TE structures, mechanisms of transposition, and phylogenetic relationships (Wicker et al. 2007). Class I retroelements are classified according to the presence or absence of long terminal repeats (LTRs). LTR elements include (endogenous) retroviruses, which have an envelope gene, and LTR retrotransposons (Ty3/gypsy, BEL, and Ty1/copia elements), which do not encode any envelope protein most of the time. Classical LTR elements produce among others an integrase and a reverse transcriptase. In addition, DIRS1-like retrotransposons present atypical inverted and/or internal long repeats and use a tyrosine recombinase instead of an integrase for their insertion. Non-LTR retrotransposons, which encode

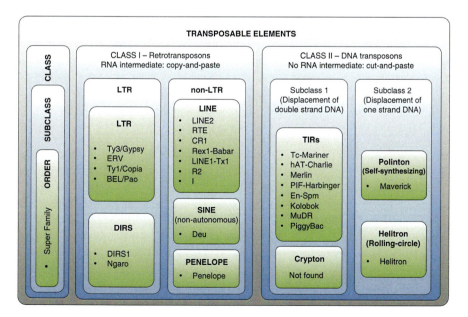

Fig. 7.1 Overview of the different transposable elements using RepeatMasker found in this analysis of the transcriptome of *A. baerii*

an endonuclease and a reverse transcriptase, include the Penelope and the long interspersed nuclear elements (LINEs, e.g., L1 and L2). Non-autonomous noncoding retroelements called SINEs (for short interspersed nuclear elements) like the Alu, B2, and B4 sequences, can retrotranspose using as parasite the enzymatic machinery from the coding sequence of the autonomous LINE sequences (Dewannieux et al. 2003). Class II DNA transposons are subdivided into DDE transposons (e.g., PiggyBac, Tc1/mariner, MuDR/Foldback, hAT, Transib, PIF, Merlin, and P), Helitrons, Polintons/Mavericks, and noncoding non-autonomous DNA transposons (MITEs for miniature inverted transposable elements, e.g., MADE1). DDE transposons are named after the highly conserved aspartate-aspartate-glutamate catalytic triad in the transposase. Helitrons use a rolling-circle-based mechanism to transpose. Polinton/Maverick transposons are complex TEs encoding an integrase and other proteins. Finally, MITE elements are short noncoding sequences with a palindromic structure that use the transposase encoded by autonomous DNA transposons (Feschotte and Pritham 2007).

Through their ability to integrate into genomes, TEs can disrupt functional regulatory and coding sequences of the host, this leading to mutations and genetic diseases which are amply characterized in humans (Hancks and Kazazian 2012). The multiple copies of TEs in a genome also provide the substrate for ectopic homologous recombination. This can promote genomic rearrangements such as deletions, duplications, inversions, and translocations, which can be deleterious to the host (Kazazian 2004; Burns and Boeke 2012). On the other hand, TEs occasionally provide new genetic material for the evolution of genes and genomes, for instance new regulatory and coding sequences for host genes (Nekrutenko and Li 2001; Feschotte 2008; Lynch et al. 2011; Rebollo et al. 2012; Ellison and Bachtrog 2013; Pé et al. 2013). In addition, novel protein-coding and RNA genes have been repeatedly formed from TEs during evolution in vertebrates and other organisms, an evolutionary process called "molecular domestication" (Volff 2006; Sinzelle et al. 2009; Kapusta et al. 2013). TEs are therefore considered as major drivers of gene and genome evolution, with roles in biological diversity and speciation (McClintock 1956; Coyne and Orr 1998; Kraaijeveld 2010). Lineage-specific activity/extinction and molecular domestication of TEs have contributed to genomic and biological diversification in vertebrates (Böhne et al. 2008; Warren et al. 2015). In a recent study, Chalopin et al. (2015) showed in a comparative analysis of 23 genomes that TE content, diversity, and age can drastically vary between and even within major vertebrate lineages. As an example, there is an order of magnitude more TEs in mammals compared to birds, or in the zebra-fish compared to the pufferfish *Tetraodon*. The diversity of TE superfamilies is much higher in sea lamprey, cartilaginous fish, teleost fish, and coelacanth, decreases in amphibians and non-avian sauropsids, and shows its lowest level in mammals and birds. Competition between TEs could maintain them in low copy numbers, and loss of one family could allow the expansion of others. There might also be some differences in the rate of DNA elimination pending of the vertebrate lineages. For example, mammalian genomes harbor more of older TE copies as faster riddance seems to have operated in teleosts. Also, some types of TE are widespread in vertebrates, while others show a more patchy distribution probably driven by lineage-specific extinction and/or horizontal transfer (Volff et al. 2003; Chalopin et al. 2015). Even widespread superfamilies can show strong lineage-specific differences in their diversity and copy number. For instance, there is only one family of LINE1 retrotransposons that fills 20% of the human and mouse genomes, as opposed to 30 different families of LINE1 with a much lower content in zebra-fish (Furano et al. 2004). Competition between TE families has been proposed to modify TE content between genomes of closely related species (Le Rouzic and Capy 2006). As an example, the invasion of endogenous retroviruses (ERVs) may have been responsible for the loss of the LINE1 elements in the genome of a group of muroid rodents (Erickson et al. 2011).

In order to better understand the molecular basis of biological diversity and speciation in ray-finned fish, it is of the highest importance to analyze the genomes of non-teleost fish species, which represent only 0.2% of actinopterygian species. These species include gars (the genome of the spotted gar *Lepisosteus oculatus* will be published soon), bowfins (*Amia calva*), paddlefishes, sturgeons, and bichirs (including the *Polypterus* genus). Sturgeons belong to the order of Acipenseriformes and to the Acipenseridae family. There are 25 extant species of sturgeons (27 when subspecies are considered) subdivided in four genera, *Acipenser, Huso, Pseudoscaphirhynchus*, and *Scaphirhynchus*, with several species critically endangered (Birstein et al. 1997; Pikitch et al., 2006; Nelson 2006; Froese and Pauly 2008). Acipenseridae present a high number of chromosomes (from $2n = 100$ to 500), which contrasts with the lower number observed in other species like *Amia calva* ($2n = 46$ chromosomes) (Fontana et al. 2004, 2008; Peng et al. 2007). This is strongly indicative of WGD events (Ludwig et al. 2001), the big size of sturgeon genomes making them difficult to sequence for now. Consequently, almost nothing is known about the structure and evolution of sturgeon genomes, and, to the best of our knowledge, only very few TEs have been described in sturgeons and other basal Actinopterygii fish (Pujolar et al. 2013; Zhang et al. 2014). These sturgeon-specific duplication events have been estimated 53 Mya for the sturgeon Atlantic lineage and 70 Mya for the sturgeon Pacific lineage. The youngest of these dates has been correlated with the closing of the Tethys Sea, which will lead to the Mediterranean Sea and birth of new water bodies that are the Caspian, the Black, and the Aral seas (Peng et al. 2007). Whole genome duplications are expected to be correlated with a burst of TEs, as they may play some role in the process of rediploidization favoring chromosome rearrangements and divergence among homologous chromosomes (Lien et al. 2016).

To contribute to a better understanding of the evolution and evolutionary impact of TEs in sturgeons and other fish lineages, we searched for TEs through the analysis of transcriptome data from the Siberian sturgeon *Acipenser baerii* (Vizziano-Cantonnet and Klopp, submitted) and the lake sturgeon *Acipenser fulvescens* (Hale et al. 2010). In the absence of genomic data, which are necessary to perform an exhaustive analysis of the mobilome (a common neologism referring to mobile TE information collected from large-scale data-rich biological technics) of a species, transcriptome analysis offers the possibility to identify expressed TEs, even at basal expression level, that might therefore be potentially active—but without providing precise insights into the structure of the elements. In this paper, we present two types of TEs that are expressed in the Siberian sturgeon: class 2 Tc1/mariner transposons and a class 1 *DIRS*-like retrotransposon, which display completely different evolutionary dynamics.

7.1 Material and Methods

7.1.1 Sequence Data

For this analysis, we used an *Acipenser baerii* transcriptome of gonads from non-differentiated and differentiated animals (testis and ovaries) ranging in size from 13.4 to 77.5 cm and aged between 2.5 and 17 months (Vizziano-Cantonnet and Klopp, submitted). The sequence data produced consisted in 91,582 contigs (the overlapping sequence reads) with a total length of 136,616,422 base pairs (bp). Contig lengths ranged from 200 bp to 21,420 bp, with a median of 1010 bp. Additional transcriptome data from gonadal biopsies of the lake sturgeon *Acipenser fulvescens* were also used (one non-mature female with immature eggs and one mature male that produced milt; Hale et al. 2010). A total of 32,630 contigs for the female and 12,792 contigs for the male were downloaded (http://datadryad.org/handle/10255/dryad.37812), for a total number of 18,139,815 and 2,624,330 bp, respectively.

7.1.2 Relevant Websites

NCBI-BLAST (http://blast.ncbi.nlm.nih.gov) National Center for Biotechnology Information-Basic Local Alignment Search Tool. BLASTN searches a nucleotide database using a nucleotide query; BLASTP searches a protein database using a protein query; TBLASTN searches a translated nucleotide database using a protein query.

RepeatMasker http://www.repeatmasker.org Smit, AFA, Hubley, R and Green, P. RepeatMasker Open-4.0. 2013–2015.

E!ensembl http://www.ensembl.org/Ensembl genome browser.

7.1.3 TE Identification

The method used here for the TE identification is shown in Fig. 7.2. Transcriptome sequence datasets were locally masked using RepeatMasker version 3.3.0 (Smit et al. 2013). Two TE libraries were used: REPBASE 20.05 with either the default library or a fish-specific TE database (Chalopin et al. 2015). TEs were retrieved according to their categories among the repName (and thus repClass) subdivision. A total of 121 different subcategories (RepName) that corresponded to at least 30 different contigs were selected. Redundant sequences were excluded, then a consensus sequence was generated based on a >50% majority rule using an option implemented in SeaView (Gouy et al. 2010). Further manual adjustments were made to get a potential in-frame coding sequence in order to recreate a full-length

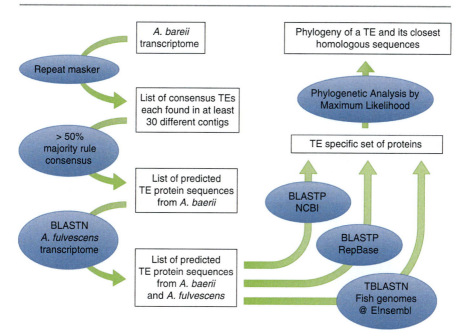

Fig. 7.2 Schematic representation of the main steps provided in detail in the material and method section

protein sequence. *A. baerii* TE consensus sequences were used as queries to retrieve similar elements in the transcriptomic data of *A. fulvescens* (Hale et al. 2010) using BLASTN (Altschul et al. 1990). Contigs from both sturgeon species were then aligned using Clustal Omega (Sievers and Higgins 2014). Predicted TE protein sequences from sturgeons were used as query for an online BLASTP (Altschul et al. 1990) against the protein NCBI database, and a local BLASTP search against the protein sequence database of TEs in Eukaryotes retrieved from REPBASE 20.05.

Finally, the E!nsembl website was used to screen the available fish genomes with TE protein sequences from sturgeon using TBLASTN (Altschul et al. 1990) in order to retrieve related elements from other species.

7.1.4 Phylogenetic Analyses

TE proteins were aligned with Clustal Omega (Sievers and Higgins 2014). After visualizing the alignments under SeaView (Gouy et al. 2010) and trimming out the redundancies, maximum likelihood (ML) trees were generated using PhyML (v2.4.4) (Guindon et al. 2010) under a WAG model with estimated gamma distribution parameter and 1000 bootstrap replicates. Phylogenetic trees were subsequently processed using FigTree (v1.4.2) (http://tree.bio.ed.ac.uk/software/figtree/).

7.2 Results

7.2.1 Many Types of Vertebrate TEs are Found in the Gonadal Transcriptome of the Sturgeon

RepeatMasker analysis of the transcriptome contigs from the Siberian sturgeon *A. baerii* revealed a total coverage of 2.3% (TE coverage) of the total length of the contigs, which is lower than the value that could be expected from the direct analysis of large fish genomes (e.g., 6% in the green spotted pufferfish, and 67% in humans). Among the 2.3% of TE coverage, all main types of vertebrate TEs were detected (Fig. 7.1), including LINEs (~13.5% of TE coverage), LTRs (~15.6%), SINEs (~0.4%), and DNA transposons (~10.7%). 60% of sequences masked by TEs corresponded to still non-annotated elements present in the fish-specific database.

Among the LINEs, all major vertebrate superfamilies were detected, including L2 (44% of LINE coverage), L1 (20%), Rex3 (13%), CR1 (12%), and Rex1/Babar (6%). Most LTR elements were Gypsy/Ty3 retrotransposons (70% of LTR coverage), with a significant contribution of DIRS-like elements (17%) and endogenous retroviruses (7%). Other LTR elements (BEL, Copia/Ty1) were minority. The low level of SINE elements is expected as they are filtered out in the preparation of the transcriptome. Most SINE elements were identified as DEU sequences. These sequences are also highly represented in the transcriptome of the coelacanth (Forconi et al. 2014). However, the analysis of the contribution of SINEs must be taken with caution since some of these sequences are often lineage-specific. Sturgeon-specific elements might therefore be present in the *Acipenser* genomes but absent from the databases used to mask the transcriptome datasets. Within DNA transposons, 42% of masked sequences corresponded to Tc1/mariner, 29% to hAT, and 13% to EnSpm elements. Other elements like Merlin, PIF-Harbinger, and Helitron transposons were marginally detected. Taken together, this analysis uncovers a high diversity of TEs in sturgeons. Most main TE families are expressed in gonads, suggesting that at least some of them might be active in germ cells.

7.2.2 Diversity of Tc1/Mariner DNA Transposons in Sturgeon

Nine different consensus sequences of Tc1/mariner DNA transposons were obtained from the transcriptome data of *A. baerii*. The degree of nucleotide identity between these elements ranged from 50 to 83%, indicating that most of them clearly corresponded to very divergent elements. The predicted transposases encoded by these sequences show sizes ranging from 328 to 363 aa, which are in range with lengths reported for Tc1/mariner transposases, and which display the canonical transposase composed of the three aa DDE that function together in these enzymes and are called the catalytic triad (Fig. 7.3). This motif allows the positioning of two divalent

7 Analysis of Transposable Elements Expressed in the Gonads of the Siberian

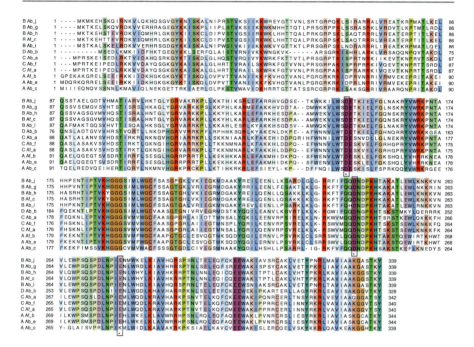

Fig. 7.3 Multiple protein sequence alignments of sturgeon Tc1/mariner transposases. The *A. baerii* (Ab) and *A. fulvescens* (Af) sequences are presented. The catalytic DD35E triad is shown by asterisks. The dataset of Tc1/mariner elements available in Yuana and Wessler (2011) was used to determine the location of the DDE catalytic motif, shown with an asterisk for each of the three aa. The color code refers to that of clustalX implemented in Jalview v2 (Waterhouse et al. 2009). Names refer to the four groups observed in Fig. 7.3. Ab and Af stand for *A. baerii* and *A. fulvescens*, respectively. Numbers refer to the aa position of each sequence in this alignment

ions involved in the strand cleavage and strand transfer during the transposition process. As observed in other Tc1/mariner transposases, 35 aa were found between the second and third catalytic residues (the catalytic triad harbor a DD35E motif, Fig. 7.3). Phylogenetic analysis of *A. baerii* transposase sequences showed that the Tc1/mariner elements identified grouped into at least four distinct phylogenetic groups, termed A to D (Fig. 7.4). For three of these groups (A–C), related sequences were detected in the lake sturgeon *A. fulvescens*. Extensive transcriptome and genome sequencing of these two sturgeon species and all others might reveal more sequences in all groups. Group A elements were also related but not identical to the Tana1 sturgeon DNA transposon previously described (Pujolar et al. 2013). In all cases, sturgeon elements were more related to teleost fish and sometime tetrapod sequences from a same group than to sturgeon sequences from another group, indicating distinct evolutionary transposon lineages. Hence, sturgeon genomes contain divergent lineages of Tc1/mariner transposons, which are transcribed in gonads and potentially active in germ cells.

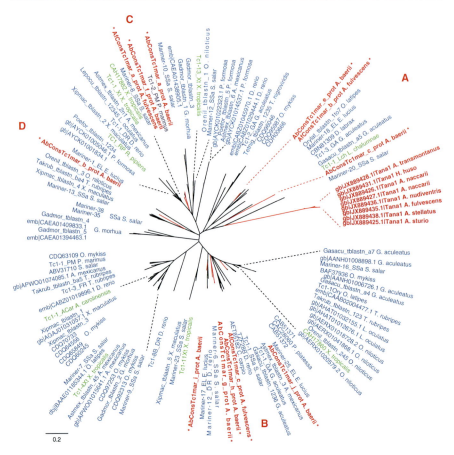

Fig. 7.4 Radial phylogenetic tree of the *Tc1/mariner* transposases. This phylogeny was generated using a maximum likelihood method with a WAG model calculated by the PhyML program (Guindon et al. 2010) (247 aa in length, 1000 bootstrap replicates). Transposases from lobe-finned fishes are in green, from ray-finned fishes in *blue* except for sturgeon elements that are in bold and red characters. Asterisk points out at data generated in this work. Tana1 proteins have been previously described in various sturgeons (Pujolar et al. 2013). Protein ID refers to the database screened (emb or gb, for embl or genbank, respectively). Others refer at translated DNA sequences that we screened from genomes ("Cons" when sequences are from a consensus of sequences). Each protein ID is followed by the first letter of the genus and species name. Groups defined here (in red) refer to the *Tc1/mariner* found in sturgeons which are phylogenetically closely related

7.2.3 A Single DIRS-Like Retrotransposon Lineage Transcribed in Sturgeon

At least three reverse transcriptase consensus sequences (514–515 aa in length) identified as belonging to DIRS-like retrotransposons were identified, showing 60–98% identity at the nucleotide level. Phylogenetic analysis showed that all four sequences grouped into a single phylogenetic group (Fig. 7.5). DIRS-like sequences

7 Analysis of Transposable Elements Expressed in the Gonads of the Siberian

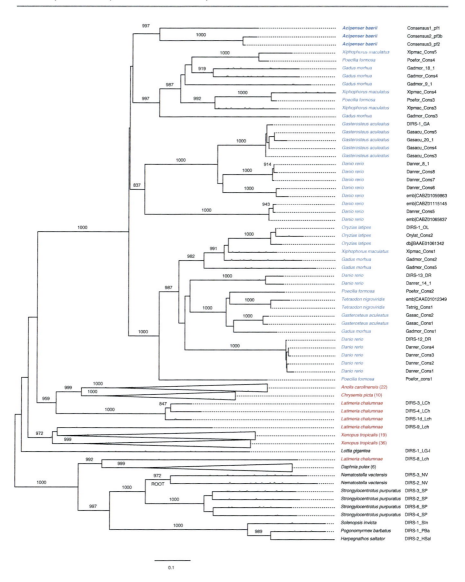

Fig. 7.5 Phylogenetic tree of DIRS-like reverse transcriptases. This analysis was performed on a 573 aa protein alignment of the reverse transcriptase domain using an unrooted PhyML (Guindon et al. 2010) tree based on 1000 bootstrap replicates (a potential root is indicated at the base of the DIRS2 and DIRS3 found in the starlet sea anemone *Nematostella vectensis*). Bootstrap values below 800 are not shown. For species searched in the Ensembl genomes, we used the consensus sequence of similar sequences (labeled as "Cons"). Ray-finned fish species are in *blue* (sturgeon in **bold**) and lobe-finned fish species in *red* (DIRS are absent from mammals and birds). Numbers in brackets refer to the number of sequences that have been collapsed

were detected in teleosts as well as in lobe-finned fish including amphibians and reptiles, but neither in birds nor in mammals. Sturgeon DIRS-like sequences were more related to other ray-finned fish sequences, and ray- and lobe-finned fish elements formed two clearly distinct phylogenetic groups. Hence, the data support the presence of DIRS-like retrotransposons in the last common ancestor of ray-finned and lobe-finned fishes, with subsequent lineage-specific independent evolution and losses in birds and mammals.

Conclusions

In this paper, we report a first global analysis of transposable element diversity in a divergent non-teleost ray-finned fish, the sturgeon. We show that many families of TEs are represented in the germ cell-containing gonad transcriptome of the Siberian sturgeon *Acipenser baerii*. This suggests that the genome of this fish contains many families of active TEs, which might contribute to its evolution. Interestingly, even if genome-wide analyses will be required to more exhaustively analyze TE content in sturgeons, the use of transcriptome data was sufficient to identify many TE families that are also found in teleost fish, coelacanth, and amphibians, but are absent from birds and mammals (Chalopin et al. 2015). Hence, the data obtained here confirms the probable high TE diversity in the ancestral vertebrate mobilome, which has been maintained in fish, coelacanth, and amphibians but strongly reduced in birds and mammals (Volff et al. 2003; Chalopin et al. 2015). We have further analyzed in more details two types of sturgeon transposable elements, namely, Tc1/mariner DNA transposons and DIRS-like retrotransposons. Even if more sequences need to be analyzed, both types of elements apparently show different evolutionary dynamics. Sturgeon DIRS-like elements were all grouped in a single monophyletic group, which was included within a ray-finned fish-specific group clearly distinct from elements of the lobe-finned fish. This might suggest an evolution by lineage-specific classical vertical transmission. In contrast, we found evidence of at least four distinct Tc1/mariner DNA transposon lineages in sturgeon, with elements more related to teleost and sometime lobe-finned fish sequences than to other sturgeon elements. Even if we could not find any strong evidence of "horizontal" interspecific transfer (strongly similar elements in divergent species), such a mode of transmission might be involved in the distribution of Tc1/mariner elements we have observed. Accordingly, Tc1/mariner transposons have been shown to be able to "jump" between species, for example, between teleost fish and lampreys (Kuraku et al. 2012). Transcriptome analyses have also previously been useful to determine their coding sequences and to identify transcribed TEs, which mRNA could potentially produce an active transposase (e.g., the coelacanth; Forconi et al. 2014). Therefore, it helps for a first characterization of some TEs. However, only TEs producing polyA mRNA at a level sufficient to allow detection have been found in this work. Poorly or not expressed elements, or elements expressed in non-gonadal somatic tissues, have therefore escaped our analysis. In addition, TE structures that are not embedded in the mRNA are not accessible through transcriptome analysis. Whole sequencing of the large genome of sturgeons will be necessary to characterize their complete mobilome. Such an analysis will help to understand the evolution and evolutionary impact of TEs in fish and other vertebrates. In addition, sturgeons look like the first Acipenseriformes fishes recorded among fossils (~200 Mya), with seemingly not much morphological changes (Gardiner, 1984). The same controversial considerations were also made for the coelacanth species (reviewed in Casane and Laurenti 2013). As such, it was suggested that the genome evolution could have been lowered. The description of the TE repertoire in these species has led to think otherwise (Naville et al. 2015). The description of these TE repertoires could help to tell whether the previous finding in coelacanths can also be found in these other "living fossils" and generalize that genomes continue to harbor a strong genome plasticity due to TE activity even though the phenotype barely changed over a long time. Finally, sturgeon caviar has become seafood production with strong economic production in aquaculture (Diana 2009). Along with any economy raising, illegal and deceitful products might be encountered, e.g., mix caviar of different species sold to the price of the most expensive one. Knowing the TE repertoire can be used as a strategy to figure out whether caviar is made of only one or various sturgeon species. At least three approaches can be considered. The first implies the classical phylogenetic approach as showed here. The second will use the inventory of non-active TE occurrences at species-specific positions along the orthologous genomic regions of sturgeons. The third would use a mere TE PCR amplification and sequencing, looking for TEs characterizing one species and not the other or, at least, using the previous description of species-specific signatures within one common TE. The Russian sturgeon (*A. gueldenstaedtii*) and the Siberian sturgeon are estimated to have diverged since 25 Mya (Peng et al.), a time frame amply sufficient for TE divergence and species-specific signatures, as found between the Siberian sturgeon and the Lake sturgeon *A. fulvescens*, which are ~55 Mya old. TEs that show characteristics of burst amplification in the genomes would be best as minute amount of DNA in the sample analyzed would be more favorably amplified. As such, the sequencing of multiple sturgeon species and characterization of their TE repertoires will have a dual interest: the academic knowledge, and economically with the development of molecular strategies to ensure the quality of the caviar origin being sold worldwide.

Glossary

TEs   Transposable elements
mobilome The TE repertoire in a genome
cDNA   Complementary DNA
bp   Base pair
aa   Amino acids

References

Altschul SF, Gish W, Miller W et al (1990) Basic local alignment search tool. J Mol Biol 215:403–410

Berthelot C, Brunet F, Chalopin D et al (2014) The rainbow trout genome provides novel insights into evolution after whole-genome duplication in vertebrates. Nat Commun 5:3657

Biémont C, Vieira C (2006) Genetics: junk DNA as an evolutionary force. Nature 443:521–524

Birstein VJ, Bemis WE, Waldman JR (1997) The threatened status of acipenseriformes species: a summary. Environ Biol Fishes 48:427–435

Böhne A, Brunet F, Galiana-Arnoux D et al (2008) Transposable elements as drivers of genomic and biological diversity in vertebrates. Chromosome Res 16:203–215

Burns KH, Boeke JD (2012) Human transposon tectonics. Cell 149:740–752

Casane D, Laurenti P (2013) Why coelacanths are not 'living fossils': a review of molecular and morphological data. Bioessays 35:332–338

Chalopin D, Naville M, Plard F et al (2015) Comparative analysis of transposable elements highlights mobilome diversity and evolution in vertebrates. Genome Biol Evol 7:567–580

Coyne JA, Orr HA (1998) The evolutionary genetics of speciation. Philos Trans R Soc Lond B Biol Sci 353:287–305

de Koning APJ, Gu W, Castoe TA, Batzer MA, Pollock DD (2011) Repetitive elements may comprise over two-thirds of the human genome. PLoS Genet. 7:e1002384

Deininger PL, Moran JV, Batzer MA et al (2003) Mobile elements and mammalian genome evolution. Curr Opin Genet Dev 13:651–658

Dewannieux M, Esnault C, Heidmann T (2003) LINE-mediated retrotransposition of marked Alu sequences. Nat Genet 35:41–48

Diana JS (2009) Aquaculture production and biodiversity conservation. Bio Science 59:27–38

Ellison CE, Bachtrog D (2013) Dosage compensation via transposable element mediated rewiring of a regulatory network. Science 342:846–850

Erickson IK, Cantrell MA, Scott L et al (2011) Retrofitting the genome: L1 extinction follows endogenous retroviral expansion in a group of muroid rodents. J Virol 85:12315–12323

Feschotte C (2008) Transposable elements and the evolution of regulatory networks. Nat Rev Genet 9:397–405

Feschotte C, Pritham EJ (2007) DNA transposons and the evolution of eukaryotic genomes. Annu Rev Genet 41:331–368

Fontana F, Bruch RM, Binkowski FP (2004) Karyotype characterization of the lake sturgeon, *Acipenser fulvescens* (Rafinesque 1817) by chromosome banding and fluorescent *in situ* hybridization. Gen Natl Res Counc Can 47:742–746

Fontana F, Congiu L, Mudrak VA et al (2008) Evidence of hexaploid karyotype in shortnose sturgeon. Gen Natl Res Counc Can 51:113–119

Forconi M, Chalopin D, Barucca M et al (2014) Transcriptional activity of transposable elements in Coelacanth. J Exp Zool (Mol Dev Evol) 322B:379–389

Froese R, Pauly D (2008) Fish Base. Available at http://www.fishbase.org

Furano AV, Duvernell DD, Boissinot S (2004) L1 (LINE-1) retrotransposon diversity differs dramatically between mammals and fish. Trends Genet 20:9–14

Gardiner BG (1984) Sturgeons as Living Fossils. In: Eldredge N, Stanley SM (eds) Living Fossils. Springer, New-York, pp 148–152

Gouy M, Guindon S, Gascuel O (2010) SeaView version 4: A multiplatform graphical user interface for sequence alignment and phylogenetic tree building. Mol Biol Evol 27:221–224

Guindon S, Dufayard JF, Lefort V et al (2010) New algorithms and methods to estimate maximum-likelihood phylogenies: assessing the performance of PhyML 3.0. Syst Biol 59:307–321

Hancks DC, Kazazian HH Jr (2012) Active human retrotransposons: variation and disease. Curr Opin Genet Dev 22:191–203

Hale MC, Jackson JR, DeWoody JA (2010) Discovery and evaluation of candidate sex-determining genes and xenobiotics in the gonads of lake sturgeon (*Acipenser fulvescens*). Genetica 138:745–756

Hurley IA, Mueller RL, Dunn KA et al (2007) A new time-scale for ray-finned fish evolution. Proc Biol Sci 274:489–498

Jaillon O, Aury JM, Brunet F et al (2004) Genome duplication in the teleost fish *Tetraodon nigroviridis* reveals the early vertebrate protokaryotype. Nature 431:946–957

Kapusta A, Kronenberg Z, Lynch VJ et al (2013) Transposable elements are major contributors to the origin, diversification, and regulation of vertebrate long noncoding RNAs. PLoS Genet 9:e1003470

Kazazian HH Jr (2004) Mobile elements: drivers of genome evolution. Science 303:1626–1632

Kraaijeveld K (2010) Genome size and species diversification. Evol Biol 37:227–233

Kuraku S, Qiu H, Meyer A (2012) Horizontal transfers of Tc1 elements between teleost fishes and their vertebrate parasites, lampreys. Genome Biol Evol 4:929–936

Le Rouzic A, Capy P (2006) Population genetics models of competition between transposable element subfamilies. Genetics 174:785–793

Lien S, Koop BF, Sandve SR et al (2016) The Atlantic salmon genome provides insights into rediploidization. Nature 533:200–205

Ludwig A, Belfiore NM, Pitra C et al (2001) Genome duplication events and functional reduction of ploidy levels in sturgeon (*Acipenser, Huso* and *Scaphirhynchus*). Genetics 158:1203–1215

Lynch VJ, Leclerc RD, May G et al (2011) Transposon-mediated rewiring of gene regulatory networks contributed to the evolution of pregnancy in mammals. Nat Genet 43:1154–1159

McClintock B (1956) Controlling elements and the gene. Cold Spring Harb Symp Quant Biol 21:197–216

Meyer A, Schartl M (1999) Gene and genome duplications in vertebrates: The one-to-four (−to-eight in fish) rule and the evolution of novel gene functions. Curr Opin Cell Biol 11:699–704

Naville M, Chalopin D, Casane D, Laurenti P, Volff JN (2015) The coelacanth: Can a "living fossil" have active transposable elements in its genome? Mob Genet Elements. 5:55–59

Nekrutenko A, Li WH (2001) Transposable elements are found in a large number of human protein-coding genes. Trends Genet 17:619–621

Nelson JS (2006) Fishes of the world, 4th edn. Wiley, Hoboken, New Jersey

Ohno S (1970) Evolution of gene duplication. Springer, New-York

Pé J, Jeyakani J, Bourque G (2013) The majority of primate-specific regulatory sequences are derived from transposable elements. PLoS Genet 9:e1003504

Peng Z, Ludwig A, Wang D, Diogo R, Wei Q, He S (2007) Age and biogeography of major clades in sturgeons and paddleWshes (Pisces: Acipenseriformes). Mol Phyl Evol 42:854–862

Pikitch EK, Doukakis P, Lauck L et al (2006) Status, trends and management of sturgeon and paddlefish fisheries. Fish and Fisheries 6:233–265

Pujolar JM, Astolfi L, Boscari E et al (2013) Tana1, a new putatively active Tc1-like transposable element in the genome of sturgeons. Mol Phylogenet Evol 66:223–232

Ravi V, Venkatesh B (2008) Rapidly evolving fish genomes and teleost diversity. Curr Opin Genet Dev 18:544–550

Rebollo R, Romanish MT, Mager DL (2012) Transposable elements: an abundant and natural source of regulatory sequences for host genes. Annu Rev Genet 46:21–42

Santini F, Harmon LJ, Carnevale G et al (2009) Did genome duplication drive the origin of teleosts? A comparative study of diversification in ray-finned fishes. BMC Evol Biol 9:194

Sievers F, Higgins DG (2014) Clustal Omega, accurate alignment of very large numbers of sequences. Methods Mol Biol 1079:105–116

Sinzelle L, Izsvák Z, Ivics Z (2009) Molecular domestication of transposable elements: from detrimental parasites to useful host genes. Cell Mol Life Sci 66:1073–1093

Smit AFA, Hubley R, Green P. RepeatMasker Open-4.0. 2013–2015. http://www.repeatmasker.org

Taylor JS, Braasch I, Frickey T et al (2003) Genome duplication, a trait shared by 22000 species of ray-finned fish. Genome Res 13:382–390

Volff JN (2005) Genome evolution and biodiversity in teleost fish. Heredity 94:280–294

Volff JN (2006) Turning junk into gold: domestication of transposable elements and the creation of new genes in eukaryotes. Bioessays 28:913–922

Volff JN, Bouneau L, Ozouf-Costaz C et al (2003) Diversity of retrotransposable elements in compact pufferfish genomes. Trends Genet 19:674–678

Warren IA, Naville M, Chalopin D et al (2015) Evolutionary impact of transposable elements on genomic diversity and lineage-specific innovation in vertebrates. Chromosome Res 23:505–531

Waterhouse AM, Procter JB, Martin DMA et al (2009) Jalview version 2: a multiple sequence alignment and analysis workbench. Bioinformatics 25:1189–1191

Wicker T, Sabot F, Hua-Van A et al (2007) A unified classification system for eukaryotic transposable elements. Nat Rev Genet 8:973–982

Yuana Y-W, Wessler SR (2011) The catalytic domain of all eukaryotic cut-and-paste transposase superfamilies. Proc Nat Acad Sci USA 108:7884–7889

Zhang HH, Feschotte C, Han MJ et al (2014) Recurrent horizontal transfers of Chapaev transposons in diverse invertebrate and vertebrate animals. Genome Biol Evol 6:1375–1386

Early Ontogeny in the Siberian Sturgeon

8

Enric Gisbert and Yoon Kwon Nam

Abstract

This chapter is a synthesis of published information describing the development of Siberian sturgeon during the embryonic (from fertilization to hatching), prelarval (from hatching to the onset of exogenous feeding) and larval (from first feeding to the juvenile stage) periods. Siberian sturgeon embryos undergo holoblastic cleavage in which each cleavage divides the entire egg cytoplasm, including the yolk. Early cleavage, the end of gastrulation and the beginning of neurulation are the most critical stages during sturgeon embryogenesis, a process that is influenced by water temperature. The higher incubation temperature, the shorter time from fertilization to hatching; the optimal incubation water temperature in Siberian sturgeon is comprised between 12 and 20 °C. Morphogenesis and differentiation are more intense during the prelarval than the larval and early juvenile stages of development. Characteristically, during the prelarval period, embryonic adaptations and functions are replaced by definitive ones, such as gill respiration, exogenous feeding and active swimming. Such modifications involved dramatic alterations in the relationship of the developing fish with the environment that were reflected in morphological and morphometric changes. In addition, relevant information regarding the histological development of the digestive system and olfactory and visual systems is presented and their significance discussed in terms of early development and feeding behaviour.

Keywords

Embryo • Prelarva • Larva • Early development • Digestive system • Sensory system

---

E. Gisbert (✉)
IRTA-Sant Carles de la Ràpita, Unitat de Cultius Experimentals,
Crta. Poble Nou km 5.5, 43540 Sant Carles de la Ràpita, Tarragona, Spain
e-mail: enric.gisbert@irta.cat, enric.gisbert@irta.es

Y.K. Nam
Department of Marine Bio-Materials and Aquaculture, Pukyong National University,
Busan 608737, South Korea

© Springer International Publishing AG, part of Springer Nature 2018
P. Williot et al. (eds.), *The Siberian Sturgeon (Acipenser baerii, Brandt, 1869)*
Volume 1 - Biology, https://doi.org/10.1007/978-3-319-61664-3_8

Introduction

Development of the embryo starts from fertilization, which transforms, under favourable conditions, through a series of sequential changes, into a prelarva (yolk sac larva), larva, juvenile and, finally, adult fish. Embryonic development of sturgeons can be subdivided into five successive periods: (a) fertilization; (b) cleavage; (c) blastulation and gastrulation; (d) early organogenesis characterized by the appearance of rudiments of the most important systems of organs such as nervous, excretory, muscular, and circulatory and the onset of heart beating; and (e) from the onset of heart beating to hatching. The newly hatched prelarvae conduct a free way of life and use reserves of nutrients which are contained in the yolk sac. Later, it starts to actively capture food and becomes a larva, which grows and gradually transforms into a juvenile, already similar in the main structural features with the adult fish. As a consequence, morphogenesis and differentiation are very intense processes during the early life stages of Siberian sturgeon. These processes lead not only to the formation of characteristic developmental patterns, but also to the appearance of characteristic structural defects under unfavourable rearing conditions that may directly affect the growth performance, survival and quality of young fish. In this context, a description of the stages and timing of normal development of fish during early ontogeny is of practical importance, since it also allows the comparison of specimens from different batches and reared under different conditions. In addition, it also allows estimating the quality of specimens and their suitability for further on-growing or restocking purposes. Although the early development is very similar among different sturgeon species, there are interspecific variations in the timing of organ formation, development and functionality, so it is necessary to conduct studies of organogenesis for individual species. Thus, the aim of this chapter is to provide a comprehensive description of the embryonic, prelarval and larval development of Siberian sturgeon from morphological and allometric points of view and completed with a detailed histological description of the digestive, visual and olfactory systems.

8.1 Embryonic Development

The information presented in this section is a synthesis of data reported by Park et al. (2013a) obtained from incubating Siberian sturgeon embryos at the constant temperature of 18 °C. Embryonic staging is based on the descriptions conducted by Dettlaff et al. (1993) in Russian sturgeon, which comprised 36 stages from the unfertilized egg to mass hatching.

The mature and *unfertilized egg* (stage 1) is already distinctly polarized, with more yolk at the vegetal pole. Concentric rings of differential pigmentation (generally lighter rings on a darker background) are centred on the animal pole of the slightly ovoid egg. Pigmentation may vary significantly between different females, though it is generally quite uniform within a single batch of eggs. Fertilization is accomplished by the entry of a sperm through the micropyles present in the chorion above the animal pole, and between 15 and 20 min after fertilization, the egg rotates 90° and the animal pole is located in an upward position that retains this orientation. Rapid series of cortical contractions accompanied by shifts in the pigmentation of the concentric rings around the animal pole are detected within a few minutes of fertilization, resulting in the formation of a single light spot at the animal pole, centred within a dark area that is in turn surrounded by a broad zone of reduced pigment (stage 3).

Early cleavage (Fig. 8.1). The first cleavage furrow (stage 4) appeared at the animal pore at 2 h postfertilization (hpf), whereas the second cleavage forming the typical four-cell embryo (four equal-sized blastomeres, stage 5) is observed 1 h later. Although the second cleavage furrow does not divide completely the vegetal hemisphere due to the large accumulation of yolk in this region (incomplete holoblastic cleavage), partial infiltration of the vegetal hemisphere is observed. At this stage, it is important to pay attention to embryos with more than four blastomeres, since they correspond to polyspermic embryos that are not viable (the percentage of polyspermic embryos should not exceed 4–6% under normal conditions). At 4 hpf, an eight-cell embryo (similarly sized blastomeres) is formed in the animal pole through the third cleavage event (stage 6); the cleavage infiltrated the entire vegetal hemisphere. According to Dettlaff et al. (1993), fertilization rate may be determined during the early cleavage during the first three embryo divisions when blastomeres are clearly visible and counted easily. It is undesirable to determine the percentage of fertilization at later stages of cleavage, since at this time activated unfertilized eggs usually start dividing, and it becomes more difficult to distinguish them from normal developing embryos. If it is not possible to determine the fertilization rate at stages 5–6, it is better to postpone it until the gastrula stage, when all the unfertilized eggs (both activated and nonactivated) degenerate and differ distinctly from gastrulating embryos.

When the 16 blastomeres are formed in the animal pole as a consequence of the fourth cleavage at 5 hpf (stage 7), one-half (newly formed) of the 16 blastomeres in the centre are smaller than the remaining half, which were formed at earlier cleavages. The sizes and shapes of the blastomeres that result from subsequent cleavages (stages 8–10) are different, and the furrows in the vegetal hemisphere were also formed in an irregular manner.

Blastulation (Fig. 8.2). As cleavage continued, small blastomeres proliferate in the animal hemisphere, concomitantly with irregular divisions in the vegetal hemisphere. Blastomeres are barely distinguishable from one to another in the animal pole due to their small size, and the blastocoel begins to form at the apex of the animal hemisphere at 9 hpf (stage 11). The primordial cleavage cavity increases in size and becomes more evident in the animal hemisphere at 11 hpf (stage 12), whereas cell division continues to generate smaller blastomeres in the vegetal

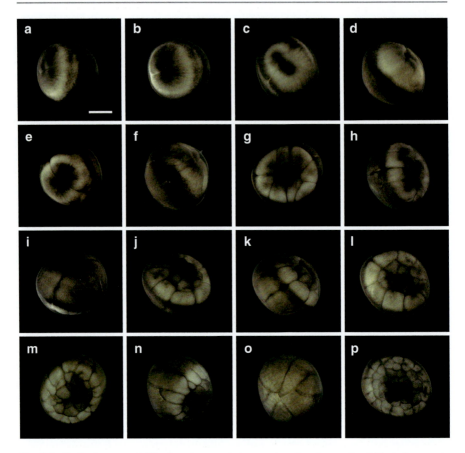

Fig. 8.1 Early cleavage of Siberian sturgeon *Acipenser baerii* embryos. (**a–c**) First cleavage to form two blastomeres. (**d, e**) Second cleavage in animal hemisphere (*blastomeres*). (**f**) Lateral view of four-celled embryo showing the partial infiltration of cleavage furrow into the vegetal hemisphere. (**g, h**) Eight blastomeres in animal hemisphere. (**i**) Vegetal view of the embryos showing eight blastomeres in the animal hemisphere. (**j–l**) Embryos showing 16 cells in animal hemisphere. (**m–o**) Irregular blastomeres formed after the fifth cleavage in the animal hemisphere (*animal view*, *lateral view* and *vegetal view*, respectively). (**p**) Continued cleavages in animal hemisphere. Scale bar: *a* = 1 mm (**a–p**) (Reprinted with permission from Park et al. (2013a))

hemisphere. Nearing the completion of blastulation at 18–19 hpf, the smooth appearance of the animal hemisphere changed to a milky white blastula roof.

Gastrulation (Fig. 8.3). The gastrulation is the process by which the structurally simple, hollow blastula forms a multilayered embryo with an internal cavity (archenteron) and a distinct anterior–posterior axis, and clearly delineated germ layers, the endoderm, ectoderm and mesoderm, are formed. Upon the onset of gastrulation (19–20 hpf, stage 12), a band is formed close to the equator of the embryo, and this is followed by the formation of a "dorsal blastopore lip" between the animal and vegetal hemispheres (stage 13). At this time point, the vegetal hemisphere contains a number of divided blastomeres of differing sizes, whereby relatively large and

8 Early Ontogeny in the Siberian Sturgeon

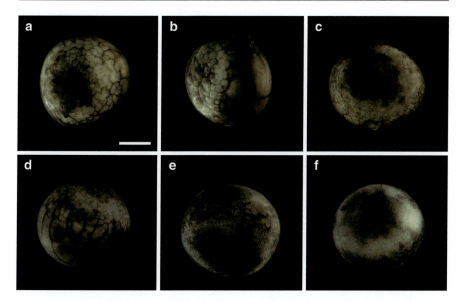

Fig. 8.2 Embryogenesis of Siberian sturgeon *Acipenser baerii* from the early blastula stage (**a–d**) to the late blastula stage (**e**). The onset of gastrulation is shown in (**f**). Scale bar: $a = 1$ mm (**a–f**) (Reprinted with permission from Park et al. (2013a))

distinguishable blastomeres are present in the region close to the apex; smaller blastomeres are detected in the region close to the marginal zone. The blastula roof of the animal hemisphere envelopes progressively the vegetal hemisphere. At 25 hpf, approximately two-thirds of the embryo are covered by the blastoderm (stage 15). As gastrulation continues, epiboly covers more than two-thirds of the embryos, and at 28 hpf, a large yolk plug is formed (stage 16). About 2 h later, the size of the yolk plug is further reduced to less than one-fifth the diameter of the embryo (small yolk plug formation, stage 17). The size of the yolk plug gradually decreases until the blastopore appeared only as a small circle at the apex of the vegetal pore (stage 18). As gastrulation neared completion (32 hpf), the blastopore has a slitlike appearance that signals the onset of neurulation.

As Dettlaff et al. (1993) reviewed, most of the defects observed in sturgeon embryos arise as a result of disturbance of morphogenetic movements during gastrulation. If epiboly of the vegetal hemisphere of the blastula is inhibited at the early stages of gastrulation, the embryo generally dies. When such disturbances are not very important and a more or less large yolk plug remains naked by the moment of the neural plate formation, the embryos can live much longer and reach the stage of hatching but with important morphological defects. Gastrulation defects may also arise due to abnormal embryo cleavage (incomplete cleavage) and unfavourable incubating conditions. In good-quality eggs, gastrulation proceeds simultaneously, and at its end only single embryos with large yolk plugs of irregular shape can be detected. If an egg batch contains a large quantity of embryos with a wide range of yolk plug sizes, this indicates insufficiently favourable incubation conditions for normal embryogenesis.

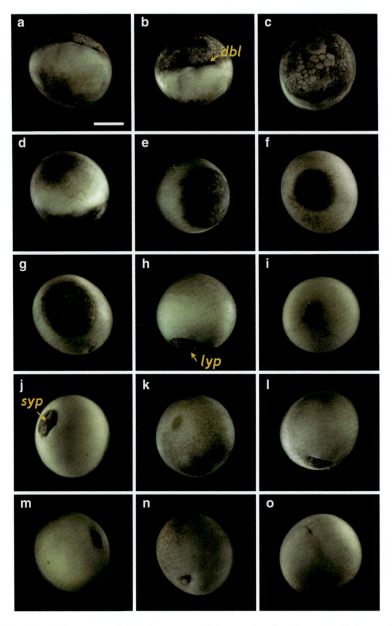

Fig. 8.3 Gastrulation stages of Siberian sturgeon *Acipenser baerii* embryos. (**a**, **b**) Onset of gastrulation with the formation of dorsal blastopore lip (*dbl*). (**c**) Vegetal view of dbl.-formed embryo. (**d–f**) The epiboly of the blastoderm covers two-thirds of the embryo in lateral, vegetal and animal views, respectively. (**g**) Further covering of vegetal hemisphere by the blastoderm. (**h**) Large yolk plug (*lyp*) formation at the vegetal pole. (**i**) Animal view of the embryo with a large yolk plug. (**j**) Small yolk plug (*syp*) formation. (**k**) Animal view of the embryo with a small yolk plug. (**l–n**) Blastopore formation. (**o**) Completion of the gastrulation. Scale bar: *a* = 1 mm (**a–o**) (Reprinted with permission from Park et al. (2013a))

8 Early Ontogeny in the Siberian Sturgeon

Neurulation (Fig. 8.4). Neurulation begins with the appearance of a slit-shaped neural groove in the blastopore (stage 19), followed by the formation of the neural plate at the dorsal surface with the folded structure in the head region (33 hpf). With the progression of development (33–35 hpf), the neural plate in the head region widens and the neural folds rises and thickens (stage 20). In the dorsal region, the neural groove is more evident, and the excretory system rudiments are faintly visible parallel to the neural groove (37 hpf, stage 21); a folded shape is seen in the caudal region of the embryo (tail). As neurulation continues, the folding of the head region is slightly reorganized, and the pronephros (head kidney) rudiments become evident as cords running parallel to the neural groove (40 hpf, stage 23). Thereafter, the pronephros becomes elongated and thickens along with the risen neural folds.

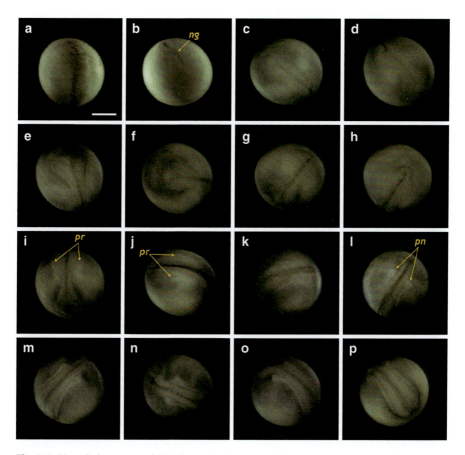

Fig. 8.4 Neurulation stages of Siberian sturgeon *Acipenser baerii* embryos. (**a, b**) Onset of neurulation with a slitlike neural groove (*ng*). (**c, d**) Formation of the neural plate. (**e, f**) Folding of the head region. (**g**) Appearance of excretory rudiments. (**h**) Folded structure in the tail region. (**i, j**) Pronephros rudiments (*pr*) running in parallel to the neural groove. (**k, n**) Elongation of the pronephros (*pn*). (**o, p**) Thickened of the tail region and the pronephros becomes perpendicular to the neural tube. Scale bar: *a* = 1 mm (**a–p**) (Reprinted with permission from Park et al. (2013a))

The tail region continues to thicken as development progresses. At 46 dpf, the pronephros are located almost perpendicular to the neural tube and the neural tube is mostly closed (stage 25).

Developmental defects arising during this period are most frequently due to disturbances of epiboly (Dettlaff et al. 1993). If the yolk plug is large, a shortened and deformed neural plate forms; the embryo develops abnormally and soon dies. Rarely, embryos occur with bifurcated axial rudiments, asymmetrical neural tube and one row of somites and one pronephros. In most of the cases, a unique complex defect is found where the head, trunk and heart develop abnormally at the same time, which is a result of unfavourable environmental conditions at the end of gastrulation and during the neural plate formation or just before the heart rudiment formation.

Organogenesis and hatching (Figs. 8.5, 8.6, and 8.7). At 55 hpf, the lateral plates are fused to the prosencephalon (forebrain), and the head region appears as a round-shaped object in the dorsal view, in which the rudimentary (unpigmented) eyes and undeveloped mouth are easily visible (stage 26). At this stage, the dorsal region of

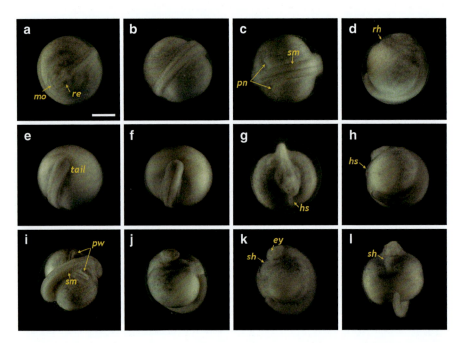

Fig. 8.5 Organogenesis of Siberian sturgeon *Acipenser baerii* embryos. (**a–c**) Embryos at stage 26 showing the round-shaped head, rudimentary eye (*re*), underdeveloped mouth (*mo*) and somite (*sm*) formation. A pair of pronephric ducts (*pn*) become v-shaped. (**d–f**) Embryos at stage 27 characterized by the formation of the rudimentary heart as well as the rod-shaped tail that begins to separate from the yolk. (**g, i**) Embryos at stage 28 displaying the heart straightened (*hs*) and a well-developed pronephros wing (*pw*). (**j–l**) Embryos at stage 29 showing the s-shaped heart (*sh*) and evident eye caps (*ey*). Scale bar: *a* = 1 mm (**a–l**) (Reprinted with permission from Park et al. (2013a))

8 Early Ontogeny in the Siberian Sturgeon

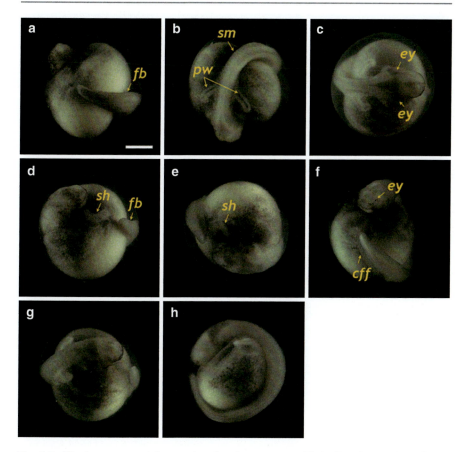

Fig. 8.6 Siberian sturgeon *Acipenser baerii* embryos at stage 30 (**a–d**) and stage 31 (**e–h**). In stage 30, the tail was transformed to be straightened structure with the rudimentary fin bud (*fb*) in caudal region. Somites (*sm*), pronephros wings (*pw*), eye caps (*ey*) and s-heart (*sh*) were more developed than previous stages. In stage 31, rudimentary fin bud was further developed to caudal finfold (*cff*) and tail approached the s-heart. Scale bar: $a = 1$ mm (**a–h**) (Reprinted with permission from Park et al. (2013a))

the embryo has a rod-shaped appearance with the development of somites. A rudimentary heart is formed in the embryo at 59 hpf. In the caudal region, the flattened tail is transformed to a rod-shaped structure, and the tail begins to separate from the yolk sac. Between 60 and 70 hpf, the head region thickens and begins to separate from the yolk, in which the heart is visible as a short straight tube (stage 27). The tail continues to lengthen and somites are visible over the entire embryonic body. At 73 hpf, the heart achieves an s-shape and cardiac beatings begin (stage 29). At 86 hpf, the round, rod-shaped structure of the tail straightens, and fin-bud rudiments are visible in the caudal region; the anterior part of the head separates from the yolk sac; and the heart gets a more pronounced s-shape (stage 30). At 94 hpf, the tail reaches the heart and finfolds appear in the caudal region (stage 31). Blood circulation is noticeable, and olfactory organs are also visible. As the tail continues to

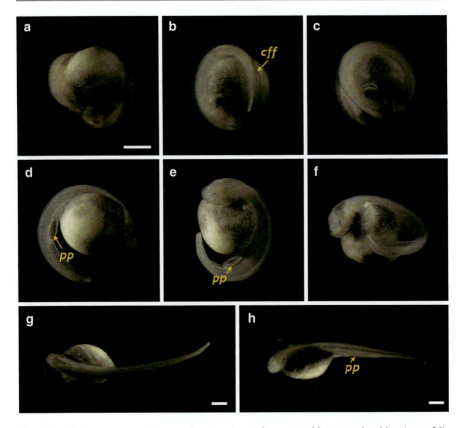

Fig. 8.7 Siberian sturgeon *Acipenser baerii* embryos from stage 32 to mass hatching (stage 36). (**a–c**) Embryos at prehatching stage (stage 32) with well-developed caudal finfold (*cff*). (**d–f**) Advanced hatchling (stage 35) with the pigment plug in the rudimentary intestine (*pp*) and the trunk still curled. (**g, h**) Newly hatched prelarva (stage 36). Scale bars: *a* = 1 mm (**a–f**), **g, h** = 1 mm (Reprinted with permission from Park et al. (2013a))

lengthen, it approaches the head at 101 hpf (stage 32). Eye caps are evident and caudal finfolds are easily distinguishable. At this stage, embryos perform circular movements within the egg chorion. Between stages 32 and 35, the finfolds that surround the tail rapidly widen. From 119 hpf afterwards, several advanced embryos start to hatch, and mass hatching is observed at 130 hpf (stage 36). Newly hatched prelarvae measure in average between 10.5 and 12.5 mm in total length (TL); prelarval size at hatching is affected by the egg size, and the larger the egg diameter, the larger the prelarvae size (Gisbert et al. 2000).

Abnormal embryos cannot free themselves of their membranes; the less abnormal ones are able to hatch but usually with some delay compared to the normal embryos. If the incubation conditions are favourable, only single deformed embryos occur in batches of good-quality eggs and up to 25–30% in those of poor-quality eggs (Dettlaff et al. 1993). The former authors were amazed when detecting alive embryos with severe abnormalities after hatching; if these prelarvae were separated and put in clean water, they were able to survive several days using their endogenous reserves, whereas if they were left in the incubators together with rests of chorions and dead embryos, they soon died. The underdevelopment of the anterior regions of the body is the most common developmental defects in hatchlings. In some cases, these abnormalities consist on the simply reduction of the head region, brain and sensory organs, whereas in more severe cases, the anterior region of the embryo is completely absent, i.e. there is no prosencephalon, diencephalon, olfactory sacs or eyes. In some other cases, embryos have no anterior trunk region and shortened and bent tails. Embryos with underdeveloped anterior regions of the body move their tail spasmodically, but usually they are unable to swim, whereas embryos without the anterior region of the head and more deformed ones remain immobile in the bottom of the tank and even do not respond to needle pricking. The majority of the above-mentioned deformities are caused by disturbances at the earliest stages of development, i.e. during oocyte maturation and fertilization. When these abnormalities appear at later stages of development as a consequence of abnormal embryogenesis, they are due to abnormal environmental incubation conditions or undesirable water quality, as data from several sturgeon species revealed (Dettlaff et al. 1993; Van Eenennaam et al. 2005; Park et al. 2013b).

Effects of water temperature on the embryonic development. Temperature is considered as the environmental factor with the largest effect on the development of fish due to their poikilothermic condition. The developmental rate and embryonic survival in Siberian sturgeon were affected by the water temperature at which embryos were incubated (Fig. 8.8), as it was shown by Park et al. (2013a). In particular, the time lapse comprised between fertilization and the formation of the small yolk plug (stage 17) ranged from 19 h after fertilization (haf) at 24 °C to 58 haf at 12 °C. Embryo survival

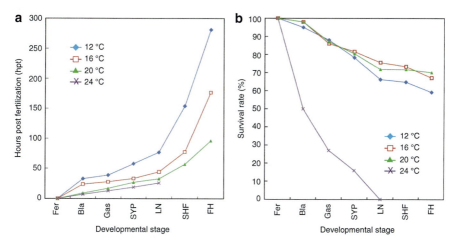

Fig. 8.8 Effects of water temperature on developmental progress (**a**) and survival rates (**b**) of Siberian sturgeon *Acipenser baerii* embryos incubated at different temperatures ranging from 12 to 24 °C. None embryo survived after the completing of the neurulation in incubation water temperature of 24 °C. Abbreviation of developmental stages: *Fer* fertilization; *Bla* blastula; *Gas* gastrulation; *SYP* small yolk plug formation; *LN* late neurulation; *SHF* s-heart formation; *FH* onset of hatching (Reprinted with permission from Park et al. (2013b))

rates (78–81% in average) at the small yolk plug stage did not differ among the groups that were incubated at 12, 16 or 20 °C. However, embryos that were incubated at 24 °C had the lowest survival (16%), and none of them survived until the completion of neurulation (stage 22). The time to reach the end of neurulation ranged from 33 haf (at 20 °C) to 77 haf (at 12 °C), whereas the s-heart formation stage (stage 25) was observed at 57 haf (at 20 °C), 77.5 haf (at 16 °C) and 154 haf (at 12 °C) after fertilization.

The onset of hatching (stage 36) was also affected by water temperature, as well as the survival of embryos. In embryos incubated at 12 °C, hatching was observed to start at 281 haf, while hatching started at 176.5 haf and 96 haf in those embryos incubated at 16 °C and 20 °C, respectively. The percent survival in each of the temperature groups until the first occurrence of hatching was $59.0 \pm 3.8\%$ (at 12 °C), $67.0 \pm 4.0\%$ (at 16 °C) and $70.0 \pm 1.8\%$ (at 20 °C). It should be noted that hatching rates in Siberian sturgeon tend to be generally higher (Williot 1997) than the above-mentioned ones (Park et al. 2013a), which might be linked to different egg quality and incubation conditions between different studies and hatcheries. No differences in the incidence of morphological abnormalities (9.0–11.0%) were found in newly hatched prelarvae incubated at water temperatures ranging from 12 to 20 °C. According to Park et al. (2013a), the incubation of Siberian sturgeon embryos at temperatures close to 20 °C could provide advantages over low-temperature incubation, because there was a much more synchronized timing window for hatching. At 20 °C, the hatching events of almost all of the tail-beating embryos were completed within only 3 days, whereas lower incubation water temperatures resulted in longer hatching times, at least 5–9 days at 16 °C and 12 °C, respectively. Furthermore, hatchability was significantly higher at 20 °C than at the two lower temperatures, without any notable signs of additional morphological abnormalities. Under hatchery conditions, large variation in the timing of hatching inevitably leads to non-uniform larval populations, which consequently results in the unavoidable and unwanted loss of a portion of the larvae that show either advanced or delayed development. However, higher incubation temperatures may enhance the growth and proliferation of fungus (*Saprolegnia* sp.) in egg incubators, affecting the performance of the incubation process. Thus, the selection of the water incubation temperature should be a choice of each hatchery manager depending on their incubation systems, water quality and production strategy.

8.2 External Morphological Development

The early development of the Siberian sturgeon from hatching to the juvenile stage can be divided into two periods according to Dettlaff et al. (1993). The prelarval stage comprised between hatching and the onset of the exogenous feeding, and the larval stage between the beginning of external feeding and metamorphosis.

During the prelarval stage, embryonic adaptations and functions were replaced by definitive ones, such as gill respiration, exogenous feeding and active swimming, which involved dramatic alterations in the relationship of the developing fish with the environment, that were reflected in morphological and morphometric changes that enhanced the performance and survival chances of the developing specimens. Those changes were initially described by Gisbert (1999) and Gisbert et al. (1998a)

8 Early Ontogeny in the Siberian Sturgeon

and recently confirmed by Park et al. (2013b) in specimens reared at *ca*.18 °C, and they may be summarized as follows:

Newly hatched (0 days post hatching, dph) Siberian sturgeon prelarvae measured between 10.4 and 11.1 mm in total length (TL) had the hatching gland still visible on the lower head surface (Fig. 8.9a). The mouth opening and gill clefts were absent; the tail was proterocercal; the trunk and tail were bordered by a wide primordial finfold, especially in the caudal region; sense organs were poorly differentiated, rudiments of olfactory organs were rounded with only one external opening, eyes were dark pigmented spots and the seismosensory system was present only in the head region. Pairs of distinct veins were visible on the surface of the yolk, and the pigment plug was evenly spread inside the digestive tract; pronephric ducts (head kidney) were visible. Very little pigmentation (scattered melanocytes) in the embryonic body was observed in most prelarvae. Between 1 and 2 dph (12.5–13.7 mm TL), pectoral fin

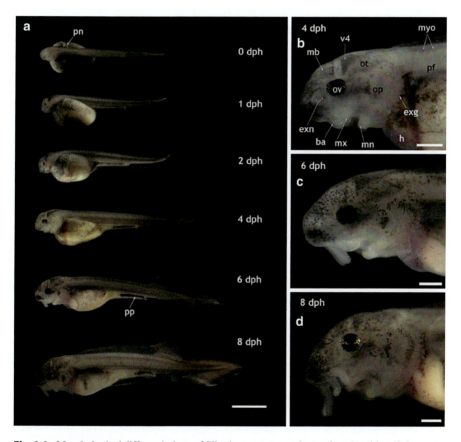

Fig. 8.9 Morphological differentiations of Siberian sturgeon prelarvae from hatching (0 days post hatching, dph) to 8 dph reared at 18 °C. Detailed view of head region at selected ages is shown at the right side. Abbreviations: *pn* pronephros; *pp.* pigment plug; *mb* midbrain; *v4* fourth ventricle; *myo* myotomes; *ov* optic vesicle; *ot* otocyte; *op* operculum; *pf* pectoral fin; *exg* external gills; *exn* external nares; *ba* barbels; *mn* mandibular process; *mx* maxillary process; *h* heart. Scale bars: A = 3 mm, B–D = 0.5 mm (Reprinted with permission from Park et al. (2013b))

rudiments appeared as little folds, the mouth opened, barbel buds were apparent and the eye pigmentation was more pronounced; rudiments of gill filaments could be seen behind the transparent opercula, and the lateral line grew in length, reaching the Cuvier duct zone (Fig. 8.9a). At 3 dph (13.9–15.1 mm TL), the eyes were darkly pigmented. Lower and upper lips were distinguished as small folds surrounding the mouth opening. The finfold started to get pigmented and pigmentation indicated where future rays would appear along the dorsal region of the finfold. The finfold was wider on the ventral side of the trunk and narrowed at the caudal peduncle and protruded slightly where the future dorsal, caudal and anal fins would develop.

At 4 dph (14.8–15.6 mm LT), the level of pigmentation in the cephalic region and abdominal cavity increased; external gill filaments developed and were not covered by the operculum, the lobes dividing the olfactory opening made contact and barbel rudiments increased in size; the midbrain, the fourth ventricle and the otocysts could be distinguished in the lateral view (Fig. 8.9b). The lateral line system reached the middle part of the stomach. Rudiments of dorsal fin rays (18–19) appeared in the dorsal finfold, and rudiments of pelvic fins appeared and could be differentiated as narrow folds of epidermis. The yolk sac appeared as bipartite, with a furrow that separated the anterior stomach rudiment to the mid and posterior intestine, whereas the pigment plug became concentrated and moved to the posterior part of the digestive tract (spiral valve).

Between 5 and 6 dph (17.5–18.3 mm TL), the network of blood vessels that covered the middle and posterior part of the yolk sac decreased and almost disappeared, and pectoral fins developed (Fig. 8.9a). The lobes of the olfactory organ fused, some neuroepithelial follicles appeared on the ventral surface of the rostrum, barbels elongated, tooth rudiments were seen in both the upper (maxillary) and lower (mandibular) lips and the gills were still not covered by the operculum (Fig. 8.9c). The lateral line extended to the posterior part of the digestive tract. Pigmentation levels increased along the entire body of prelarvae, especially in the cranial and caudal sections. The posterior margin of the yolk sac had been markedly reduced, the yolk sac had separated into presumptive organs and organs in the abdominal cavity could be visualized. The dorsal fin was separated from the finfold and rudimentary fin rays extended from the fin base, and the anal and pelvic fins also began to appear as ridges in the finfold. At 7–8 dph (17.9–19.5 mm LT), the snout became horizontal and ventral, and barbels were longer and extended beyond the rostrum. Rudiments of pectoral (7–9) and anal (9–10) fin rays appeared in their respective finfolds. Canine-like teeth were more obvious and sharpened in the upper and lower jaws. Food seizure and sense organs were completely developed for exogenous feeding. The forth ventricle was less contrasted than in earlier larvae due to the thickened cranial cover, which had significant pigmentation on its surface. Although the external gills that had been prominent in earlier larval stages were largely covered by the extended operculum, the coverage was not yet completed (Fig. 8.9d). The lateral line reached the posterior margin of the pelvic fin. Dorsal and anal fins were separated from the tail fin, rudiments of pelvic fin rays (8–10) were visible and the heterocercal structure of the caudal fin was clear. At this period, the yolk sac was difficult to discern clearly. The pigment plug had moved further in the anal direction and eventually started to be evacuated, indicating that the larvae were transitioning to exogenous feeding (Gisbert and Williot 1997).

At the beginning of the larval phase (9–10 dph, 19.7–21.0 mm LT), new morphological traits appeared; the yolk sac was apparently depleted and most larvae had the artificial diet in their digestive tracts. The teeth lengthened and sharpened with age, and a shark tooth-like shape peaked between 9 and 11 dph. Upper teeth were sharper and more tapered than lower teeth. However, the numbers of both upper and lower teeth were not uniform among individuals, with the number of upper teeth in most individuals ranging from 8 to 10 and the number of lower teeth ranging from 6 to 8 (Fig. 8.10). Rudiments of dorsal scutes were distinguishable in the finfold, and the external gill filaments were almost covered by the operculum.

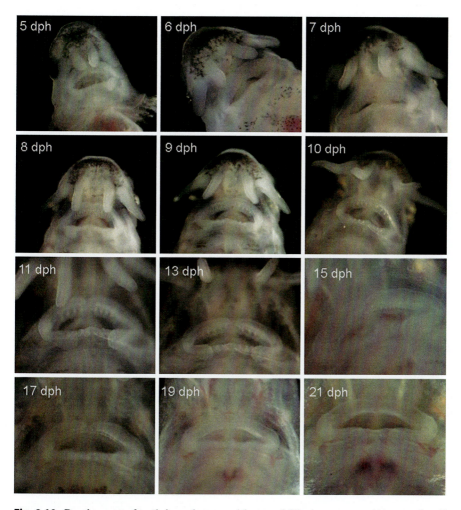

Fig. 8.10 Development of teeth in prelarvae and larvae of Siberian sturgeon *Acipenser baerii* from 5 days post hatching (*dph*) to 21 dph reared at 18 °C. Teeth rudiments were firstly detected in both maxillary and mandibular at 5 dph. Afterwards, the development of teeth was peaked at 9–11 dph and then began to degenerate. At 21 dph, teeth were completely diminished (Reprinted with permission from Park et al. 2013b)

Few morphological changes took place from days 11 to 18 (Fig. 8.11). At 11 dph (21.2–22.4 mm TL), rudimentary spikes on the dorsal head surface and on the pectoral zone became visible. At 13 dph (22.5–23.1 mm TL), several rows of sensory pits were evident on the dorsal side of the snout, which increased with age. After the complete transition to exogenous feeding, teeth began to degenerate; the upper teeth became thinner while the lower teeth took on a blunted shape, and both upper and lower teeth became shorter. Between 15 and 17 dph, teeth were less than half the size of those observed at 9 dph, whereas at 19 dph, teeth were only vestigial and they had disappeared completely by 21 dph. Between 18 and 19 dph (27.2–29.7 mm TL), rudiments of lateral and ventral scutes could be differentiated as four rows of little spikes running laterally (2) and ventrally (2) on the body surface. Between 20 and 40 dph, metamorphosis took place and some new juvenile traits appeared. Dorsal, lateral and ventral rows of scutes developed greatly, and rudiments of caudal fin rays (19–21) appeared and achieved their definitive structure; head spikes increased in number and the tip of the snout became definitively sharp. Between

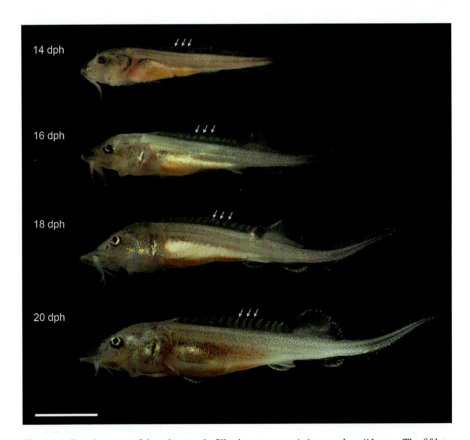

Fig. 8.11 Development of dorsal scutes in Siberian sturgeon *Acipenser baerii* larvae. The fifth to seventh bony scutes from the dorsal fin were indicated by arrows. Scale bar = 5 mm (Reprinted with permission from Park et al. 2013b)

8 Early Ontogeny in the Siberian Sturgeon 147

days 36 and 40, the rudiment of the preanal finfold disappeared completely, and specimens resembled miniature adult Siberian sturgeon.

It is important to mention that there exist an important gap of knowledge on the skeletal development in Siberian sturgeon (Leprévost and Sire 2014), as well as in other Acipenseriformes, which may be of special relevance since the axial skeleton is involved in body support among other important functions and skeletal disorders are quite common in juvenile specimens of this species.

8.3 Digestive System

Sturgeon embryos have intraembryonic yolk, and their alimentary tract develops from the yolk sac endoderm (Dettlaff et al. 1993). This is a result of holoblastic cleavage in Acipenseriformes, contrary to the teleost fish in which the egg is subjected to meroblastic cleavage and in which the alimentary tract develops independently from the extraembryonic yolk sac. When Siberian sturgeon prelarvae emerge from the egg membranes, they possess a primordial alimentary canal subdivided into two different regions, which does not communicate with the exterior as the mouth and anus are not yet opened. The anterior region presents a closed buccopharynx and a gastric cavity filled with yolk and lined by endodermal cells that will develop into the stomach. The posterior region of the alimentary tract consists of a partially differentiated hindgut, which will develop into the spiral intestine and the anus (Gisbert et al. 1998b, 1999a).

The development of the alimentary tract and accessory digestive glands (liver and pancreas) is an asynchronous and very intense phenomenon that takes place during the prelarval stage, just before the onset of exogenous feeding. Differentiation proceeds from the posterior towards the anterior part, being the spiral intestine the first to develop and the glandular stomach the last, whereas after the onset of the exogenous feeding, the organization and differentiation of the digestive tract do not undergo any remarkable modification. The histological development of the digestive system in Siberian sturgeon can be summarized as follows:

At hatching (10.4–11.2 mm TL), the *buccopharynx* is not opened, lined by a two- to three-cell-long pseudostratified cuboidal epithelium, surrounded by connective tissue and a thin lamina propria; epithelial cells covering the buccopharynx present some vacuoles filled with rests of acidophilic yolk and pigmented granules. The posterior region of the buccopharynx is undifferentiated. There is no cellular differentiation, nor connection with the yolk sac cavity. One day after hatching, the buccopharynx (mouth) opens, and two well-differentiated oral valves with a thick stratified cuboidal epithelium are visible in prelarvae aged 3 dph (13.0–13.5 mm TL). Taste buds appeared in the buccopharyngeal epithelium and become more numerous as specimens grow. Between 5 and 7 dph, mucous cells started to secrete a combination of neutral and acidic mucopolysaccharides. Canine teeth develop in the connective tissue underlying the epithelium, erupting in the lips and pharyngeal regions during the period of transition to exogenous feeding.

The *oesophagus* is not differentiated in prelarvae aged 1 dph. At 3 dph, the oesophagus appears as a simple columnar epithelium with a narrow and short lumen. A large number of oesophageal cells present supranuclear vacuoles with yolk inclusions and pigment granules, as a consequence of the holoblastic embryo cleavage. At this age, the oesophagus is in differentiation and its connection with future stomach is not yet established. At 5–7 dph, mucous cells were already visible in the oesophageal mucosa containing a combination of neutral and acid mucopolysaccharides, although mucous cells containing neutral mucins were more abundant than acid ones. Between 8 and 9 dph (19.0–20.2 mm TL), the connection between the oesophagus and the cardiac stomach takes place due to the resorption of a yolk mass that separated them. At this moment, the oesophagus has elongated, and two differentiated regions associated with secretion (anterior) and food transport functions (posterior) can be distinguished. The anterior region is characterized by abundant mucous cells secreting a combination of neutral and acidic mucins, whereas the posterior oesophageal region has a lower number of mucous cells and is mainly characterized by a ciliated epithelium.

The *stomach* is undifferentiated and filled with yolk at hatching. At 3 dph, an oblique furrow of the digestive tube wall starts to develop on the dorsal–posterior region of the yolk sac and divides the alimentary canal into two differentiated regions: the future stomach and intermediate intestine. The anterior part of this furrow, which consists of squamous cells, will become the lower wall of the stomach, while its posterior part lined with a ciliated columnar epithelium will constitute the upper wall of the intermediate intestine. At 5 dph (14.6–15.6 mm TL), a non-glandular stomach surrounded by a prominent muscular layer appears in the anterior–ventral region of the body. Between 6 and 7 dph (17.2–18.4 mm TL), mucosal folds increase in size and number, and the non-glandular stomach is separated from the anterior intestine by an epithelial fold that will develop into the pyloric sphincter. During this period, yolk sac reserves are considerably reduced, and the glandular stomach starts to develop and few scattered gastric glands are visible. At 9 dph, numerous simple and tubular gastric glands are visible arranged along numerous longitudinal folds; the lumen of the gastric stomach was still filled with remains of yolk. The glandular region of the stomach has a cubical simple epithelium with basal nuclei and prominent microvilli with mucous cells secreting a large quantity of neutral mucosubstances. Between 12 and 13 dph (21.6–23.1 mm TL), the only noticeable change in the stomach is the disappearance of yolk granules from the glandular region, whereas until 21 dph (28.1–31.1 mm TL), the non-glandular and glandular regions of the stomach do not show any further noticeable modification, rather than their increase in size with age.

In 1 dph prelarvae, the *intestine* is lined by a simple columnar ciliated epithelium in differentiation. Enterocytes have basal nuclei and apical vacuoles with yolk granules inside that are generally smaller than those from the anterior region of the alimentary canal. Between 2 and 3 days dph, the spiral valve starts to differentiate in a simple columnar epithelium with prominent microvilli, and pigment granules accumulate in the spiral valve lumen forming the melanin plug. As fish grow, goblet cells containing neutral mucosubstances appear in the posterior region of the spiral valve. Anteriorly to the spiral valve, a large number of epithelial folds are clearly distinguishable in the intermediate intestine. At 5 dph, the intestine becomes more elongated and its anterior region appears; the epithelium of the anterior gut is similar to the intermediate with exception of the number and size of mucosal folds, which are less numerous and shorter, and lipids start to accumulate inside enterocytes. Goblet cells containing a mixture of neutral and acid mucins are detected along the intestine, which increase in number as fish grows, although these secretory-like cells were more abundant in the intermediate intestine and hindgut than in the anterior intestine, and their number increased with larval development. Between 7 and 9 dph, the terminal section of the digestive tract is differentiated into a short rectum lined with a cuboidal epithelium. From 9 to 11 dph, the melanin plug is ejected from the lumen of the spiral valve.

At the onset of exogenous feeding (9–11 dph), the general anatomy and histology of the digestive system in Siberian sturgeon are similar to that in juvenile or adult specimens. The buccopharynx is lined with a stratified squamous epithelium with numerous fungiform and filiform papillae, canine-like teeth in the pharyngeal region and taste buds scattered along the buccopharyngeal epithelium. The lumen of the oesophagus is lined by abundant mucous cells secreting neutral and acidic mucins. The large number of mucous cells and their abundant mucins secreted may not be simply explained as a lubricant function to compensate the lack of salivary glands; this region of the alimentary canal is also able to respond to changes in environmental conditions and maintain osmotic balance when gill arches are not completely functional. In addition, neutral mucins secreted by oesophageal mucous cells are considered to cooperate in the digestion of food and its transformation into chyme, as well as in the absorption of disaccharides and short-chain fatty acids, whereas acid mucosubstances, especially those containing sialic acid, may serve to prevent the oesophageal mucosa from sialidase produced by bacteria and virus recognition sites (Sarasquete et al. 2001). The epithelium of the cardiac stomach is composed of cuboidal cells, with numerous simple and tubular gastric glands secreting a mixture of hydrochloric acid and pepsin. Neutral secretory products present in this region of the digestive tract may serve to protect the epithelium of the stomach from autodigestion processes. The pyloric stomach is lined with a simple columnar ciliated epithelium containing supranuclear vacuoles filled with neutral mucins and organized in folds surrounded by a muscular layer. The secretion of neutral mucosubstances, in conjunction with thick mucosa of pyloric region, may serve to protect the underlying layers from chemical and physical damages during food trituration processes. The histological organization of the intestine is generally similar in different regions, with a simple columnar ciliated epithelium and numerous goblet cells that secreted neutral and acidic mucosubstances. At the onset of feeding, the cells of intestinal mucosa were filled with large lipid vacuoles that gradually disappeared after onset of feeding.

8.4   Olfactory System

The chemosensory organs play an important role in the feeding behaviour in sturgeon larval and juvenile stages. During development, the ability of Siberian sturgeon to react to food extracts appears immediately after changing to exclusively exogenous feeding (Kasumyan and Kazhlayev 1993). This observation fits the process of cellular differentiation of the olfactory epithelium described by Zeiske et al. (2003) in Siberian sturgeon embryos incubated at ca. 14 °C, which seems to be finished well before onset of exogenous feeding.

The anlage of the olfactory organ became microscopically visible as placodal ectodermal thickenings after closure of the neural folds and a little before first heart contractions of the embryo were observed. Externally, it remains in the state of invaginated olfactory pits even after hatching, whereas further development of the olfactory apparatus slowly occurred during the prelarval stage. In particular, the formation of the olfactory system begins with the formation of the olfactory pits at 64 haf (stage 25), which are visible by scanning electron microscopy. The paired pits are located between the developing eyes and a deep recess at the tip of the head, which obliterates, and the olfactory pits deepen during embryogenesis until hatching. Semi-thin and ultrathin sections through the head of embryos at stage 25 show a line of more or less cuboidal cells between the epidermis and the nerve cord, the so-called subepidermal layer that thickens on either side of the developing forebrain by adding cells. These thickenings represent the earliest visible sign of the olfactory placodes. At 72 haf, embryos show increased cell division in a region on the top of the head, where the surface epithelium will invaginate to form the primordia of the olfactory pits. In embryos aged 74 haf, connective tissue completely separates the brain from the olfactory placodes. As shown by transmission electron microscopy, the cells of the placodes and those of the brain contain vesicles filled with yolk or products of its digestion, which is a consequence of the holoblastic development of embryos.

Sections of embryos 84–90 haf show that the cells bearing microvillus-like protrusions (already detected at 80 haf) in the central part of the placode have stretched in proximal–distal direction, whereas the surface cells covering the deeper, more advanced layer of the placode appear structurally unchanged during this stage. During the following hours, a junctional complex consisting of tight junction, intermediate junction and desmosomes develops in the cell membranes of adjacent superficial cells slightly below their free surface, indicating their further development. The first few free endings of dendrites (olfactory knobs) bearing a single short cilium appear at the surface of the olfactory pit of embryos around 93 haf, whereas only a few hours later, these knobs may have up to 20 short cilia and microvillus-like protrusions in various numbers.

The basic structure of the olfactory epithelium is achieved between 108 and 126 haf. These changes are characterized by a pronounced development of dendrites that reach the free epithelial surface with their distal endings (knobs) that bear microvillus-like protrusions; axon bundles of receptor cells develop in the basal region of the placode during this period. The loose connective tissue, which separates olfactory placode and brain, has increased in thickness, and the axons within the proximal part of the placode have increased in number. In cells of the mid-region of the placode, single primary cilia are found. Primary cilia are also present in the region of the brain which is close to the olfactory placode and in fibroblasts in the gap between the brain and placode. At 3 dph (stage 39), the opening of the deeply invaginated pit elongates in the rostro-caudal direction,

8 Early Ontogeny in the Siberian Sturgeon

and 1 day later, superficial extensions appear on either side of each pit. At 144 haf (6 daf), the endings of dendrites from sensory cells appear at the free epithelial surface in the olfactory pit. Two different sensory cell types can be distinguished in the differentiated olfactory organ of Siberian sturgeon, ciliated and microvillus receptor cells, being the first ones being more numerous than the second ones. Further details about these groups of sensory cells and supporting nonsensory cells may be found in Zeiske et al. (2003). The anterior and posterior openings are separated from each other at 6 dph. At 8 dph, the olfactory organ is completed from a morphological point of view, and both nasal openings already tend to acquire the characteristic shape of the juvenile and adult: the anterior being smaller and the posterior larger, slitlike and vertically oriented. The division of the primary olfactory opening into the anterior incurrent and posterior excurrent opening in front of the eye is completed 6 dph, before the onset of exogenous feeding. The formation of the nasal cavity and the anterior (incurrent) and posterior (excurrent) nostrils from an original single pit as described by Zeiske et al. (2003) is considered the phylogenetically primitive (plesiomorphic) type among actinopterygians. The completion of the organogenesis of the olfactory system in Siberian sturgeon, as well as in other sturgeon species studies so far, indicates the importance of olfaction at first feeding and larval behaviour at early life stages.

8.5   Visual System

The range of activity in fish larvae increases with growth, as the function of sensory organs becomes more refined and locomotor structures develop (Osse and van den Boogaart 1995). In this context, the eye is one of the major sensory organs in fish. It detects light stimuli, forms images and shows a wide range of structural adaptations to the visual environment. In most species, vision is considered the dominant sense during early life stages of development, as it is required for feeding, orientation, schooling and avoiding potential predators (Olla et al. 1995). In Siberian sturgeon, light seems to be a crucial stimulus during the first stages of development (prelarval and larval stages), as it directly affects the reaction of prelarvae to water currents, their activity patterns and the development of schooling behaviour just before the completion of yolk, thus influencing fish distribution, recruitment and survival (Gisbert et al. 1999b). Several ontogenetic changes occur in the eye that are particularly related to retinal morphology and associated with alterations in the photic environment as a result of fish migration or habitat shifts (Loew and Sillman 1993).

According to Rodriguez and Gisbert (2001, 2002), the most relevant morphological changes occurred between hatching (10.4–11.1 mm TL) and 5–6 dph (17.5–18.3 mm TL) at 18 °C, and it can be summarized as follows: between hatching and 2 dph (12.5–13.7 mm TL), the eye was scantly pigmented with a hemispherical rudimentary retina lined by neuroblastic cells, a pseudostratified and undifferentiated epithelium and a lens. At 3 dph (13.9–15.1 mm TL), the eye consisted of three parts: an external basophilic layer of cylindrical stratified epithelium that will differentiate into the choroid gland and sclera, a basophilic rudimentary retina in differentiation with some scattered melanosomes and a simple eosinophilic tissue which constituted the crystalline lens. Between 3 and 4 dph (13.9–15.6 mm TL), the first retinal layer was observed as a thick, simple pigmented epithelium; the rest of the retina was not yet differentiated completely, and only a neuroblastic layer could be distinguished.

Between 5 and 6 dph (17.5–18.3 mm TL), the eyes of Siberian sturgeon prelarvae were completely differentiated. The crystalline lens was a spherical ball consisting of two well-differentiated tissues: an encapsulating spherical sheath with a thick stratified cuboidal epithelium and the inner tissue consisting of acidophilic, large, nucleated fibre cells with a blunt edge. The lens was almost in contact with the cornea, which was comprised of compact basophilic connective tissue and numerous layers of lamellar cells. The sclera consisted of a fibre–collagen lax support tissue, which contained some scattered melanocytes and lymphocytes. The iris consisted of a smooth epithelium on the outer edge and a stratified, slightly prominent cuboidal epithelium on the inner edge. The choroid gland was observed to be in contact with the marginal border of the iris and appeared as a conjunctive highly vascularized tissue, with two thin pigmented layers. All retinal layers could be observed: the pigment epithelium was distinguished as a single layer of cuboidal cells. The amount of melanin granules in the pigment epithelium increased considerably in relation to other retinal layer components, corresponding to a noticeable increase in eye pigmentation. The outer nuclear layer consisted of a regular row of two types of photoreceptor cells, rods and single cones. A thick cylindrical outer segment, a small ovoid ellipsoid and a long filamentous myoid formed the eosinophilic part of the rods, while the nuclei were basophilic and almost round. In contrast, cone photoreceptors presented a shorter eosinophilic tapered outer segment, large ovoid ellipsoids containing an oil droplet and nuclei of a homogeneous shape and size. Between the photoreceptor cells and the inner nuclear layer, a thin basophilic outer plexiform layer was identified. The sclera – most part of the inner nuclear layer – consisted of a single row of cells with large flattened nuclei, indicating the presence of horizontal cells. The vitread segment of the inner nuclear layer contained numerous round cells with large basophilic nuclei and a reduced cytoplasm, resembling amacrine cells. Scattered between amacrine cells, bipolar cells were distinguished as elongated cells with ovoid, slightly basophilic stained nuclei. The inner plexiform layer consisted of a broad layer of reticular tissue. The ganglion cell layer was composed of a single row of spherical cells with large basophilic nuclei and a reduced cytoplasm. The nerve fibre layer was visible on the vitreal surface of the retina.

From 7 dph (17.9–18.5 mm TL) to the end of the study (30 dph, 26.5–35.0 mm TL), the only noticeable changes detected in the eye of Siberian sturgeon were an increase in the number and size of cells, particularly in the retina where there was an increase in photoreceptor cells and melanin granules in the pigmented epithelium.

8.6 Allometric Growth During Prelarval and Larval Development

Allometry describes how the characteristics of an organism scale with each other and with the body size (Fuiman 1983). For a morphological characteristic, allometry can be visualized as plots of character's size against the size of the body, reflecting ontogenetic shifts in morphology and phenotypic plasticity in specific environments and rearing conditions (Simonovic et al. 1999). Morphogenesis and differentiation are more intense during the prelarval than the larval and early juvenile stages of development in Siberian sturgeon. Characteristically, during the prelarval period embryonic adaptations and functions are replaced by definitive ones, such as branchial respiration, exogenous feeding and active swimming. Such modifications involved dramatic alterations in the relationship of the developing fish with the environment that are reflected in morphological and morphometric changes, as well as in the growth profiles of the head and tail (Fig. 8.12). Several morphological characters show positive allometric growth soon after hatching, and after a short period of time, these patterns become isometric; such abrupt changes in growth patterns are linked with changes in respiration, feeding and swimming ability (Gisbert 1999).

According to Gisbert (1999), the head growth in length was positively allometric from hatching to 9 dph in Siberian sturgeon when larvae measured ca. 20.0 mm LT. During this period, sensorial, respiratory and feeding systems developed rapidly and prepared the fish for the larval phase. The positive allometric growth of the head would allow prelarvae to improve their feeding and seizing organs, as well as their respiratory capabilities. In natural environments, starvation and predation are the main factors for high mortality rates in early stages of development (Chambers and Trippel 1997); thus, a rapid development of sensorial organs and fins might enhance both prey detection and predator evasion. In particular, during prelarval stages, morphogenesis and differentiation of sensorial structures, such as barbels covered with taste buds, neuroepithelial follicles on the ventral side of the head and olfactory organ, were very intense and allowed larvae to search for food when yolk sac reserves are depleted. Another explanation for such rapid head length growth is the fact that as yolk sac becomes depleted, the larvae must switch to exogenous feeding and thus they need a functional food intake apparatus. Furthermore, as the prelarva hatches with a relatively underdeveloped head, head length increases rapidly in order to allow the uptake of food particles of increasing size, which are energetically more favourable (Osse et al. 1997). In this context, the positive allometric growth of the mouth width changed to isometry at 14 dph (24.2 mm LT), coinciding with the complete transition to exogenous feeding and confirming that morphological and morphometric changes during early ontogeny are concomitant with physiological modifications to optimize food uptake and match expected feeding priorities. This fact is supported by the complete development of the olfactory, tactile and gustatory sensory systems during the transition to exogenous feeding that participate in feeding processes (Kasumyan and Kazhlayev 1993).

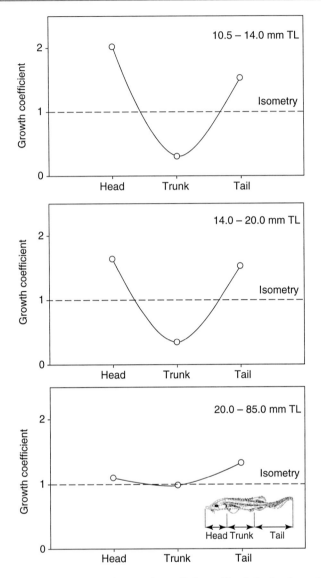

Fig. 8.12 Changes in the allometric growth coefficients (*b*) of the head, trunk and tail in Siberian sturgeon *Acipenser baerii* during early ontogeny. Growth coefficients were calculated according to Fuiman (1983), using the formula $Y = aX^b$, where X is the independent variable (total length, TL), Y the dependent variable (head, trunk and tail length), *a* the intercept and *b* the growth coefficient. Values of $b > 1$ indicate a positive allometric growth; $b = 1$ indicates isometry (both variable growths at the same rate); $b < 1$ indicates a negative allometric growth (the specimen grows faster in TL with regard to the considered body region) (Data modified from Gisbert and Ruban (2003))

The inflexion point of tail growth in length detected at 9 dph (20.2 mm LT) coincides with the timing of finfold differentiation into unpaired fins and is concomitant with the improvement of fish swimming ability, enabling larvae to disperse actively over a wide area and allowing them to use a major number of feeding zones and rearing habitats, to search for food, to avoid predators, to attack prey and so to enhance their survival (Gisbert and Solovyev 2016). These results indicate that feeding, sensory, respiratory and locomotor systems develop allometrically in mutual balance, which is of special relevance in terms of prelarval and larval fitness. These findings are quite similar to other sturgeon species, although species-specific differences exist, which may be attributed to their different size at hatching, heterochrony in development, behaviour and environmental cues (Gisbert and Doroshov 2006; Gisbert et al. 2014).

8.7 Concluding Remarks

Effective culture of larval fish requires alignment of rearing strategies with the ontogenetic status of larvae, *i.e.* with the developing abilities of the specimen to cope with different environmental and nutritional conditions and/or synchronizing the stage of fish development with zootechnical procedures. Thus, comprehension of normal larval development and growth patterns might be considered useful tools for monitoring and optimizing hatchery production, evaluating the suitability of produced fish for stocking or further rearing and assessing the quality of the produced young fish (Koumoundouros et al. 1999), as well as infer the biology, ecology and behaviour of early life stages of development in natural environments (Gisbert 1999; Gisbert et al. 1999b). The information compiled in this chapter indicates that during embryogenesis gastrulation and neurulation are two of the most critical developmental periods, whereas morphogenesis and cell differentiation are more intense during the prelarval than the larval period. This information may serve as a reference guide for further studies on the early development in Siberian sturgeon, as well as to hatchery managers.

References

Chambers RC, Trippel EA (1997) Early life history and recruitment in fish populations, 1st edn. Chapman & Hall, London, p 596

Dettlaff TA, Ginsburg AS, Schmalhausen O (1993) Sturgeon fishes. Developmental biology and Aquaculture. Springer-Verlag, Berlin

Fuiman LA (1983) Growth gradients in fish larvae. J Fish Biol 23:117–123

Gisbert E (1999) Early development and allometric growth patterns in Siberian sturgeon and their ecological significance. J Fish Biol 54:852–862

Gisbert E, Williot P (1997) Larval behaviour and effect of the timing of initial feeding on growth and survival of Siberian sturgeon larvae under small scale hatchery production. Aquaculture 156:63–76

Gisbert E, Ruban GI (2003) Ontogenetic behavior of Siberian sturgeon, *Acipenser baerii*: a synthesis between laboratory tests and field data. Env Biol Fish 67:311–319

Gisbert E, Doroshov SI (2006) Allometric growth in green sturgeon larvae. J Appl Ichthyol 22:202–207

Gisbert E, Solovyev M (2017) Behaviour of early life stages in the Siberian sturgeon. In: Williot P, Nonnotte G, Vizziano D, Chebanov M, (eds), Springer Verlag

Gisbert E, Williot P, Castelló-Orvay F (1998a) Morphological development of Siberian sturgeon (*Acipenser baeri*, Brandt) during prelarval and larval stages. Rivista Italiana di Acuacoltura 33:121–131

Gisbert E, Rodríguez A, Williot P, Castelló-Orvay F (1998b) A histological study of the development of the digestive tract of Siberian sturgeon (*Acipenser baerii*) during early ontogeny. Aquaculture 167:195–209

Gisbert E, Sarasquete MC, Williot P, Castelló-Orvay F (1999a) Histochemistry of the development of the digestive system of Siberian sturgeon (*Acipenser baerii*, Brandt) during early ontogeny. J Fish Biol 55:596–616

Gisbert E, Williot P, Castelló-Orvay F (1999b) Behavioural modifications of Siberian sturgeon (*Acipenser baeri*, Brandt) during early life stages of development: their significance and use. J Appl Ichthyol 15:237–242

Gisbert E, Williot P, Castelló-Orvay F (2000) Influence of egg size on growth and survival of early stages of Siberian sturgeon (*Acipenser baerii*) under small scale hatchery conditions. Aquaculture 183:83–94

Gisbert E, Asgari R, Rafiee G, Agh N, Eagderi S, Eshaghzadeh H, Alcaraz C (2014) Early development and allometric growth patterns of beluga *Huso huso* (Linnaeus, 1758). J Appl Ichthyol 30:1264–1272

Kasumyan AO, Kazhlayev AA (1993) Formation of searching behavioral reaction and olfactory sensitivity of food chemical signals during ontogeny of sturgeons (*Acipenseridae*). J Ichthyol 33:51–65

Koumoundouros G, Divanach P, Kentouri M (1999) Ontogeny and allometric plasticity of *Dentex dentex* in rearing conditions. Mar Biol 135:561–572

Leprévost A, Sire JY (2014) Architecture, mineralization and development of the axial skeleton in Acipenseriformes, and occurrences of axial anomalies in rearing conditions; can current knowledge in teleost fish help? J Appl Ichthyol 30:767–776

Loew ER, Sillman AJ (1993) Age-related changes in the visual pigments of white sturgeon (*Acipenser transmontanus*). Can J Zool 71:1552–1557

Olla BL, Davis MW, Ryer CH, Sogard SM (1995) Behavioural responses of larval and juvenile walleye pollock (*Theragra chalcogramma*): possible mechanisms controlling distribution and recruitment. ICES Mar Sci Symp 201:3–15

Osse JWM, van den Boogaart JGM (1995) Fish larvae, development, allometric growth, and the aquatic environment. ICES Mar Sci Symp 201:21–34

Osse JWM, van den Boogart JGM, van Snik GMJ, van der Sluys L (1997) Priorities during early growth of fish larvae. Aquaculture 155:249–258

Park C, Lee SY, Kim DS, Nam YK (2013a) Embryonic development of Siberian sturgeon *Acipenser baerii* under hatchery conditions: an image guide with embryological descriptions. Fish Aquat Sci 16:15–23

Park C, Lee SY, Kim DS, Nam YK (2013b) Effects of incubation temperature on egg development, hatching and pigment plug evacuation in farmed Siberian sturgeon *Acipenser baerii*. Fish Aquat Sci 16:25–34

Rodriguez A, Gisbert E (2001) Morphogenesis of the eye of Siberian sturgeon. J Fish Biol 59:1427–1429

Rodriguez A, Gisbert E (2002) Eye development and the role of vision during Siberian sturgeon early ontogeny. J Appl Ichthyol 18:280–285

Sarasquete C, Gisbert E, Ribeiro L, Vieira L, Dinis MT (2001) Glycoconjugates in epidermal, branchial and digestive mucous cells and gastric glands of gilthead sea bream, *Sparus aurata*, Senegal sole, *Solea senegalensis* and Siberian sturgeon, *Acipenser baerii* development. Eur J Histochem 45:267–278

Simonovic PD, Garner P, Eastwood EA, Kovac V, Copp GH (1999) Correspondence between ontogenetic shifts in morphology and habitat use in minnow *Phoxinus phoxinus*. Env Biol Fish 56:117–128

Van Eenennaam JP, Linares-Casenave J, Deng X, Doroshov SI (2005) Effect of incubation temperature on green sturgeon embryos, *Acipenser medirostris*. Env Biol Fish 72:145–154

Williot P (1997) Reproduction de l'esturgeon sibérien (*Acipenser baerii* Brandt) en élevage: gestion des génitrices, compétence à la maturation in vitro de follicules ovariens et caractéristiques plasmatiques durant l'induction de la ponte. Doctoral Thesis n° 1822, University of Bordeaux I, Bordeaux, France, p 227.

Zeiske E, Kasumyan A, Bartsch P, Hansen A (2003) Early development of the olfactory organ in sturgeons of the genus Acipenser: a comparative and electron microscopic study. Anat Embryol 206:357–372

Behaviour of Early Life Stages in the Siberian Sturgeon

9

Enric Gisbert and Mikhail Solovyev

Abstract

In this chapter, field data and laboratory studies on the behaviour and development of Siberian sturgeon at early-life intervals have been reviewed and the available information assessed under two different points of view, the biological/ecological approach and the aquaculture one. In this context, data on sturgeon behaviour have been correlated to different ecologically relevant environmental factors that may play a role in the distribution, recruitment and survival of young fish in the wild. In addition, behavioural data have been also considered under a hatchery scope, since fish behaviour may help to define an explicit criterion to assess the quality and fitness of fish, detect abnormal situations during rearing and optimise culture conditions. Four behavioural phases (swimming-up, rheotactism, schooling and, foraging and active dispersal) are observed from hatching to the juvenile phase. Different behavioural patterns are associated with an early-life interval and might allow fish to occupy different river habitats, directly influencing their distribution, survival and recruitment. In hatchery conditions, these phases may be used for evaluating larval quality and the stage of development.

Keywords

Prelarva • Larva • Early juvenile • Behavioural patterns • Swimming

---

E. Gisbert (✉)
IRTA-Sant Carles de la Ràpita, Unitat de Cultius Experimentals Crta,
Poble Nou km 5.5, 43540 Sant Carles de la Ràpita, Tarragona, Spain
e-mail: enric.gisbert@irta.cat; enric.gisbert@irta.es

M. Solovyev
Institute of Systematics and Ecology of Animals, Russian Academy of Sciences,
Frunze Str. 11, 630091 Novosibirsk, Russia

© Springer International Publishing AG, part of Springer Nature 2018
P. Williot et al. (eds.), *The Siberian Sturgeon (Acipenser baerii,* Brandt, 1869)
Volume 1 - Biology, https://doi.org/10.1007/978-3-319-61664-3_9

Introduction

Behavioural studies during the early life stages of fish have a great importance due to its several applications. Behavioural laboratory tests can provide useful tools in aquaculture in order to define an explicit criterion to assess the quality and fitness of specimens, detect abnormal situations during rearing and optimise culture conditions. However, behavioural assessments are generally considered difficult to use, because they have to be referred to general standards that need to be developed for each species of interest under controlled and well-standardised conditions. In addition, laboratory tests can also provide insight into probable behaviour of fish in their natural environments (different types of substrate, water velocity, turbidity, etc.) and in response to different environmental cues that are especially useful for determining values for parameters of models used in behavioural ecology and management plans (Noakes and Baylis 1990). Studying fish swim performance under various biotic and abiotic conditions may help in understanding their ecology (Olla et al. 1995; Plaut 2001; Kynard and Horgan 2002; Wolter and Arlinghaus 2003) and physiology (Milligan 1996; Kieffer 2000; McKenzie et al. 2007) and may also provide an explanation for specific morphological traits (i.e. shape and size of body and fins; Long 1995; Wilga and Lauder 1999; Liao and Lauder 2000; Grunbaum et al. 2007).

Field data and laboratory studies on the behaviour and development of Siberian sturgeon at early-life intervals have been reviewed and the available information assessed under two different scopes, the biological and ecological point of view and the aquaculture one. In this context, data on sturgeon behaviour have been correlated to different ecologically relevant environmental factors that may play a role in the distribution, recruitment and survival of young fish in the wild, information that may be of guidance for protection and stock enhancement programmes of early-life intervals of the Siberian sturgeon. In addition, available information regarding the behaviour of this species under aquaculture conditions may serve as a guide to hatchery managers to synchronise the state of development of specimens with the rearing process, as well as evaluate the fitness and quality of hatchery-raised fish, regardless of their final use, aquaculture rearing for commercial purposes or restocking programmes.

The information here summarised about the behaviour during prelarval, larval and early juvenile periods was mainly obtained from laboratory studies of Siberian sturgeon kept at 18 °C (Gisbert and Williot 1997; Gisbert et al. 1999). This fact is of special relevance since water temperature directly affects the rate of ontogenesis and, consequently, changes in water temperature may advance or delay the following behavioural characteristics in Siberian sturgeon. In this context, during Siberian sturgeon early life stages (prelarval, larval and early juvenile development), several periods could be clearly distinguished and characterised by different behavioural patterns, as it is summarised in Table 9.1.

Table 9.1 Summary of the major behavioural changes during the early development of Siberian sturgeon (*Acipenser baerii*) described under laboratory conditions in fish reared at 18 °C according to Gisbert and Williot (1997), Gisbert et al. (1999), and Gisbert and Ruban (2003)

	Prelarvae				Larvae	
Behavioural pattern	<1 dph	1–3 dph	4–6 dph	7–8 dph	9–10 dph	11–25 dph
Swimming-up	Yes	Yes	Yes	No	No	No
Phototaxis	Positive	Positive	Positive	No	No	No
Rheotaxis	No	No	Positive	Positive	No	No
Schooling	No	No	No	Yes	No	No
Swimming ability	No	No	No	Yes	Yes	Yes
Pelagic swimming	Yes	Yes	Yes	No	No	No
Benthic swimming	No	No	Yes	Yes	Yes	Yes
Foraging behaviour	No	No	No	No	Yes	Yes

Ages are expressed as days post hatch (dph)

9.1 Behavioural Patterns During the Prelarval Stage

The prelarval stage is the developmental period comprised between hatching and the onset of exogenous feeding (Dettlaff et al. 1993). Laboratory studies revealed that during the first 3 days after hatching, prelarvae exhibit vertical swim-up behaviour and are unable to swim for a long time in a certain direction. Prelarvae perform alternate movements: they move upwards by active movement of the posterior trunk region and tail and then passively sink to the bottom where they rest on their side or yolk sac before active swimming-up starts again. The time that newly hatched prelarvae spend swimming and resting on the bottom of the tank is quite similar, while between 1 and 3 days post hatch (dph), the time that they spend swimming in the water column increases considerably due to the reduction of the yolk sac volume and increase in trunk musculature. Deviations from the above-described swimming pattern may be used in aquaculture practices to detect specimens with severe morphological defects, i.e. underdevelopment of the cephalic region, trunk myomeres, twisted body and/or hydrocele in the abdominal region, among others, as prelarvae showing these types of developmental defects would show impaired swimming capabilities or swim in circles at the bottom of the tank without being able to reach the water surface.

During this period (1–3 dph), prelarvae are positively phototactic (strongly attracted to light) and show a preference for white substrates in comparison to greyish and blackish ones, as it was shown in laboratory studies by Gisbert et al. (1999). In some sturgeon species like shortnose sturgeon (*A. brevirostrum*) and white sturgeon (*A. transmontanus*), swimming-up behaviour has been observed when bottom concealment was not provided to prelarvae (Richmond and Kynard 1995; Loew and Sillman 1993); thus, this behavioural pattern in those species was attributed to seeking for cover in those species with negative phototactism. From an ecological perspective, the swim-up and drift behaviour of free embryos inhabiting inadequate substrate is interpreted as an evasive strategy that allows hatchlings to move short distances mostly by passive transport in moderate currents, thereby increasing survival by dispersal and improved shelter. Nevertheless, substrate availability may limit the downstream migration fidelity of the free embryo and the intensity of migration during the first days of the prelarval stage of free embryonic development (Gessner et al. 2009). However, the swimming-up behaviour observed in Siberian sturgeon prelarvae could not be attributed to the absence of bottom cover and shelter, since prelarvae exhibited such behaviour trait even though bottom concealment was provided in laboratory studies. In natural environments, the above-mentioned behavioural patterns in Siberian sturgeon, i.e. swimming-up and drift behaviour and positive phototactism, may allow newly hatched prelarvae to disperse downstream and abandon the spawning grounds (Gisbert and Ruban 2003). Laboratory results are in accordance with field captures in the upper Ob River (Votinov and Kas'yanov 1978), where prelarvae were found far away from the spawning areas. The time spent by prelarvae in the water column may affect the dispersal of prelarvae from spawning grounds, as well as the water depth and current speed; the deeper and the faster the water current is, the further the prelarvae may be dispersed from spawning grounds. Field data on the diel patterns of downstream migration at early life stages of postembryonic development in Siberian sturgeon are scarce. However, data from other sturgeon species exhibiting the same behavioural patterns during the prelarval stage may indicate that downstream migration of prelarvae with a lower water transparency continues all day long and is independent of light intensity (Khodorevskaya et al. 2009). Downstream migration velocities of the beluga, Russian sturgeon and stellate sturgeon free embryos were about 80–90% of the river flow velocity (Khodorevskaya 2002). Thus, since water velocity in the spawning grounds of the Lena River is 140 cm/s (Sokolov and Malyutin 1977), free Siberian sturgeon embryos might passively migrate downstream at about 100 km/day and possibly up to 300–440 km during the swimming-up behavioural stage (Gisbert and Ruban 2003). The velocity of migration in the Tobol–Irtysh river system has been determined to range between 5 and 70 km/day depending on river flow conditions (Gundrizer et al. 1983). The migration of prelarvae is a passive process with some regulation with regard to fish behaviour and always coincides with the direction of the river flow. Some sturgeon species tend to hide under pebbles for reducing their downstream migration, as well as regulate their vertical movements in the water column (Khodorevskaya 1977; Khodorevskaya 2002). In sterlet (*A. ruthenus*), the velocity of downstream migration decreases during development, thus regulating dispersal distance from the spawning grounds and preventing fish from reaching estuarine and marine areas (Khodorevskaya 2002). Dispersal of hatchlings and free embryos from spawning grounds may be important in maintaining a stable population of fish, since it allows them to utilise more rearing habitats. On the other hand, such early movement of newly hatched prelarvae and young prelarvae from spawning sites could be an important source of mortality for young Siberian sturgeon, as during this period specimens seem to be extremely vulnerable due to their reckless swimming in the water column and poor avoidance capacity from potential predators (Gisbert et al. 1999). Similarly to other sturgeon species, Siberian sturgeon hatchings are likely to be highly susceptible to entrainment into water-diversion structures during their passive dispersal phase, because these fish have virtually no ability to man-made water-diversion structures during this life stage (Verhille et al. 2014). In addition, the dispersal rate of sturgeon prelarvae in natural environments seems to be also related to size (and perhaps other characteristics) of bottom substrate, making modelling of prelarval dispersal difficult (Kynard et al. 2013).

According to Nasirov (1982), the effects of different light levels in sturgeon behaviour during the prelarval stage are genetically determined, whereas after the transition to active feeding, they mainly depend on environmental cues. Siberian sturgeon prelarvae are strongly photopositive and capable of distinguishing bottoms of different brightness (Gisbert et al. 1999). Such responses to light stimuli and the positive allometric growth of the eye diameter in relation to body length, coupled with the first stages of differentiation of the retina (Rodriguez and Gisbert 2002), observed from hatching to 3 dph, suggest that light signals are a significant information source in the early life stages of Siberian sturgeon. In this sense, prelarvae would be able to detect light intensity, signal when it is night and detect the contrast between a potential predator and the water, enabling prelarvae to select a suitable habitat (Kasumyan and Kazhlayev 1993; Loew and Sillman 1993). According to Dettlaff et al. (1993), the photobehaviour is established in sturgeons at the mass hatching stage, enabling prelarvae to adapt to different ecological conditions. The strong positive phototactic response to light stimuli showed by Siberian sturgeon prelarvae may be used to evaluate their fitness in hatchery conditions; the placement of a light focus in one corner of the tank would attract normal prelarvae, while those with malformations or developmental defects would not be able to reach the illuminated zone and could be easily detected and eliminated from the group. The optimal illumination levels for Siberian sturgeon rearing have been reported to be comprised between 30 and 800 lux, whereas higher light intensities are not recommended since this species inhabits in muddy waters and in the polar night condition (Ruchin 2008). In contrast, prelarvae of Russian sturgeon (*A. gueldenstaedtii*) and ship sturgeon (*Acipenser nudiventris*) generally try to avoid areas with illumination values higher than 10 lux, whereas prelarvae of beluga (*Huso huso*) prefer 2–60 lux and stellate sturgeon (*A. stellatus*) between 80 and 100 lux (Ruchin 2008).

At 4 dph, swimming-up behaviour starts to decrease and about the 50% of the prelarvae become benthic and are actively swimming along the bottom and positively rheotactic, whereas between 5 and 6 dph, swimming-up completely disappears and the most part of prelarvae are observed to swim against the water current (positive rheotactism), while at 7–8 dph, specimens aggregated into schools of different shapes as a consequence of water turbulence until the end of the endogenous feeding phase and the onset of exogenous feeding. These new behavioural patterns are a result of the major development of sensory organs (i.e. eye, lateral line, and sensory follicles on the rostrum), swimming structures (i.e. development of pectoral fins and unpaired fins, reduction of the embryonic fin fold) and a considerable reduction of yolk sac volume that allows prelarvae to vigorously swim against the water current (Gisbert 1999). In addition, during this phase of the prelarval development, specimens are still positively phototactic, even though a small proportion of fish shows preference for greyish and dark substrates (25–30%) and seeks for cover (15–20%). The development of schooling behaviour in Siberian sturgeon prelarvae may correspond with a major maturation of the visual and the ear–lateral line systems (Dettlaff et al. 1993), as it has been described in many teleost fish species. Although sturgeons are described as having not very well-developed visual reception (Kasumyan and Kazhlayev 1993), vision during this period seemed to play a more important role than tactile system in maintaining the schooling behavioural pattern, as it disappeared during night hours like other fish species (Gallego and Heath 1994). This hypothesis is also supported by the positive allometric growth of the eye diameter observed during this period (Gisbert 1999). Normal schooling behaviour at the end of prelarval stage under daylight conditions may be used as a quality criterion in Siberian sturgeon (Gisbert and Williot 1997; Gisbert et al. 1999). In natural environments, schooling is a common behaviour in fish to prevent injury or death caused by reckless swimming ability and poor avoidance from predators (Gallego and Heath 1994). This behaviour in Siberian sturgeon prelarvae may be related to the ending of their endogenous reserves contained in the yolk sac and the preparation of specimens for the larval phase and the onset of exogenous feeding, which would highly reduce their vulnerability to potential predators. These fish aggregations would allow specimens to remain concealed between rocks and in crevices protecting them from predators during daylight hours, which would also suppose a reduction of the energy wasted in swimming against the current that could be derived to finish their morphological and physiological development before the larval stage. Schooling behaviour disappears during the night, and prelarvae disperse actively, which suggested that during this period active dispersal downstream may possibly occur at night, reducing the risk of predation. Such changes in behaviour during day and night hours may determine the downstream dispersal distance, as laboratory tests seem to indicate that prelarvae may only disperse only during the night hours (Gisbert et al. 1999). Such active dispersal pattern during the ending of the prelarval stage, added to the drift behaviour observed after hatching, may be responsible for the dispersal of Siberian sturgeon over a wide area, allowing the utilisation of a major number of feeding areas and rearing habitats by larval stages and minimisation of competition for those resources (Gisbert and Ruban 2003). From an aquaculture point of view, normal schooling behaviour of prelarvae at this stage under daylight conditions could be used as a quality criterion like in reared coregonid species (Zitov and Millard 1988), as well as a swimming pattern for assessing the stage of development of prelarvae (Gisbert and Williot 1997).

The timing and length of above-mentioned behavioural patterns observed during the prelarval development (swimming-up and drifting behaviour, benthic swimming, positive rheotactism and schooling behaviour) are likely dependent on a suite of temperature-dependent processes involving prelarval development, yolk sac absorption and the transition from endogenous to exogenous feeding, as it has been reported in other sturgeon species (Braaten et al. 2012).

9.2 Behavioural Patterns at Larval and Early Juvenile Stages

The larval period is comprised between the beginning of exogenous feeding and metamorphosis when the specimen reaches the juvenile stage and resembles an adult in miniature. At the beginning of the larval phase, schooling behaviour disappears and larvae scatters along the bottom coinciding with the transition to exogenous feeding and the expelling of melanin plug from the spiral valve (Gisbert and Williot 1997). Thus, the end of the schooling behaviour may be considered as a useful visual guide for hatchery managers to determine the best moment for starting feeding Siberian sturgeon larvae.

During this period, larval activity increases and swimming ability improves compared to prelarvae, since larvae change from an anguilliform to a subcarangiform swimming mode (Gisbert 1999). At this stage, starvation is reported to be an important factor affecting growth and survival of Siberian sturgeon larvae due to the completion of endogenous reserves contained in the yolk sac (Gisbert and Williot 1997); hence, an improved swimming performance might allow larvae not only to disperse actively over a wide rearing area, thus improving their foraging capacities and chances to avoid predators (Gisbert and Ruban 2003).

Contrary to the closely related Russian sturgeon (*A. gueldenstaedtii*), the ontogenetic behaviour of the Siberian sturgeon is characterised by the almost complete absence of pelagic feeding in the water column and a more rapid morphological development (Malyutin 1980 quoted by Gisbert and Ruban 2003). These facts are associated with the shorter length of Siberian summers, which limits planktonic food resource availability, thus enabling first feeding Siberian sturgeon larvae to exploit as many food resources as possible. Although no specific laboratory tests have been conducted in order to evaluate the preference of Siberian sturgeon larvae to different types of substrates, considering the preference of larvae for light-coloured and open substrates (Gisbert et al. 1999) and substrates with fine granulometry (Jatteau pers. commun. quoted in Gisbert and Ruban 2003), the ideal Siberian sturgeon larval nursery habitats most likely would be sandy and fine gravel bottoms. Similar observations have been reported for other sturgeon species, such as the Chinese sturgeon *A. sinensis* (Zhuang et al. 2002), the lake sturgeon *A. fulvescens* (Peake 1999), the white sturgeon *A. transmontanus* (Parsley and Beckman 1994; Kynard et al. 2013), the stellate sturgeon *A. stellatus* (Sbikin and Khomenkov 1980) and the Russian sturgeon (Sbikin and Bibikov 1988). Although sturgeon larvae seem to prefer sandy bottoms for foraging, behavioural laboratory tests using artificial stream tanks (see design in Kynard et al. 2013) revealed that larvae foraging on sandy bottoms may be stationary in areas of water velocity refuges downstream or at the edge of rocks of all sizes, but particularly beside large rubbles. This behaviour may conserve energy and enable fish to forage on drifting food (Kynard et al. 2013).

At the juvenile stage, the Siberian sturgeon disperse into different sections of the tributary system of the river (side channels, inlets, oxbow lakes) (Votinov and Kas'yanov 1978) and seek for covered bottom habitats (Gisbert et al. 1999). This behaviour differs from that of Ponto–Caspian anadromous sturgeon species (giant sturgeon, Russian sturgeon and stellate sturgeon), whose juveniles continue to migrate downstream and forage in the riverbeds (Khodorevskaya 2002). In the Indigirka and Kolyma Rivers, Siberian sturgeon juveniles forage in sandy and pebble grounds in river channels and silty grounds in the river arms and bays (Ruban and Konoplya 1994). In nearly all instances, this is an opportunistic, nonselective benthivorous species, whose diet composition reflects the zoobenthic diversity of the foraging grounds (Ruban and Konoplya 1994).

Olfaction is the main long-distance sensory system in sturgeon and might be crucial for their homing performances. The ability to smell is part of the male–female communication system during the spawning period. Ripe females release postovulatory sexual pheromones that evoke behavioural responses in males (Kasumyan and Kazhlayev 1993). Natural food odour emanated by food organisms and watery solution of some pure chemical substances evoke specific food search behaviour in sturgeon species that attracts them to the zone where the highest concentration of food chemical stimuli is present (Kasumyan 2002). Their behavioural responses are fast, enabling these virtually blind fish to inspect quickly and carefully an area with the highest concentration of food chemical stimuli (Kasumyan and Kazhlayev 1993; Kasumyan 1999). Laboratory studies have shown that sturgeons are very sensitive to free amino acids and raised the idea that chemical stimulus might be the primary reliable sensory mechanism in the food seeking behaviour of this group of species (Kasumyan 2002; Jafari-Shamushaki et al. 2008; Zhuang et al. 2008). In this sense, olfactory sensitivity to food odours appears immediately after the completion of the transition to the exogenous feeding [11–12 dph at 18 °C (Gisbert et al. 1998) and 12–14 dph at 14–15 °C]. At this time, the larva has many receptor cells of microvillar and flagellar types in the olfactory epithelium comparable to the ratio observed in the adult. In addition, at this time, connection of the axon of the receptor cells with mitral cells of the differentiated olfactory bulb takes place, which indicates the formation of a functional olfactory sensory system as Kasumyan and Kazhlayev (1993) validated in laboratory studies exposing stellate, Russian and Siberian sturgeon larvae of different ages to different prey extracts. According to the former authors, in spite of the appearance of reaction to food extracts during the period of transition to exogenous feeding (mixed nutrition phase based on endogenous and exogenous nutrients), it is not olfactory sensitivity, but tactile and taste (extraoral) sensitivity and electroreception that are of importance for searching for food and foraging behaviour, as the ability to recognise different taste chemical signals develops during the first days after changing to active feeding.

Olfactory sensitivity to food extracts in different sturgeon species has been evaluated by exposing them to different sources (i.e. live preys, amino acids) and levels of food extracts and evaluating fish response in terms of swimming and feeding behaviours (Kasumyan 2002). According to Kasumyan and Kazhlayev (1993), the behavioural reaction to food chemical stimuli develops during the third and fourth weeks of life, and it is characterised by the following stereotypical reaction: during the first two days of exogenous feeding, most larvae swim along the bottom, whereas few specimens were reported to swim vertically and then after a short period of swimming descend to the bottom. During this period, supplying zooplankton extracts at 10^{-4} g/L did not change the above-mentioned behaviour, indicating that larvae were not able to detect preys by the odour. In contrast, after the complete transition to exogenous feeding, larvae show typical elements of searching behaviour when they are exposed to the same olfactory stimulus, consisting on repeated swimming movements in the zone with the maximum concentration of the chemical attractant based on circular and S-shaped swimming trajectories and scouring movements that consist of the displacement of the fish head to both sides from the main trajectory. During the searching reaction, inspection of the bottom is provided by continuous tight contact with the substrate using the tip of the barbels, which have external taste buds, and also by frequent and repeated seizing mouth movements (bites). When sturgeon does not receive conformation about the presence of food objects by non-olfactory sensory systems, they rapidly leave the zone of search and resume their background behaviour. The level of olfactory sensitivity to food extracts in Siberian sturgeon juveniles is 0.1 mg/L; this low sensitivity to food odours is compensated by the high speed of swimming maintained over long periods, allowing fish to actively patrol large areas and discover olfaction zones with high concentration of food organisms (Kasumyan and Kazhlayev 1993). The above-described foraging behaviour may be used in sturgeon hatcheries to evaluate the response of larvae to different diets and test their preference in terms of their organoleptic and odour properties. In addition to taste and odour, the large number of electrosensory organs indicates the importance of electrosense in the life of the sturgeons. The common structural features of all genera of sturgeon include four barbels and about 5000 electroreceptors distributed on the ventrally flattened rostrum in fish of 20–30 cm total length, TL (Zhang et al. 2012). The development of the ampullary system in Siberian sturgeon from the Lena River population has been described to occur during the first 3–4 days after hatching, whereas at the beginning of the larval stage, ca. 86% of ampoules are already formed and located on the ventral part of the head. The distribution of ampoules have species-specific mode. In natural environments, the stronger bioelectric intensities would represent larger preys or schools of prey, and in this sense, electrosensitivity helps fish to find more prey in turbid environments full of mixed cues. Different studies have shown that aquatic animals are dipolar, a negative bioelectric field at the level of the head and a positive one at the tail (Peters et al. 2002), a trait that may help sturgeons to distinguish or to chase individuals from schools of prey in combination with the mechanosensory lateral line system that may also be used in orientation of objects (Zhang et al. 2012).

Regarding the swimming performance of Siberian sturgeon juveniles, the body and tail shape of sturgeons differs from that of the typical, faster-swimming actinopterygians and more closely resembles that of chondrichthyans. Differing from the more derived actinopterygian homocercal caudal fin, the sturgeon caudal fin is heterocercal, which, for lake sturgeon, has been shown to generate 66% less thrust than for the homocercal caudal fin of trout (Webb 1986). Further reductions in acipenserid swimming efficiency arise from increased drag due to the sturgeon spindle-shaped body form, the presence of bony scutes and roughened skin (Webb 1986), in comparison to the smooth surfaces and streamlined fusiform body shape of more derived fishes. This inefficiency in body design results in reduced burst, prolonged and sustained swimming capacities compared with those of salmonids (Peake et al. 1997; Deslauriers and Kieffer 2012). The behavioural tendency of sturgeons to station hold at high water velocities (Webb 1986; Parsons et al. 2003; Hoover et al. 2011; Kieffer et al. 2009; Deslauriers and Kieffer 2012) is of special relevance for controlling their downstream migration of this potamodromous species and preventing juveniles to reach estuarine or marine habitats before achieving their osmoregulatory capacities (Khodorevskaya et al. 2009). As Gisbert and Ruban (2003) reviewed, juvenile and adult Siberian sturgeon migratory patterns depend on the ecological characteristics of each river basin. Because of the lack of anadromous forms, in all large Siberian rivers, this species is represented by population continuums, in which in many cases, the foraging range also includes the spawning area (Ruban 1997). The term 'population continuum', introduced by Mayr (1970), was first applied to the Siberian sturgeon by Ruban (1998a), since this sturgeon species presents a clinal variation in morphometric and meristic characters and different spawning times and sites within the same river (northern and southern reaches) that reveals that distinct populations of this species occur within the same river without significant mixture between them. This condition differentiates the Siberian sturgeon from anadromous sturgeon species. In the Ob–Irtysh basin, juveniles migrate downstream from the middle and lower parts of those rivers to the Ob Bay due to oxygen depletion in winter (November–January), while in spring, after the ice melts, they migrate upstream. The Ob Bay and the delta of the river is the main foraging area for almost all fishes inhabiting the Ob River including Siberian sturgeon. Siberian sturgeons that inhabit the head waters of the Ob and Irtysh Rivers do not perform extensive migrations (Votinov 1963). The range size of migration fry in Ob River is comprised between 5 and 52 cm in TL, among them 90% of Siberian sturgeons are 1-year-old specimens, between 5 and 10% are fish aged 2 years, and less than 5% are older fish. Some reports have indicated that ca. 93% of migrating fish belonged to the 2+ year class. Active migration of Siberian sturgeon fry is absent in the part of the middle Ob River (Popkov 2014). The separation of foraging and spawning habitats is also characteristic of the Baikal Lake population of the Siberian sturgeon. In Lake Baikal, conditions for Siberian sturgeon reproduction are not appropriate; thus, fish inhabiting this area must undertake extended potamodromous migrations in the Selenga, Bargusin and Verkhnyaya Angara rivers to reproduce. However, these rivers are very poor in food resources and cannot support stable populations of sturgeon (Ruban 1997, 1998b).

9.3 Concluding Remarks

In natural environments, ontogenetic changes in behaviour might allow each early-life interval (prelarva, larva and juvenile) to disperse and occupy different river habitats, thus directly influencing their distribution, survival and recruitment. River current intensity, substrate typology, food resources and predation pressure seem to be the most important factors affecting the distribution of Siberian sturgeon prelarvae and larvae, while juvenile and adult fish disperse and migrate according to food abundance and reproduction. Migratory movements of juvenile and adult Siberian sturgeons also depend on the river basin that they inhabit. Most parts of the rivers inhabited by Siberian sturgeon, with the exception of Lake Baikal (Yenisey River basin), show a population continuum that shares both spawning and foraging grounds. Ontogenetic changes in Siberian sturgeon behaviour are interpreted as a species-specific mechanism to maintain the population continuums described in this species without significant mixture of local populations within the river. This information can be used to give guidance for protection and stock enhancement programmes of early-life intervals of the Siberian sturgeon. In addition, by integrating laboratory-based findings on the ontogeny of Siberian sturgeon with their swimming capacities and behavioural information, season-specific recommendations for water flow limitations in different river basins around water-diversion structures can be obtained for managers seeking to balance water demands with restoration and conservation of sturgeons (Verhille et al. 2014). From an aquaculture point of view, different behavioural patterns (i.e. swimming-up and drifting behaviour, benthic swimming, positive phototactism and rheotactism, schooling behaviour and active foraging) of Siberian sturgeon prelarvae and larvae along their early ontogeny might be useful visual guides to evaluate their fitness and quality, as well as indicate changes in their development that may serve hatchery managers to improve and optimise the rearing process by means of synchronising the state of development of developing fish with the rearing protocol.

The revision of available literature for preparing this chapter revealed that there exist importance gaps of knowledge regarding the behaviour of Siberian sturgeon during the prelarval and larval developmental periods in natural environments in order to better characterise their preferences for different types of substrates, foraging areas and distribution patterns depending on water currents and environmental parameters. This information may be obtained from Siberian rivers where this species inhabits, but it can be gathered by means of laboratory test using experimental structures for studying behaviour at prelarval, larval and early juvenile stages (SCOLA device consisting of an annular flume of 8 m long, 0.3 m wide, 0.6 m deep, 2 m^3 of volume) like those available at the IRSTEA in the research station of the Saint Seurin sur l'Isle in France (Trancart et al. 2014) or the artificial stream tanks (Kynard et al. 2013). Similarly to early life stages of development, available information regarding Siberian sturgeon behaviour at older ages is also scarce. Monitoring larger juveniles is typically achieved through conventional methods that involve direct capture and handling of individuals (e.g. mark recapture, telemetry). However, these methods can be logistically challenging for small populations residing in large systems, especially when data pertaining to distributional trends are lacking. The available technology for monitoring fishes in non-invasive ways has recently advanced, both at the individual and population levels. Examples include underwater videography, split beam sonar, hydroacoustics and dual-frequency *id*entification *son*ar (DIDSON) (Crossman et al. 2011).

References

Braaten PJ, Fuller DB, Lott RD, Ruggles MP, Brandt TF, Legare RG, Holm RJ (2012) An experimental test and models of drift and dispersal processes of pallid sturgeon (*Scaphirhynchus albus*) free embryos in the Missouri River. Env Biol Fish 93:377–392

Crossman JA, Martel G, Johnson PN, Bray K (2011) The use of dual-frequency IDentification SONar (DIDSON) to document white sturgeon activity in the Columbia River, Canada. J Appl Ichthyol 27:53–57

Deslauriers D, Kieffer JD (2012) The effects of temperature on swimming performance of juvenile shortnose sturgeon (*Acipenser brevirostrum*). J Appl Ichthyol 28:176–181

Dettlaff TA, Ginsburg AS, Schmalhausen O (1993) Sturgeon fishes. Developmental biology and aquaculture. Springer-Verlag, Berlin

Gallego A, Heath MR (1994) The development of schooling behaviour in Atlantic herring *Clupea harengus*. J Fish Biol 45:569–588

Gessner J, Kamerichs CM, Kloas W, Wuertz S (2009) Behavioural and physiological responses in early life phases of Atlantic sturgeon (*Acipenser oxyrinchus* Mitchill 1815) towards different substrates. J Appl Ichthyol 25:83–90

Gisbert E (1999) Early development and allometric growth patterns in Siberian sturgeon and their ecological significance. J Fish Biol 54:852–862

Gisbert E, Williot P (1997) Larval behaviour and effect of the timing of initial feeding on growth and survival of Siberian sturgeon larvae under small scale hatchery production. Aquaculture 156:63–76

Gisbert E, Ruban GI (2003) Ontogenetic behavior of Siberian sturgeon, *Acipenser baerii*: a synthesis between laboratory tests and field data. Env Biol Fishes 67:311–319

Gisbert E, Rodríguez A, Williot P, Castelló-Orvay F (1998) A histological study of the development of the digestive tract of Siberian sturgeon (*Acipenser baeri*) during early ontogeny. Aquaculture 167:195–209

Gisbert E, Williot P, Castelló-Orvay F (1999) Behavioural modifications of Siberian sturgeon (*Acipenser baeri*, Brandt) during early life stages of development: their significance and use. J Appl Ichthyol 15:237–242

Grunbaum T, Cloutier R, Mabee PM, Le Francois NR (2007) Early developmental plasticity and integrative responses in arctic charr (*Salvelinus alpinus*): effects of water velocity on body size and shape. J Exp Zool B Mol Dev Evol 308:396–408

Gundrizer AN, Egorov AG, Afanasieva VG (1983) Perspectives of reproduction of Acipenseridae in Siberia. In: Altukhov JP (ed) Biological basis of sturgeon culture. Nauka, Moscow, pp 241–258

Hoover JJ, Collins J, Boysen KA, Katzenmeyer AW, Killgore KJ (2011) Critical swimming speeds of adult shovelnose sturgeon in rectilinear and boundary-layer flow. J Appl Ichthyol 27:226–230

Jafari-Shamushaki VA, Abtahi B, Kasumyan AO, Abedian Kenari A, Ghorbani R (2008) Taste attractiveness of free amino acids for juveniles of Persian sturgeon *Acipenser persicus*. J Ichthyol 48:124–133

Kasumyan AO, Kazhlayev AA (1993) Formation of searching behavioral reaction and olfactory sensitivity of food chemical signals during ontogeny of sturgeons (Acipenseridae). J Ichthyol 33:51–65

Kasumyan AO (1999) Olfaction and taste senses in sturgeon behavior. J Appl Ichthyol 15:228–232

Kasumyan AO (2002) Sturgeon food searching behaviour evoked by chemical stimuli: a reliable sensory mechanism. J Appl Ichthyol 18:685–690

Khodorevskaya RP (1977) Osobennosti povedeniya molodi osetrovyih v svyazi s ih pokatnoy migratsiey. Moskva. Author's abstract of dissertation. [In Russian] Institut Evolucionnoi morfologii i ecologii jivotnih im A.N. Severcova. p 20

Khodorevskaya RP (2002) Behavior, distribution and migrations of sturgeon fishes inVolgo-Caspian basin. Doctors dissertation, Severtsov Institute of Ecology and Evolution, Russian Academy of Sciences, Moscow. A. N., p 446 (in Russian)

Khodorevskaya RP, Ruban GI, Pavlov DS (2009) Behaviour, migrations, distribution and stocks of sturgeons in the Volga-Caspian basin. World Sturgeon Conservation Society: Special Publication 3, Nordesrstedt, Germany, p 244

Kieffer JD (2000) Limits to exhaustive exercise in fish. Comp Biochem Physiol A Mol Integr Physiol 126:161–179

Kynard B, Horgan M (2002) Ontogenetic behavior and migration of Atlantic sturgeon, *Acipenser oxyrhinchus oxyrhinchus*, and shortnose sturgeon, *A. brevirostrum*, with notes on social behavior. Env Biol Fish 63:137–150

Kieffer JD, Arsenault LM, Litvak MK (2009) Behaviour and performance of juvenile shortnose sturgeon *Acipenser brevirostrum* at different water velocities. J Fish Biol 74:674–682

Kynard B, Parker E, Kynard B, Horgan M (2013) Behavioural response of Kootenai white sturgeon (*Acipenser transmontanus*, Richardson, 1836) early life stages to gravel, pebble, and rubble substrates: guidelines for rearing substrate size. J Appl Ichthyol 29:951–957

Liao J, Lauder GV (2000) Function of the heterocercal tail in white sturgeon: flow visualization during steady swimming and vertical maneuvering. J Exp Biol 203:3585–3594

Loew E, Sillman AJ (1993) Age-related changes in the visual pigments of the white sturgeon (*Acipenser transmontanus*). Can J Zool 71:1552–1557

Long JH Jr (1995) Morphology, mechanics, and locomotion: the relation between the notochord and swimming motions in sturgeon. Env Biol Fishes 44:199–211

Mayr E (1970) Populations, species, and evolution. The Belknap Press of Harvard University Press Cambridge, Massachusets, p 460

McKenzie DJ, Hale ME, Domenici P (2007) Locomotion in primitive fishes. Fish Physiol 26:319–380

Milligan CL (1996) Metabolic recovery from exhaustive exercise in rainbow trout. Comp Biochem Physiol A 113:51–60

Noakes DLG, Baylis JR (1990) Behaviour. In: Schreck CB, Moyle PB (eds) Methods in fish biology. American Fisheries Society, Bethesda, Maryland, pp 555–584

Nasirov NV (1982) Sravnitelnaya harakteritika povedencheskih, morfofiziologicheskih i biohimicheskih parametrov molodi osetrovyih ryib v rannem ontogeneze. Author's abstract of dissertation. [In Russian] Baku. Institut fiziologii im. A.I. Karaeva. p 18

Olla BL, Davis MW, Ryer CH, Sogard SM (1995) Behavioural responses of larval and juvenile walleye pollock (*Theragra chalcogramma*): possible mechanisms controlling distribution and recruitment. ICES Mar Sci Symp 201:3–15

Parsley MJ, Beckman LG (1994) White sturgeon spawning and rearing habitat in the lower Columbia River. N Am J Fish Manage 14:8412–8827

Parsons GR, Hoover JJ, Killgore KJ (2003) Effect of pectoral fin ray removal on station-holding ability of shovelnose sturgeon. N Am J Fish Manage 23:742–747

Peake S (1999) Substrate preferences of juvenile hatchery reared lake sturgeon, *Acipenser fulvescens*. Env Biol Fish 56:367–374

Peake S, Beamish FWH, McKinley RS, Scruton DA, Katopodis C (1997) Relating swimming performance of lake sturgeon, *Acipenser fulvescens*, to fishway design. Can J Fish Aquat Sci 54:1361–1366

Peters RC, van Wessel T, van den Wollenberg BJ, Bretschneider F, Olijslagers AE (2002) The bioelectric field of the catfish *Ictalurus nebulosus*. J Physiol 96:397–404

Popkov VK (2014) Population structure and condition of Siberian sturgeon stock: *Acipenser baerii* Brandt in the River Basin of the Ob. Int J Environ Stud 71:707–715

Plaut I (2001) Critical swimming speed: its ecological relevance. Comp Biochem Physiol A 131:41–50

Richmond AM, Kynard B (1995) Ontogenetic behaviour of shortnose sturgeon, *Acipenser brevirostium*. Copeia 1995:172–182

Rodriguez A, Gisbert E (2002) Eye development and the role of vision during Siberian sturgeon early ontogeny. J Appl Ichthyol 18:280–285

Ruban GI (1997) Species structure, contemporary distribution and status of the Siberian sturgeon *Acipenser baerii*. Env Biol Fish 48:221–230

Ruban GI (1998a) On the species structure of the Siberian sturgeon *Acipenser baerii* Brandt (Acipenseridae). J Ichthyol 38:345–365

Ruban GI (1998b) An analysis of adaptive features of the Siberian sturgeon *Acipenser baerii* Brandt. Russ J Aquatic Ecol 7:59–65

Ruban GI, Konoplya LA (1994) Diet of the Siberian sturgeon, *Acipenser baeri*, in the Indirgka and Kolyma rivers. J Ichthyol 34:154–158

Ruchin AB (2008) The effects of permanent and variable illumination on the growth, physiological and hematological parameters of the Siberian sturgeon (*Acipenser baerii*) juveniles. Zoologichesky Zhurnal 87:964–972

Sbikin YN, Bibikov NI (1988) The reaction of juvenile sturgeons to elements of bottom topography. J Ichthyol 28:155–160

Sbikin YN, Khomenkov A (1980) Effect of the ground character and current on the behavior of Acipenserid fry in the experiment. Zool Zhurn 59:1661–1670. (in Russian)

Sokolov LI, Malyutin VS (1977) Features of the population structure characteristics of spawners of the Siberian sturgeon, *Acipenser baeri*, in the spawning grounds of the Lena River. J Ichthyol 17:210–218

Trancart T, Lambert P, Daverat F, Rochard E (2014) From selective tidal transport to countercurrent swimming during watershed colonisation: an impossible step for young-of-the-year catadromous fish? Knowl Managt Aquatic Ecosyst 412:04

Verhille CE, Poletto JB, Cocherell DE, DeCourten B, Baird S, Cech JJ Jr, Fangue NA (2014) Larval green and white sturgeon swimming performance in relation to water-diversion flows. Conserv Physiol 2:1–4

Votinov NP (1963) The biological basis of the artificial breeding of the Ob River sturgeon. Trudy Ob Tazovskogo otdeleniya Gosudarstvennogo nauchno-issledovatelskogo institute ozernogo i rechnogo rybnogo khozyaistva III, p 5–102 (in Russian)

Votinov NP, Kas'yanov VP (1978) Ecology and reproductive efficiency of the Siberian sturgeon *Acipenser baeri* Brandt in the Ob River as affected by hydraulic structure. Vopr Ikhtiol 18:25–35. (in Russian)

Webb PW (1986) Kinematics of lake sturgeon, *Acipenser fulvescens*, at cruising speeds. Can J Zool 64:2137–2141

Wilga CD, Lauder GV (1999) Locomotion in sturgeon: function of the pectoral fins. J Exp Biol 202:2413–2432

Wolter C, Arlinghaus R (2003) Navigation impacts on freshwater fish assemblages: the ecological relevance of swimming performance. Rev Fish Biol Fish 13:63–89

Zitov RE, Millard JL (1988) Survival and growth of lake withefish (*Coregonus clupeaformis*) larvae fed only formulated dry diets. Aquaculture 69, 105–113.

Zhuang P, Kynard B, Zhang L, Zhang T, Cao W (2002) Ontogenetic behavior and migration of Chinese sturgeon, *Acipenser sinensis*. Env Biol Fish 65:83–97

Zhuang P, Zhang LZ, Luo G, Zhang T, Feng GP, Liu J (2008) Function of sense organs to the feeding behavior juveniles Chinese sturgeon captured from the Yangtze estuary. Acta Hydrobiol Sin 32:475–481

Zhang X, Song J, Fan C, Guo H, Wang X, Bleckman H (2012) Use of electrosense in the feeding behavior of sturgeons. Integr Zool 7:74–82

Olfaction and Gustation in *Acipenseridae*, with Special References to the Siberian Sturgeon

10

Alexander Kasumyan

Abstract

In sturgeons our knowledge about chemosensory systems is restricted by olfaction and gustation which are well developed in these fish. Data about the common chemical sense in Acipenseriformes is totally absent up to now. Basic morphology of olfactory organ is nearly identical among genera and species. It is paired structure and lies on the dorsal side of rostrum. Cuplike olfactory rosette comprises radially arranged primary lamellae (up to several dozens). Primary lamellae have secondary lamellae which vary in number. Olfactory sensory neurons are presented by ciliated, microvillous and crypt receptor cells. Olfaction is the main distant sensory system in sturgeons and plays a significant role in feeding, reproduction and, as supposed, homing. Ability to respond to food odour appears just after the beginning of exogenous feeding. Gustatory system is presented by two distinct subsystems, oral and extraoral. Taste buds are situated not only within the oral cavity, pharynx and gills but also on the lips and barbels and are absent in the rostrum and fins and over the entire trunk surface. Morphology of oral and extraoral taste buds is similar. Organic and inorganic chemicals and food extracts are highly effective taste stimuli. Oral gustatory system has a narrower spectrum of effective substances than extraoral one and has a higher specificity. Taste buds develop later than the olfactory organ. Extraoral taste buds appear first, then come oral taste buds and then larvae start feeding. The significance of basic knowledge in sturgeon olfaction and gustation for aquaculture is emphasised.

Keywords

Olfaction · Gustation · Chemoreception · Taste buds · Feeding behaviour · Feeding stimulants · Sex pheromones · Taste preferences

A. Kasumyan
Department of Ichthyology, Faculty of Biology, Moscow State University,
Leninskie Gory, Moscow 119992, Russia
e-mail: alex_kasumyan@mail.ru

© Springer International Publishing AG, part of Springer Nature 2018
P. Williot et al. (eds.), *The Siberian Sturgeon* (*Acipenser baerii*, Brandt, 1869)
Volume 1 - Biology, https://doi.org/10.1007/978-3-319-61664-3_10

Introduction

The aquatic environment is rich in chemical signals which provide fish with various information. There are three chemosensory systems, olfaction, taste and common chemical sense, which evolved in fish as well as in all other vertebrates for reception of signal substances. Olfactory system can detect substances in nano- or picomolar concentrations and is one of the most distant sensory system in fish. The sense of smell elicits behaviours related to essential life processes like reproduction, feeding, defence and migration (Døving 1986). In contrast to olfaction, the gustatory system serves a role in feeding behaviour only. Based on information from this contact sensory system, fish snap and swallow appropriate food items or neglect and reject unpalatable ones. Thus, gustation takes part in the ultimate sensory evaluation in the feeding process and serves the consummatory phase of feeding behaviour (Atema 1980; Pavlov and Kasumyan 1998). The common chemical sense is presented in fish by free nerve endings of the cranial and spinal nerves and by solitary chemosensory cells usually distributed over the entire body surface. The function of this system in fish remains mostly mysterious.

Sturgeons have well developed both olfaction and gustation (Kasumyan 1999; Billard and Lecointre 2000), but data about the common chemical sense in these fishes is totally absent up to now. In this chapter, structure and function of both olfactory and gustatory systems in Acipenseridae are briefly reviewed mainly by the example of the Siberian sturgeon. The recent data about morphology of olfactory organ and taste buds as well as their cell composition, innervation and development in sturgeon ontogeny is provided. Special emphasis to the functional characteristics of olfactory and gustatory systems and their role in sturgeon behaviour related to feeding, spawning and migration is attended. The evidences of interrelationship between olfactory and gustatory systems and the ability of gustation which specifically compensate the loss of olfaction are provided. Usefulness of knowledge about olfactory and taste preferences for sturgeon aquaculture is shortly discussed.

10.1 Olfaction

10.1.1 Olfactory Organ

10.1.1.1 Gross Morphology

Although many investigators have studied the morphology, development and function of olfactory organs in teleosts (see, for review, Yamamoto 1982; Zeiske et al. 1992; Kasumyan 2004), studies of olfactory organs in Acipenseriformes and other primitive actinopterygians are few. The position and structure of the olfactory organ vary among actinopterygians. But the basic morphology of the olfactory organ of Acipenseriformes is nearly identical among genera and species except for the minor differences in some details (Chen and Arratia 1994).

The gross morphology of the adult olfactory organ of sturgeons was described as early as 1887 (Dogiel 1887) followed by several studies on various aspects of the

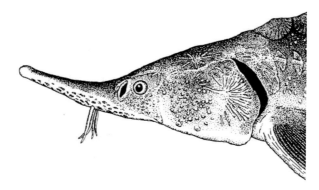

Fig. 10.1 The head of Siberian sturgeon, *A. baerii*, with the view on the excurrent nasal opening (by the courtesy of Paul Vescei)

peripheral parts of olfactory system (Pyatkina 1975, 1976; Chen and Arratia 1994) and central (Rustamov 1987; Obukhov 1999; Palatnikov 1989; Hofmann and Bleckmann 1997; Huesa et al. 2000). The olfactory organ in sturgeons is paired structure and lies on the dorsal side of the cartilaginous rostrum, relatively far away from its anterior end and just in front and laterally of the eyes (Fig. 10.1). The pre-olfactory organ length is less than the 50% of the head length (Chen and Arratia 1994). Each olfactory organ consists of an olfactory cavity, which connects with the exterior through two openings, the anterior (incurrent) and the posterior (excurrent) nostrils, and separated by a narrow nasal bridge. Numerous flat, stellate bones that articulate with each other covered laterally the olfactory organ. These bones vary in shape and size; the posterodorsal corner of the olfactory organ is covered by the dermosphenotic (Chen and Arratia 1994).

The anterior nostril is smaller and more dorsal than the posterior one. Both nostrils are elongate, adjacent and evident in dorsal view. The nasal bridge has relatively small flap rather ridge that helps water to pass into the olfactory cavity through the anterior nostril. The nasal bridge is usually a soft structure, e.g. mainly skin, separating the nostrils in many actinopterygians. In contrast, the ossicles bearing the anterior portion of the supraorbital sensory canal are positioned in the nasal bridge (Chen and Arratia 1994). Protheolytic bacteria such as *Spherotilus natans* (Chlamydobacteriacea) were shown as capable of completely destroying the nasal bridge in sturgeon juveniles (Schmalhausen 1962b; Goryunova et al. 2000).

A nasal diverticulum and accessory nasal sacs are lacking in the olfactory organ of Acipenseriformes (Chen and Arratia 1994).

10.1.1.2 Olfactory Rosette

On the floor of the olfactory cavity, the cuplike olfactory rosette is located. In sturgeons, rosette comprises radially arranged primary folds or lamellae surround an almost round, slightly eccentric axis. The axis is closer to the posterior margin of the olfactory rosette than the anterior one (Fig. 10.2). The highest numbers of primary lamellae are found in *Acipenser brevirostrum*, 29 (TL 46.9 cm);

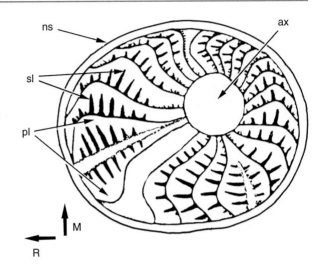

Fig. 10.2 Pattern of olfactory rosette in Acipenseridae: *pl*, primary lamellas; *sl*, secondary lamellas; *ax*, axes; *ns*, nasal suck; R, M, rostral and medial directions, respectively (modified after Chen and Arratia 1994)

Scaphirhynchus albus, 29 (TL 60.9 cm); and *A. sinensis*, 32 (TL 42.5 cm) (Chen and Arratia 1994). The Siberian sturgeon has 18 primary lamellas at the age 38 days after hatching (TL 5.1 cm) (Devitsina and Kazhlayev 1993). The shape of the primary lamellae is similar in all acipenserids studied. There are numerous secondary lamellae also which vary in number among primary lamellae, but their orientation is constant and almost perpendicular to the bases of the primary lamellae (Chen and Arratia 1994). Tertiary lamellae are found in *Acipenser oxyrhynchus* only (Chen and Arratia 1994).

10.1.1.3 Olfactory Epithelium

The largest part of each secondary lamella is covered by a pseudostratified sensory epithelium of around 50–60 μm thickness. Sensory epithelium contains olfactory receptor cells and several other types of cell as well: supporting cells, ciliated nonsensory cells, goblet cells and basal cells (Fig. 10.3). At the apex of lamellae, the epithelium is nonsensory (Pyatkina 1975 1976; Devitsina and Kazhlayev 1993; Zeiske et al. 2003).

Olfactory receptor cells that detect the chemical substances are bipolar primary sensory neurons and are presented by three distinct cell types: ciliated, microvillous and crypt receptor cells. They withdraw deep into the epithelium to form axons that cross the basal lamina and joint in olfactory nerve (*nervus olfactorius*, I). The axons of olfactory nerve carry external information directly to the brain and converge to a specific region of the olfactory bulb where they connect to a set of secondary or relay neurones. The axons of these relay neurones leave in several bundles of olfactory tract to the telencephalon (Døving 1986).

The dendrites of ciliated receptor cells and microvillous receptor cells end by a superficial apical ending, the olfactory knob, bearing up to 30 cilia or several dozen microvilli (Devitsina and Kazhlayev 1993). Olfactory knobs of ciliated receptor cells and microvillous receptor cells have diameter around 2.0–2.5 μm; diameter and length of cilia and microvilli of this cells are 0.33 and 0.14 μm and

Fig. 10.3 Scanning electron microscopic survey of the olfactory epithelium in Siberian sturgeon *A. baerii* juveniles, 5 cm TL: (**a**) Secondary lamella; SC, supporting cell; SNSC, ciliated nonsensory cells, cilia; *asterisks*, olfactory knobs. *Arrow* indicates the opening of goblet cell; scale bar = 5 µm. (**b**) Olfactory knob of the olfactory receptor cell with cilia (*arrows*) and microvilli (*arrowheads*); scale bar = 2 µm (by the courtesy of Galina Devitsina)

6.0 and 0.4 µm, respectively. In Siberian sturgeon, almost all ciliated olfactory sensory neurons bear both cilia and microvilli or at least microvillous-like protrusions on their dendritic endings (Fig. 10.3) (Devitsina and Kazhlayev 1993; Zeiske et al. 2003). The egg-shaped crypt cells only recently discovered in the olfactory epithelium of *Acipenser* species have no olfactory knob, but bear special microvilli bordering a deep invagination into which several short cilia protrude (Hansen and Finger 2000; Hansen and Zielinski 2005; Camacho et al. 2010).

Cellular composition of the olfactory epithelium of acipenserids has not yet been adequately studied, but it can be supposed that all three morphological types of sensory neurones spread in a random arrangement. The density of olfactory sensory neurons population in the Siberian sturgeon olfactory epithelium is estimated as 40.000 per mm^2 (Devitsina and Kazhlayev 1993). In an average teleost, there are approximately 5–10 million olfactory sensory neurons in each olfactory organ (Yamamoto 1982). These cells, unlike other neurons of the vertebrate central nervous system, are replaced every few weeks (Hara 1994a).

Olfactory sensory neurons are surrounded by supporting cells: the nonsensory ciliated cells and mucus (goblet) cells located usually out of the sensory epithelium. In sturgeons, nonsensory ciliated cells have the broad, flat apex and bear up to 50 cilia that are typical kinocilia. Nonsensory ciliated cells move water across the epithelium. Goblet cells produce mucus which can help to remove particles entering the olfactory cavity or aid the water transport (Døving et al. 1977). It has been hypothesised that mucus also plays an important role in chemoreception and that molecules of odorants are first introduced into the mucus and only then reach the receptor areas in the cellular membrane of cilia and microvilli of sensory neurons (Sorensen and Caprio 1998). The deepest part of the epithelium is occupied by basal cells lying adjacent to the basal lamina. These cells are undifferentiated cells and are assumed to be the progenitors of the sensory and supporting cells (Hara 1994b).

Fish have around 100 distinct genes encoding for odorant receptors (Barth et al. 1996). This is about one tenth of that found in mammals (Buck and Axel 1991). Nothing is known concerning genes encoding odorant receptors in sturgeon fish.

10.1.2 Behaviour

In most fishes, the main source of information about environment is vision. In Acipenseriformes the visual system is functionally poorly developed and allows these fish to distinguish only large and moving contrast objects. In such circumstance, olfaction becomes the main distant sense and plays a significant role in Acipenseriformes behaviour and orientation (Pavlov et al. 1970; Palatnikov 1983; Kasumyan 1999). The activation of the sense of smell can evoke in sturgeons behaviours related to feeding, reproduction and, as supposed, homing also.

10.1.2.1 Feeding Behaviour

The smell of food evokes specific and easily recognisable behavioural response in sturgeons. Several seconds after the start of food odour injection into the water, fishes sink to the bottom and crowd in the zone where the maximum concentration of odorant is created. Fish hover close to the bottom, move along the circular and "S"-shaped trajectories, or loops and trail the barbels at the bottom surface. The size of trajectories is of the same as the fish body length. While moving fish shift from side to side and check a wide area of bottom surface. The fish often attack the

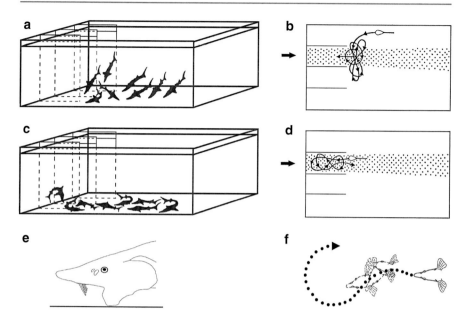

Fig. 10.4 Scheme of sturgeon bottom food searching behaviour: (**a**) Beginning of the response, side view; (**b**) Beginning of the response, individual fish trajectory, top view; (**c**) The maximum of the response, side view; (**d**) The maximum of the response, individual fish trajectory, top view; (**e**) Bottom snaps in the maximum of the response; (**f**) Head oscillation in sturgeon while swimming along the circular and "S"-shaped bottom food search trajectories. Fish trajectory is indicated by dotted line and fish oscillation by sequence of fish head outlines. Spotted zone, area of distribution of food water extract in the fluvarium

bottom in attempts to snap up food pieces which were not here (Fig. 10.4). Siberian sturgeon and all other sturgeon species tested—Russian sturgeon *Acipenser gueldenstaedtii*, stellate sturgeon *A. stellatus*, green sturgeon *A. medirostris*, Persian sturgeon *A. persicus* and beluga *Huso huso*—have the same behavioural response to food odour. Feeding response is fast and enables these virtually blind fish to inspect a feeding area quickly and carefully (Kasumyan and Kazhlaev 1993a; Kasumyan 1999). Anosmic sturgeons after destroying of both olfactory rosettes do not respond to food odour and do not change the background behaviour even if the high concentration of food extract was used for fish stimulation (Pavlov et al. 1970; Kasumyan 1999).

Free amino acids. Between various chemical substances, free amino acids are widely present in the excretory products of aquatic animals and plants and are potent olfactory stimuli for fish (Williams and Poulet 1986). Many of the free amino acids are used as common stimuli in electrophysiological and behavioural studies of fish olfactory system (Caprio 1982; Hara 1994b). Free amino acids were used as odorants for studies in sturgeon chemoreception also (Kasimov and Mamedov 1988; Kasumyan and Taufik 1994; Kasumyan 1999; Mamedov et al. 2009).

In Siberian sturgeon juveniles, only two and the simplest free amino acids, glycine and L-alanine, evoke typical food searching behaviour. The glycine is much more effective than L-alanine but weaker than those caused by *Daphnia* extract. Both amino acids stimulate the same pattern of searching behaviour as the water extract of natural food organisms did. The threshold concentration of glycine is 1 µM for Siberian sturgeon. The same or similar spectra of amino acids induce feeding behaviour in other sturgeon species studied (Table 10.1) (Kasumyan and Taufik

Table 10.1 The intensity of bottom food searching behaviour evoked by water solutions of free amino acids and betaine in juveniles of six species of Acipenseridae: Siberian sturgeon *Acipenser baerii* (4–6 cm), Russian sturgeon *A. gueldenstaedtii* (6–7 cm), stellate sturgeon *A. stellatus* (6–8 cm), green sturgeon *A. medirostris* (30–40 cm), and beluga *Huso huso* (15–20 cm)

Amino acids	Concentration (µM)	Siberian sturgeon	Russian sturgeon	Stellate sturgeon	Green sturgeon	Beluga
l-alanine	100			1.7 ± 0.2		0
	10	0.5 ± 0.1	0.8 ± 0.2			
	1	0.2 ± 0.2	0.5 ± 0.1			
	0.1	0	0			
l-arginine	100	0		0		
l-asparagine	100	0		0		
l-aspartic acid	100	0		0		
l-cysteine	100	0		0		
l-glutamine	100	0		0		
l-glutamic acid	100	0		0		
Glycine	100	2.6 ± 0.3		1.2 ± 0.2	1.4 ± 0.2	0.5 ± 0.2
	10	2.0 ± 0.2	2.5 ± 0.5			
	1	1.1 ± 0.2	0.9 ± 0.2			
	0.1	0	0			
l-histidine	100	0		0		
l-isoleucine	100	0		0		
l-leucine	100	0		0		
l-lysine	100	0		0		
d,l-methionine	100	0		0		
l-norvaline	100	0		0		0
d,l-phenylalanine	100	0		0		
l-proline	100	0		0		
l-serine	100	0		0		
d,l-threonine	100	0		0		
l-tyrosine	100	0		0		
d,l-valine	100	0		0		
Betaine	100	0		0		

The scale of 5 was used to estimate the intensity of bottom food searching behaviour—from "0" (no response) to "4" (maximum response). Values are as mean ± standard error (modified from Kasumyan and Taufik 1994)

1994; Shamushaki et al. 2011). It supports the conclusion of previous studies that olfactory spectra have weak species specificity in closely related fish (Caprio 1982; Hara 1994a, b). Anosmia totally blocked the ability of sturgeons to respond to glycine and L-alanine (Kasumyan and Taufik 1994).

According to these data, glycine and *l*-alanine can be considered as feeding attractants for Siberian sturgeon and for other acipenserids as well. Amino acids with repellent effect were not found in sturgeons, while some of the free amino acids such as glutamic acid evoke weak avoidance response in Persian sturgeon (Shamushaki et al. 2007). A spectra of free amino acids induced food searching behaviour which is usually far broader in teleosts and in many fish may include glycine and *l*-alanine also. However, for some species, these two amino acids have weak activity or even are inert, and their solutions do not evoke obvious behavioural response (Jones 1992).

Based on the sturgeon phylogeny (Vasil'ev 1985), it is possible to suggest that the olfactory spectra of free amino acids in sturgeon were not changed for a time more than 80 Mya. The last revision of molecular phylogeny of the order Acipenseriformes found that the date of 170 Mya is appropriate to the origin of the sturgeon lineage (Krieger et al. 2008). It gives possibility to suppose that olfactory spectra in sturgeon were stable during a longer period than 80–100 Mya. It is interesting to compare olfactory sensitivity to free amino acids in sturgeon species, belonging to genus *Acipenser* and *Scaphirhynchus* or belonging to groups with different chromosome numbers and ploidy levels (Vasil'ev 2009), or compare species from Atlantic and Pacific phylogenetic groups (Krieger et al. 2008).

10.1.2.2 Spawning Behaviour

Hormonal pheromones (steroids, prostaglandins and/or their metabolites) are generally involved in the mating and reproductive activities of teleost fishes (see for review Stacey and Sorensen 2006), but knowledge about chemical signals in reproductive behaviour of sturgeons is scarce. It was shown that sturgeons have releaser postovulatory sex pheromone(s) which is released by ripe females during spawning time. In bioassay males of Russian sturgeon, stellate sturgeon and Persian sturgeon respond to water solution of ovarian liquid obtained from ripe sturgeon females. Males highly increase motion activity and swim near the place where test solution was injected into the basin. In this case, short (2–4 s), spasmodic, convulsive body trembling are observed in males. Some males, on leaving the odour zone, immediately perform an abrupt turn and rapidly return along a short circular trajectory (trajectory radius of approximately 1.0–1.5 m) to the site of delivery of ovarian fluid. Fishes that responded in such a way do not stay long in a zone where odour concentration is the highest; therefore, no obvious aggregation of males at the site of delivery of ovarian fluid solution occurs. Fish swimming rate increases by a factor of three to five with respect to the background (according to visual estimation) (Kasumyan 1993; Kasumyan and Mamedov 2011).

With odour distribution, almost all males in the basin pass from slow quiet swimming to rapid movements over the entire basin. At this time, fishes can drastically change their direction of movement, suddenly and rapidly increase their swimming rate and even perform darts forward. As a result, the general pattern of fish behaviour in the basin becomes utterly different: slow and monotonous swimming and long stops are replaced by the activation of most fish males, increase in their locomotion and uneven swimming (accelerations and darts, turns). Excitation of fish is retained for a short time, and quiet background behaviour of males is completely restored after 5–10 min. The response is similar in males of Russian sturgeon, Persian sturgeon and starred sturgeon, and no differences in their behaviour were found (Kasumyan and Mamedov 2011). The reproductive behaviour and its olfactory control were not studied in Siberian sturgeon.

These results show that the regulation of reproductive behaviour using sex pheromones is typical not only of Teleostei but also of an ancient group of fish such as Chondrostei. All previous attempts to reveal, using electrophysiological methods, sensitivity of acipenserids to prostaglandins (F2α) and to various hormonal steroid substances – highly efficient olfactory stimuli for bony fish—were unsuccessful (Kitamura et al. 1994; Stacey and Cardwell 1997). As in most other studied fish, sex pheromones in acipenserids are produced by mature females, most probably, at the final stage of their reproductive cycle, apparently, before or during ovulation. Such pheromones, according to the adopted modern classification of chemical signals, should be assigned to releasing postovulatory sex pheromones. To answer the question whether releasing preovulatory and primer pheromones exist in acipenserids, additional studies are necessary.

10.1.2.3 Homing

The sense of smell has been implied in the homing performance of many fish species, in particular anadromous fish like salmonids (Stabell 1984; Dittman et al. 1994; Døving and Stabell 2003; Kasumyan 2004). It was suggested that olfaction seems essential for homing migration of sturgeons also. The critical or sensitive period during which sturgeon may become olfactory imprinted has been suggested to be the time when sturgeon larvae changes over to exogenous feeding (Boiko et al. 1993). In salmonids, the sensitive period is connected with smoltification and an increase of the thyroid hormone level (Hasler and Sholz 1983). It seems that the thyroid hormones participate in the processes of imprinting of chemical signals in sturgeon as well (Boiko and Grigor'yan 2002). Undoubtedly, olfactory imprinting and homing hypotheses for sturgeons are needed in more strong evidences.

10.1.3 Ontogeny

10.1.3.1 Morphogenesis

Development of the olfactory organ of the Siberian sturgeon was followed in details by means of histology, scanning and transmission electron microscopy (Devitsina and Kazhlayev 1993; Zeiske et al. 2003). The olfactory placodes, the anlage of the olfactory organ, become microscopically visible as ectodermal thickenings after closure of the neural folds and a little before first heart contractions of the embryo. The olfactory placodes become two layered early in embryonic development, and both the superficial epidermal and the subepidermal layer can be easily distinguished by ultrastructural properties. The morphogenesis of the olfactory epithelium in acipenserids follows a different pattern than that seen in the teleost. In sturgeons, only olfactory receptor neurons develop from cells of the subepidermal layer. Supporting and ciliated nonsensory cells derive from epithelial surface cells. In this respect, acipenserids clearly demonstrate close resemblance to the morphogenetic process found in the tetrapod *Xenopus* (Anura) (Zeiske et al. 2003).

Externally olfactory organ remains in the state of invaginated olfactory pits even after hatching. Further development of the olfactory organ slowly occurs during the yolk sac larval phase. The development of the primary olfactory pit by invagination initially forming a shallow pit represents a widespread and probably phylogenetically a plesiomorphic pattern seen in many other species of fish including teleosts like cyprinids (Kleerekoper 1969; Zeiske et al. 1992; Pashchenko and Kasumyan 2015). The division of the primary olfactory opening into the anterior incurrent and posterior excurrent opening is completed 6 days AH[1] (12 days AF[2]), still well before onset of exogenous feeding (at Detlaff stage 45) (Fig. 10.5). Differentiation of the basic cellular composition of the olfactory epithelium is far advanced at the time of onset of exogenous feeding (Fig. 10.6) (Zeiske et al. 2003).

10.1.3.2 Function

The ability of fishes to respond to olfactory stimuli develops hours or days after the sensory neurons first appear in the olfactory organ (Døving et al. 1994; Kasumyan et al. 1998). In Siberian sturgeon, and in other acipenserids studied, the response to food odour appears just after the beginning of exogenous feeding (Kasumyan and Kazhlaev 1993a). This observation fits the process of cellular differentiation of the olfactory epithelium (Zeiske et al. 2003). Already when the juveniles are 1.5 months old, they display the same thresholds to natural food stimuli and to free amino acids as found in older fish (Kasumyan and Kazhlaev 1993b; Kasumyan and Taufik 1994; Kasumyan 1999). In comparison with sturgeons, the food odour perception develops less rapidly in the bottom-feeding cyprinid fish and even slower in the pelagic plankton-feeding fish (Fig. 10.7) (Kasumyan and Ponomarev 1990).

10.2 Gustation

The morphology of the taste buds in fish has been studied extensively since the nineteenth century. Phylogenetically, the basic patterns of the taste bud are well conserved among fishes, although there are many fine-scale peculiarities and

---

[1] AH—after hatching

[2] AF—after fertilisation

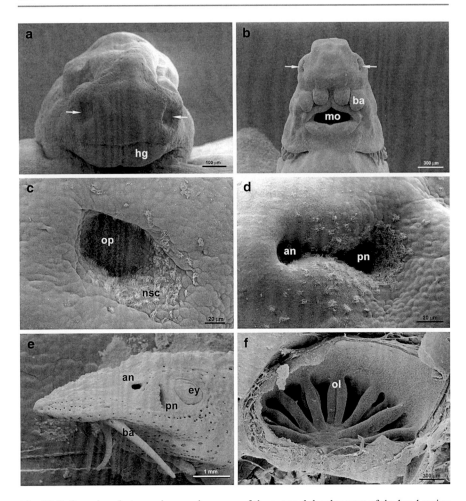

Fig. 10.5 Scanning electron microscopic survey of the external development of the head region and olfactory placode up to hatching and the formation of the olfactory openings and olfactory rosette in Siberian sturgeon *A. baerii*: (**a**) Late embryo, 120 h (5 days) AF, 8.7 mm total length, one day before hatching, showing deep olfactory pit (*arrows*) and hatching glands. Scale bar = 100 μm. (**b**) Larva, 10 days AF (4 days AH), 14 mm total length, olfactory pit elongated (*arrows*). Scale bar = 300 μm. (**c**) Late embryo, 132 h AF, lateral view of the primary olfactory pit surrounded by ciliated nonsensory cells. Scale bar = 20 μm. (**d**) Late yolk sac larva, 11 days AF (5 days AH), 14 mm total length, lateral view, division of anterior (incurrent) (*an*) and posterior (excurrent) (*pn*) nostril incomplete. Note the ingrowth of ciliated nonsensory cells. *Scale bar* = 20 μm. (**e**) Advanced juvenile, 33 days AF (27 days AH), 35 mm total length. *Scale bar* = 1 mm. (**f**) Juvenile, 52 days AF (46 days AH), 85 mm total length, opened olfactory cavity containing olfactory lamellae arranged in a rosette. Scale bar = 300 μm. *An*, anterior (incurrent) nostril; *ba*, barbel; *ey*, eye; *hg*, hatching gland; *mo*, mouth; *nsc*, ciliated nonsensory cell; *ol*, olfactory lamellae; *op*, olfactory pit; *pn*, posterior (*excurrent*) nostril; *arrows* point to the developing olfactory organs (from Zeiske et al. 2003)

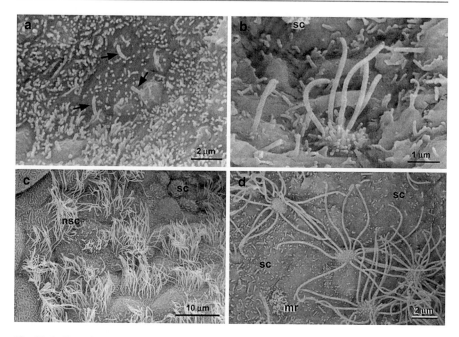

Fig. 10.6 Scanning electron microscopic survey of the olfactory pit in Siberian sturgeon, *A. baerii*: (**a**) Embryo, 96 h AF. Early dendritic endings (olfactory knobs) of ciliated olfactory receptor cells showing their first cilium (*arrows*) appear at the surface of the olfactory pit. *Scale bar* = 2 µm. (**b**) Late embryo, 120 h (5 days) AF. Olfactory knob in advanced development showing cilia grown in length. Note the less pronounced olfactory knob bearing microvilli. Surrounding supporting cells bear microvillous-like protrusions. *Scale bar* = 1 µm. (**c**) Late embryo, 132 h (5 days 12 h) AF. Ciliated nonsensory cells populate the olfactory pit. *Scale bar* = 10 µm. (**d**) Young larva, 8 days AF (2 days AH), 11 mm total length. Mature olfactory knobs of ciliated olfactory receptor cells. Scale bar = 2 µm. *mr* microvillous olfactory receptor cell, *nsc* ciliated nonsensory cell, *sc* supporting cell (from Zeiske et al. 2003)

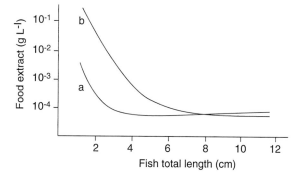

Fig. 10.7 Development of sensitivity to food odour in acipenserid and cyprinid fish. The graph illustrates changes in threshold concentration of food odour (water extract of Chironomidae larvae) for Siberian sturgeon *A. baerii* (**a**) and common carp *C. carpio* (**b**) during their first months of life (modified after Kasumyan 2004)

differences between groups (Kapoor et al. 1975; Reutter 1986; Jakubowski and Whitear 1990). A typical taste bud has an ovoid-shaped form; its long axis is oriented vertically to the surface of the epithelium, and the base is often situated on a small ascending papilla of the corium (Reutter et al. 1974). Marginal cells form the border or margin between a taste bud and the neighbouring stratified squamous epithelium. Taste buds consist of two types of gustatory sensory cells, the electron-lucent light cells and the electron-denser dark cells; their number in taste buds greatly varies from a few to several dozens (Jakubowski and Whitear 1990; Reutter 1992). Each sensory cell bears apical microvilli which together form the taste bud pore exposed to the environment. Between light cells, it is possible to distinguish cells which end in one large apical microvillus and cells which show different numbers of distinctly shaped microvilli. The most common dark cell subtype normally bears small and undivided microvilli. The micromorphology of microvilli is species specific (Reutter and Hansen 2005). Up to five basal cells are situated at the bottom of the taste bud. It has been suggested that both light and dark cells as well as basal cells have synaptic associations with unmyelinated and myelinated nerve fibres (Fig. 10.8) (Reutter 1986; Finger et al. 1991).

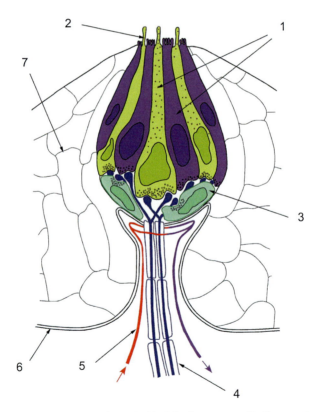

Fig. 10.8 The gross morphology of taste bud in fish: *1*, receptor cells; *2*, protrusions of receptor cells in taste pore; *3*, basal cell; *4*, nerve fibres with synaptic contacts; *5*, blood vessel; *6*, epidermal basal lamina; *7*, epidermal cell (modified after Døving and Kasumyan 2008)

Nerve fibres innervating taste buds are derived from the facial (VII), glossopharyngeal (IX) and vagal (X) cranial nerves. The facial nerve supplies extraoral taste buds of the barbels and lips (*nervus facialis, r. maxillaris*), fins and body surface (*n. facialis, r. recurrens*) and oral taste buds of rostral palate (*n. facialis, r. palatinus*). The vagal nerve innervates most of the orobranchial taste buds, whereas the glossopharyngeal nerve plays a minor role in the gustatory supply to the oral cavity (Herrick 1901; Atema 1971; Finger 1976). A single nerve fibre may innervate several taste buds, and each sensory cell forms synapses with several nerve fibres (Jakubowski and Whitear 1990). The primary gustatory nuclei lie in the *medulla oblongata* and form the facial, vagal and glossopharyngeal lobes. Within the medulla, local gustato-motor networks and descending pathways form the neural substrate for reflexive movements such as food pick up or biting, and swallowing or reject (Kanwal and Finger 1992).

10.2.1 Taste bud Morphology and Distribution

As in other fishes, the taste buds in sturgeons have the light and dark sensory cells which are distinguished by the smooth endoplasmic reticulum, number of mitochondria and shape of apical microvilli. Single microvilli in light receptor cells has length up to 1 µm and diameter around 0.5 µm; microvilli in dark receptor cells are much shorter and thinner (Pevzner 1981).

In Siberian sturgeon, as in teleost fish, taste buds can be submerged in the surrounding epithelium or are on the same level with the epithelium surface (Fig. 10.9). As a rule, the two to five coating cells surround the pore of submerged taste buds. Using scanning electron microscope, several types of cell differing by the shapes of their apical protrusions were found on the receptor area. The polyvillous cells with

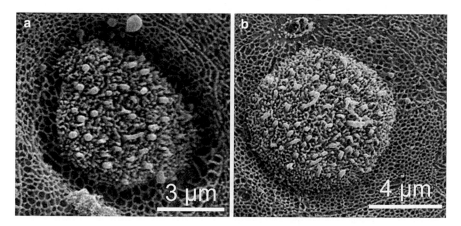

Fig. 10.9 Scanning electron microscopy survey of the taste buds in the hard palate of Siberian sturgeon *A. baerii* juveniles (85–320 mm body length): (**a**). Taste bud of the submerged type, scale bar = 3 µm. (**b**) Taste bud of the nonsubmerged type, scale bar = 4 µm (from Devitsina et al. 2011)

numerous microvilli (up to 20) may have short microvilli of uniform length or bunch of elongated microvilli. The monovillous cells may have a single long and a thin microvillus (common) or a thick microvillus (rare). In total, six different shapes of the microvillar protrusions in the taste receptor cells are described in Siberian sturgeon, but the special patterns of distribution of these cells in the receptor area are not found (Devitsina et al. 2011). Some of these receptor cells have been found previously in common carp *Cyprinus carpio*, rainbow trout *Oncorhynchus mykiss* and Atlantic cod *Gadus morhua* and in other teleosts (Reutter et al. 1974; Reutter and Hansen 2005; Devitsina and Golovkina 2008; Devitsina et al. 2011). It was found that the taste buds located on the pharyngeal surfaces of the gill arches consisted of three types of cells: (1) cells with numerous microtubules associated with mitochondria and (2) cells containing many dense vesicles and packs of microfilaments (The apical tip of these two types of cells was characterised by small microvilli.) and (3) basal cells containing small electron-dense vesicles. A lot of afferent myelin-free nervous fibres were concentrated at this level and were closed together with sensory cells. These fibres contained numerous mitochondria, synaptic vesicles and some electron-dense vesicles. They joined to each other and made up a nerve in the underlying conjunctive tissue. Despite differences in gill organisation between the chondrostean and the teleost species, the taste buds of the sturgeon themselves ultrastructurally and closely resembled those described in other fishes and other vertebrates (Nonnotte et al. 1991).

The taste buds are not smoothly distributed in the oral cavity of Siberian sturgeon juveniles. Between 14 zones, only four in ventral side and five in dorsal side have taste buds. The largest zone is the soft palate located close to the branchial cavity and supposed to be an analogous to the palatal organ of cyprinids (Fig. 10.10). Zonal distribution of taste buds is probably associated with the functional differences of the taste receptors (Devitsina et al. 2011).

In acipenserids, taste buds are situated not only within the oral cavity, pharynx and gills but also on the lips and barbels and are absent in the rostrum and fins and

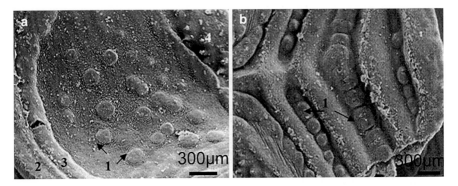

Fig. 10.10 Scanning electron microscopy survey of the dorsal surface of oral cavity of Siberian sturgeon *A. baerii* juveniles (85–320 mm body length): (**a**) hard palate, scale bar = 300 μm; (**b**) soft palate, scale bar = 300 μm. *1*, taste papillae; *2*, the upper lip; *3*, the maxillary zone (from Devitsina et al. 2011)

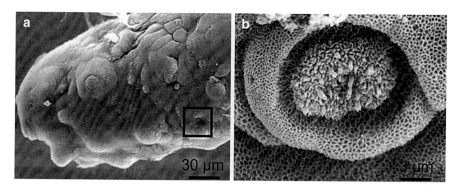

Fig. 10.11 Scanning electron microscopy survey of the lateral barbel of Siberian sturgeon *A. baerii* juveniles (85–320 mm body length): (**a**) tip of the barbel with taste buds on the papillae, *scale bar* = 30 μm; (**b**) taste bud of the submerged type designated by the frame in "*a*", scale bar = 3 μm (from Devitsina et al. 2011)

over the entire trunk surface (Fig. 10.11) (Devitsina and Kazhlayev 1993; Devitsina and Gadzhieva 1994, 1996; Boglione et al. 2006). In Siberian sturgeon, the density of taste buds on the caudal side of barbels is higher than on the rostral side. Morphology of oral and extraoral taste buds is similar (Devitsina et al. 2011). In fish like acipenserids, the gustatory system may be divided into two distinct gustatory subsystems or systems, extraoral and oral ones (Finger and Morita 1985).

10.2.2 Function Characteristics and Behaviour

The functional characteristics of the fish gustatory system have mainly been studied by electrophysiological means (Kiyohara et al. 1981; Marui and Caprio 1992; Hara 2007). The threshold concentration for the most potent substances is less than 10^{-9} M. The oral taste buds are generally less sensitive by 1.5–2 orders of magnitude than extraoral taste buds, although they have similar amino acid specificity (Kanwal and Caprio 1983). Using behavioural approach, it was shown that taste preferences in fish are highly species specific and the differences among fish species are apparent when comparing the width and composition of spectra for both taste stimulants and taste deterrents. There is strong similarity in taste preferences between geographically isolated fish populations of the same species, and taste preferences are similar in males and females although taste preferences at the individual level vary among conspecifics. Taste preferences in fish show low plasticity (in relation to the diet) and are determined genetically (for review see Kasumyan and Døving 2003).

Our knowledge about functional peculiarities of gustatory system in sturgeons is based on data obtained from behavioural assays only. It was supported that gustatory system in sturgeons is highly responsive to organic and inorganic chemicals as well as to natural food extracts. Fishes are able to recognise the differences between pellets with the same size, shape, colour and texture but having different taste substances inside (Kasumyan 1999; Lari et al. 2013). Extraoral taste buds are used by sturgeons for preliminary evaluation of food. A rapid touch of a food item by a barbels covered by numerous taste buds is sufficient to assess food quality. This preliminary assessment of food takes place in a millisecond and is essential for inducing food grasping in sturgeon. The final acceptability of food is based on the oral taste receptors which regulate swallowing of appropriate food and rejection of aversive one (Kasumyan and Kazhlaev 1993b; Kasumyan 1999).

10.2.2.1 Classical Taste Substances

As was shown that the classical taste substances which are considered to be sweet, sour, bitter and salty for human are tastants for Siberian sturgeon too. These substances are perceived by fish as different and evoke different taste responses. Siberian sturgeons much often grasp pellets with citric acid (0.26 M) and have decreased grasping rate for the pellets with calcium chloride (0.9 M); grasping of pellets with sodium chloride and sucrose (0.29 M and 1.73 M, respectively) was the same as for the blank pellets (control). Citric acid, sodium chloride and calcium chloride are taste deterrents and significantly decreased the pellet consumption, but sucrose does not have effect on the pellet consumption in Siberian sturgeon (Kasumyan and Kazhlaev 1993b).

10.2.2.2 Free Amino Acids

Free amino acids are highly effective stimuli for taste receptors of various freshwater and marine fishes (for review see Marui and Caprio 1992; Jones 1989, 1990). In some species, the mixtures of free amino acids alone had the same level of palatability as the whole extract of preferable food organisms (Mackie 1982; Takeda et al. 1984; Johnsen and Adams 1986; Mearns et al. 1987). In 4–6 cm-long Siberian sturgeon juveniles, 14 of 19 amino acids were incitants; they evoke a significant increase in the number of food pellets caught (Table 10.2). The most potent incitants were L-asparagine, L-threonine and L-methionine (all—in 0.1 M). Eleven of such amino acids were common to the four sturgeon species studied, although the Spearman rank correlation coefficient was low and insignificant for Siberian sturgeon versus all other three species (Fig. 10.12) (Kasumyan 1999).

The oral gustatory system has a narrower spectrum of effective amino acids than the extraoral one. There were only seven amino acids found to be stimulants for the Siberian sturgeon, and L-alanine had a deterrent effect. For comparison, there were six amino acids found to be stimulants for the Russian sturgeon, three for the stellate sturgeon and 15 for the Persian sturgeon (Tables 10.3 and 10.4). Among the stimulant amino acids, none were common to all four sturgeon species studied, and there was no significant correlation between sturgeon species concerning the oral taste responses (Fig. 10.12). These data show that the oral taste receptors in sturgeons have a higher specificity than the extraoral receptors (Kasumyan 1999; Shamushaki et al. 2008). Specificity of taste preferences in sturgeon species is supported by observations that the palatability of natural prey organisms varies

Table 10.2 Extraoral taste preferences to L-amino acids. Four species of *Acipenser* sturgeons, Siberian sturgeon, *A. baerii* (4–6 cm body length); Russian sturgeon, *A. gueldenstaedtii* (6–7 cm body length); stellate sturgeon, *A. stellatus* (6–8 cm body length); and Persian sturgeon (9–13 cm body length), were exposed to pellets containing amino acids

Amino acids	Concentration, M	Siberian sturgeon	Russian sturgeon	Stellate sturgeon	Persian sturgeon
Alanine	0.1	30.4***	20.6**	23.8***	20.8**
Arginine	0.1	16.5	28.7***	25.3***	31.2***
Asparagine	0.1	46.6***	27.7***	21.4***	29.3***
Valine	0.1	30.1***	9.6	12.3**	23.7***
Histidine	0.1	23.7*	30.7***	26.5***	32.4***
Glycine	0.1	36.0***	13.7	19.1***	26.6**
Glutamine	0.1	29.2***	25.4***	23.3***	27.2***
Lysine	0.1	21.5	28.7***	27.7***	21.5**
Methionine	0.1	41.0***	15.9*	20.3***	18.4*
Norvaline	0.1	–	7.1	11.0	–
Proline	0.1	15.3	7.1	10.3	19.6
Serine	0.1	28.2***	23.6***	27.9***	22.3**
Threonine	0.1	42.0***	29.6***	21.9***	43.1***
Phenylalanine	0.1	22.0*	22.9**	18.7***	29.2**
Cysteine	0.1	–	35.1***	35.1***	28.3***
Aspartic acid	0.01	25.9**	37.9***	35.9***	22.8***
Glutamic acid	0.01	32.2***	35.6***	28.1***	14.5*
Isoleucine	0.01	13.9	14.4	13.8***	11.8
Leucine	0.01	30.3***	15.0*	11.5**	9.9
Tryptophan	0.01	32.2***	23.2***	12.5***	16.8*
Tyrosine	0.001	19.0	13.6	9.4	3.7
Control	–	19.0	13.6	7.4	6.3

Extraoral taste preference is the mean number of snaps of 50 agar pellets by five fishes (Siberian sturgeon, stellate sturgeon, Russian sturgeon) or by three fishes (Persian sturgeon) during 5-min exposure time. Controls were pellets containing Ponceau 4R (*Red colour dye*) or 0.3% Cr_2O_3. *, ** and *** indicate that value are significantly different from the blank pellets $P < 0.05$, $P < 0.01$ and $P < 0.001$ (student's t-test), respectively (modified from Kasumyan 1999; Shamushaki et al. 2008, 2011)

among sturgeon species. In aquaria, Russian sturgeon and sterlet *A. ruthenus* consume the amphipod *Gammarus* and refuse the polychaete *Nereis*. Under similar conditions, stellate sturgeon prefers *Nereis* and ignores *Gammarus* (Stroganov 1968). Selective feeding is also typical for other species (Soriguer et al. 2002; Pyka and Kolman 2003).

In contrast, there is the close correlation between the number of pellet grasps and the number of pellets swallowed by the sturgeons. In other words, there is a correspondence between oral and extraoral taste preferences in the same fish species. Comparison of the extraoral and oral taste responses within one species revealed a significant correlation in the Siberian sturgeon ($r_s = 0.48$; $p < 0.05$) and in the Russian

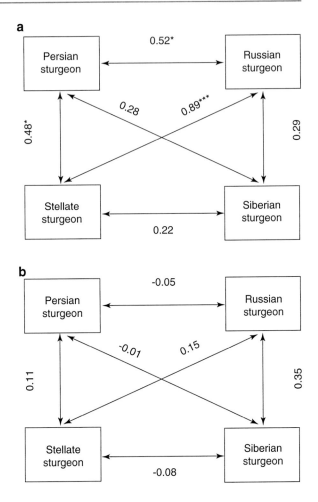

Fig. 10.12 The Spearman rank correlation coefficients between taste ranges of free amino acids mediated by extraoral (**a**) and oral (**b**) gustatory systems in four sturgeon species belonging to genus *Acipenser*. Correlation coefficient was calculated for Russian sturgeon *A. gueldenstaedtii* and stellate sturgeon *A. stellatus* by gustatory responses of fish towards 21 free amino acids, for Siberian sturgeon *A. baerii* by gustatory responses to 19 free amino acids and for Persian sturgeon *A. persicus* by gustatory responses to 20 free amino acids. Correlation coefficients were calculated taken into account control. * and *** indicate significance levels, respectively, $p < 0.05$ and $p < 0.001$ (from Shamushaki et al. 2011)

sturgeon, stellate sturgeon and Persian sturgeon as well (Table 10.5) (Shamushaki et al. 2011). The correlation between oral and extraoral taste spectra has been found not only in sturgeons but also in other species (Caprio and Derby 2008). It is indicated that there is a consistency and coordination between sensory information provided by an extraoral gustatory apparatus for the initial acceptance test and by the oral taste buds making the final judgement of acceptance. It is noteworthy that some substances have an opposite effect on extraoral and oral taste receptors. For example, citric acid is a strong incitant for Siberian sturgeon juveniles and increased the frequency of pellets caught in comparison with blank pellets; but grasped pellets with citric acid are immediately rejected by fish (Kasumyan and Kazhlaev 1993b).

10 Olfaction and Gustation in *Acipenseridae*

Table 10.3 Oral taste preferences to L-amino acids. Four species of *Acipenser* sturgeons, Siberian sturgeon, *A. baerii* (4–6 cm body length); Russian sturgeon, *A. gueldenstaedtii* (6–7 cm body length); stellate sturgeon, *A. stellatus* (6–8 cm body length); and Persian sturgeon (9–13 cm body length), were exposed to pellets containing amino acids. Oral taste preference is the mean consumption of pellets in relation to the number of grasps (%) of 50 agar pellets by five fishes (Siberian sturgeon, stellate sturgeon, Russian sturgeon) or by three fishes (Persian sturgeon) during 5-min exposure time

Amino acids	Concentration, M	Siberian sturgeon	Russian sturgeon	Stellate sturgeon	Persian sturgeon
Alanine	0.1	34.9*	15.6	59.9***	9.4**
Arginine	0.1	62.5*	41.9*	11.5	18.2 **
Asparagine	0.1	62.4*	35.0	20.9	12.6***
Valine	0.1	65.1*	34.0	11.6	7.6
Histidine	0.1	64.3*	27.6	57.3***	19.6**
Glycine	0.1	71.1***	34.2	33.0	7.0*
Glutamine	0.1	41.4	35.9	11.9	11.0**
Lysine	0.1	38.8	39.0	29.5	2.2
Methionine	0.1	55.5	25.0	14.7	6.0
Norvaline	0.1	–	15.1	14.8	–
Proline	0.1	45.6	19.0	14.8	3.8*
Serine	0.1	47.0	42.3	23.6	12.2*
Threonine	0.1	86.6***	43.4*	17.0	11.6**
Phenylalanine	0.1	54.7	40.7	16.0	6.0**
Cysteine	0.1	–	80.3***	55.9***	15.4***
Aspartic acid	0.01	54.6	68.6***	24.4	5.0*
Glutamic acid	0.01	69.8**	50.1*	7.2	3.8*
Isoleucine	0.01	52.3	20.4	16.2	2.4
Leucine	0.01	60.6	26.1	15.1	4.8
Tryptophan	0.01	52.8	53.1**	15.3	4.4*
Tyrosine	0.001	47.5	21.1	14.3	0.2
Control	–	47.5	21.1	15.7	1.0

Controls were pellets containing Ponceau 4R or 0.3% Cr_2O_3. *, ** and *** indicate that value are significantly different from the blank pellets $P < 0.05$, $P < 0.01$ and $P < 0.001$ (student's t-test), respectively (modified from Kasumyan 1999; Shamushaki et al. 2008, 2011)

Table 10.4 The number of free amino acids evoked positive behavioural response mediated by extraoral and oral gustatory systems in sturgeons belonging to genus *Acipenser*. Percentage of efficient amino acids from the number of amino acids tested is given in parentheses

Species	Gustatory system		Ratio[a]	The number of free amino acids tested
	extraoral	oral		
Siberian sturgeon	14 (73.7%)	8 (42.1%)	1.75	19
Russian sturgeon	15 (71.4%)	6 (28.6%)	2.50	21
Stellate sturgeon	18 (85.7%)	3 (14.3%)	6.33	21
Persian sturgeon	16 (80%)	14 (70%)	1.14	20

[a]The ratio between number of effective amino acids for extraoral and oral gustatory systems

Table 10.5 The value of Spearman rank correlation coefficient (r_s) for gustatory preferences to free amino acids between extraoral and oral gustatory systems in sturgeons belonging to genus *Acipenser*

Species	r_s	p
Siberian sturgeon	0.47	0.040
Russian sturgeon	0.78	0.00003
Stellate sturgeon	0.40	0.074
Persian sturgeon	0.81	0.00001

The data for Siberian sturgeon, Russian sturgeon and stellate sturgeon are given by Kasumyan (1999); the data for Persian sturgeon are given by Shamushaki et al. (2011). Correlation coefficients are calculated taken into account control

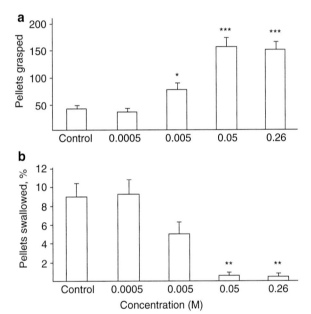

Fig. 10.13 Oral (**a**) and extraoral (**b**) taste preferences for different concentrations of citric acid in juveniles of Siberian sturgeon *A. baerii*. Control, blank pellets. Vertical columns and T-bars are the mean and the SEM (modified after Kasumyan and Kazhlaev 1993b)

10.2.2.3 Thresholds

An important parameter of the fish gustatory system is the minimum concentration needed to give a positive or negative response of a fish towards taste stimuli. In Siberian sturgeon juveniles (at 38 days old, 49–52 mm), the threshold concentrations were $5 \cdot 10^{-2}$ M for citric acid for extraoral taste, 1.7 *M* for sodium chloride for oral taste and 9×10^{-2} *M* for calcium chloride for both oral and extraoral taste (Fig. 10.13) (Kasumyan and Kazhlaev 1993b).

Based on threshold concentrations determined by behavioural assays, it is possible to estimate the amount of substance in one pellet which is sufficient to release a significant taste response. These amounts were 16 μg for citric acid, 169 μg for sodium chloride and 17 μg for calcium chloride. Because only substance that are situated in the surface layer of pellet is available to taste receptors, the actual quantities of the substance sufficient for eliciting a taste response would be lower than calculated by one to two orders of magnitude (Kasumyan and Døving 2003).

10.2.3 Ontogeny

Many studies concerning ontogeny of the gustatory system in sturgeons have been focused on the morphology of the taste buds. In general, taste buds develop later than the olfactory organ. The first taste buds appear several days or even hours before fish larvae start to catch food items, which is at the beginning of exogenous feeding, the time when the estimation of taste quality of food becomes essential and necessary (Dragomirov 1954; Schmalhausen 1962a; Pevzner 1985; Devitsina and Gadzhieva 1996).

In sturgeons, the rate of taste bud development seems to be species specific; taste bud formation is most rapid in beluga, followed by ship sturgeon *Acipenser nudiventris* and Russian sturgeon. The taste pore is formed by moving apart the epithelial cells. The first receptor cells to appear in the open taste pore are the light cells with a single apical microvillus. The apical microvilli begin to form in receptor cells after the taste buds reach the epithelium surface and the receptor area is exposed to the environment. The first synaptic contacts between the receptor cells and the nerve endings begin to form one day after formation of the taste bud receptor area, just 1–2 days before first feeding. Basal cells are the last cell type to appear in taste bud differentiation (Pevzner 1985).

In Siberian sturgeon and in other acipenserids, extraoral taste buds appear first, then come the oral taste buds and then the larvae start feeding (Table 10.6). The first extraoral taste buds appear on the rostral tip of the barbels and are seldom found in the middle or basal region of barbel (Devitsina and Kazhlayev 1993; Devitsina and Gadzhieva 1996). In Adriatic sturgeon (*Acipenser naccarii*), the extraoral and pharyngeal taste buds are the first to appear, followed by the palatal and lingual ones (Boglione et al. 1999). In Persian sturgeon extraoral and oral taste buds appear at 8 days AF, i.e. one day before first feeding (Devitsina and Gadzhieva 1996). The advanced development of the extraoral gustatory system gives an obvious advantage for these fish in searching behaviour: touching a food item by barbels is the first step in the feeding sequence by sturgeons (Pavlov et al. 1970; Kasumyan and Kazhlaev 1993b; Pavlov and Kasumyan 1998).

The size of taste buds, the height and number of cells, as well as the diameter of the taste pore increase during fish development (Pevzner 1985; Devitsina et al. 2011). Starvation caused different changes in taste bud morphology and abundance. In the first-feeding larvae of Russian sturgeon, starvation caused an increase in the size of

Table 10.6 The sequence of oral and external taste bud appearance in fish ontogeny, in relation to the first feeding (FF)

Species	Sequence	References
Siberian sturgeon, *Acipenser baerii*	External → oral → FF	Devitsina and Kazhlayev (1993)
Stellate sturgeon, *Acipenser stellatus*	External → oral → FF	Devitsina and Kazhlayev (1993)
Adriatic sturgeon, *Acipenser naccarii*	External → oral → FF	Boglione et al. (1999)
Ship, *Acipenser nudiventris*	External → oral = FF	Devitsina and Gadzhieva (1996)
Persian sturgeon, *Acipenser persicus*	FF → oral = external	Devitsina and Gadzhieva (1996)
Paddlefish, *Polyodon spathula*	Oral = FF → external	Devitsina and Kazhlayev (1993)

taste buds on the barbels and palate, whereas in non-starving larvae, the number of taste buds remained the same. The taste buds of the tongue and gill arches remained the same size, but the number increased. Following a 10-day starvation period, taste buds on the gill arches disappeared completely and the number on the palate, tongue and on the lips decreased significantly (Devitsina and Gadzhieva 1996).

The combination of morphological studies and behavioural assays has shown that there is good correspondence between the development of the gustatory system in sturgeons and the ability of fish to discriminate taste properties of food. Larvae at the start feeding stage can discriminate only some taste substances and have a more restricted spectrum of effective taste stimuli than the juveniles. For Russian sturgeon larvae (10–15 after hatching), 11 amino acids were incitants and only 1 amino acid a stimulant, and the extraoral and oral taste preferences for amino acids were not significantly correlated ($r_s = 0.36$; $p > 0.05$). In contrast, for juveniles 16 amino acids were incitants, and 6 amino acids were stimulants, and the juveniles showed a positive correlation between oral and extraoral preferences of the same amino acids ($r_s = 0.79$; $p < 0.001$). The extraoral taste preferences were similar in larvae and juveniles ($r_s = 0.69$; $p < 0.01$), while the oral taste preferences changes dramatically ($r_s = 0.04$; $p > 0.05$) (Fig. 10.14).

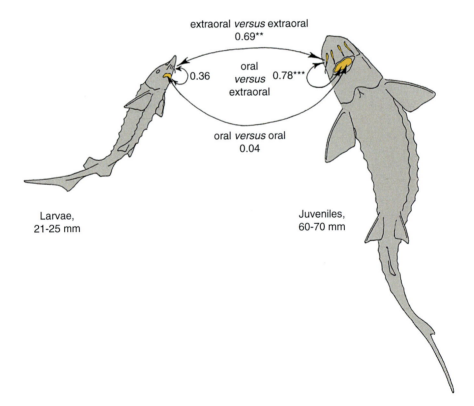

Fig. 10.14 Development of oral and extraoral taste preferences in Russian sturgeon *A. gueldenstaedtii*: values of Spearman rank correlation coefficient between oral and extraoral taste preferences of free amino acids in larvae and juveniles (modified after Kasumyan 1999)

10.3 Interrelationship of Olfaction and Gustation

Parasites such as protozoa, microsporidia and crustaceans may seriously damage the olfactory sensory epithelium in fish (Devitsina 1977; Doroshenko and Pinchuk 1978; Morrison and Plumb 1994; Burdukovskaya and Pronin 2010). Following a viral infection (*Iridoviridae*), the olfactory lamellae in infected white sturgeon juveniles *A. transmontanus* became swollen and truncated, and the olfactory epithelium became hyperplastic, dysplastic and focally necrotic (Watson et al. 1998). Protheolitic bacteria can completely destroy the nasal bridge between the incurrent and excurrent nostrils in up to 20–50% on farmed sturgeon juveniles (Fig. 10.15) (Goryunova et al. 2000). Various pollutants such as heavy metals and detergents may inhibit olfactory responses, and fish sensitivity to olfactory stimuli decreased dramatically or becomes blocked (Baatrup and Døving 1990; Klaprat et al. 1992).

Recently it was found that gustatory system in sturgeons can specifically compensate the loss of olfaction, in particular when this main distant sensory system in acipenserids is, either totally or partially, eliminated (Kasumyan 2002). Anosmic sturgeons partly regain their ability to respond to food odour several months after the bilateral cauterisation of olfactory rosettes (Fig. 10.16). This might be related to

Fig. 10.15 Head of stellate sturgeon *A. stellatus* (12–15 cm in body length) without nasal bridge: 1, remainder parts of the nasal bridge; 2, olfactory rosette and olfactory primary lamellas

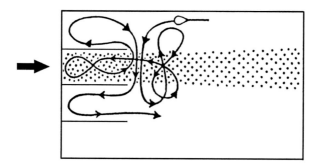

Fig. 10.16 Trajectory of the movements for anosmic stellate sturgeon *A. stellatus* (12–15 cm in body length) 6 months after bilateral cauterisation of olfactory rosettes. Spotted zone—area of distribution of food water extract in the fluvarium

the compensatory hypertrophied development of the extraoral gustatory system in anosmic sturgeons. The density of external taste buds on the surface of the barbels increases in sturgeons after cauterisation, the most pronounced in the apical section of the barbels. The taste buds become also larger and had greater variability than in intact fish (Kasumyan and Devitsina 1997). It is possible to suggest that the extraoral gustatory system is responsible for restoration of the ability of anozmized sturgeons to respond to food odour. This suggestion is supported by results obtained for Siberian sturgeon: juveniles 3 months after bilateral cauterisation revealed extraoral taste responses for a higher number of free amino acids (17) than intact fish (14) out of 19 free amino acids tested (Kasumyan 2002).

These findings also show that there is a close functional relationship between the olfactory and gustatory systems in sturgeon. The vicariation of one chemosensory system by another is considered a sensory mechanism aimed at the reliability of food search behaviour in sturgeon. Redundancy of receptor cells in the olfactory organ and their ability for renewal is another sensory mechanism responsible for the safety of sturgeon chemocommunication and chemoorientation under conditions where olfactory receptor cells might be destroyed.

10.4 Application

Information about fish smell and taste preferences, and about attractive and palatable substances might be of significant value when applied to sturgeon cultivation. The eventual incorporation of feeding stimulants into diets consisting of cheap and normally non-palatable protein sources may be of practical significance to the fish farming industry (Jobling et al. 2001). Feeding stimulant supplementation masks deterrent ingredients that lower the palatability of feeds and improves feed intake, weight gain and feed efficiency of reared fish (Tandler et al. 1982; Gomes et al. 1997). For instance, certain antibiotics have aversive taste for cultivated fish and reduce the palatability of feed (Poe and Wilson 1989; Robinson et al. 1990; Hustvedt et al. 1991; Toften et al. 1995; Maklakova et al. 2011).

Palatability of artificial food might also be raised by removing the ingredients with bad smell or/and taste or by replacing those ingredients with equivalents, which do not provoke negative chemosensory-mediated responses. Ingredients of artificial feeds have strongly different olfactory and gustatory qualities for sturgeon fish. Nothing from 19 ingredients which are often used for artificial fish diets have attractive smell or taste as it has natural food like mosquito larvae (Chironomidae) (Kasumyan et al. 1995). Such highly palatable animals can be recommended as potent additives for enhancing the taste attractiveness of artificial feeds for sturgeon. The palatability of extracts of these animals are long-term stable (Lari et al. 2013). It should be noted that some of invertebrate and vertebrate animals can contain unpalatable compounds with strong deterrent effect (Kasumyan and Døving 2003). The evaluation of palatability of feed ingredients becomes crucial to nutritional research and feed development for aquaculture species. In evaluating ingredients for use in aquaculture feeds, the information about ingredient palatability is one of

10 Olfaction and Gustation in *Acipenseridae*

important knowledge that should be estimated to enable the judicious use of a particular ingredient in feed formulation (Glencross et al. 2007; Jobling 2015).

10.5 Concluding Remarks and Future Perspectives

Chemosensory systems in sturgeons, olfaction and gustation, are well investigated in comparison with some other sensory systems like lateral line, electroreception, vision and hearing. Many studies in sturgeon chemoreception were performed on Siberian sturgeon that substantially improve our knowledge about the chemosensory systems and their role in behaviour of this fish and other acipenserids as well. The morphology of olfactory organ and its morphogenesis by means of histology, scanning and transmission electron microscopy was followed in details in Siberian sturgeon ontogeny. The olfactory epithelium and olfactory receptor cell fine structure and distribution is well studied. Pattern of food search behaviour evoked by smell of natural food organisms is easily recognisable in sturgeons and is described for several species including *A. baerii*. Spectra of olfactory effective free amino acids which attract sturgeon fish and evoke typical food search is ascertain also. The morphology and distribution of the taste buds in Siberian sturgeon larvae and juveniles is investigated, and the taste preferences to free amino acids and other substances are defined. The threshold concentration for some effective taste substances is founded. The rate of extraoral and oral taste bud development and distribution and sturgeon responsivity to taste stimuli is well studied.

Nevertheless, functional modes in olfaction and gustation of sturgeons are still poorly understood. Recent and widely used electrophysiological approach has not been applied to explore the width and composition of olfactory and taste spectra, dose-response relationships, sensitivity level, differential sensitivity, adaptation rate and other basic physiological characteristics of sturgeon chemosensory systems. Many aspects of sturgeon chemocommunication are still needed to be studied; homing and olfactory imprinting are needed in proof. Our knowledge about the sex pheromones and olfactory control of sturgeon reproduction is very poor, and what is done so far is only a beginning of an extended series of experiments that will provide a basis for our understanding of the subject. Behavioural data concerning the role of chemical signals in antipredator behaviour and social interactions in acipenserid fishes are extremely scarce (Kasimov and Mamedov 1998). The morphological structure and functional peculiarities of the primary olfactory and gustatory centres are still weakly investigated. The common chemical sense has not been studied yet, and any data about the common chemical sense in sturgeons are absent. The list of unsolved problems in sturgeon chemoreception can be easily extended, but many of them will be undoubtedly uncovered in the nearest future. Such studies may give the results highly useful for sturgeon aquaculture and for the improving and restoration of their natural populations.

Acknowledgments The study on sturgeon fish chemoreception was supported by the Russian Foundation for Basic Research (grant No. 13-04-00711). Manuscript preparing was supported by the Russian Science Foundation (grant No. 14-50-00029).

References

Atema J (1971) Structures and functions of the sense of taste in the catfish (*Ictalurus natalis*). Brain Behav Evol 4:273–294

Atema J (1980) Chemical senses, chemical signals and feeding behaviour in fishes. Fish behaviour and its use in the capture and culture of fishes, Manila, pp 57–101

Baatrup E, Doving KB (1990) Histochemical demonstration of mercury in the olfactory system of salmon (*Salmo salar L.*) following treatments with dietary methylmercuric chloride and dissolved mercuric chloride. Ecotoxicol Environ Saf 20:277–289

Barth AL, Justice NJ, Ngai J (1996) Asynchronous onset of odorant receptor expression in the developing zebrafish olfactory system. Neuron 16:23–34

Billard R, Lecointre G (2000) Biology and conservation of sturgeon and paddlefish. Rev Fish Biol Fish 10:355–392

Boglione C, Bronzi P, Cataldi E et al (1999) Aspects of early development in the Adriatic sturgeon *Acipenser naccarii*. J Appl Ichthyol 15:207–213

Boglione C, Cataldi E, Sighicelli M et al (2006) Contribution to the trophic ecology of the adriatic sturgeon *Acipenser naccarii*: morphological observations on mouth and head sensorial equipment. J Appl Ichthyol 22:208–212

Boiko NE, Grigor'yan RA (2002) Effect of thyroid hormones on imprinting of chemical signals at early ontogenesis of the sturgeon *Acipenser gueldenstaedtii*. J Evol Biochem Physiol 38(2):218–222

Boiko NE, Grigor'yan RA, Chichachev AS (1993) Obonyatel'niy imprinting u molodi russkogo osetra *Acipenser gueldenstaedtii* (olfactory imprinting in juveniles of Russian sturgeon, *Acipenser gueldenstaedtii*). Zhurnal Evolyutsionnoi Biokhimii i Fiziologii 29:509–514. (in Russian)

Buck L, Axel R (1991) A novel multigene family may encode odorant receptors: a molecular basis for odor recognition. Cell 65:175–187

Burdukovskaya TG, Pronin NM (2010) Novie vidi paraziticheskih copepod roda *Salmincola* (Copepoda, Lernaeopodidae) iz obonyatel'nih yamok hariusovih (Thymallidae) i sigovih (Coregonidae) rib basseina oz. Baikal (new species of parasitic copepods of genus *Salmincola* (Copepoda, Lernaeopodidae) from the olfactory pits of Thymallidae and Coregonidae fish from Baikal basin). Izvestia Irkutskogo gosudarstvennogo universiteta, seria "Biologia. Ecologia" 3(2):20–29. (in Russian)

Camacho S, Ostos-Garrido MV, Domezain A et al (2010) Study of the olfactory epithelium in the developing sturgeon characterization of the crypt cells. Chem Senses 35:147–156

Caprio J (1982) High sensitivity and specificity of olfactory and gustatory receptors of catfish to amino acids. In: Hara TJ (ed) Chemoreception in fishes. Elsevier Scientific Publishing Comp, Amsterdam, pp 109–134

Caprio J, Derby CD (2008) Aquatic animal models in the study of chemoreception. In: Basbaum AI, Kaneko A, Shepherd GM, Westheimer G, Firestein S, Beauchamp GK (eds) The senses: a comprehensive reference, 4: olfaction and taste. Academic Press, San Diego, pp 97–134

Chen X-Y, Arratia G (1994) Olfactory organ of Acipenseriformes and comparison with other Actinopterygians: patterns of diversity. J Morphol 222:241–267

Devitsina GV, Golovkina TV (2008) Morfologia vkusovogo apparata rotovoi polosti navagi *Eleginus navaga* L. i treski *Gadus morhua marisalbi* der. (Gadidae, Teleostei) (morphology of the gustatory apparatus of the oral cavity of navaga, *Eleginus navaga* L. and cod *Gadus morhua marisalbi* der. (Gadidae, Teleostei)). Sensornie sistemi 22(1):70–85. (in Russian)

Devitsina GV, Golovkina TV, Rod'kin MM (2011) Features of gustatory system morphology in early juveniles of Siberian sturgeon *Acipenser baerii* (Acipenseridae, Acipenseriformes). J Ichthyol 51:1104–1116

Devitsina GV, Kazhlayev AA (1993) Development of chemosensory organs in Siberian sturgeon, *Acipenser baeri*, and stellate sturgeon, *A stellatus*. J Ichthyol 33:9–19

Devitsina GV, Gadzhieva AR (1994) Morphological features of development of taste receptors in early ontogeny of the Russian sturgeon *Acipenser gueldenstaedtii* Brandt. Sens Syst 8(2):65–70

Devitsina GV, Gadzhieva AR (1996) Dynamics of morphological development of gustatory system during the early ontogenesis in two representatives of acipenserids *Acipenser nudiventris* and *A. persicus*. J Ichthyol 36(8):642–653

Devitsina GV (1977) Sravnitel'noe issledovanie morfologii obony'atel'nogo analizatora rib (comparative study of the morphology of the olfactory analyzer of fishes). Voprosy Ikhtiologii 17:129–133. (in Russian)

Dittman AH, Quinn TP, Dickhoff WW et al (1994) Interactions between novel water, thyroxine and olfactory imprinting in underyearling coho salmon (*Oncorhynchus kisutch* Walbaum). Aquacult Fish Manag 25:157–169

Dogiel A (1887) Ueber den Bau des Geruchsorgans bei Ganoiden, Knochenfischen und Amphibien. Arch Mikrosk Anat 29:74–139

Doroshenko MA, Pinchuk LE (1978) Osobennosti morfologii porazhennogo mikrosporidiami obony'atel'nogo epitelia morskih rib (morphology of the olfactory epithelium in marine fishes affected with microsporidia). Biologia Moria 3:88. (in Russian)

Doving KB (1986) Functional properties of the fish olfactory system. In: Autrum H et al (eds) Progress in sensory physiology. Springer-Verlag, Berlin, pp 39–104

Døving KB, Kasumyan A (2008) Chemoreception. In: Finn RN, Kapoor BG (eds) Fish larval physiology. Science Publishers, Enfield, pp 331–394

Døving KB, Mårstøl M, Andersen JR et al (1994) Experimental evidence of chemokinesis in newly hatched cod larvae (*Gadus morhua* L.) Mar Biol 120:351–358

Døving KB, Stabell OB (2003) Trails in open waters: sensory cues in salmon migration. In: Collin SP, Marshall NJ (eds) Sensory processing in the aquatic environment. Springer-Verlag, New York, pp 39–52

Døving KB, Dubois-Dauphin M, Holley A et al (1977) Functional anatomy of the olfactory organ of fish and the ciliary mechanisms of water transport. Acta Zool 58:245–255

Dragomirov NI (1954) Razvitie kozhnih receptorov na nizhney storone golovi u lichinok osetra, perehodiashih k pridonnomu obrazu zhizni (development of skin receptors on the lower side of head in sturgeon larvae passing over to bottom mode of life). Dokladi Akademii Nauk SSSR 97(l):173–176. (in Russian)

Finger TE (1976) Gustatory pathways in the bullhead catfish. I. Connections of the anterior ganglion. J Comp Neurol 165:513–526

Finger TE, Drake SK, Kotrschal K et al (1991) Postlarval growth of the peripheral gustatory system in the channel catfish, *Ictalurus punctatus*. J Comp Neurol 314:55–66

Finger TE, Morita Y (1985) Two gustatory systems: facial and vagal gustatory nuclei have different brainstem connections. Science 227:776–778

Glencross BD, Booth M, Allasn GL (2007) A feed is only as good as its ingredients – a review of ingredient evaluation strategies for aquaculture feeds. Aquac Nutr 13:17–34

Gomes E, Dias J, Kaushik SJ (1997) Improvement of feed intake through supplementation with an attractant mix in European sea bass fed plant protein rich diets. Aquat Living Resour 10:385–389

Goryunova VB, Shagaeva VG, Nikol'skaya MP (2000) Analysis of anomalies in the structure of larval and young acipenseridae in the Volga-Caspian basin under artificial reproduction. J Ichthyol 40:762–766

Hansen A, Finger TE (2000) Phyletic distribution of crypt-type olfactory receptor neurons in fishes. Brain Behav Evol 55:100–110

Hansen A, Zielinski BS (2005) Diversity in the olfactory epithelium of bony fishes: development, lamellar arrangement, sensory neuron cell types and transduction components. J Neurocytol 34:183–208

Hara TJ (1994a) Olfaction and gustation: an overview. Acta Physiol Scand 152:207–217

Hara TJ (1994b) The diversity of chemical stimulation of fish olfaction and gustation. Rev Fish Biol Fish 4(1):1–35

Hara TJ (2007) Gustation fish physiology. In: Farrell AP, Brauner CJ (eds) Sensory systems neuroscience. Elsevier Press, New York, pp 45–96

Hasler AD, Sholz AT (1983) Olfactory imprinting in salmon. Springer, New York

Herrick CJ (1901) The cranial nerves and cutaneous sense organs of the north American silurid fishes. J Comp Neurol Psychol 11:177–249

Hofmann MH, Bleckmann H (1997) Subdivision of the olfactory system in the sterlet *Acipenser ruthenus*. Neurosci Lett 233:154–156

Huesa G, Anadón R, Yáñez J (2000) Olfactory projections in a chondrostean fish, *Acipenser baerii*: an experimental study. J Comp Neurol 428:145–158

Hustvedt SO, Storebakken T, Salte R (1991) Does oral administration of oxolinic acid or tetracycline affect feed intake of rainbow trout? Aquaculture 92:109–113

Jakubowski M, Whitear M (1990) Comparative morphology and cytology of taste buds in teleosts. Z mikrosk-anat Forsch 104(4):529–560

Jobling M (2015) Fish nutrition research: past, present and future. Aquacult Int 23. doi:10.1007/s10499-014-9875-2

Jobling M, Gomes E, Dias J (2001) Feed types, manufacture and ingredients. In: Houlihan D, Boujard T, Jobling M (eds) Food intake in fish: feed types, manufacture and ingredients. Blackwell Science, Oxford, pp 25–48

Johnsen PB, Adams MF (1986) Chemical feeding stimulants for the herbivorous fish, *Tilapia zillii*. Comp Biochem Physiol 83A(1):109–112

Jones KA (1989) The palatability of amino acids and related compounds to rainbow trout, *Salmo gairdneri* Richardson. J Fish Biol 34:149–160. doi:10.1111/j.1095-8649.1989.tb02964.x

Jones KA (1990) Chemical requirements of feeding in rainbow trout, *Oncorhynchus mykiss* (Walbaum): palatability studies on amino acids, amides, amines, alcohols, aldehydes, saccharides, and other compounds. J Fish Biol 37:413–423. doi:10.1111/j.1095-8649.1990.tb05872.x

Jones KA (1992) Food search behaviour in fish and the use of chemical lures in commercial and sports fishing. In: Hara TJ (ed) Fish chemoreception. Chapman and Hall, London, pp 288–320

Kanwal JS, Caprio J (1983) An electrophysiological investigation of the oro-pharyngeal (IX-X) taste system in the channel catfish, *Ictalurus punctatus*. J Comp Physiol 150A:345–357

Kanwal JS, Finger TE (1992) Central representation and projections of gustatory systems. In: Hara TJ (ed) Fish chemoreception. Chapman and Hall, London, pp 79–102

Kapoor BG, Evans HE, Pevzner RA (1975) The gustatory system in fish. Adv Mar Biol 13:53–108

Kasimov RY, Mamedov CA (1988) Hemoreceptornie povedencheskie reakcii u kurinskogo osetra I beluga v rannem ontogense (chemoreceptor behavioral responses in Kura river sturgeon and beluga in early ontogenesis). Izvestia Akademii Nauk Az SSR, Seria Biologicheskie Nauki 1:68–75. (in Russian)

Kasimov RY, Mamedov CA (1998) Vnutrividovie himicheskie signali y molodi kurinskogo osetra (interspecific chemical signals in Kura sturgeon juveniles). Izvestia Akademii Nauk Az SSR, Seria Fiziologia cheloveka i zhivotnih 54(3–4):118–121. (in Russian)

Kasumyan AO (1999) Olfaction and taste in sturgeon behaviour. J Appl Ichthyol 15:228–232

Kasumyan AO (2004) The olfactory system in fish: structure, function, and role in behaviour. J Ichthyol 44(Suppl 2):S180–S223

Kasumyan AO, Mamedov CA (2011) Behavioral response of mature males of Acipenseridae to female sex pheromone. J Ichthyol 51(6):457–465

Kasumyan AO, Ryg M, Døving KB (1998) Effect of amino acids on the swimming activity of newly hatched turbot larvae (*Scophthalmus maximus*). Mar Biol 131:189–194

Kasumyan AO, Taufik LR (1994) Behavioral reaction of juvenile sturgeons (Acipenseridae) to amino acids. J Ichthyol 34(1):90–103

Kasumyan AO (1993) Behavioral reaction of male sturgeons to the releaser postovulatory sex pheromone of females. Doklady Biol Sci 333(1–6):439–441

Kasumyan AO (2002) Sturgeon food searching behaviour evoked by chemical stimuli: a reliable sensory mechanism. J Appl Ichthyol 18:685–690

Kasumyan AO, Døving KB (2003) Taste preferences in fish. Fish Fish 4:289–347. doi:10.1046/j.1467-2979.2003.00121.x

Kasumyan AO, Devitsina GV (1997) The effect of olfactory deprivation on chemosensory sensitivity and the state of taste receptors of acipenserids. J Ichthyol 37:786–798

Kasumyan AO, Kazhlaev AA (1993a) Formation of searching behavioural reaction and olfactory sensitivity to food chemical signals during ontogeny of sturgeons (*Acipenseridae*). J Ichthyol 33:51–65

Kasumyan AO, Kazhlaev AA (1993b) Behavioral responses of early juveniles of Siberian sturgeon *Acipenser baerii* and stellate sturgeon *A. stellatus* (*Acipenseridae*) to gustatory stimulating substances. J Ichthyol 33:85–97

Kasumyan AO, Ponomarev VY (1990) The ontogeny of feeding behavior in relation to natural chemical signals in cyprinid fishes. J Ichthyology 30(5):89–100

Kasumyan AO, Sidorov SS, Kazhlaev AA, Pashchenko NI (1995) Behavioral responses of young stellate sturgeon to smell and taste of artificial feeds and their components. In: Gershanovich AD, Smith TIJ (eds) Proceeding of international symposium on Acipenserids. VNIRO Publishing, Moscow, pp 278–288

Kitamura S, Ogata H, Takashima F (1994) Olfactory responses of several species of teleost to F-prostaglandins. Comp Biochem Physiol 107A:463–467

Kiyohara S, Yamashita S, Harada S (1981) High sensitivity of minnow gustatory receptors to amino acids. Physiol Behav 26(6):1103–1108

Klaprat DA, Evans RE, Hara TJ (1992) Environmental contaminants and chemoreception in fishes. In: Hara TJ (ed) Fish chemoreception. Chapman and Hall, London, pp 321–341

Kleerekoper H (1969) Olfaction in fishes. Indiana University Press, Bloomington

Krieger J, Hett AK, Fuerst PA et al (2008) The molecular phylogeny of the order Acipenseriformes revisited. J Appl Ichthyol 24(Suppl. 1):36–45. doi:10.1111/j.1439-0426.2008.01088.x

Lari E, Kasumyan A, Falahat F et al (2013) Palatability of food animals for stellate sturgeon *Acipenser stellatus* Pallas, 1771. J Appl Ichthyol 29:1222–1224. doi:10.1111/jai.12324

Mackie AM (1982) Identification of the gustatory feeding stimulants. In: Hara TJ (ed) Chemoreception in fishes. Elsevier Scientific Publishing Comp, Amsterdam, pp 275–291

Maklakova ME, Kondratieva IA, Mikhailova ES et al (2011) Effect of antibiotics on immunophysiological status and their taste attractiveness for rainbow trout *Parasalmo (=Oncorhynchus) mykiss* (Salmoniformes, Salmonidae). J Ichthyol 51(11):1133–1142. doi:10.1134/S0032945211110063

Mamedov CA, Gadzhiev RY, Akhundov MM (2009) Novie tehnologii osetrovodstva v Azerbaydzhane (new technologies of sturgeon cultivation in Azerbaijan). Elm, Baku. (in Russian)

Marui T, Caprio J (1992) Teleost gustation. In: Hara TJ (ed) Fish chemoreception. Chapman and Hall, London, pp 171–198

Mearns KJ, Ellingsen OF, Døving KB et al (1987) Feeding behaviour in adult rainbow trout and Atlantic salmon parr, elicited by chemical fractions and mixtures of compounds identified in shrimp extract. Aquaculture 64(1):47–63

Morrison EE, Plumb JA (1994) Olfactory organ of channel catfish as a site of experimental *Edwardsiella ictaluri* infection. J Aquat Anim Health 6:101–109

Nonnotte L, Nonnotte G, Truchot JP, Williot P (1991) Gill chemoreceptor ultrastructure in a Chondrostean fish : *Acipenser baerii*. In: Williot P (ed) Acipenser. Cemagref Publ, Anthony, France, p 169

Obukhov DK (1999) Cytoarchitectonics and neuronal organization of the sturgeon telencephalon. J Appl Ichthyol 15(4–5):92–95

Palatnikov GM (1983) Rol' obonyania v pishedobivatel'nom povedii osetra *Acipenser güldenstädti* (role of olfaction in foraging behavior in sturgeon *Acipenser güldenstädti*). In: Kreps EM (ed) Funkcional'naya evolucia central'noy nervnoy sistemi. Izdatel'stvo Nauka, Leningradskoe otdelenie, pp 37–41. (in Russian)

Palatnikov GM (1989) Elektrofisiologicheskoe izuchenie oboniatel'nih proekciy v promezhutochnom I srednem mozgu osetra *Acipenser güldenstädti* (electrophysiological study of olfactory projections in medulla oblongata and midbrain of sturgeon *Acipenser güldenstädti*). Zhurnal Evolyutsionnoi Biokhimii i Fiziologii 25(3):354–359. (in Russian)

Pashchenko NI, Kasumyan AO (2015) Scanning electron microscopy of olfactory organ development in grass carp *Ctenopharyngodon idella* ontogeny. J Ichthyol 55(6)

Pavlov DS, Kasumyan AO (1998) The structure of the feeding behaviour of fishes. J Ichthyol 38:116–128

Pavlov DS, Sbikin YN, Popova IK (1970) Rol' organov chuvstv pri pitanii molodi osetrovih rib (the significance of sense organs in feeding behaviour of young sturgeons). Zoologicheskii Zhurnal 49:872–880. (in Russian)

Pevzner RA (1981) Ul'trastrukturnaia organizacia vkusovih receptorov kostno-hriashevih rib. I. Vzroslie osetrovie ribi (the fine structure of taste buds of the ganoid fishes. I. Adult Acipenseridae). Tsitologia 23(7):760–765. (in Russian with English summary)

Pevzner RA (1985) Ul'trastrukturnaia organizaciya vkusovih receptorov kostno-hriashevih rib. III. Lichinki v period zheltochnogo pitania (ultrastructure of taste receptors in Chondrichthyes. III. Sturgeon larvae in yolk-feeding period). Tsitologia 27(11):1240–1246. (in Russian with English summary)

Poe WE, Wilson RP (1989) Palatability of diets containing sulfadimethoxine, ormetoprim, and Romet 30 to channel catfish fingerlings. Progress Fish Cult 51:226–228

Pyatkina GA (1975) Elektronnomikroskopicheskoe issledovanie organa oboniania sterlyadi (Acipenser ruthenus) (Electronnomicroscopic study of the olfactory organ in sterlet (*Acipenser ruthenus*)). Arhiv Anatomii, Gistologii i Embriologii 68(5):85–93. (in Russian)

Pyatkina GA (1976) Receptornie kletki razlichnih tipov I kolichestvennoe sootnoshenie mezhdu nimi v organe oboniania lichinok I polovozrelih osobey osetrovih rib (receptor cells of various types and their proportional interrelation in the olfactory organ of larvae and adults of acipenserid fishes). Tsitologia 18:1444–1449. (in Russian)

Pyka J, Kolman R (2003) Feeding intensity and growth of Siberian sturgeon *Acipenser baeri* Brandt in pond cultivation. Arch Pol Fish 11:287–294

Reutter K (1986) Chemoreceptors. In: Bereiter-Hahn J, Matoltsy AG, Richards KS (eds) Biology of the integument, II. Springer, Berlin, pp 586–604

Reutter K (1992) Sturcture of the peripheral gustatory organ, represented by the siluroid fish Plotosus lineatus (Thunberg). In: Hara TJ (ed) Fish chemoreception. Chapman and Hall, London, pp 60–78

Reutter K, Breipohl W, Bijvank GJ (1974) Taste bud types in fishes. II. Scanning electron microscopical investigations on *Xiphophorus helleri* Heckel (Poeciliidae, Cyprinodontiformes, Teleostei). Cell Tissue Res 153:151–165

Reutter K, Hansen A (2005) Subtypes of light and dark elongated taste bud cells in fishes. In: Reutter K, Kapoor BG (eds) Fish chemosenses. Sci. Publ. Inc., Enfield, pp 211–230

Robinson EH, Brent JR, Crabtree JT et al (1990) Improved palatability of channel catfish feeds containing Romet-30®. J Aquatic Animal Health 2:43–48

Rustamov EK (1987) Razvitie konechnogo mozga osetrovih rib v rannem ontogenese (development of telencephalon in early ontogeny in sturgeons). Zhurnal Evolyutsionnoi Biokhimii i Fiziologii 23(2):253–257. (in Russian)

Schmalhausen OI (1962a) Morfologicheskoe issledovanie oboniatel'nih organov rib (morphological study of olfactory organs in fishes). Trudi Instituta morfologii i ekologii zhivotnih imeni AN Severtzova AN SSSR 40:157–187

Schmalhausen OI (1962b) Narushenie razvitiya oboniatel'nogo organa u osetrovih rib pri opredelennih usloviah (disruption of olfactory organ development in specific conditions). In: Proceedings of AN Severzev's Institute of Animal Evolutionary Morphology and Ecology, Russian Acad Sci, vol 40, pp 188–218

Shamushaki VAJ, Kasumyan AO, Abedian A et al (2007) Behavioral response of the Persian sturgeon (*Acipenser persicus*) juveniles to free amino acid solutions. Mar Freshwater Behav Physiol 40(3):219–224

Shamushaki VAJ, Abtahi B, Kasumyan AO et al (2008) Taste attractiveness of free amino acids for juveniles sturgeon *Acipenser persicus*. J Ichthyol 48:130–140. doi:10.1134/S0032945208010116

10 Olfaction and Gustation in *Acipenseridae*

Shamushaki VAJ, Abtahi B, Kasumyan AO (2011) Olfactory and taste attractiveness of free amino acids for Persian sturgeon juveniles, *Acipenser persicus*: a comparison with other acipenserids. J Appl Ichthyol 27:241–245. doi:10.1111/j.1439-0426.2011.01687.x

Sorensen PW, Caprio J (1998) Chemoreception. In: Evans DH (ed) The physiology of fishes. CRC Press, Boca Raton, pp 375–405

Soriguer MC, Domezain A, Aragonés J et al (2002) Feeding preference in juveniles of *Acipenser naccarii* Bonaparte 1836. J Appl Ichthyol 18:691–694

Stabell OB (1984) Homing and olfaction in salmonids: a critical review with special reference to the Atlantic salmon. Biol Rev Camb Philos Soc 59:333–388

Stacey N, Sorensen P (2006) Reproductive pheromones. In: Sloman KA, Wilson RW, Balshine S (eds) Behaviour and physiology of fish. Academic Press, London, pp 359–412

Stacey NE, Cardwell JR (1997) Hormonally derived pheromones in fish: new approaches to controlled reproduction. In: Fingerman M, Nagabhushanam R, Thompson MF (eds) Recent advances in marine biotechnology, vol V. 1. Science Pub, Enfield, pp 407–454

Stroganov NS (1968) Akklimatizacia i virashivanie osetrovih rib v prudah (acclimatization and cultivation of sturgeon fishes in punds). Moscow University Press, Moscow. (in Russian)

Takeda M, Takii K, Matsui K (1984) Identification of feeding stimulants for juvenile eel. Bull Jap Soc Sci Fish 50(4):645–651

Tandler A, Berg BA, Kissil GW et al (1982) Effect of food attractants on appetite and growth rate of gilthead bream, *Sparus aurata* L. J Fish Biol 20:673–681

Toften H, Jorgensen EH, Jobling M (1995) The study of feeding preferences using radiography: oxytetracycline as a feeding deterren and squid extract as a stimulant in diets for Atlantic salmon. Aquac Nutr 1:145–149

Vasil'ev VP (1985) Evolucionnaia kariologia rib (evolutionary karyology of fish). Nauka, Moscow. (in Russian)

Vasil'ev VP (2009) Mechanisms of polyploid evolution in fish: polyploidy in sturgeons. In: Carmona R, Domezain A, Garsia Gallego M et al (eds) Biology, conservation and sustainable development of sturgeons. Springer, NY, pp 97–117

Watson LR, Groff JM, Hedrick RP (1998) Replication and pathogenesis of white sturgeon iridovirus (WSIV) in experimentally infected white sturgeon *Acipenser transmontanus* juveniles and sturgeon cell lines. Dis Aquat Org 32:173–184

Williams R, Poulet SA (1986) Relationship between the zooplankton phytoplankton particulate matter and dissolved free amino acids in the Celtic Sea. Mar Biol 90:279–284

Yamamoto M (1982) Comparative morphology of the peripheral olfactory organ in teleosts. In: Hara TJ (ed) Chemoreeeption in fishes. Elsevier, Amsterdam, pp 39–59

Zeiske E, Theisen B, Breucker H (1992) Structure, development, and evolutionary aspects of the peripheral olfactory system. In: Hara TJ (ed) Fish chemoreception. Chapman and Hall, London, New York, pp 13–39

Zeiske E, Kasumyan A, Bartsch P et al (2003) Early development of the olfactory organ in sturgeons of the genus *Acipenser*, a comparative and electron microscopic study. Morphol Embryol 206(5):357–372

Nutritional Requirements of the Siberian Sturgeon: An Updated Synthesis

11

Bahram Falahatkar

Abstract

Sturgeon aquaculture grows fast with 465% increasing through 2003–2013. Unfortunately, there is no special culture system or nutritional requirements for these valuable species. Among 27 different sturgeon species, Siberian sturgeon *Acipenser baerii* is one of the best sturgeon fish for culture in different environments with high growth rate and adaptation to captivity. With growing Siberian sturgeon aquaculture, nutritional requirements at any life stage and culture condition should be considered. This chapter provides basic information and recent data on nutrient requirements of Siberian sturgeon to illuminate the way for sturgeon aquaculturists, feed factories, and researchers to improve and provide appropriate specialized feeds for the fish. Hence, this chapter is focused on macro- and micronutrients, feeding practices, and starvation effect of Siberian sturgeon.

Keywords

Siberian sturgeon • Feeding • Requirement • Macro- and micronutrients • Aquaculture

B. Falahatkar
Fisheries Department, Faculty of Natural Resources, University of Guilan, Sowmeh Sara, 1144, Guilan, Iran
e-mail: falahatkar@guilan.ac.ir

Introduction

As livestock and poultry production, nutrition plays an important role in aquaculture industry by affecting growth, product quality, health, production cost, and environmental issue. Feed costs are around 50% of variable costs of aquaculture activities and therefore influence economic profits (NRC 2011). Improvement of nutritious competence and cost-effective diets depends on knowing fish nutritional requirements and matching those requirements with balanced diet formulations and proper feeding practices.

Nutrients are necessary for any organism to maintain life developments and set aside steady state, activity, growth, and reproduction. Nutrients provide precursors for the biosynthesis of structural or storage molecules, enzymes, metabolic intermediates, and molecules' surplus.

Many sturgeon species are endangered or critically endangered because of dam construction, cutting the reproduction migration, poaching and overfishing, and environmental pollution (Birstein 1993; Williot et al. 2002; Falahatkar and Nasrollahzadeh 2011). Sturgeon culture has grown worldwide owing to their meat quality and market value of caviar (Williot et al. 2001; Bronzi et al. 2011; Bronzi and Rosenthal 2014). During the last two decades, sturgeon culture is favored by many countries because of high growth rate, better feed efficiency, and resistance to stressors and diseases (Falahatkar et al. 2009; Poursaeid et al. 2015). Sturgeon farming is important not only for producing delicious boneless meat and precious caviar but also for stock rehabilitation and conservation programs. Sturgeon aquaculture has taken over fisheries because there is almost no sturgeon fishery, and wild caviar is banned for export to protect wild stocks. China is the leader of sturgeon production with the value of 85% of the total sturgeon production worldwide (FAO 2014). After China, Russia and EU are the main producers. In 2013, 367.7 tons of Siberian sturgeon were recorded for inland aquaculture production, only in freshwater, and Uruguay, Spain, Argentina, Belarus, Bulgaria, Cyprus, France, and Italy are the main producers, respectively (FAO 2014).

Among 27 sturgeon species in Acipenseridae family, Siberian sturgeon is one of the best species for aquaculture because of fast growing, stress resistance, and adaptation to aquaculture condition (Eslamloo et al. 2012; Falahatkar et al. 2014b). In the 1970s, Siberian sturgeon was first introduced into the farming system in the former Soviet Union. Also, the first specimens, more or less at the same time, were introduced into France as part of scientific program (Williot et al. 1997).

Siberian sturgeon can be reared monoculture in raceways, circular tanks, artificial ponds, recirculation systems, nets, cages, and earthen ponds under temperate conditions (Williot et al. 2001; Adámek et al. 2007). Although this is a Eurasian fish and lives in temperate waters, the experience shows that Siberian sturgeon grows well in high temperature as well. The average time to obtain meat is 2–3 years, but for caviar production, it takes 6–8 years (Falahatkar et al. 2014a). So females remain in culture ponds for many years under optimal growth and development conditions; males are reared for meat production (Falahatkar et al. 2011). It is recommended to separate males and females with appropriate methods at 2.5–3 kg in size to sell males on the market (Steffens 2008; Falahatkar and Poursaeid 2014).

This fish is typically considered a benthophagous species (Ruban and Konoplja 1994), but in individual cases, it can be shifted into the predatory habit (Ruban 2005). They are considered excellent bottom-feeders because of sensitive barbells underside the long snout to find food items and a protruding lip for sucking up the prey. They have a unique gastrointestinal tract with pyloric stomach wall which is hypertrophied to gizzard-like organ (Hung and Deng 2002). Also, the hindgut is morphologically changed to spiral valve.

There are some information regarding the developmental growth and feeding biology in Holčik (1989), Gisbert and Williot (2002), and Ruban (2005). In culture condition, they can easily adapt to formulated diets and be fed with pellets with fish meal and fish oil and vegetable ingredients. After more than four decades of sturgeon rearing, the lack of nutrition and feeding information is irritating. Therefore, due to increased interest in sturgeon aquaculture and restocking programs, it is necessary to find nutritional requirements of these fish including Siberian sturgeon.

11.1 Protein and Amino Acids

Proteins and amino acids are vital molecules because of their role in the structure and metabolism of animals. It is assumed that sturgeon like the other fish species needs the same ten essential amino acids (arginine, lysine, histidine, leucine, isoleucine, valine, phenylalanine, threonine, methionine, and tryptophan; Table 11.1) due to the inability to synthesize them. Thus, protein is an essential component for all cell types in the body. Body tissues are continuously formed and broken down. In the process of animal growth, protein synthesis exceeds degradation, and the balance between these two routes consequences in protein deposition or degradation.

Besides ten essential amino acids, fish have a general need to a source of amino groups for synthesis of nonessential amino acids. Most nonessential amino acids can be synthesized by simple routes.

Free amino acids as feed stimulants were evaluated in many sturgeon fish and in Siberian sturgeon (Kasumyan 1994; Kasumyan and Taufik 1994). From 20 tested amino acids, glycine and L-alanine in the minimum level of 1 μM induced feed searching behavior.

When the Siberian sturgeon larvae were fed with live food and semi-purified casein-gelatin diet, amino acid absorption was more efficient when fish were fed with live food compared with semi-purified diet (Dabrowski et al. 1987). Fauconneau et al. (1986) analyzed protein metabolism in the same species. During short period of fasting, protein synthesis decreased to a large extent, and arginine oxidation increase was small. Fish fed with live food showed a better protein metabolism efficiency, incorporated with a lower rate of protein synthesis and a lower arginine oxidation compared to fish fed with artificial feeds. Further studies are compulsory on the digestibility of different protein sources, the rate of protein synthesis, and excretion in different age classes.

The maximum 50% dietary protein requirement was determined in Siberian sturgeon (Médale et al. 1995). Protein-to-energy ratio was reported to be 20–22 mg kJ^{-1}. Kaushik et al. (1994) reported that apparent digestibility of protein and growth performance in Siberian sturgeon are higher when the fish was fed with diet containing casein or casein and soybean diets compared with fish fed with a fish meal-based diet. It should be considered in the research with sturgeon that this kind of fish grows poorly when fed with crystalline amino acids (Ng et al. 1996). In a study by Kaushik et al. (1991), essential amino acid requirement was provided for 22 g of Siberian sturgeon, showing higher requirement of lysine, leucine, and arginine, respectively (Table 11.1). The results showed the best performance in diet containing 40 ± 2% crude protein.

Kaushik et al. (1989) found that growth performance of Siberian sturgeon with 90–400 g wet weight utilizing different carbohydrates was better with a diet containing 36–38% crude protein compared with the diet containing 42% crude protein (Table 11.2).

Table 11.1 Whole body amino acid composition and estimated essential amino acid requirements for the Siberian sturgeon (Kaushik et al. 1991)

Amino acid[a]	Body weight (g) 44.7	1000	Requirement as % of dietary protein
Arginine	6.7 ± 0.3	7.7 ± 0.4	2.8 ± 0.6
Lysine	11.7 ± 1.8	11.1 ± 1.1	6.3 ± 2.0
Histidine	2.0 ± 0.1	2.4 ± 0.1	1.1 ± 0.4
Isoleucine	5.1 ± 0.3	5.4 ± 1.2	2.1 ± 0.6
Leucine	6.8 ± 0.8	7.4 ± 1.8	3.2 ± 1.8
Valine	5.3 ± 0.2	6.0 ± 1.8	2.3 ± 0.6
Phenylalanine	4.1 ± 0.1	3.8 ± 0.5	1.5 ± 0.4
Threonine	5.0 ± 0.2	5.4 ± 0.8	2.2 ± 0.6
Tyrosine	4.0 ± 0.2	5.3 ± 0.8	–
Methionine	NR[b]	NR	–
Tryptophan	NR	NR	–
Aspartic acid	8.4 ± 3.1	10.2 ± 2.1	–
Glutamic acid	18.5 ± 0.1	18.3 ± 1.6	–
Serine	4.8 ± 0.1	5.6 ± 0.7	–
Glycine	9.2 ± 0.4	11.9 ± 1.7	–
Alanine	7.6 ± 0.1	7.3 ± 2.7	–
Proline	NR	NR	–

[a]Expressed as g/16 g nitrogen
[b]Not reported

Table 11.2 The effect of macro and micro-nutrients in Siberian sturgeon at different sizes

Ingredient	Dose and duration of experiment	Fish weight (g)	Best results	References
Protein	29–52%, 12 weeks	22	40 ± 2%	Kaushik et al. (1991)
Lipid	12.5 or 21.8%, 8 weeks	49	38% extruded starch	Médale et al. (1991)
Carbohydrate	30% gelatinized starch or 38% crude starch, gelatinized starch, or extruded starch, 12 weeks	90 and 400		Kaushik et al. (1989)
Vitamin C	0 or 300 mg ascorbic acid kg^{-1}, 4 months	25.5 ± 0.5	300 mg kg^{-1} as ascorbyl-2-polyphosphate	Moreau et al. (1996)
Vitamin A	22,500 or 772,500 IU retinyl acetate kg^{-1}, 27 days	30 mg	Not significant	Fontagné et al. (2006)
Fluoride	75.2–1478.3 mg fluoride kg^{-1}, 12 weeks	11.9 ± 0.1	<162.6 mg kg^{-1}	Shi et al. (2013)
Phosphorus	0–1.38%, 56 days	14.6 ± 2.3	0.5–0.87%	Xu et al. (2011)
Choline	0–7.5 g kg^{-1}, 12 weeks	37.7 ± 2.9	1.5 g kg^{-1}	Yazdani Sadati et al. (2014)

11.2 Lipid and Fatty Acids

While most lipids are complex, fatty acids typically consist of dietary lipid intake. The requirement for specific fatty acids depends upon their differing functional roles and whether they can be synthesized endogenously. All fatty acids can be stored as energy sources, but some specific long-chain polyunsaturated fatty acids have a number of essential roles in metabolism.

Siberian sturgeon requires dietary lipid levels for growth, energy, and other physiological activities. Dietary lipid levels for sturgeon vary based on life stage, source of ingredient, and rearing condition. The quantity of dietary lipid level is also affected by dietary protein and carbohydrate levels, which can supply the energy need of fish. Protein and carbohydrate can be sources of energy through lipogenesis with amino acids and pyruvate serving as the main carbon source. As protein sources are the most costly ingredients in formulated diets for commercial use, the goal is to minimize dietary protein used as a source of energy. Thus, with an appropriate amount of energy supplied by lipid, protein requirements can be reduced or speared. In this case, protein will be used for growth, and other ingredients such as lipids and carbohydrates will be utilized as main sources of energy.

Studies on the utilization of dietary nonprotein energy sources in a juvenile Siberian sturgeon revealed that apparent digestibility coefficient of lipid was lower in fish fed with diet containing 22% lipid with a raw starch compared with diet containing 12.5% lipid with gelatinized starch (Médale et al. 1991). They suggested that dietary lipid can be more influential in sparing protein than dietary carbohydrates, and lipid appeared as the best energy source in sturgeon.

Fatty acids composition in diverse tissues of farmed Siberian sturgeon (2.53 kg body weight) was analyzed, revealing that fatty acids profile of diet differed from fatty acids of tissues where sturgeon particularly accumulated highly unsaturated fatty acids. They suggested that they are probably derived from the diet (Nieminen et al. 2014), but it seems that fish can also have desaturase enzymes for elongation of fatty acids (Xu et al. 1996). In a study on Siberian broodstocks, 6 months of feeding with high levels of eicosapentaenoic acid (EPA) and docosahexaenoic acid (DHA) showed higher egg mass weight, fecundity, and fertilization rate in fish fed with high level of DHA as well as higher weight, length, weight gain, and survival in 35 days post-hatch larvae (Lou et al. 2015).

Based on the studies on the other species, it seems that sturgeon needs to have both n-3 and n-6 fatty acids in the diet (Deng et al. 1998). Although studies on other sturgeon species have determined the appropriate levels of lipid in 25.8–35.7% for sub-yearling white sturgeon, *Acipenser transmontanus* (Hung et al. 1997b), there is no certain study showing the best level of dietary lipid in Siberian sturgeon in all life stages.

11.3 Carbohydrates

Aquatic animals do not have a specific requirement for dietary carbohydrates. Many studies show that when fish are fed diets without carbohydrates, they can survive and grow. This is probably because glucose is synthesized efficiently from nonglucose precursors through gluconeogenesis, mainly amino acids, the major substrates for gluconeogenesis (Walton and Cowey 1982; Cowey and Walton 1989).

Carbohydrates, including low-molecular-weight sugars, starch, and different cell wall and storage nonstarch polysaccharides, supply the most important part of energy for many organisms. They use the least glucose coming up from carbohydrate digestion for ATP production to provide cellular energy needs. The requirement of a suitable quantity of digestible carbohydrates in formulated diets for farmed fish is significant to spare the use of lipids and carbohydrates as sources of energy. However, dietary carbohydrate utilization by fish merits interest because of efforts to efficiently replace fish meal with plant protein sources, including considerable amounts of various carbohydrate components. The capability of aquatic animals to use dietary carbohydrates for growth suggests a difference along with fish species and fundamentally matches the feeding habits of the species; however, sturgeon are not equipped to directly digest carbohydrates.

The sturgeon's ability to utilize carbohydrates varies from teleosts. Lin et al. (1997) showed more efficient use of D-glucose and maltose than dextrin, fructose, lactose, and raw starch in white sturgeon. Furthermore, continuous feeding improves utilization of D-glucose and raw maize starch compared with hand-feeding. The low digestion ability of sucrose and lactose is related to the low activities of digestive enzymes in brush-border membrane (Hung et al. 1989). Also, because of lacking fructose active transport system in brush-border membrane, fructose can only be absorbed through brush-border membrane by passive diffusion (Hung and Deng 2002). Similar to white sturgeon, juvenile Siberian sturgeon did not use raw starch appropriately; however, it can be improved by gelatinization of starch or extruding feeds (Kaushik et al. 1989), also showing that high level of carbohydrate negatively impacts the liver size, morphology, and function. Using 15–30% of each glucose and corn starch in juvenile Siberian sturgeon, diet showed higher liver size in fish fed with 15% glucose, but no significant effect on growth performance after 8 weeks of rearing (Taati et al. 2015).

Although there are studies on the effects of various carbohydrate sources on different sturgeon species, data is rarely available on Siberian sturgeon in terms of utilization, function, transportation, and renal threshold of D-glucose through urinary excretion.

11.4 Energy

Sustaining life processes, activity, growth, and reproduction requires nutrients. Nutrients provide basic materials for the biosynthesis of structural molecules, enzymes, metabolic intermediates, and a surplus of molecules.

There is limit data on energy value and expenditure in sturgeon at different ages and culture conditions. In a study of Médale and Kaushik (1991) at different age classes of Siberian sturgeon (40, 230, and 1500 g body weight), voluntary feed intake decreased with age from 1.9% in small fish to 0.5% body weight daily in large fish at 18 °C. Energy retention was higher in large fish (55% as lipid) compared to other classes, retaining energy primarily as protein. In different studies on sturgeon, the results clearly demonstrated that these kinds of fish can utilize lipids more efficiently than carbohydrates as energy sources (Médale et al. 1991; Hung and Deng 2002). Because of difficulty in continuous feeding (nibbling) in habitats of sturgeon, diet (energy) intake is impossible to measure. Therefore, the experiments on Siberian sturgeon (Médale and Kaushik 1991) and white sturgeon (Cui et al. 1996) had difficulty to determine the energy requirement because of slaughter method and inaccurate measurement of oxygen consumption and ammonia excretion in the indirect calorimetry methods they used. Future experiments should be focused on the use of developed methods.

Despite the limitations, it is needed to embrace explicit and integrative utilization approaches of nutrients in terms of energy sources and different aspects of energy budget in different growth and reproduction phases.

11.5 Minerals

Information regarding minerals in fish feeds is limited compared to terrestrial animals because the fish can absorb some minerals from the aquatic media they live in, in addition to the diets. Problems related to the quantification of mineral needs consist of identification of the contribution of minerals from water, leakage of minerals from the diets before consumption, availability of appropriate test diets with low concentration of targeted mineral, and limited data on bioavailability of mineral (NRC 2011).

There is some general information regarding fish and shrimp showing the functions and deficiency signs but limited information regarding sturgeon. Selenium requirement of juvenile white sturgeon was studied; no growth changes appeared after 14 weeks of rearing with purified diet, except for fish fed with diet with no selenium compared to 0.3, 1.0, and 1.2 mg selenium/kg diet (from Hung and Deng 2002). Shi et al. (2013) reported that growth was inhibited at the higher doses when juvenile Siberian sturgeon are fed with diets containing 75.2 (control)–1478.3 mg fluoride kg^{-1} diet. They concluded that fluoride can be considered in the Siberian sturgeon diet of 360.8 mg kg^{-1} without any adverse effect on growth, but because of fluoride accumulation in scute and cartilage and dangerous possibility of fluoride that are fed in fish as low as 162.6 mg kg^{-1}, consumption of cartilage in sturgeon should be avoided.

Xu et al. (2011) determined the dietary phosphorus requirement of juvenile Siberian sturgeon. The highest concentration of 0.6% phosphorus/kg diet showed the highest weight gain after 56 days of rearing. They recommended that diet can be supplemented with 0.5–0.87% phosphorus/kg diet based on weight gain data and phosphorus content in the whole body.

Bagherzadeh Lakani (2016) used copper nanoparticle (0, 250, 500, and 1000 mg kg^{-1} diet) on the growth and physiology of juvenile Siberian sturgeon. The highest concentration of dietary copper nanoparticle reduced the weight and length of fish after 84 days of rearing. In addition, the highest concentration of copper nanoparticle was observed in the intestine, liver, gills, and muscle, respectively. Toxic effect of this nanoparticle was reported in the diet of sturgeon which is not a suitable nanomineral to be used in commercial scale of feed factory.

Now, many farmers and feed producers use designed mineral premix for salmonids in Siberian sturgeon diet. More investigations are needed to find the specific requirement for each mineral in purified or semi-purified diets in sturgeon.

11.6 Vitamins

Vitamins are classified into two groups of water- and fat-soluble forms. Eight different water-soluble vitamins are needed in small quantities in the diet, with coenzyme functions (B complex) as the main role. Three water-soluble vitamins, vitamin C, choline, and inositol, are required at larger amounts with more functions than coenzymes. Fat-soluble vitamins (A, D, E, and K) have independent function as enzymes and somehow as coenzymes such as vitamin K. Some vitamins may be synthesized from other essential nutrients to spare the dietary requirement; for example, choline has been proved to form methionine in channel catfish (*Ictalurus punctatus*) (Wilson and Poe 1988), ascorbic acid can form glucose in lake sturgeon *A. fulvescens* through *de novo* synthesis (Moreau et al. 1999a), and some water-soluble vitamins can be derived from microorganisms in the gastrointestinal tract in certain warmwater fish, e.g., tilapia (*Tilapia nilotica*) and channel catfish (Lovell and Limsuwan 1982; Burtle and Lovell 1989; Shiau and Lung 1993).

Quantitative and qualitative vitamin requirements of many farmed species are determined (NRC 2011). In case of sturgeon, although some information is available on vitamin C (Moreau et al. 1999a; Falahatkar et al. 2006, 2015), vitamin E (Agradi et al. 1993; Moreau et al. 1999b; Safarpour Amlashi et al. 2011), vitamin A (Fontagné et al. 2006; Wen et al. 2008), and vitamin B1 (Ghiasi et al. 2014), there is lack of information on Siberian sturgeon.

Regarding vitamin C, although sturgeon can synthesize it because of L-guluno-1,4-lactone oxidase activity in the kidneys, the results of Moreau et al. (1996) on juvenile Siberian sturgeon showed that the growth and other physiological performances were not affected by treatments (0 or 300 mg of vitamin C kg^{-1} diet) and showed no need of dietary vitamin C in this fish. Xie et al. (2006) revealed that dietary vitamin C may be necessary in different conditions (e.g., diseases) to achieve optimal immune response, especially in early developmental stages. Also, after 8 weeks of rearing juvenile Siberian sturgeon by different graded levels of vitamin C from 0 to 1600 mg kg^{-1}, the number of white blood cells and lysozyme activity improved when fish were fed 100 and 800 mg vitamin C kg^{-1} diet, respectively (Pourgholam et al. 2015).

Fontagné et al. (2006) found that when diet contained higher level of vitamin A (772,500 IU kg^{-1} vitamin A as retinyl acetate), deformation of Siberian sturgeon larvae will be decreased in the oxidized lipid diets. The effect of various levels of dietary choline was determined in juvenile Siberian sturgeon (initial weight 37.7 g) after 12 weeks of experiment. The results revealed the best growth performance was observed in fish fed with 1500 mg kg^{-1} diet (Yazdani-Sadati et al. 2014), while the best growth rate was observed in white sturgeon (initial weight 34.1 ± 0.2 g) fed with 400–600 mg choline chloride kg^{-1} diet (Hung 1989).

In sturgeon broodstocks, however, it is recommended to use double or triple amount of dietary vitamin premix to support the offspring quality. Although vitamin premix designed for salmonids is decent to be used in sturgeon diets, further studies should be focused to find specific requirement of vitamins at different stages and conditions for sturgeon.

11.7 Feed Additives

Feed or diet additives are included into the formulations as nonnutritive ingredients to affect both physical and chemical properties of the diet or influence the growth performance, body composition, and physiological-immunological features. Some like preservatives and binders can enhance the pellet quality, and others such as acidifiers, immunostimulants, pigments, chemotherapeutants, exogenous enzymes, growth promoters, probiotics, and prebiotics may directly affect fish performance and product quality. The use of different food additives in Siberian sturgeon diet is summarized in Table 11.3.

Table 11.3 Use of food additives in Siberian sturgeon diet

Ingredient	Dose and length of administration	Fish weight (g)	Results	Reference
Spirulina	40, 50, 60%—12 weeks	92.1 ± 3.6	↑ growth rate, protein efficiency ↑ Palmitic and linoleic acids ↓ FCR, myristic acid	palmegiano et al. (2005)
Vitaton	0.7 g Kg⁻¹, 60 days	3.9.6 ± 12.3	↑ carotenoid	Czeczuga et al. (2006)
Lecithin	0, 2.5, 5, 7.5, 10%—8 weeks	32.9 ± 0.3	↑ growth rate, protein efficiency ↑ HUFA, EPA, DHA ↓ FCR, hemoglobin → PUFA	Najafipour Moghadam et al. (2011, 2015)
Lactoferrin	0, 100, 200, 400, 800, 1600 mg Kg⁻¹—8 weeks	26.3 ± 0.2	→ growth rate → stress indicators ↑ liver iron content, total iron binding capacity ↓ plasma iron ↑ mucus secretion and serum bactericidal activity → non-specific immune responses	Eslamloo et al. (2012)
Arabinoxylan oligosaccharides	1, 2, 4%—12 weeks	25 and 30	→ growth rate → survival ↑ abundance of bacteria in the gut ↑ concentration of acetate, butyrate, and total short-chain fatty acids	Geraylou et al. (2013a)
Arabinoxylan oligosaccharides or *L. lactis* and *B. circulans*	0, 2%—4 weeks	48.4 ± 1.4	↑ growth rate ↓ FCR ↑ innate immune responses ↑ colonization capacity of *L. lactis* → colonization capacity of *B. circulans*	Geraylou et al. (2013b)
Lactoferrin	0, 100, 200, 400, 800, 1600 mg kg⁻¹; 70 days	26.3 ± 0.2	→ growth rate ↓ stress indicators	Falahatkar et al. (2014b)
Lactobacillus plantarum	10⁷, 10⁸, 10⁹ cfu g⁻¹—8 weeks	14.6 ± 2.3	↑ growth rate ↑ immune responses	Pourgholam et al. (2015)
Pediococcus pentosaceus	10⁷, 10⁸, 10⁹ cfu g⁻¹—8 weeks	14.3 ± 0.01	↑ growth rate ↓ FCR ↑ immune responses	Moslehi et al. (2015)

Symbols represent an increase (↑), decrease (↓), or no influence (→) on the response

Effect of poly-β-hydroxybutyrate, as a natural polymer that can be depolymerized into the water-soluble short-chain fatty acids monomers, at 2–5% levels, was investigated regarding the growth and microbial control in Siberian sturgeon fingerling. After 12 weeks, the results showed better growth performance, survival, richness, and diversity of microbial species in the gastrointestinal tract at 2% level (Najdegerami et al. 2012).

Vitaton as a natural component containing β-carotene, which is derived from a biotechnological process of corn by fungus *Blakeslea trispora*, was supplemented into the trout commercial diet for Siberian sturgeon. The muscle contained all carotenoids and provitamin A, suggesting that vitaton can be used as feed supplementation of sturgeon (Czeczuga et al. 2006).

Lecithin as a source of phospholipids was used in juvenile Siberian sturgeon diet at 0, 2.5, 5, 7.5, and 10% for 8 weeks. The results showed some growth improvements and hematological features in fish fed with 7.5% lecithin (Najafipour Moghadam et al. 2011). Fatty acids composition including C14:0, EPA, DHA, saturated fatty acids, and high unsaturated fatty acids (HUFA) increased in fish fed with 10% lecithin, but monounsaturated fatty acids (MUFA) and polyunsaturated fatty acids (PUFA) were not affected. C18:2n-6 showed the highest in fish fed with diet containing 2.5% lecithin (Najafipour Moghadam et al. 2015). Overall, these findings demonstrate the positive effect of lecithin on growth, immunity, and fatty acid composition in Siberian sturgeon.

Eslamloo et al. (2012) determined the effect of dietary bovine lactoferrin at 0, 100, 200, 400, 800, and 1600 mg kg^{-1} in juvenile Siberian sturgeon after 8 weeks of rearing. Growth performance, hematological factors, serum protein, and hepatic enzymes were unchanged, but total iron binding capacity increased significantly in fish fed with 800 mg lactoferrin kg^{-1}. Also, some immune parameters and iron content in the liver were affected by dietary lactoferrin. In addition, after 10 weeks of feeding with the same diets, dietary lactoferrin levels affected cortisol and lactate post-stress levels, demonstrating that dietary lactoferrin can suppress stress response in Siberian sturgeon fed with 400 mg lactoferrin kg diet (Falahatkar et al. 2014b).

During recent years, many studies have been conducted to determine the best probiotic, prebiotic, and synbiotic in sturgeon commercial feeds (see Hoseinifar et al. 2014). Mahious and Ollevier (2005) reported significant growth enhancement in juvenile Siberian sturgeon fed with diet containing 20 g kg^{-1} inulin and oligofructose. When juvenile Siberian sturgeon fed with diet containing 2 g *Bacillus* sp. per kg diet, growth and digestibility were positively affected by dietary probiotics (Gao et al. 2009). Geraylou et al. (2012) isolated putative probiotics from the gut of Siberian sturgeon and introduced *Lactococcus lactis* spp., *lactis* STG45 and STG81, as candidate probiotics. Geraylou et al. (2013a) used arabinoxylan oligosaccharides (ANOX) as probiotics, and the results showed that growth did not change, but gut microbia were affected by dietary ANOX. They reported that this probiotics can be a good candidate to be used in Siberian sturgeon diet. They also found that when fish are fed with *L. lactis* STG45 + 2% ANOX, significant alternations in the gut microbiota increased the growth and immunity, while bacterial diversity decreased. In the study of Delaedt et al. (2008), ANOX was used in Siberian sturgeon diet for 18 weeks, showing suitable bacterial community in the gut (Geraylou et al. 2013b). Rurangwa et al. (2008) used 1% and 2% ANOX in Siberian sturgeon diet and found no effect on growth, while significant production of short-chain fatty acids was reported in the hindgut. The effect of *Lactobacillus plantarum* as probiotics was investigated in juvenile Siberian sturgeon diet. Growth and immune parameters including lysozyme activity, immunoglobulin M, and complement component 3 (as a protein of the immune system) increased in fish fed with diet containing 1×10^8 colony-forming unit g^{-1} (Pourgholam et al. 2015). The effect of *Pediococcus pentosaceus* as a probiotic was studied on Siberian sturgeon for 8 weeks, revealing better performance and immunity in fish fed with probiotics than the control (Moslehi et al. 2015).

Single cell biomass such as Spirulina (as a freshwater microalgae) seems as a good protein source for aquatic diets. Spirulina meal was tested at 40%, 50%, and 60% in the diet of Siberian sturgeon (42% crude protein) with initial weight of 92 g for 12 weeks. Fish fed with 50% Spirulina showed the highest weight gain and the best FCR and the highest MUFA and PUFA contents in the muscle of fish compared to control diet, but with increasing the level of Spirulina, apparent digestibility coefficient of protein was reduced from 91% in control diet to 80.5% in Spirulina 60% (Palmegiano et al. 2005). So, the results revealed the possibility of using Spirulina in the diet of sturgeon.

The effect of ovine growth hormone on growth performance of juvenile Siberian sturgeon at different injection doses of 0, 2, 4, and 8 $\mu g \ g^{-1}$ body weight for 8 weeks showed the best significant performance on fish injected with the highest dose, and the fish response was a dose-dependent manner (Falahatkar and Poursaeid 2015).

Despite many studies on the effects of additives in sturgeon, further studies are required to investigate the best additives and doses and the physiological-immunological effects on fish and the possible effect on sturgeon reproduction and more nutritive value production of meat and caviar.

11.8 Replacement of Fish Meal by Other Protein Sources

Fish meal is the most costly ingredient in formulated diets. Fish nutritionists have tried to use less-expensive plant or other protein sources for partial or total fish meal replacement. Extensive effort has been performed to make out and assess alternatives to fish meal to be used as protein sources in diets of many species. In sturgeon, less activity has been carried out in this field. Partial replacement of animal protein with full-fat soybean meal in juvenile Siberian sturgeon demonstrated that diets containing fish meal (13%), meat meal (14.7%), full-fat soybean meal (15%), and also diet supplemented with 1.12% L-lysine and 0.85% DL-methionine gave the higher weight and better FCR (Ronyai et al. 2002). Replacement of fish meal by soy protein concentrate and rapeseed meal in juvenile Siberian sturgeon for 50 days of rearing showed that both can be used partially instead of fish meal with amounts of 23.6% and 7% in formulated diets, respectively (Mazurkiewicz et al. 2009). An 8-week experiment of fish meal replacement by blend of rendered animal protein (including 40% meat and bone meal, 40% poultry by-product meal, and 20% hydrolyzed feather meal) in juvenile Siberian sturgeon revealed the best weight gain in fish fed with diet containing 50% blend of rendered animal protein with crystallized amino acids. The results also suggest that sturgeon can efficiently utilize crystalline amino acids as properly as spray-dried blood meal (Zhu et al. 2011). Xue et al. (2012) showed no significant effect of dietary rendered animal protein blend replacing fish meal and protein levels in juvenile Siberian sturgeon on growth performance and amino acid profile after 8-week trial. Fish meal replacement by soybean concentrate in Siberian sturgeon (mean weight 186 g) for 8 weeks showed that it can be replaced by 20% soybean concentrate without any adverse effect on growth and body composition (Yazdani Sadati and Rezaii 2015). Replacement of fish meal by plant protein blend (soybean meal and wheat gluten meal) at 36% and 40% protein levels in Siberian sturgeon suggested that total replacement of fish meal by plant proteins with supplemented crystalline essential amino acids is possible without any effect on growth performance (Yun et al. 2014). Based on this result, phosphorus excretion is reduced in plant protein diets both at 36% and 40% dietary crude protein levels. Using rice protein concentrate in juvenile Siberian sturgeon (initial weight 19.1 g) for 4 months showed the capability of 53% replacement with no adverse effect on growth and body composition (Sicuro et al. 2015). A recent study on replacement of fish meal by earthworm meal in juvenile Siberian sturgeon (initial weight 21.3 g) for 12 weeks showed that using 10–20% of earthworm meal instead of fish meal is possible with no adverse effect on growth and body composition (Soleimani et al. 2016). A study showed that when fish meal is replaced partially with gammarus meal in 14.2 g juvenile Siberian sturgeon, growth performance is influenced by dietary levels and the best level observed in fish fed with diet (42.1% crude protein) containing 10–20% gammarus meal, while no effect was observed on proximate body composition 12 weeks after the experiment (Razgardani Sharahi et al. 2016).

With increasing the cost and limiting fish meal sources, it is essential to find less-expensive protein sources for fish feeds. Therefore, by increasing the production of sturgeon worldwide, it is important to produce low-cost feed with high efficiency in the future.

11.9 Starvation and Compensatory Growth

Starvation and/or fasting happens in both natural environments because of food limitation, migration or reproduction, and culture conditions because of stressors, pathogens, turbidity, and other unlikely conditions including low or high temperatures (Falahatkar et al. 2013). Feeding protocols that limit the amount of food have the potential to lower the production cost by decreasing the quantity of food used to produce a unit of fish via reducing maximum daily feeding rates or decreasing the feeding frequency. The ability of animals to adjust to food deprivation and resume growth after limited periods of fasting is well established (Hornick et al. 2000; McCue 2010).

There have been few studies on the effects of starvation in sturgeon. Only short-term effects of starvation (4–5 weeks) on white epaxial muscle have been studied in white sturgeon (Kiessling et al. 1993). In addition, some other studies on lake sturgeon (Gillis and Ballantyne 1996), white sturgeon (Hung et al. 1997a; Georgiadis et al. 2000a, b), and Adriatic sturgeon (*A. naccarii*) (Furné et al. 2012) showed the starvation's negative impact on some morphophysiological parameters. In a study on beluga sturgeon (*Huso huso*), the fasting and feeding strategies showed some partial compensatory growth (Falahatkar 2012).

Regarding Siberian sturgeon, a few researches have been carried out to find the effect of starvation on growth and physiological performance. A study on juvenile fish for 40 days of fasting and refeeding in different treatments exhibited weight loss during starvation (Morshedi et al. 2013). However, they can adjust the metabolic activity to overcome the fasting period in terms of energy sources (Ashouri et al. 2013). Another study showed that when juvenile fish (mean weight 54 g) were exposed to starvation, feeding and satiation were restricted for 8 weeks; weight loss and gain were −11.7% and +110.1% in fish faced to starvation and satiation, respectively (Shirvan et al. 2013). Also, this fish showed better conservation of body protein stores compared with lipids during starvation. Feeding various levels of dietary protein, fat, and carbohydrate for 3 weeks followed with starvation (2 weeks) and refeeding (5 weeks) in juvenile Siberian sturgeon (30 ± 5 g) showed an increasing superoxide dismutase activity in starved fish, while fish fed with low protein diets (38%) represented increase in catalysis activity (Babaei et al. 2016). Further studies should be focused on finding the best strategies for catch-up growth after a short period of food restriction and the mechanism involved to supply the energy need during fasting/starvation.

11.10 Antinutritional Factors and Toxins in Feeds

Some ingredients especially plants or plant-derived containing antinutrients may cause poisoning signs, decrease growth and feed efficiency, liver dysfunction and pancreatic hypertrophy, goitrogenesis, hyperglycemia, and immune suppression. Various animal by-products also contain antinutrients. The effects of such antinutrients vary with species, age, size, stage of development, sex and maturity stage, environmental condition, health state, and stress factors which can cause different responses of the animal (NRC 2011).

In commonly used potential fish feed ingredients, antinutrients in plant feedstuffs such as enzyme inhibitors, lectins, saponins, phytoestrogens, glucosinolates/goitrogens, tannins, phytic acid, gossypol, cyclopropene fatty acids – sterculic and malvalic acid – glycoalkaloids, arginase inhibitors, quinolizidine alkaloids, cyanide-releasing compound, erucic acid, and allergens have shown different effects of growth retardation, low resistance to disease, immunosuppression, and abnormalities in the gastrointestinal tract (NRC 2011). Moreover, some antinutrients in animal feedstuffs such as thiaminase, chitin, fluorine, and some toxins in fish feed ingredients such as oxidized lipid, biogenic amines, mycotoxins, and other contaminants may influence fish growth and survival.

A 4-week trial with vitamin A and oxidized lipid in Siberian sturgeon larvae showed 25% deformed larvae in fish fed with diet containing 80 g kg⁻¹ oxidized lipid, but higher vitamin A prevented this malformation (Fontagné et al. 2006). Sex steroids in commercial diets of Siberian sturgeon were measured and revealed high amount of sex steroids in the diet which can affect plasma sex steroid and vitellogenin levels in fish (Pelissero et al. 1989). Pelissero et al. (1991) showed that there is lower influence of various phytoestrogens present in vegetable ingredients than that of estradiol-17β in Siberian sturgeon.

Nevertheless, research is needed to find various antinutrients' effect on sturgeon growth at different ages and different functions such as microbiota in the gastrointestinal tract. Long-term effects of different antinutrients' advantage examination help predict how combination of ingredients may affect Siberian sturgeon and recognize steps needed to reduce or prevent negative health effects.

11.11 Feeding Practices

Feeding and feeding strategies are significant features in aquaculture and are widely influenced by different factors such as fish species, age and life stage, strain, feed characteristics, and rearing conditions and practices. Characteristics of the diets including type (live, fresh, frozen, or compounded feeds) and physical and chemical characteristics (e.g., shape of feed, size, durability, density, and palatability) should be considered for different life steps and species. Furthermore, feed allowance, delivery system, ration size, feeding frequency, and strategy are important issues affecting nutrient uptake and retention and, therefore, growth, fitness, health, and quality of the fish.

Information on optimum feeding rate is very important in sturgeon farming because the best feeding rate provides the greatest fish performance and lower production cost. Different feed delivery methods such as hand-feeding, automated feeders, and demand feeders are used to culture sturgeon. Hand and automated feeders are mainly used in larval hatcheries. Although there is some information on the feeding rate of white sturgeon in larval stage, no information is available on Siberian sturgeon. It seems that they can be fed 25–30% of body weight per day as initial feeding (9 days post-hatch in Siberian sturgeon). Now, many hatcheries use live foods as the first feeding of Siberian sturgeon larvae, *Artemia* nauplii, followed by *Daphnia* or chironomids. Some formulated diets are available for this stage, and feed can be divided into equal portions and fed to larvae every 2–3 h using time-controlled automated feeders or belt feeders. Most farmers, however, prefer to hand-feed larvae to satiation every 2 h. Some other farmers hand-feed the fish during working hours and use automated feeders during the rest of the day. Excess feed in tanks is removed or siphoned once or twice a day based on the amount of remaining feeds and water quality. Feeding is decreased during the next 30 days of rearing to 3–4% body weight per day (Hung 1991a). Automated or demand feeders can be used for fingerlings and grow-out phases. There is a 3-h delay in ammonia excretion after the feeding of Siberian sturgeon at different weights, and continuous feeding is suitable for sturgeon farms regarding ammonia loading (Jatteau 1997). Since there is no difference regarding light or dark times in feed consumption in sturgeon (Najafi et al. 2017), feeding with hand method, automated, or demand feeders should be considered through the day-night hours using filled feeders in early morning and late afternoon.

Different feeding rates from 0.75 to 1.5% body weight/day were evaluated in Siberian sturgeon (mean weight 1736 g) for 2 weeks at 19–22 °C. The best growth performance in order to SGR and FCR was observed in fish fed 1% body weight/day (Rad et al. 2003).

Larval behavior and initial feeding time were determined in small hatchery production of Siberian sturgeon at 18 °C (Gisbert and Williot 1997), showing that larvae at 1–3 days old were strongly photopositive, but at 4–8 days old, they showed rheotactic and benthic behavior. At age 9–10 years, fish were active swimmers, and yolk sac was depleted, and exogenous feeding occurred. They concluded that expulsion of melanin plug is not a good indicator for transition of feeding form yolk sac to exogenous feeding.

Siberian sturgeon larvae were fed with *Artemia* nauplii, *Limnodrillus* sp., and commercial feeds for 30 days. The results revealed the highest survival rate in fish fed with *Artemia* nauplii, while the highest growth rate was achieved in fish fed with *Limnodrillus*. The authors suggest using *Artemia* nauplii first and then *Limnodrillus* for better survival rate and growth performance in larval stage (Zhang et al. 2009).

Live *Tubifex* and three different dry diets were examined in larval stage of Siberian sturgeon (Dabrowski et al. 1985). A dry diet-based single cell protein and freeze-dried liver showed better results than the salmon starter feed and casein-based diet. After 30 days of rearing, final weights of 960–1450 mg in larvae-fed freeze-dried liver and *Tubifex* were observed, respectively. They concluded that live feed can be totally eliminated from larval rearing of Siberian sturgeon, but nutritional quality of dry feed should be studied, and more attention is needed on making a complete starter diet to address all requirements.

Diets of +1 year with Siberian sturgeon in concrete storage ponds were evaluated by Adámek et al. (2007). Twenty-five feed items were found in the gut with the highest amount of chironomids (40.4–52.8%), cladocereans (19.1–28.8%), and detritus (16.3–19.4%). Monoculture of first year Siberian sturgeon in earthen ponds exhibited diptera and cladocereans as dominant food items in the gut (Pyka and Kolman 2003).

The growth rate of 9.2 g Siberian sturgeon was evaluated in concrete raceways for 135 days, and the results showed that this species has a good potential to be cultured in specialized rainbow trout farms, and the growth is even higher than that of rainbow trout, sea bass, and sea bream (Köksal et al. 2000).

In Siberian sturgeon, there is no information on the optimum feeding rates in terms of water temperature, fish weight, and energy content of the diet. In addition, feeding under different light regimes and feeding systems should be carried out to find the optimum practical feeding at different ages. Because of the large size, small numbers, late puberty, and long life span, researches on sturgeon broodstocks' nutritional requirements are scarce. Some special requirements are needed to be considered in the diet of brooders such as higher vitamin levels and fatty acids.

Future examinations should try to find the best macro- and micronutrient levels and the effects on reproductive and offspring performance.

Conclusion

Siberian sturgeon aquaculture grows fast and makes this species as one of the best sturgeon species for meat and caviar production. To ensure the success and expansion of intensive sturgeon farming, more data are needed to find nutritional requirements, digestibility of feedstuffs, and adequate balance of nutrients leading to suitable growth rate, high feed efficiency, limiting water pollution, and producing high-quality meat and caviar. Also, more studies are necessary to find the best feeding practices to improve feeding frequency, feeding method, and metabolic utilization of digestible nutrients and optimum level of each dietary ingredient.

References

Adámek A, Prokeš M, Baruš V, Sukop I (2007) Diet and growth of 1+ Siberian sturgeon, *Acipenser baerii* in alternative pond culture. Turk J Fish Aquat Sci 7:153–160

Agradi E, Abrami G, Serrini G, Mckenzie D, Bolis C, Bronzi P (1993) The role of dietary n-3 fatty acid and vitamin E supplements in growth of sturgeon (*Acipenser naccarii*). Comp Biochem Physiol 105A:187–195

Ashouri G, Yavari V, Bahmani MA, Yazdani M, Kazemi R, Morshedi V, Fatollahi M (2013) The effect of short-term starvation on some physiological and morphological parameters in juvenile Siberian sturgeon, *Acipenser baerii* (Actinopterygii: Acipenseriformes: Acipenseridae). Acta Ichthyol Pisc 43:145–150

Babaei S, Abedian Kenari A, Hedayati M, Yazdani Sadati MA, Metón I (2016) Effect of diet composition on growth performance, hepatic metabolism and antioxidant activities in Siberian sturgeon (*Acipenser baerii*, Brandt, 1869) submitted to starvation and refeeding. Fish Physiol Biochem 42:1509–1520

Bagherzadeh Lakani F (2016) Effects of cooper nanoparticle on physiological and pathological responses of Siberian sturgeon (*Acipenser baerii*). PhD Thesis, Urmia University, p 132 (in Persian)

Birstein VJ (1993) Sturgeons and paddlefishes: threatened species in need of conservation. Cons Biol 7:773–787

Bronzi P, Rosenthal H (2014) Present and future sturgeon and caviar production and marketing: a global market overview. J Appl Ichthyol 30:1536–1546

Bronzi P, Rosenthal H, Gessner J (2011) Global sturgeon aquaculture production: an overview. J Appl Ichthyol 27:169–175

Burtle GJ, Lovell RT (1989) Lack of response of channel catfish (*Ictalurus punctatus*) to dietary myo-inositol. Can J Fish Aquat Sci 46:218–222

Cowey CB, Walton MJ (1989) Intermediary metabolism. In: Halver JE (ed) Fish nutrition. Academic Press, San Diego, CA, pp 260–329

Cui Y, Hung SSO, Zhu X (1996) Effect of ration and body size on energy budget of juvenile white sturgeon. J Fish Biol 49:863–876

Czeczuga B, Kolman R, Czeczuga-Semeniuk E, Szczepkowski M, Semeniuk A, Kosielinski P, Sidorov N (2006) Carotenoid composition in the muscles of Siberian sturgeon (*Acipenser baerii* Br.) and Sterlet (*Acipenser ruthenus* L.) juveniles fed feed supplemented with vitaton. Arch Polish Fish 14:213–224

Dabrowski K, Kaushik S, Fauconneau B (1985) Rearing of sturgeon (*Acipenser baerii* Brandt) larvae: I. Feeding trial. Aquaculture 47:185–192

Dabrowski K, Kaushik S, Fauconneau B (1987) Rearing of sturgeon (*Acipenser baerii* Brandt) larvae: III. Nitrogen and energy metabolism and amino acid absorption. Aquaculture 65:31–41

Delaedt Y, Diallo MD, Rurangwa E, Courtin CM, Delcour JA, Ollevier F (2008) Impact of arabinoxylooligosaccharides on microbial community composition and diversity in the gut of Siberian sturgeon (*Acipenser baerii*). Aquaculture Europe, Krakow, Poland

Deng DF, Hung SSO, Conklin DE (1998) White sturgeon (*Acipenser transmontanus*) require both n-3 and n-6 fatty acids. Aquaculture 163:333–335

Eslamloo K, Falahatkar B, Yokoyama S (2012) Effects of dietary bovine lactoferrin on growth, physiological performance, iron metabolism and non-specific immune responses of Siberian sturgeon *Acipenser baerii*. Fish Shellfish Immunol 32:976–985

Falahatkar B (2012) The metabolic effects of feeding and fasting in beluga *Huso huso*. Mar Environ Res 82:69–75

Falahatkar B, Nasrollahzadeh A (2011) Caspian Sea and the sturgeon catch in Iran. Doc Nat 60:1–12. (In German)

Falahatkar B, Soltani M, Abtahi B, Kalbassi M, Pourkazemi M (2006) Effects of dietary vitamin C supplementation on performance, tissue chemical composition and alkaline phosphatase activity in great sturgeon, *Huso huso*. J Appl Ichthyol 26:283–286

Falahatkar B, Soltani M, Abtahi B, Kalbassi M, Pourkazemi M (2015) The role of dietary L-ascorbyl-2-polyphosphate on the growth and physiological functions of beluga, *Huso huso* (Linnaeus, 1758). Aquac Res 46:3056–3069

Falahatkar B, Poursaeid S, Shakoorian M, Barton B (2009) Responses to handling and confinement stressors in juvenile great sturgeon *Huso huso*. J Fish Biol 75:784–796

Falahatkar B, Tolouei MH, Falahatkar S, Abbasalizadeh A (2011) Laparoscopy, a minimally-invasive technique for sex identification in cultured great sturgeon *Huso huso*. Aquaculture 321:273–279

Falahatkar B, Akhavan SR, Efatpanah I, Meknatkhah B (2013) Effect of feeding and starvation during the winter period on the growth performance of young-of-year (YOY) great sturgeon, *Huso huso*. J Appl Ichthyol 29:26–30

Falahatkar B, Akhavan SR, Poursaeid S, Hasirbaf I (2014a) Use of sex steroid profile and hematological indices to identify perinucleolus and migratory gonadal stages of captive Siberian sturgeon *Acipenser baerii* (Brandt, 1869) females. J Appl Ichthyol 30:1578–1584

Falahatkar B, Eslamloo K, Yokoyama S (2014b) Suppression of stress responses in Siberian sturgeon, *Acipenser baerii*, juveniles by the dietary administration of bovine lactoferrin. J World Aqua Soc 45:699–708

Falahatkar B, Poursaeid S (2014) Gender identification in great sturgeon (*Huso huso*) using morphology, sex steroids, histology and endoscopy. Anat Histol Embryol 43:81–89

Falahatkar B, Poursaeid S (2015) Effects of ovine growth hormone on growth performance and body composition in Siberian sturgeon *Acipenser baerii*. Middle East and Central Asia Aquaculture 2015, December 14–16, Tehran, Iran

FAO (2014) The state of world fisheries and aquaculture. Fisheries and Aquaculture Department, Rome, Italy

Fauconneau B, Aguirre P, Dabrowski K, Kaushik S (1986) Rearing of sturgeon (*Acipenser baerii* Brandt) larvae: 2. Protein metabolism: influence of fasting and diet quality. Aquaculture 51:117–131

Fontagné S, Bazin D, Bréque J, Vachot C, Bernarde C, Rouault T, Bergot P (2006) Effects of dietary oxidized lipid and vitamin a on the early development and antioxidant status of Siberian sturgeon (*Acipenser baerii*) larvae. Aquaculture 257:400–411

Furné M, Morales AE, Trenzado CE, García-Gallego M, Hidalgo MC, Domezain A, Rus AS (2012) The metabolic effects of prolonged starvation and refeeding in sturgeon and rainbow trout. J Comp Physiol 182B:63–76

Gao X, Ge L-Q, Li M-Y, Guo X-X, An R-Y (2009) Effects of *Bacillus* spp. on the growth performance and digestibility of juvenile *Acipenser baerii*. J Hebei Norm Univ 33:377–382

Geraylou Z, Souffreau C, Rurangwa E, D'Hondt S, Callewaert L, Courtin CM, Delcour JA, Buyse J, Ollevier F (2012) Effects of arabinoxylan-oligosaccharides (AXOS) on juvenile Siberian sturgeon (*Acipenser baerii*) performance, immune responses and gastrointestinal microbial community. Fish Shellfish Immunol 33:718–724

Geraylou Z, Souffreau C, Rurangwa E, De Meester L, Courtin CM, Delcour JA, Buyse J, Ollevier F (2013a) Effects of dietary arabinoxylan-oligosaccharides (AXOS) and endogenous probiotics on the growth performance, non-specific immunity and gut microbiota of juvenile Siberian sturgeon (*Acipenser baerii*). Fish Shellfish Immunol 35:766–775

Geraylou Z, Souffreau C, Rurangwa E, Maes GE, Spanier KI, Courtin CM, Delcour JA, Buyse J, Ollevier F (2013b) Prebiotic effects of arabinoxylan oligosaccharides on juvenile Siberian sturgeon (*Acipenser baerii*) with emphasis on the modulation of the gut microbiota using 454 pyrosequencing. FEMS Microbiol Ecol 86:357–371

Gisbert E, Williot P (1997) Larval behaviour and effect of the timing of initial feeding on growth and survival of Siberian sturgeon (*Acipenser baerii*) larvae under small scale hatchery production. Aquaculture 156:63–76

Gisbert E, Williot P (2002) Influence of storage duration of ovulated eggs prior to fertilisation on the early ontogenesis of sterlet (*Acipenser ruthenus*) and Siberian sturgeon (*Acipenser baerii*). Internat Rev Hydrobiol 87:605–612

Ghiasi S, Falahatkar B, Dabrowski K, Abbasalizadeh A, Arslan M (2014) Effect of thiamine injection on growth performance, hematology and germinal vesicle migration in sterlet sturgeon *Acipenser ruthenus* L. Aquacult Int 22:1563–1576

Gillis TE, Ballantyne JS (1996) The effects of starvation on plasma free amino acid and glucose concentrations in lake sturgeon. J Fish Biol 49:1306–1316

Georgiadis MP, Hedrick RP, Johnson WO, Gardner IA (2000a) Growth of white sturgeon (*Acipenser transmontanus*) following recovery from stunted stage in a commercial farm in California, USA. Prev Vet Med 43:283–291

Georgiadis MP, Hedrick RP, Johnson WO, Gardner IA (2000b) Mortality and recovery of runt white sturgeon (*Acipenser transmontanus*) in a commercial farm in California, USA. Prev Vet Med 43:269–281

Holčík J (1989) The freshwater fishes of Europe. General introduction to fishes acipenseriforms, vol 1–2. AULA Verlag, Wiesbaden, Germany

Hornick JL, Van Eenaeme C, Gérard O, Dufrasne I, Istasse L (2000) Mechanisms of reduced and compensatory growth. Dom Anim Endocrinol 19:121–132

Hoseinifar SH, Ringø E, Shenavar Masouleh A, Esteban MÁ (2014) Probiotic, prebiotic and synbiotic supplements in sturgeon aquaculture: a review. Rev Aquacult 6:1–14

Hung SSO (1991a) Sturgeon, *Acipenser* spp. In: Wilson RP (ed) Handbook of nutrient requirements of finfish. CRC Press, Boca Raton, FL, pp 153–160

Hung SSO (1991b) Choline requirement of hatchery-produced juvenile white sturgeon. Aquaculture 78:183–194

Hung SSO, Fynn-Aikins FK, Lutes PB, Xu R (1989) Ability of juvenile white sturgeon (*Acipenser transmontanus*) to utilize different carbohydrate sources. J Nut 119:727–733

Hung SS, Deng D-F (2002) Sturgeon, *Acipenser* spp. In: Webster CD, Lim C (eds) Nutrient requirements and feeding of finfish for aquaculture. CAB International, England, pp 344–357

Hung SSO, Liu W, Li H, Storebakken T, Cui Y (1997a) Effect of starvation on some morphological and biochemical parameters in white sturgeon, *Acipenser transmontanus*. Aquaculture 151:357–363

Hung SSO, Storebakken T, Cui Y, Tian L, Einen O (1997b) High energy diets for white sturgeon, *Acipenser transmontanus*, Richardson. Aquacult Nut 3:281–286

Jatteau P (1997) Daily patterns of ammonia nitrogen output of Siberian sturgeon *Acipenser baerii* (Brandt) of different body weights. Aquac Res 28:551–557

Kasumyan AO (1994) Olfactory sensitivity of the sturgeon to free amino acids. Biophysics 39:519–522

Kasumyan AO, Taufik LR (1994) Behaviour reaction of juvenile sturgeons (Acipenseridae) to amino acids. J Appl Ichthyol 34:90–103

Kaushik S, Breque J, Blanc D (1991) Requirements for protein and essential amino acids and their utilization by Siberian sturgeon (*Acipenser baerii*). In: Williot P (ed) Acipenser. Proceedings of the first International Symposium on Sturgeon. CEMAGREF, Antony, France, pp 25–37

Kaushik S, Breque J, Blanc D (1994) Apparent amino acid availability and plasma free amino acid levels in Siberian Sturgeon (*Acipenser baerii*). Comp Biochem Physiol 107A:433–438

Kaushik S, Luquet P, Blanc D, Paba A (1989) Studies on the nutrition of Siberian sturgeon, *Acipenser baerii*: I. Utilization of digestible carbohydrates by sturgeon. Aquaculture 76:97–107

Kiessling A, Hung SSO, Storebakken T (1993) Differences in protein mobilization between ventral and dorsal parts of white epaxial muscle from fed, fasted and re-fed white sturgeon (*Acipenser transmontanus*). J Fish Biol 43:401–408

Köksal G, Rad F, Kindir M (2000) Growth performance and feed conversion efficiency of Siberian sturgeon juveniles (*Acipenser baerii*) reared in concrete raceways. Turk J Vet Anim Sci 24:435–442

Lin JH, Cui Y, Hung SSO, Shiau SY (1997) Effect of feeding strategy and carbohydrate source on carbohydrate utilization by white sturgeon and hybrid tilapia. Aquaculture 148:201–211

Lou L, Ai L, Li T, Xue M, Wang J, Li W, Wu X, Liang X (2015) The impact of dietary DHA/EPA ratio on spawning performance, egg and offspring quality in Siberian sturgeon (*Acipenser baeri*). Aquaculture 437:140–145

Lovell RT, Limsuwan T (1982) Intestinal synthesis and dietary nonessentiality of vitamin B_{12} for *Tilapia nilotica*. Trans Am Fish Soc 111:485–490

Mahious AS, Ollevier F (2005) Probiotics and prebiotics in aquaculture: a review. In: Agh N, Sorgeloos P (eds). 1st Regional Workshop on Techniques for Enrichment of Live Food for Use in Larviculture. Urima, Iran

Mazurkiewicz J, Przybył A, Golski J (2009) Usability of some plant protein ingredients in the diets of Siberian sturgeon *Acipenser baerii* Brandt. Arch Polish Fish 17:45–52

McCue MD (2010) Starvation physiology: reviewing the different strategies animals use to survive a common challenge. Comp Bioch Physiol 156A:1–18

Médale F, Kaushik S (1991) Energy utilization by farmed Siberian sturgeon (*Acipenser baerii*) from 3 age classes. In: Williot P (ed) Acipenser. Proceedings of the first International Symposium on Sturgeon. CEMAGREF, Antony, France, pp 13–23

Médale F, Blanc D, Kaushik S (1991) Studies on the nutrition of Siberian sturgeon, *Acipenser baerii*. II. Utilization of dietary non-protein energy by sturgeon. Aquaculture 93:143–154

Médale F, Corraze G, Kaushik S (1995) Nutrition of farmed Siberian sturgeon. In: Gershanovich AD, Smith TIJ (eds) Proceeding of the third International Symposium on Sturgeons. VNIRO Publishing, Moscow, Russia

Moreau R, Kaushik SJ, Dabrowski K (1996) Ascorbic acid status as affected by dietary treatment in the Siberian sturgeon (*Acipenser baerii* Brandt): tissue concentration, mobilisation and L-gulonolactone oxidase activity. Fish Physiol Biochem 15:431–438

Moreau R, Dabrowski K, Sato PH (1999a) Renal L-gulono-1, 4- lactone oxidase activity as affected by dietary ascorbic acid in lake sturgeon (*Acipenser fulvescens*). Aquaculture 180:250–257

Moreau R, Dabrowski K, Czesny S, Chila F (1999b) Vitamin C-vitamin E interaction in juvenile lake sturgeon (*Acipenser fulvescens*), a fish able synthesize ascorbic acid. J Appl Ichthyol 15:252–257

Morshedi V, Kochanian P, Bahmani M, Yazdani-Sadati M, Pourali H, Ashouri G, Pasha-Zanoosi H, Azodi M (2013) Compensatory growth in sub-yearling Siberian sturgeon, *Acipenser baerii* Brandt, 1869: effects of starvation and refeeding on growth, feed utilization and body composition. J Appl Ichthyol 29:978–983

Moslehi F, Sattari M, Khoshkholgh M, Shenavar Masule A, Abbasalizade A (2015) The effect of *Pediococcus pentosaceus* as a probiotic on growth and immune factors of Siberian sturgeon (*Acipenser baerii*). Fish Sci Technol 3:81–92. (in Persian)

Najafi M, Falahatkar B, Safarpour Amlashi A, Tolouei Gilani MH (2017) The combined effects of feeding time and dietary lipid levels on growth performance in juvenile beluga sturgeon *Huso huso*. Aquacult Int 25:31–45

Najafipour Moghadam E, Falahatkar B, Kalbassi M (2011) Effects of lecithin on growth and hematological indices in juveniles of Siberian sturgeon (*Acipenser baerii* Brandt 1869). Iran Sci Fish J 20:143–154. (in Persian)

Najafipour Moghadam E, Falahatkar B, Kalbassi Masjed Shahi M (2015) Changes in dietary and muscle fatty acids composition in Siberian sturgeon (*Acipenser baerii* Brandt 1869) fed with different levels of lecithin. J Oceanograph 6:97–105. (in Persian)

Najdegerami EH, Tran TN, Defoirdt T, Marzorati M, Sorgeloos P, Boon N, Bossier P (2012) Effects of poly-β-hydroxybutyrate (PHB) on Siberian sturgeon (*Acipenser baerii*) fingerlings performance and its gastrointestinal tract microbial community. FEMS Microbiol Ecol 79:25–33

Ng WK, Hung SSO, Herold MA (1996) Poor utilization of dietary free amino acids by white sturgeon. Fish Physiology Biochem 15:131–142

Nieminen P, Westenius E, Halonen T, Mustonen A-M (2014) Fatty acid composition in tissues of the farmed Siberian sturgeon (*Acipenser baerii*). Food Chem 159:80–84

NRC (2011) Nutrient requirements of fish and shrimp. National Research Council, The National Academies Press, Washington, DC, p 376

Palmegiano GB, Agradi E, Forneris G, Gai F, Gasco L, Rigamonti E, Sicuro B, Zoccarato I (2005) Spirulina as a nutrient source in diets for growing sturgeon (*Acipenser baerii*). Aquac Res 36:188–195

Pelissero C, Cuisset B, Le Menn F (1989) The influence of sex steroids in commercial fish meals and fish diets on plasma concentration of estrogens and vitellogenin in cultured Siberian sturgeon *Acipenser baerii*. Aquatic Liv Res 2:161–168

Pelissero C, Le Menn F, Kaushik S (1991) Estrogenic effect of dietary soya bean meal on vitellogenesis in cultured Siberian sturgeon *Acipenser baerii*. Gen Comp Endocrinol 83:447–457

Pourgholam MA, Khara H, Safari R, Sadati MAY, Aramli MS (2015) Dietary administration of *Lactobacillus plantarum* enhanced growth performance and innate immune response of Siberian sturgeon, *Acipenser baerii*. Probiot Antimicrob Prot 8:1–7

Poursaeid S, Falahatkar B, Van Der Kraak G (2015) Short-term effects of cortisol implantation on blood biochemistry and thyroid hormones in previtellogenic great sturgeon *Huso huso*. Comp Biochem Physiol 179A:197–203

Pyka J, Kolman R (2003) Feeding intensity and growth of Siberian sturgeon *Acipenser baerii* Brandt in pond cultivation. Arch Ryb Polsk 11:287–294

Rad F, Köksal G, Kindir M (2003) Growth performance and food conversion ratio of Siberian sturgeon (*Acipenser baerii* Brandt) at different daily feeding rates. Turk J Vet Anim Sci 27:1085–1090

Razgardani Sharahi A, Falahatkar B, Efatpanah I (2016) Replacement of fish meal with gammarus meal and its effects on growth and body composition of juvenile Siberian sturgeon, *Acipenser baerii* (Brandt, 1869). J Aquatic Ecol 6:102–113 (in Persian)

Ronyai A, Csengeri I, Varadi L (2002) Partial substitution of animal protein with full fat soybean meal and amino acid supplementation in the diet of Siberian sturgeon (*Acipenser baerii*). J Appl Ichthyol 18:682–684

Rurangwa E, Delaedt Y, Geraylou Z, Van De Wiele T, Courtin CM, Delcour JA, Ollevier F (2008) Dietary effect of arabinoxylan oligosaccharides on zootechnical performance and hindgut microbial fermentation in Siberian sturgeon and African catfish. Aquaculture Europe, Krakow, Poland

Ruban GI (2005) The Siberian sturgeon *Acipenser baerii* Brandt: species structure and ecology. World Sturgeon Conservation Society-Special Publication, No.1, Norderstedt, Germany, p 203

Ruban GI, Konoplja LA (1994) Food of Siberian sturgeon *Acipenser baerii* from the Indigirka and Kolyma rivers. Vop Ichtiol 34:130–132. (in Russian)

Safarpour Amlashi A, Falahatkar B, Sattari M, Tolouei MH (2011) Effect of dietary vitamin E on growth, muscle composition, hematological and immunological parameters in sub-yearling beluga, *Huso huso*. Fish Shellfish Immunol 30:807–814

Shi X, Wang R, Zhuang P, Zhang L, Feng G (2013) Fluoride retention after dietary fluoride exposure in Siberian sturgeon *Acipenser baerii*. Aquac Res 44:176–181

Shiau SY, Lung CQ (1993) No dietary vitamin B_{12} required for juvenile tilapia *Oreochromis niloticus* × *O. aureus*. Comp Biochem Physiol 105A:147–150

Shirvan S, Falahatkar B, Noveirian H, Abasalizadeh A (2013) Effect of long-term starvation and restricted feeding on growth performance and body composition of juvenile Siberian sturgeon (*Acipenser baerii* Brandt 1869). Iran Sci Fish J 22:91–102. (in Persian)

Sicuro B, Piccinno M, Daprà F, Gai F, Vilella S (2015) Utilization of rice protein concentrate in Siberian sturgeon (*Acipenser baerii* Brandt) nutrition. Turk J Fish Aquatic Sci 15:313–319

Soleimani SM, Sajjadi MM, Falahatkar B, Yazdani MA (2016) Replacement fish meal with earthworm powder (*Eisenia foetida*) in juvenile *Acipenser baerii* and the effect on growth performance, feed efficiency and body composition. J Aquatic Ecol 5:21–30. (in Persian)

Steffens W (2008) Significance of aquaculture for the conservation and restoration of sturgeon populations. Bulgar J Agri Sci 14:155–164. (in Persian)

Taati R, Mohseni M, Khoshsima S (2015) Effect of different sources and levels of carbohydrate (glucose and corn starch) on feed efficiency and carcass composition of juvenile Siberian sturgeon (*Acipenser baerii*). Fish Sci Technol 4:77–88

Walton MJ, Cowey CB (1982) Aspects of intermediary metabolism in salmonid fish. Comp Biochem Physiol 73B:59–79

Wen H, Yan AS, Gao Q, Jiang M, Wei QW (2008) Dietary vitamin a requirement of juvenile Amur sturgeon (*Acipenser schrenckii*). J Appl Ichthyol 24:534–538

Williot P, Rochard E, Castelnaud G, Rouault T, Brun R, Lepage M, Elie P (1997) Biological characteristics of European Atlantic sturgeon, *Acipenser sturio*, as the basis for a restoration program in France. Environ Biol Fish 48:359–370

Williot P, Sabeau L, Gessner J, Arlati G, Bronzi P, Gulyas T, Berni P (2001) Sturgeon farming in Western Europe: recent developments and perspectives. Aqua Living Res 14:367–374

Williot P, Arlati G, Chebanov M, Gulyas T, Kasimov R, Kirschbaum F, Patriche N, Pavlovskaya L, Poliakova L, Pourkazemi M, Yu K, Zhuang P, Zholdasova IM (2002) Status and management of Eurasian sturgeon: an overview. Int Rev Hydrobiol 87:483–506

Wilson RP, Poe WE (1988) Choline nutrition of fingerling channel catfish. Aquaculture 68:65–71

Xie Z, Niu C, Zhang Z, Bao L (2006) Dietary ascorbic acid may be necessary for enhancing the immune response in Siberian sturgeon (*Acipenser baerii*), a species capable of ascorbic acid biosynthesis. Comp Biochem Physiol 145A:152–157

Xu R, Hung SSO, German JB (1996) Effects of dietary lipids on the fatty acid composition of triglycerides and phospholipids in tissues of white sturgeon. Aquac Nutr 2:101–109

Xu Q, Xu H, Wang C, Zheng Q, Sun D (2011) Studies on dietary phosphorus requirement of juvenile Siberian sturgeon *Acipenser baerii*. J Appl Ichthyol 27:709–714

Xue M, Yun B, Wang J, Sheng H, Zheng Y, Wu X, Qin Y, Li P (2012) Performance, body compositions, input and output of nitrogen and phosphorus in Siberian sturgeon, *Acipenser baerii* Brandt, as affected by dietary animal protein blend replacing fishmeal and protein levels. Aquac Nutr 18:493–501

Yazdani Sadati M, Rezaii E (2015) The effect of processed soy bean meal (SPH) on growth and body composition of juvenile fish *Acipenser baerii*. Iran Sci Fish J 23:73–84. (in Persian)

Yazdani-Sadati MA, Sayed Hassani M, Pourkazemi M, Shakourian M, Pourasadi M (2014) Influence of different levels of dietary choline on growth rate, body composition, hematological indices and liver lipid of juvenile Siberian sturgeon *Acipenser baerii* Brandt, 1869. J Appl Ichthyol 30:1632–1636

Yun B, Xue M, Wang J, Sheng H, Zheng Y, Wu X, Li J (2014) Fishmeal can be totally replaced by plant protein blend at two protein levels in diets of juvenile Siberian sturgeon, *Acipenser baerii* Brandt. Aquac Nutr 20:69–78

Zhang T, Zhuang P, Zhang LZ, Wang B, Gao LJ, Xia YT, Tian MP (2009) Effects of initial feeding on the growth, survival, and body biochemical composition of Siberian sturgeon (*Acipenser baerii*) larvae. Ying Yong Sheng Tai Xue Bao 20:358–362. (In Chinese)

Zhu H, Gong G, Wang J, Wu X, Xue M, Niu C, Guo L, Yu Y (2011) Replacement of fish meal with blend of rendered animal protein in diets for Siberian sturgeon (*Acipenser baerii* Brandt), results in performance equal to fish meal fed fish. Aquac Nutr 17:389–395

Swimming Characteristics of the Siberian Sturgeon

12

Ming Duan, Yi Qu, and Ping Zhuang

Abstract

The Siberian sturgeon, *Acipenser baerii* Brandt, widely distributed throughout all major rivers in Siberia, inhabits the largest range compared to other sturgeon species. Nowadays, the Siberian sturgeon is the main species for sturgeon aquaculture in several countries. Swimming and respiratory characteristics of fish can be modulated, affected, and constrained by morphological, physiological, and environmental variables. Compared with other sturgeon species, Siberian sturgeon exhibits similar behaviors including station-holding under high water velocity. However, this species has some special swimming characteristics, including relatively strong swimming performance and a relative low drag coefficient incurred by their unique body form and scarce bony scutes and a high ability to cope with the lack of ambient oxygen. This chapter will try to summarize and discuss existing knowledge on behavior and kinematics, capacity, metabolism, and functional morphology in swimming performance of this species and to give enlightenment for potential application on sturgeon aquaculture.

Keywords

Siberian sturgeon · Behavior · Kinematics · Swimming performance · Metabolism · Functional morphology · Aquaculture

---

M. Duan
Institute of Hydrobiology, Chinese Academy of Sciences, Wuhan 430072, People's Republic of China

East China Sea Fisheries Research Institute, Chinese Academy of Fishery Sciences, Shanghai 200090, People's Republic of China
e-mail: duanming@ihb.ac.cn

Y. Qu · P. Zhuang (✉)
East China Sea Fisheries Research Institute, Chinese Academy of Fishery Sciences, Shanghai 200090, People's Republic of China
e-mail: pzhuang@ecsf.ac.cn

Introduction

Swimming of fish has gained the attention of researchers for more than 50 years. Influenced by technology (e.g., availability of telemetry devices and access to swim flumes), research has focused on the swimming behavior, swimming performance, metabolism, morphological effect on locomotion, and so on. Most fishes lack other weapons against predators, and thus swimming is their main way to avoid and survive their attack (Videler 1993; Reidy et al. 2000; Watkins 1996). Moreover, it is assumed that the maximal swimming performance may strongly influence the ability of a fish to obtain food, find a mate, avoid unfavorable conditions, etc. (Drucker 1996). The range of swimming speeds over which a fish can operate has traditionally been divided into three distinct categories: sustained, prolonged, and burst (Beamish 1978). The highest swimming speed that can be maintained aerobically is called maximum sustained speed and is often estimated by measuring critical swimming speed, or U_{crit}, in a swim tunnel respirometer. Swimming capacity and respiratory characteristics of fishes can be modulated, affected, and constrained by morphological, physiological, and environmental variables. For sturgeon, relatively poor swimming performance is generally attributed to their unique combination of physiological and morphological characteristics, which include a slower metabolism (Singer et al. 1990), a heterocercal tail which sacrifices thrust for dynamic lift, and a high drag coefficient incurred by their protective bony scutes and rough skin (Webb 1986).

Siberian sturgeon *Acipenser baerii* Brandt (Fig. 12.1) is of particular interest to evolutionary biologists because its morphology and physiology are quite unique and, in many ways, intermediate between sharks and the more recently evolved teleosts. At present, considerable information on species structure, ecology, distribution, status, and some ecological adaptations is summarized in Ruban (2005). However, studies conducted on swimming, and its relationship with morphology, physiology in Siberian sturgeon still have not been reviewed. This chapter will attempt to summarize and discuss existing knowledge on behavior, capacity, metabolism, and functional morphology in locomotion of this species and place this knowledge into an aquaculture context with that of other sturgeon species.

Fig. 12.1 Lateral morphology of Siberian sturgeon *Acipenser baerii*

12.1 Life History and Contemporary Status

The Siberian sturgeon, *Acipenser baerii* Brandt, inhabits the largest range compared to other sturgeon species. This potamodromous species are widely distributed throughout all major rivers in Siberia, with the most notable populations occurring in Lake Baikal, the Ob, Aldan, and Lena River systems (Sokolov and Vasilev 1989; Ruban 1997, 2005). There currently are three recognized subspecies of Siberian sturgeon: *A. b. baerii*, which occurs mainly in the Ob River basin, and *A. b. baicalensis* and *A. b. stenorhynchus*, which occur in the eastern Siberian river basins (Ruban 1997). Unlike most other sturgeon species, Siberian sturgeons are predominantly a freshwater species; however, certain populations may inhabit estuarine environments on a seasonal basis (Birstein 1993; Billard and Lecointre 2001). Inhabiting such a wide range with variable environments has caused extreme ecological and morphological plasticity based on specific adaptations (Ruban 1998, 2005) and also the high resistance of the Siberian sturgeon to environmental factors and diseases (Ruban 1997; Bauer et al. 2002). Such plasticity and resistance were the main basis for including this species in aquaculture. Nowadays, the Siberian sturgeon is the main species for sturgeon aquaculture in China, France, Russia, the Moldova Republic, the Czech Republic, Hungary, Germany, Chile, and others (Williot et al. 2001; Bronzi et al. 2011; Wei et al. 2011).

As a threatened species, the Siberian sturgeon is included in the IUCN Red Data List (Birstein 1993). All major populations of Siberian sturgeon have suffered dramatic declines (50–80%) during the last 60 years as a result of overfishing, dam construction, and water pollution (Ruban and Zhu 2010). Up to 40% of the spawning grounds of *A. b. baerii* are now inaccessible to migrating adults because of dam construction on the Ob River, and both *A. b. baerii* and *A. b. stenorhynchus* have been severely overfished (Ruban 1997). All populations of Siberian sturgeon, especially those in the Ob and Kolyma Rivers, are also hindered by reproductive abnormalities attributed to pollution, including several instances of complete sterility (Ruban and Zhu 2010). These facts make it necessary to collect as much information as possible to properly manage the natural populations and to develop aquaculture industry of this species.

12.2 Swimming Behavior and Kinematics

12.2.1 Ontogenetic Behavior

Recent laboratory studies on behavior of early-life intervals of several sturgeon species indicate that young sturgeon of all species undergo ontogenetic changes in behavior and migration (Richmond and Kynard 1995; Gisbert et al. 1999; Zhuang et al. 1999, 2002; Kynard and Horgan 2002). The behavior at early-life stage is innate and slightly affected by experimental environmental factors such as temperature and water velocity (Richmond and Kynard 1995, Kynard et al. 2002). Thus, the behavioral patterns observed in the laboratory correctly reflect young sturgeon's behavior in the river. Ontogenetic behavior of the sturgeon species observed to date reflects adaptations to river conditions (Kynard et al. 2002).

The Siberian sturgeon is capable of integrating a complex of extrinsic (e.g., water temperature, current intensity, substrate typology, predation, food resources) and intrinsic (morphophysiological development) parameters into behavioral patterns, and as a consequence, several saltatory behavioral changes in rheotaxis and swimming ability are observed during early ontogeny. Gisbert and Ruban (2003) reviewed field data and studies on the behavior and development of Siberian sturgeon at early-life intervals and related them to different ecologically relevant environmental factors that may play a role in the distribution, recruitment, and survival of young fish.

At the beginning, feeding and locomotor systems develop allometrically in mutual balance, and as a result, different saltatory behavioral patterns have been described and associated with different levels of morphological development and larval fitness in Siberian sturgeon larvae. At 3 days post-hatching, free embryos exhibit a positive phototaxis, preference for white substrates, vertical swimming, and drift behavior and are unable to swim for a long time in a certain direction. Free embryos perform an alternate set of movements. They move upward by active movements of the posterior trunk region and tail and, then, passively settle to the bottom (Gisbert et al. 1999). Siberian sturgeon embryos might passively migrate downstream at about 100 km/day and possibly up to 300–440 km during the swimming-up behavioral stage. This passive model of transport, however, would help sturgeon free embryos to conserve yolk-sac energy, by allocating those energy resources to the development of locomotor, sensorial, and feeding systems before the completion of yolk utilization. At 4 days post-hatching, there is a transition from pelagic to benthic behavior in free embryos, which gets rid of passive downstream migration in the river due to the development of pectoral fins, eye differentiation, presence of head and trunk rudimentary sensorial structures, and a considerable reduction of the yolk-sac volume (Gisbert 1999). At 5–6 days post-hatching, the fish's tactile bond with the ground is established, and free embryos become benthic swimmers that exhibit a positive rheotactic response and swim vigorously against the water current. Benthic free embryos 7–8 days old aggregate in shoals, while swimming they perform wigwag movements (anguilliform swimming style; Gisbert 1999). During this period, free embryos remain aggregated and swim against the current during the day, while at night, they do not aggregate but swim actively along the bottom of tank. These observations suggest that in their natural environments, free embryos might actively disperse from dusk to dawn, while during the daylight, fish would be concealed between the bottoms and protected from predators. This behavior supposes a reduction of the energy wasted in swimming against the current, and consequently, more energy can be derived from endogenous reserves to finish larval development before the onset of exogenous feeding. Coinciding with the end of the free embryo phase and the beginning of the larval period, shoaling behavior disappears, and fishes scatter across the bottom of the tank and along their walls. Larval activity increases, and swimming ability improves compared to free embryos, since larvae change from an anguilliform to a subcarangiform swimming mode (Gisbert 1999).

12.2.2 Behavior

Studies have also shown that sturgeons typically employ different swimming behaviors in various water velocities, allowing them to effectively move or maintain position within the lotic environment (Adams et al. 2003; Hoover et al. 2011). Observed swimming behaviors in sturgeons are characterized as station-holding (Fig. 12.2), substrate skimming, and free swimming. Swimming performance of sturgeons is typically reported to be relatively poor, compared to similarly size teleost species (Peake et al. 1995; Lee et al. 2003). In previous swim studies, researchers have found that prolonged periods of free swimming are rare in sturgeon and that the fish seem to prefer substrate skimming especially at higher current velocities (Adams et al. 1997, 2003; Boysen and Hoover 2009; Parsons et al. 2003). Most sturgeon species have an inherent ability to maintain station in flowing water either by hunkering or skimming along the substrate bottom. Both of these energy-conserving behaviors are aided by their flat ventral surface, large pectoral fins, and pointed and flattened rostrums (Adams et al. 2003). As such, these same behaviors should, presumably, allow sturgeon to efficiently maintain station along a riverine substrate even after the extreme physical exertion caused by capture and handling.

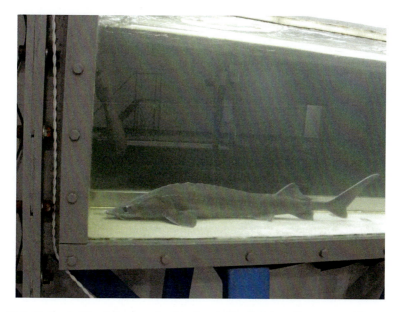

Fig. 12.2 Station-holding behavior of swimming *A. baerii*. A juvenile sturgeon (58.5 cm total length) was grasping the bottom in a modified Brett-type swim flume (Brett 1964; Beamish 1978). The swimming tank consisted of a rectangular swimming channel (length × width × height, 190 cm × 50 cm × 50 cm) made of clear acrylic. An impeller powered by a variable-speed motor (15 KW) circulated water through the flume. Fish was contained within the swim flume by 2 cm × 2 cm screening at each end. To ensure laminar flow through the tank, two rectifier grids were placed at the upstream and downstream ends of the swimming flume

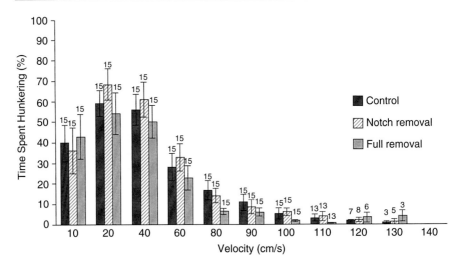

Fig. 12.3 Mean percent time spent hunkering (TSH) across a range of water velocities for *A. baerii*. *Black* bars represent results of samples without treatment. Numbers above each bar represent sample sizes. Error bars represent ± SE (Nguyen 2013)

Most swimming tests have been conducted on freshwater teleost species, especially involved the ubiquitous rainbow trout (*Oncorhynchus mykiss*). A few studies determined swimming performance of sturgeon but have provided relatively few details on swimming behavior during these trials. Compared with rainbow trout, sturgeons have been characterized as relatively docile in swim tunnels (Burggren and Randall 1978), with bouts of restlessness and violent outbursts occurring rarely, if ever. Nguyen (2013) described that most Siberian sturgeon typically spent a majority (>50%) of the time thrashing about in the tunnel at water velocity of 10 cm/s and began to exhibit a higher percentage of station-holding behavior (50–68%) at acclimation velocities of 20–40 cm/s (indicated by the time spent in hunkering, TSH, Fig. 12.3). Bottom holding is a behavior that many sturgeon species employ to hold station in a current, and individuals appear to use this strategy more readily than rainbow trout. Adams et al. (1999) reported that juvenile Pallid sturgeon (*Scaphirhynchus albus*) often held their position in a swim tunnel by pressing their bodies and pectoral fins against the bottom at intermediate speeds, while free swimming most often occurred at low (<20 cm/s) and high (>50 cm/s) water velocities. In a similar study, it was demonstrated that the frequency of bottom-holding behavior by shovelnose sturgeon increased as water velocity exceeded 40 cm/s (Adams et al. 1997). The authors hypothesized that the holding behavior of pallid and shovelnose sturgeon may allow them to occupy fast-water microhabitats with a relatively small energetic cost, thereby reducing competition from more pelagic species. The behavior has been also reported for lake sturgeon exposed to low-to-moderate water velocities (McKinley and Power 1992; Peake 1999). McKinley and Power (1992) suggested that lake sturgeon did not swim when faced with low water velocities because the flows were insufficient to lift the negatively buoyant fish off the bottom and into the water column. In addition, Nguyen (2013) found that Siberian sturgeon's ability to adhere to the tunnel bottom was reduced as velocity increased, such that the fish spent a majority of its time actively swimming at water velocities >60 cm/s. In previous swim studies, however, researchers suggested that prolonged periods of free swimming are rare in sturgeon and that the fish seem to prefer substrate skimming especially at higher current velocities (Adams et al. 1997, 2003; Boysen and Hoover 2009; Parsons et al. 2003).

12.2.3 Kinematics

The locomotion of most teleost species exhibits the following characteristics: (1) swimming speed during steady locomotion is modulated primarily by the alteration of tail beat frequency (TBF), (2) tail beat amplitude (TBA) is usually constant over most of the swimming speed range, and (3) the length of the propulsive wave is independent of speed (Webb 1986). In addition, gill beat frequency (GBF) and the distance that flow passed by the fish in the period of time that the fish spent on a type of gait (D_{gait}) have been quantified as kinematic parameters (Nguyen 2013; Qu et al. unpublished).

Nguyen (2013) reported that mean TBF of Siberian sturgeon ranged from approximately 8–17 beats/min at 10 cm/s to 198 beats/min at 140 cm/s (Fig. 12.4) during critical swimming speed test, while the mean GBF was relatively steady (Fig. 12.5). TBF had a strong positive relationship with increasing current velocity. Research has

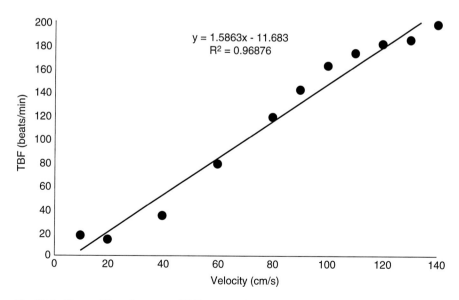

Fig. 12.4 Mean tail beat frequency (TBF) regressed against swimming velocity of *A. baerii* subjected to different fin ray sampling methods (Modified from Nguyen 2013)

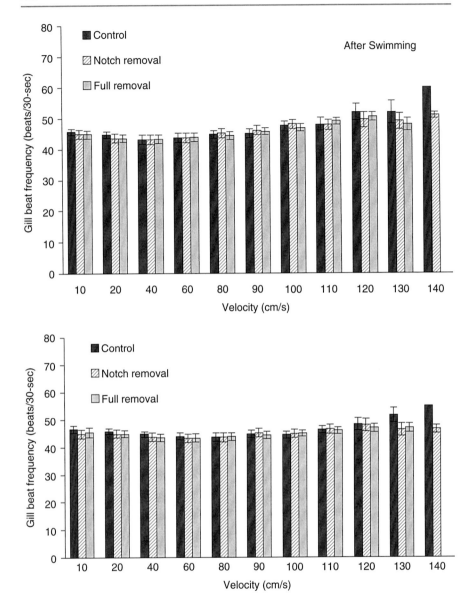

Fig. 12.5 Mean gill beat frequency across a range of velocities for *A. baerii* subjected to different fin ray sampling methods. Gill beats were recorded prior to the fish swimming at each velocity and immediately after swimming. Error bars represent ± SE (Nguyen 2013)

also indicated that sturgeon swimming has unique features but is kinematically more similar to teleosts than selachians (Webb 1986). The relationship between tail beat frequency and swimming speed is positive, linear, and similar in slope and intercept to that of rainbow trout (Webb 1986). However, other kinematic parameters have not been compared with test velocity in Siberian sturgeon.

12.3 Swimming Performance

12.3.1 Critical Swimming Speed

One method of evaluating a fish's swimming performance is to evaluate its critical swimming speed (U_{crit}), which is defined as the speed at which a fish can maintain station for a prescribed period of time (Brett 1964). By quantitatively evaluating the U_{crit} of fishes, biologists can determine how different environmental factors affect the fish's swimming performance (Plaut 2001). To evaluate U_{crit} values in fish, Brett (1964) designed a method in which fish are placed into an enclosed swim chamber or "flume" in which water velocity is increased incrementally at specific time intervals. For benthic fishes like sturgeons, Parsons et al. (2003) developed the term "critical station-holding speed" (CSHS) which accounts for benthic station-holding behaviors. Although similar to U_{crit}, critical station-holding speed is defined as the speed at which a fish can maintain station, either by swimming or adherence to the bottom, for a prescribed period of time. The U_{crit} is influenced by many factors including the size of the fish, and the value of U_{crit} is not transferable to adult, which can swim faster but at a lower rate relative to body length (BL/s).

Nguyen (2013) has determined 111 ± 1.7 cm/s of pooled 10-min CSHS for Siberian sturgeon measuring 74.5 ± 0.6 cm fork length, similar to those obtained by Qu et al. (2013) who reported a mean 10-min U_{crit} of 106 ± 2.2 cm/s for fish measuring 61.8 ± 3.1 cm total length. In addition, Cai et al. (2015) recorded a mean 20-min U_{crit} of 3.26 ± 0.11 BL/s for fish measuring 13.9 ± 0.2 cm TL. Qu et al. (2013) found that *A. baerii* is a stronger swimmer compared to *A. sinensis* with a similar body length, with an average of 25% higher U_{crit} (expressed in body lengths per second). In addition, the U_{crit}, physiology, and behavior are also influenced by environmental conditions, such as temperature, water velocity, seasonal change, and hypoxia (Deslauriers and Kieffer 2012; Killen et al. 2013). Fish may also have different swimming capability under different hydraulic conditions (Webber et al. 2007) and different reared conditions (Duan et al. 2011). Higher water temperatures may result in swimming speeds of the smaller Siberian sturgeon used by Qu et al. (2013) that were comparable to those of the larger fish used in Nguyen's trials (Nguyen 2013).

12.3.2 Speed Exponent (c)

In an exhaustive exercise for fish, the relationship between oxygen consumption rate (MO_2) and fish swimming speed (U) was described by the power function,

$$MO_2 = a + bU^c;$$

where a, b, and c are fitting constants; c, known as the speed exponent, is inversely related to swimming efficiency (Tu et al. 2012).

Videler and Nolet (1990) reviewed several studies and reported the speed exponent (c) from 1.1 to 3.0 at 10 °C–29 °C for 20 fish species. Cai et al. (2015) calculated the value of c for Siberian sturgeon which ranged from 0.966 to

1.028 in exhaustive exercise and ranged from 1.111 to 1.265 after different recovery duration, indicating this species is an efficient swimmer, even after exhaustive exercise. In addition, the value of c was also close to 1 for juvenile Amur sturgeon (*A. schrenckii*) and Chinese sturgeon (*A. sinensis*) at 20 °C and White sturgeon (*A. transmontanus*) and Lake Sturgeon (*A. fulvescens*) at 10 °C (Peake 2005; Cai et al. 2013, 2014). The similar and low values of c for these other species indicate that the swimming efficiency of sturgeons is high. The higher swimming efficiency may be a result of the distinct morphology that sturgeons have compared with other fishes (Pettersson and Hedenstrom 2000). Also, because sturgeons have the high swimming efficiencies, they can be better long-distance swimmers. Most sturgeons migrate for long distances in spring or autumn, and although researchers have reported the values of c in many fishes, those in sturgeons are limited.

12.4 Swimming Metabolism

Energy for maintaining sustained swimming is derived primarily through oxidative metabolic processes. Sustained activity is, therefore, limited by the ability of the cardiovascular system to provide oxygen to the working musculature. Several studies have examined the relationship between activity, respiration, and ambient oxygen levels for sturgeon and paddlefish.

Nonnotte et al. (1993) showed that Siberian sturgeon, like most other fish, exhibit typical O_2 regulatory behavior when subjected to graduate and moderate hypoxia. Besides, a slight lactic acidosis was developed below a critical level of 40 torr Pwo_2 and became more marked during return to normoxia. Maxime et al. (1995) subsequently testified that environmental hypoxia is a potent stimulus for catecholamines and cortisol release into blood plasma of Siberian sturgeon as in teleosts, by exposing fish to an acute and severe hypoxia followed by a rapid return to normoxic conditions. Compared with the response to hypoxia of another species of sturgeons, *Acipenser naccarii* (Randall et al. 1992), and of other studied fishes, Siberian sturgeon seems to present a high ability to cope with the lack of ambient oxygen.

Swimming metabolism and energy recovery mechanism have also been studied in exhaustive exercise. Cai et al. (2015) found that swimming capability of Siberian sturgeon (13.9 ± 0.2 cm, mean ± SE, total length) decreased significantly with exhaustive exercise and recovery required more than 1 h. The U_{crit} was reported to be ~85% with a resting time of 1 h and recovery ratio of adult sockeye salmon (*Oncorhynchus nerka*) nearly 100% with a resting time of 1.5 h at 9 °C (Jain et al. 1998). The recovery of U_{crit} in Siberian sturgeon was slower, 78% in 1 h and complete recovery in slightly more than 24 h, and even U_{crit} may increase in sturgeon exposed to exercise. The recovery pattern of maximum metabolic rate (MMR) was similar to that of U_{crit} but lower. While the recovery ratios of MMR and U_{crit} for 1-h recovery were very close (78% and 79%, respectively), the recovery ratios for 24-h recovery were diverged (88% and 93%, respectively).

12.5 Functional Morphology in Swimming

Natural selection can result in adaptations tuned to the specific environments in which the organism lives (Darwin 1859). In most fish species, an important adaptation is body morphology as it affects swimming performance (Lauder and Liem 1983). For instance, the hydrodynamic function of fins in teleost fishes has been documented recently (Lauder and Drucker 2002, 2004; Fish and Lauder 2006), providing excellent data for understanding general evolutionary patterns of form and function. *A. baerii*, a species of basal chondrostean ray-finned fish that have a slender spindle-shaped body and some plesiomorphic features such as a soft dorsal fin and heterocercal tail, clever at station-holding which is correlated closely with the structures and body form (Webb 1989; Wilga and Lauder 2001), is a good model for understanding the evolutionary pattern of functional morphology in basal ray-finned fishes.

Qu et al. (2013) compared lateral morphology and swimming performance of Siberian sturgeon (*A. baerii*) (61.82 ± 3.09 cm, mean $\pm$ SE, total length) and Chinese sturgeon (*A. sinensis*) (57.86 ± 4.66 cm) by geometric morphometrics and critical swimming speed test. Twenty-six morphological landmarks were used for recording morphological information. Then the landmark configuration was aligned by generalized Procrustes analysis (GPA), in order to minimize the sum of square distance between paired landmarks (Rohlf 1990; Rohlf and Slice 1990). All aligned configurations of specimens were compared by relative warp analysis (RWA, Rohlf 1993). This method is a principal component analysis of all body shape variations among specimens. The relative warps (RW) of the specimens were visualized on transformative grids. The relationship between the morphological variation and swimming performance was examined by regressing U_{crit} of each specimen onto its corresponding relative warp scores. Interspecies analysis (Qu et al. 2013) clearly suggested that negative scores on the first relative warp (RW1, namely, the first principal component, accounting for 71.20% of the variance between two species) correspond to a narrower and shorter snout and head, deeper anterior trunk, shortened leading edge of dorsal fin, and elongated caudal lobes (Fig. 12.6). Such key differences in morphology result in *A. baerii* that is a stronger swimmer compared to *A. sinensis*, with an average of 25% higher U_{crit} (expressed in body lengths per second, Qu et al. 2013). Intraspecific analysis on *A. baerii* and *A. sinensis* indicated negative correlations between U_{crit} and RW1 for both species (Qu et al. 2013). For *A. baerii*, the depth and length of the snout and the trailing edge length of the dorsal fin were negatively correlated with U_{crit}, whereas the height of the anterior trunk, the leading edge length of the dorsal fin and anal fin, and the length and width of the ventral lobe were positively related to U_{crit} (Figs. 12.7 and 12.8). Moreover, although the degree of upward bending of the snout of *A. baerii* was negatively related to U_{crit}, there was a positive relationship between the length of the caudal peduncle and U_{crit} as well as between the dorsal tail lobe and U_{crit} (Figs. 12.7 and 12.8).

Variations in head shape indicate that the morphology of sturgeons in this regard serve to reduce the surface area in order to save energy. The more surface area a body exposes to the water flow, the greater drag is experienced even at a high Reynolds number (Vogel 1994). A relatively smaller head and snout with less rough surface resulted in decreased friction drag on *A. baerii*. Moreover, due to the

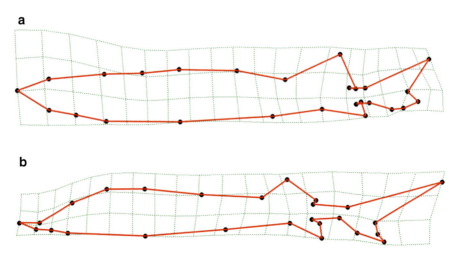

Fig. 12.6 Maps of visualized results of (**a**) positive and (**b**) negative RW1 score combinations of *A. sinensis* and *A. baerii* displayed on the deformative grids (Qu et al. 2013)

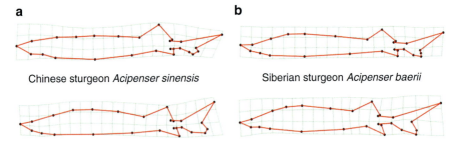

Fig. 12.7 Maps of visualized results of positive (*upper*) and negative (*lower*) RW1 scores of (**a**) *A. sinensis* and (**b**) *A. baerii* displayed on the deformative grids (Qu et al. 2013)

elongated and slightly concave downward snout of *A. baerii*, the water flow could be separated at the edge of snout ahead of time in order to avoid more friction with the frontal plane of the head and anterior segment of the trunk. This is similar to the function of the sword in the swordfish (*Xiphias gladius*) (Videler 1995). The shape of the head also affects station-holding behavior. The body form could affect efficiency of the station-holding behavior by changing the direction of pressure drag. Qu et al. (2013) suggested that the thickness and position of the snout made the frontal plane of the head useful as primary pressure plane for flow (see Fig. 12.6). The force exerted downward and forward on the head and anterior trunk was adequate to ensure enough friction generated by the ventral and pectoral fins. However, the plump ventral snout of *A. sinensis* resulted in more pressure drag, generating a clockwise torque that was inadequate to form enough friction (Qu et al. 2013).

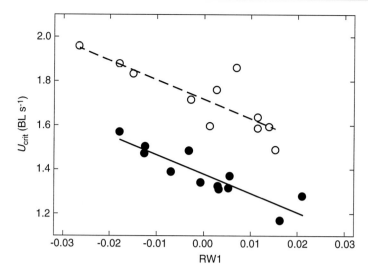

Fig. 12.8 Relationships between U_{crit} and RW1 of *A. sinensis* (*filled circle*) ($U_{crit} = 1.378 - 8.638 \times RW1$; $r^2 = 0.796$, $P < 0.001$, $n = 12$) and *A. baerii* (*open circle*) ($U_{crit} = 1.719 - 8.812 \times RW1$; $r^2 = 0.695$, $P < 0.001$, $n = 11$) (Qu et al. 2013)

The dorsal fin in sturgeons has been considered to function as a "second tail" due to its generating thrust. However, Qu et al. (2013) found that the relatively shallower dorsal fin in *A. baerii* contributes to drag reduction by determining Strouhal number (*St*, a dimensionless number describing oscillating flow mechanism) as a predictor of propulsive efficiency (Triantafyllou et al. 1993). It was concluded that thrust was developed only at low flow velocity for the oscillating dorsal fin, because strong thrust is developed from a reverse von Kármán street only when $0.2 < St < 0.5$ (Anderson et al. 1998). As a plesiomorphic feature, the soft dorsal fin cannot be controlled actively as in the rainbow trout (*Oncorhynchus mykiss*) (Drucker and Lauder 2005), such that it becomes a source of drag when the *St* is below the lower limits, especially when a fish cannot maintain intensive dorsal fin oscillation.

With regard to the caudal fin, *A. baerii* is able to develop more thrust via its deeper trailing edge, since the added mass of water accelerated by the tail is proportional to the square of the trailing edge depth (Lighthill 1971). A more vertical trailing edge of the tail fin was considered as a lift generator, which in turn might balance the torque generated from the head (Lauder *pers. comm.*).

Body streamlining, the best way to reduce pressure drag, is more significant than skin friction caused by high Reynolds numbers (Vogel 1994). Qu et al. (2013) showed that superior swimmers possessed a more streamlined body approximate to the NACA (the US National Advisory Committee for Aeronautics) 0016 airfoil shape (a type of symmetric airfoil with 16% relative thickness) by comparing the body form of *A. baerii* with NACA airfoil shape and regressing their differences

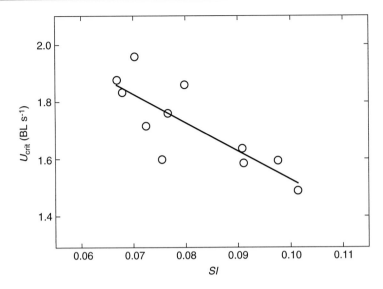

Fig. 12.9 Streamline index (*SI*) was calculated by comparing landmark coordinates on the trunk displayed in the relative warp, with its corresponding point on the NACA airfoil shape. Relationship between U_{crit} and *SI* of *A. baerii* when the reference was NACA 0016 airfoil shape (U_{crit} = 2.520– 9.907 × *SI*, r^2 = 0.656, *P* = 0.003, *n* = 11) (Qu et al. 2013)

(defined as streamline index, *SI*) on U_{crit}, while all experimental fish were swimming in a regime where a streamlined body provided a pronounced optimum for drag reduction (Fig. 12.9, Batchelor 1967; Qu et al. 2013).

Conclusions

Siberian sturgeon bears a few resemblances in swimming behavior, capacity, metabolism, and functional morphology to other sturgeon species, even to teleost, but it has its own other unique features. Swimming characteristics closely bond up with morphology and physiology from hatching to the juvenile phase of the Siberian sturgeon. As development of locomotor organ and physiological function, vertical swimming, drift behavior, benthic behavior, and rheotaxis appear successively. Ecologically, ontogenetic changes in behavior might allow each early-life interval (free embryo, larva, and juvenile) to migrate and occupy different river habitats, thus directly influencing their distribution, survival, and recruitment (Gisbert and Ruban 2003). For sturgeon aquaculture, understanding of swimming behavior in early stage is conducive to set water velocity, water depth, and substructure, which are related to larva's growth. When sturgeons grow into juvenile and adult, researchers prefer circulating water flume (often sealed and with a respiratory measurement system) for observing its swimming behavior and measuring swimming performance, respiratory metabolism, and kinematic parameter. Similar to other sturgeon species, benthic behavior (e.g., station-holding and substrate skimming) and swimming against current are two obvious features of Siberian sturgeon as well. Particularly, lower position of the narrow snout and relatively broader frontal area of the head and back might make this species better at station-holding by using pressure of current, while a big tail is beneficial to active beating, which offers staff much more choices in selecting water velocity and area. This species also seems to present a high ability to cope with the lack of ambient oxygen, indicating a possibility of intensive farming. More information might be revealed by a sort of studies on relationship between special behavior performance and morphology of corresponding locomotor organs, as what has been discussed in Chinese sturgeon (Qu et al. unpublished data).

References

Adams SR, Adams GL, Parsons GR (2003) Critical swimming speed and behavior of juvenile shovelnose sturgeon and pallid sturgeon. T Am Fish Soc 132:392–397

Adams SR, Hoover JJ, Killgore KJ (1999) Swimming endurance of juvenile pallid sturgeon, *Scaphirhynchus albus*. Copeia 3:802–807

Adams SR, Parsons GR, Hoover JJ et al (1997) Observations of swimming ability in shovelnose sturgeon (*Scaphirhynchus platorynchus*). J Freshw Ecol 12:631–633

Anderson JM, Streitlein K, Barrett DS et al (1998) Oscillating foils of high propulsive efficiency. J Fluid Mech 360:41–72

Batchelor GK (1967) An introduction to fluid dynamics. Cambridge University Press, New York

Bauer ON, Pugachev ON, Voronin VN (2002) Study of parasites and diseases of sturgeons in Russia: a review. J Appl Ichthyol 18:420–429

Beamish FWH (1978) Swimming capacity. In: Hoar WS, Randall JD (eds) Fish physiology, vol 7. Academic Press Inc, New York, pp 101–187

Billard R, Lecointre G (2001) Biology and conservation of sturgeon and paddlefish. Rev Fish Biol Fisher 10:355–392

Birstein VJ (1993) Sturgeons and paddlefishes: threatened fishes in need of conservation. Conserv Biol 7:773–787

Boysen KA, Hoover JJ (2009) Swimming performance of juvenile white sturgeon (*Acipenser transmontanus*): training and the probability of entrainment due to dredging. J Appl Ichthyol 25:54–59

Brett JR (1964) The respiratory metabolism and swimming performance of young sockeye salmon. Can J Fish Aquat Sci 21:1183–1226

Bronzi P, Rosenthal H, Gessner J (2011) Global sturgeon aquaculture production: an overview. J Appl Ichthyol 27:169–175

Burggren WW, Randall DJ (1978) Oxygen uptake and transport during hypoxic exposure in the sturgeon, *Acipenser transmontanus*. Resp Physiol 34:171–183

Cai L, Chen L, Johnson D et al (2014) Integrating pressure released from water flow, locomotor performance and respiration of Chinese sturgeon during multiple fatigue recovery cycles. PLoS One 9:e94345

Cai L, Johnson D, Mandal P et al (2015) Effect of exhaustive exercise on the swimming capability and metabolism of juvenile Siberian sturgeon. T Am Fish Soc 144:532–538

Cai L, Taupier R, Johnson D et al (2013) Swimming capability and swimming behavior of juvenile *Acipenser schrenckii*. J Exp Zool Part A 319:149–155

Darwin CR (1859) On the origin of species. John Murray, London

Deslauriers D, Kieffer JD (2012) The effects of temperature on swimming performance of juvenile Shortnose sturgeon (*Acipenser brevirostrum*). J Appl Ichthyol 28:176–181

Drucker EG (1996) The use of gait transition speed in comparative studies of fish locomotion. Am Zool 36:555–566

Drucker EG, Lauder GV (2005) Locomotor function of the dorsal fin in rainbow trout: kinematic patterns and hydrodynamic forces. J Exp Biol 208:4479–4494

Duan M, Qu Y, Yan J et al (2011) Critical swimming speed associated with body shape of Chinese sturgeon *Acipenser sinensis* under different rearing conditions. Int Aquat Res 3:83–91

Fish FE, Lauder GV (2006) Passive and active flow control by swimming fishes and mammals. Annu Rev Fluid Mech 38:193–224

Gisbert E (1999) Early development and allometric growth patterns in Siberian sturgeon and their ecological significance. J Fish Biol 54:852–862

Gisbert E, Ruban GI (2003) Ontogenetic behavior of Siberian sturgeon, *Acipenser baerii*: a synthesis between laboratory tests and field data. Environ Biol Fish 67:311–319

Gisbert E, Williot P, Castello-Orvay F (1999) Behavioral modifications in the early life stages of Siberian sturgeon (*Acipenser baerii*, Brandt). J Appl Ichthyol 15:237–242

Hoover JJ, Collins J, Boysen KA et al (2011) Critical swimming speeds of adult shovelnose sturgeon in rectilinear and boundary-layer flow. J Appl Ichthyol 27:226–230

Jain KE, Birtwell IK, Farrell AP (1998) Repeat swimming performance of mature sockeye Salmon following a brief recovery period: a proposed measure of fish health and water quality. Can J Zool 76:1488–1496

Killen SS, Marras S, Metcalfe NB et al (2013) Environmental stressors alter relationships between physiology and behaviour. Trends Ecol Evol 28:651–658

Kynard B, Henyey E, Horgan M (2002) Ontogenetic behavior, migration and social behavior of pallid sturgeon, *Scaphirhynchus albus*, and shovelnose sturgeon, *S. platorynchus*, with notes on the adaptive significance of body color. Environ Biol Fish 63:389–403

Kynard B, Horgan M (2002) Ontogenetic behavior and migration of Atlantic sturgeon, *Acipenser oxyrinchus oxyrinchus*, and shortnose sturgeon, *A. brevirostrum*, with notes on social behavior. Environ Biol Fish 63:137–150

Lauder GV, Drucker EG (2002) Forces, fishes and fluids: hydrodynamic mechanisms of aquatic locomotion. News Physiol Sci 17:235–240

Lauder GV, Drucker EG (2004) Morphology and experimental hydrodynamics of fish fin control surfaces. IEEE J Ocean Eng 29:556–571

Lauder GV, Liem KF (1983) The evolution and interrelationships of the actinopterygian fish. Bull Mus Comp Zool 150:95–197

Lee CG, Farrell AP, Lotto A et al (2003) The effect of temperature on swimming performance and oxygen consumption in adult sockeye (*Oncorhynchus nerka*) and coho (*O. kisutch*) salmon stocks. J Exp Biol 206:3239–3251

Lighthill MJ (1971) Large-amplitude elongated-body theory of fish locomotion. Proc R Soc B 179:125–138

Maxime V, Nonnotte G, Peyraud C et al (1995) Circulatory and respiratory effects of hypoxic stress in the Siberian sturgeon. Resp Physiol 100:203–212

McKinley RS, Power G (1992) Measurement of activity and oxygen consumption for adult lake sturgeon in the wild using radio-transmitted EMG signals. Ellis Horwood, West Sussex, UK

Nguyen HPL (2013) Effects of pectoral fin ray removal on white sturgeon (*Acipenser transmontanus*) and Siberian sturgeon (*Acipenser baerii*) swimming performance. Dissertation, The University of Georgia

Nonnotte G, Maxime V, Truchot JP et al (1993) Respiratory responses to progressive hypoxia in the sturgeon, *Acipenser baerii*. Resp Physiol 91:71–82

Parsons GR, Hoover JJ, Killgore KJ (2003) Effect of pectoral fin ray removal on station-holding ability of shovelnose sturgeon. N Am J Fish Manage 23:742–747

Peake S (1999) Substrate preferences of juvenile hatchery-reared lake sturgeon, *Acipenser fulvescens*. Environ Biol Fish 56:367–374

Peake S, Beamish FWH, McKinley RS et al (1995) Swimming performance of lake sturgeon, *Acipenser fulvescens*. Can Tech Rep Fish Aquat Sci 2063:26

12 Swimming Characteristics of the Siberian Sturgeon

Peake SJ (2005) Swimming and respiration. In: GTO LB, FWH B, RS MK (eds) Sturgeons and paddlefish of North America. Kluwer Academic, Dordrecht, Netherlands, pp 147–166

Pettersson LB, Hedenstrom A (2000) Energetics, cost reduction and functional consequences of fish morphology. P Roy Soc Lond B 267:759–764

Plaut I (2001) Critical swimming speed: its ecological relevance. Comp Biochem Phys A 131:41–50

Qu Y, Duan M, Yan J et al (2013) Effects of lateral morphology on swimming performance in two sturgeon species. J Appl Ichthyol 29:310–315

Randall DJ, McKenzie DJ, Abrami G et al (1992) Effects of diet on responses to hypoxia in sturgeon (*Acipenser naccarii*). J Exp Biol 170:113–125

Reidy SP, Kerr SR, Nelson JA (2000) Aerobic and anaerobic swimming performance of individual Atlantic cod. J Exp Biol 203:347–357

Richmond AM, Kynard B (1995) Ontogenetic behaviour of shortnose sturgeon, *Acipenser brevirostrum*. Copeia 1:172–182

Rohlf FJ (1990) Fitting curves to outlines. In: Rohlf FJ, Bookstein FL (eds) Proceedings of the Michigan morphometrics workshop. University of Michigan Museum of Zoology, Ann Arbor, pp 167–177

Rohlf FJ (1993) Relative warp analysis and an example of its application to mosquito wings. In: Marcus LF, Bello E, García-Valdecasas A (eds) Contributions to morphometrics. Museo Nacionalis de Ciencias Naturales, Madrid, pp 131–159

Rohlf FJ, Slice DE (1990) Extensions of the Procrustes method for the optimal superimposition of landmarks. Syst Zool 39:40–59

Ruban GI (2005) The Siberian sturgeon *Acipenser baerii* Brandt. Species structure and ecology. World Sturgeon Conservation Society. Special publication no 1, Books on Demand GmbH, Norderstedt, Germany

Ruban GI (1997) Species structure, contemporary distribution and status of the Siberian sturgeon, *Acipenser baerii*. Environ Biol Fish 48:221–230

Ruban GI (1998) On the species structure of the Siberian sturgeon *Acipenser baerii* Brandt (Acipenseridae). J Ichthyol 38:345–365

Ruban GI, Zhu B (2010) *Acipenser baerii*. In: IUCN 2012, IUCN Red List of Threatened Species

Singer TD, Mahadevappa VG, Ballantyne JS (1990) Aspects of the energy metabolism of lake sturgeon, *Acipenser fulvescens*, with special emphasis on lipid and ketone body metabolism. Can J Fish Aquat Sci 47:873–881

Sokolov LI, Vasilev VP (1989) *Acipenser baerii* Brandt, 1869. In: Holcík J (ed) The freshwater fishes of Europe, vol 1, pt. II, general introduction to fishes, Acipenseriformes. AULA-Verlag, Wiesbaden, pp 263–284

Triantafyllou GS, Triantafyllou MS, Grosenbaugh MA (1993) Optimal thrust development in oscillating foils with application to fish propulsion. J Fluids Struct 7:205–224

Tu Z, Li L, Yuan X et al (2012) Aerobic swimming performance of juvenile largemouth bronze gudgeon (*Coreius guichenoti*) in the Yangtze River. J Exp Zool Part A 317:294–302

Videler JJ (1993) Fish swimming. Chapman and Hall, London

Videler JJ (1995) Body surface adaptations to boundary-layer dynamics. In: Ellington CP, Pedley TJ (eds) Biological fluid dynamics. The Society of Biologists Ltd, Cambridge, pp 1–20

Videler JJ, Nolet BA (1990) Cost of swimming measured at optimum speed: scaling effects, differences between swimming styles, taxonomic groups and submerged and surface swimming. Comp Biochem Phys A 97:91–99

Vogel S (1994) Life in moving fluids. Princeton University Press, Princeton, NJ

Watkins TB (1996) Predator-mediated selection on burst swimming performance in tadpoles of the Pacific tree frog, *Pseudacris regilla*. Physiol Zool 69:154–167

Webb PW (1986) Kinematics of lake sturgeon, *Acipenser fulvescens*, at cruising speeds. Can J Zool 64:2137–2141

Webb PW (1989) Station-holding by three species of benthic fishes. J Exp Biol 145:303–320

Webber JD, Chun SN, MacColl TR et al (2007) Upstream swimming performance of adult white sturgeon: effects of partial baffles and a ramp. T Am Fish Soc 136:402–408

Wei Q, Zou Y, Li P et al (2011) Sturgeon aquaculture in China: progress, strategies and prospects assessed on the basis of nation-wide surveys (2007-2009). J Appl Ichthyol 27:162–168

Wilga CD, Lauder GV (2001) Functional morphology of the pectoral fins in bamboo sharks, *Chiloscyllium plagiosum*: benthic vs. pelagic station-holding. J Morphol 249:195–209

Williot P, Sabeau L, Gessner J et al (2001) Sturgeon farming in Western Europe: recent developments and perspectives. Aquat Living Resour 14:367–374

Zhuang P, Kynard B, Zhang L et al (2002) Ontogenetic behavior and migration of Chinese sturgeon, *Acipenser sinensis*. Environ Biol Fish 65:83–97

Zhuang P, Zhang L, Zhang T et al (1999) Effects of delaying first feeding on the survival and growth of Chinese sturgeon larvae. Acta Hydrobiol Sin 23:560–565. (in Chinese with English abstract)

Part II

Biology and Physiology of Reproduction

Chemical Neuroanatomy of the Hypothalamo-Hypophyseal System in Sturgeons

13

Olivier Kah and Fátima Adrio

Abstract

The preoptic-hypothalamo-hypophyseal system of sturgeons, located at the base of the brain, has a neurosecretory role exerted by hypophysiotropic neurons most of them located in the preoptic and hypothalamic periventricular region. The majority of those cells are of the cerebrospinal fluid-contacting type and exhibit short processes reaching the ventricular lumen. Moreover, the processes of those hypophysiotropic neurons course along the hypothalamic floor toward the hypophysis forming a preoptic-hypothalamo-hypophyseal tract. This chapter summarizes available data on the distribution of several hypophysiotropic factors, such as galanin, neurophysin, somatostatin, or gonadotropin-releasing hormone, in the preoptic-hypothalamo-hypophyseal system of sturgeons obtained by the use of immunohistochemical techniques. Immunoreactive neurons to those substances were found in the preoptic and hypothalamic nuclei, and immunoreactive fibers were observed along the preoptic-hypothalamo-hypophyseal tract and in the hypophysis, indicating their hypophysiotrophic role in the brain of sturgeons. Thus, most of the neuropeptides and neurohormones found in tetrapods are also present in sturgeons, suggesting that their common ancestors already possessed such regulatory systems. Unfortunately, because of the difficulty in approaching the physiology of sturgeons (size, cost, etc.), the number of experimental studies aiming at deciphering the roles of such neuropeptides and

O. Kah
Research Institute in Health, Environment and Occupation, Université de Rennes 1, Rennes cedex 35 042, France
e-mail: olivier.kah@univ-rennes1.fr

F. Adrio (✉)
Área de Bioloxía Celular, Departamento de Bioloxía Funcional,
Universidade de Santiago de Compostela, 15782 Santiago de Compostela, A Coruña, Spain
e-mail: fatima.adrio.fondevila@usc.es, fatima.adrio.fondevilla@usc.es

neurohormones is very limited, although we can speculate that part of the functions supported by these neurohormones would be similar.

Keywords

Sturgeon • Hypothalamus • Hypophysis • Galanin • Somatostatin • GnRH

13.1 The Anatomy of the Preoptic Area and Hypothalamus of Sturgeons

Studies concerning the cytoarchitecture of the diencephalon of chondrosteans, and in particular of sturgeons, are scarce, and current knowledge concerning the organization of this encephalic region is based on anatomical studies carried out at the beginning of the nineteenth century by Johnston (1901) although it has been more recently analyzed in detail by Nieuwenhuys (1998), Rupp and Northcutt (1998), and Rustamov (2006a, b).

The diencephalon is a complex brain region bounded rostrally by the telencephalon and caudally by the mesencephalon, although there is a long-lasting controversy about where exactly are those boundaries. Briefly, there are two different models of forebrain organization based on developmental studies: the *His-Herrick model* which proposed that the origin of all brain regions is in a series of longitudinal

Fig. 13.1 Micrographs of hematoxylin-eosin-stained sagittal (**a**) and transverse (**b**–**j**) sections through the brain of the Siberian sturgeon *Acipenser baerii*. (**a**) Sagittal section through the entire brain of *Acipenser*. Lines with the letters at the top indicate the levels of the figures (**b**–**j**). (**b**) Section through the rostral preoptic area showing the small cells of the parvocellular preoptic nucleus (NPOp) arranged in a few rows parallel to the ventricular surface around the preoptic recess (PR). (**c**) Detail of figure (**a**) showing the NPOp. (**d**) Section through the caudal preoptic area showing the small cells of the NPOp and the large cells of the magnocellular preoptic nucleus (NPOm) which occupy a central position in the preoptic region dorsal to the rostral portion of the optic chiasm (OC). Abundant cells are located along the ventral part of the anterior wall of the preoptic recess, in the organum vasculosum of the terminal lamina (OVLT). (**e**) Detail of figure (**d**) showing the abundant cells in the NPOp and in the OVLT. (**f**) Section through the preoptic area at a more caudal level than figure (**d**) showing the NPOp and the NPOm (*arrowheads*). (**g**) Section through the rostral hypothalamus showing the inferior hypothalamic lobes (HL), which surround the hypothalamic lateral recesses (LR). Along the ventromedial walls of the lateral recesses, the third ventricle gives rise to the infundibulum (i) in which more rostral walls the anterior tuberal nucleus (NAT) is located. (**h**) Detail of the figure (**g**) showing the major part of hypothalamic neurons located in the ventricular surface around the infundibulum and the lateral recesses. (**i**) Section through the hypothalamus showing the position of the cells of the lateral tuberal nucleus (NLT) at rostral levels. (**j**) Section through the caudal hypothalamus showing the NLT surrounding the posterior recess (POR), the median eminence (ME), the hypophysis (H), and the saccus vasculosus (SV). *Inset*: coronet cells in the floor of the hypothalamus. Abbreviations: *CB*, corpus cerebelli; *h*, habenula; *H*, hypophysis; *HL*, hypothalamic lobes; *i*, infundibulum; *LR*, lateral hypothalamic recess; *ME*, median eminence; *MO*, medulla oblongata; *NAT*, anterior tuberal nucleus; *NLT*, lateral tuberal nucleus; *NPOm*, magnocellular preoptic nucleus; *NPOp*, parvocellular preoptic nucleus; *OB*, olfactory bulb; *OC*, optic chiasm; *OT*, optic tectum; *OVLT*, organum vasculosum of the terminal lamina; *POR*, posterior recess; *PR*, preoptic recess; *PT*, posterior tubercle; *SV*, saccus vasculosus; *T*, telencephalon; *vc*, valvula cerebelli. Scale bars: 2 mm (**a**); 1 mm (**g**); 500 μm (**b**, **d**, **h**); 200 μm (**c**, **e**, **i**–**j**); 100 μm (**f**); 50 μm (inset in **j**)

13 Chemical Neuroanatomy of the Hypothalamo-Hypophyseal System in Sturgeons 251

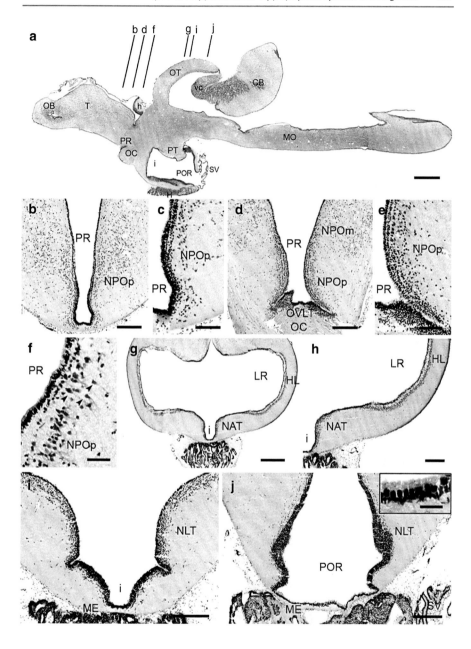

columns separated by longitudinal sulci and divides the diencephalon into four longitudinally arranged zones, epithalamus, dorsal thalamus, ventral thalamus, and hypothalamus (Herrick 1910, 1933), and the *neuromeric model* which suggests that the origin of all brain regions is in a series of transverse regions (neuromeres) separated by vertical sulci and that the diencephalon arises from a rostral parencephalic neuromere but does not include the preoptic area and the hypothalamus which together with the telencephalon are considered part of the secondary prosencephalon (Puelles and Rubenstein 1993). Thus, as in chondrosteans there are no data on neither embryological nor gene expression which could delimitate diencephalic boundaries; most authors have organized the diencephalon of sturgeons as suggested by Braford and Northcutt (1983) for the diencephalon of ray-finned fishes. Braford and Northcutt (1983) follow the His-Herrick model and thus subdivide the diencephalon into the epithalamus, thalamus, and hypothalamus and include the entire preoptic region in the diencephalon. However, Nieuwenhuys (1998), although basically following His-Herrick model for the rest of the diencephalic regions, assigns the preoptic region to the telencephalon, following classic anatomical studies by Johnston (1901). In this review we will follow the organization suggested by Braford and Northcutt (1983) for the diencephalon of ray-finned fishes, i.e., we will include the entire preoptic region in the diencephalon.

13.1.1 Preoptic Area

The preoptic area is the most rostral part of the diencephalon. It surrounds the preoptic recess of the third ventricle and lies between the anterior commissure rostrally and the optic chiasm caudally (Fig. 13.1a). This area is characterized by the periventricular position of most of its cells, containing three periventricular nuclei located around the preoptic recess: the *parvocellular preoptic nucleus*, the most rostral diencephalic nucleus with small cells arranged in a few rows parallel to the ventricular surface (Fig. 13.1b–f); the *magnocellular preoptic nucleus*, with large neurosecretory cells occupying a central position in the preoptic region dorsal to the rostral portion of the optic chiasm (Fig. 13.1f); and the *suprachiasmatic nucleus*, with small densely arranged cells situated dorsal to the optic chiasm. Moreover, there is a migrated nucleus, the *entopeduncular nucleus*, with small loosely arranged cells. In the parvocellular preoptic nucleus, Rupp and Northcutt (1998) distinguished an anterior and a posterior part. Along the ventral part of the anterior wall of the preoptic recess is located the organum vasculosum of the lamina terminalis (OVLT, Fig. 13.1d, e), one of the circumventricular organs present in all vertebrate groups.

13.1.2 Hypothalamus

During brain ontogenesis, the hypothalamus differentiates relatively late in sturgeons showing the normal histological pattern of the adult at 90 days post-hatching

13 Chemical Neuroanatomy of the Hypothalamo-Hypophyseal System in Sturgeons

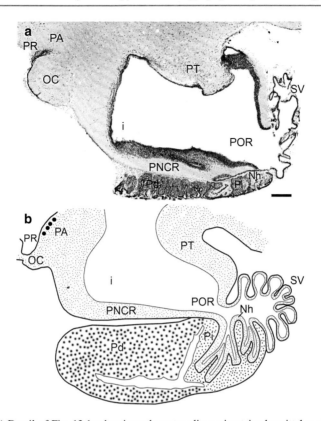

Fig. 13.2 (**a**) Detail of Fig. 13.1a showing a hematoxylin-eosin-stained sagittal section through the hypothalamus of the Siberian sturgeon *Acipenser baerii*. (**b**) Topography of the hypothalamo-hypophyseal neurosecretory system in Acipenseridae. Figure (**b**) was modified from Polenov and Garlov (1971). Abbreviations: *i*, infundibulum; *Nh*, neurohypophysis; *OC*, optic chiasm; *PA*, preoptic area; *Pd*, pars distalis of the hypophysis; *Pi*, pars intermedia of the hypophysis; *PNCR*, proximal neurosecretory contact region; *POR*, posterior recess; *PR*, preoptic recess; *PT*, posterior tubercle; *SV*, saccus vasculosus. Scale bar: 500 μm

and reaching a considerable size compared to other parts of the brain (Vázquez et al. 2002; Gómez et al. 2009). Due to its extremely enlarged ventricle, the hypothalamus forms the largest and the most ventral diencephalic structure in sturgeons (Figs. 13.1a and 13.2), and it comprises a large paired rostral region which is bordered dorsally by the preoptic area and by the posterior tubercle and a much smaller unpaired caudal portion which is continuous with the pituitary and the saccus vasculosus (Figs. 13.1a, g–j and 13.2). The major part of hypothalamic neurons of the sturgeons is located adjacent to the ventricular surface, with only a few migrated cells situated more laterally (Figs. 13.1g–j and 13.2a). Most of the neurons located in the periventricular zone are of the cerebrospinal fluid-contacting (CSF-C) type. Such neurons exhibit an apical dendrite that ends by a ventricular bulb (Fig. 13.1c, e–f). These CSF-C cells are really abundant in the hypothalamus of the sturgeons, but they represent a "primitive" character that tends to disappear in the course of evolution. Bony fish exhibit much less CSF-contacting cells.

In the rostral hypothalamus, the walls are rather thin and laterally evaginate to form the *inferior or lateral hypothalamic lobes*, which surround the *hypothalamic lateral recesses*, a pair of diverticula in the midrostral portion of the third ventricle (Fig. 13.1g–h). In sturgeons, these inferior hypothalamic lobes are much larger than in other chondrosteans and in other fish groups (Fig. 13.1g–h) (Nieuwenhuys 1998; Rupp and Northcutt 1998). Different nomenclatures for the diverse hypothalamic regions of sturgeons have been used in cytoarchitectonic studies. Thus, Nieuwenhuys (1998) divided the rostral hypothalamus of the shovelnose sturgeon (*Scaphirhynchus*) into two longitudinal periventricular regions: a hypothalamic dorsal part, extending along the dorsal and lateral walls of the lateral recesses, which corresponds to the dorsal and lateral parts of the periventricular hypothalamus of *Acipenser* (Fig. 13.1a, g–h, Rupp and Northcutt 1998), and an hypothalamic ventral part, extending along the ventromedial walls of the lateral recesses where the third ventricle gives rise to the *infundibulum* (Fig. 13.1a, g–i) in which more rostral walls the *anterior tuberal nucleus* is located (Fig. 13.1g–h). This ventral part of *Scaphirhynchus* corresponds to the ventral part of the periventricular hypothalamus in *Acipense*r (Rupp and Northcutt 1998).

The caudal portion of the hypothalamus is unpaired and surrounds the *hypothalamic posterior recess*, located in the caudal portion of the third ventricle, which displays small, laterally directed diverticula (Figs. 13.1a, j and 13.2). The caudoventral region of the posterior recess suffers narrowness in the most ventral region which leads to a cavity lined by a thin and strongly folded epithelium, named *saccus vasculosus (SV)*. This structure whose function remains unclear is present in many jawed fish, including chondrosteans (Figs. 13.1a, j and 13.2). The epithelium forming the wall of this organ contains CSF-C cells, supporting cells, and numerous *coronet cells*, a highly specialized cell type exclusive to the saccus vasculosus of jawed fishes and which is probably glial in nature (Fig. 13.1j) (Arochena et al. 2004; Sueiro et al. 2007). Coronet cells are characterized by a round or pear-shaped perikarya with a short, thick apical process that protruded into the ventricle bearing a crown of globule-tipped cilia (inset in Fig. 13.1j). Unlike other fishes, coronet cells of sturgeons are not confined to the saccus vasculosus, and they are distributed over the entire floor of the hypothalamus and also on the preoptic recess (Kotrschal et al. 1983). The function of this organ remained intriguing for a long time, and it has been suggested to be involved in sensory and transport functions as well as secretory functions (e.g., liquor pressure perception, osmoregulation, calcium homeostasis, glucose loading, and transcellular ion exchange between blood vessels and CSF). However, recently, it has been suggested that the SV is implicated in the regulation of photoperiodism in fish (Nakane et al. 2013). However, we cannot exclude the possibility that the SV serves other physiological functions such as neuroendocrine, as suggested by the localization of thyroid-stimulating hormone in the coronet cells (Nakane et al. 2013) or the expression of brain aromatase shown in trout (Menuet et al. 2003). Additionally, the SV and the hypothalamus are connected to each other by afferent and efferent fiber systems (Yáñez et al. 1997; Sueiro et al. 2007).

In the dorsolateral and ventral periventricular walls of the posterior recess of *Scaphirhynchus* lies the *lateral tuberal nucleus* (Nieuwenhuys 1998), which corresponds to the caudal zone of the periventricular hypothalamus of *Acipenser* (Fig. 13.1i–j, Rupp and Northcutt 1998). Caudally, the dorsal area progressively forms the *posterior tuberal nucleus* in both sturgeon species (Nieuwenhuys 1998; Rupp and Northcutt 1998).

The hypothalamic organization described for sturgeons, characterized by laterally expanded inferior lobes and thin hypothalamic walls, also exists in chondrichthyans (Smeets et al. 1983), but not in other ray-finned fishes, so the lateral expansions in sharks and sturgeons must be viewed as independently evolved parallel features rather than as homologous structures as it was suggested by Northcutt (1995). However, in the sturgeons, as in all ray-finned fishes, the most conserved feature of the hypothalamus is the periventricular cellular zone which exhibits basically the same type of organization in other fish groups (Rustamov 2006b).

13.1.3 Hypothalamo-Hypophyseal Relationships

In sturgeons, as in all vertebrates, the hypophysis or pituitary gland is attached to the caudoventral region of the hypothalamus (Figs. 13.1a, g–j and 13.2). The hypophysis of sturgeons, flat and elongated in the rostral-caudal direction, is composed of the adenohypophysis located rostrally and the neurohypophysis located caudally close to the transition of the infundibulum to the saccus vasculosus. The adenohypophysis is the glandular part and is mainly composed of secretory cells, while the neurohypophysis represents the neural part of the gland and consists of neurosecretory terminals originating from the hypothalamus and other brain regions (Fig. 13.2) (Grandi and Chicca 2004). During brain ontogenesis, although the hypothalamus differentiates relatively late, the adenohypophysis is already evident on the floor of the diencephalon at hatching time but reaches the adult morphology at 5 months of age, while the neurohypophysis does not begin its development until 80 days posthatching and does not reach the adult morphology before 9 months of age (Grandi and Chicca 2004; Gómez et al. 2009).

A comprehensive description of the hypothalamo-hypophyseal system was reported in the classical studies of Russian researchers (Sathyanesan and Chavin 1967; Polenov and Garlov 1971, 1973; Polenov et al. 1972, 1976, 1983, 1997; Polenov and Pavlovic 1978; Belenky et al. 1985). The adenohypophysis in sturgeons consists in a pars distalis and a pars intermedia, while the neurohypophysis is included in a neurointermediate lobe consisting in tubular rootlike processes of the bottom of the infundibular wall that penetrate among the cell cords of the pars intermedia (Fig. 13.2b). The distribution of the pituitary endocrine cells was studied by immunocytochemical techniques in the adenohypophysis of different sturgeon species, and seven types of endocrine cells were identified: the adrenocorticotropic, prolactin, growth hormone, gonadotropic and thyroid-stimulating hormone cells in the pars distalis, and the melanocyte-stimulating hormone and somatolactin cells in the pars intermedia (Hansen 1971; Hansen and Hansen 1975; Pelissero et al. 1988; Joss et al. 1990; Amemiya et al. 1999).

Moreover, in chondrosteans, the hypothalamic floor, located rostroventrally just dorsal to the pars distalis of the adenohypophysis, differentiates into a typical median eminence named *proximal neurosecretory contact region* (Fig. 13.2) (Polenov et al. 1976; Kotrschal et al. 1985). In the space between this region and the pars distalis, there are blood vessels that form part of a hypothalamo-hypophyseal portal system where numerous unmyelinated neurosecretory fibers form synaptic endings and, as in other vertebrates, discharge their products into the portal circulation. There are two types of those fibers that are also found among the processes of the neurohypophysis: Type A, peptidergic and originate in the preoptic area (magnocellular preoptic nucleus) and Type B, monoaminergic and presumably originate in the hypothalamus (lateral tuberal nucleus, Polenov et al. 1972). Thus, both the peptide and monoamine neurohormones along the way hypothalamic neurons regulate the activity of the glandular cells in the pars distalis (Polenov et al. 1976).

13.2 Hypophysiotropic Factors in the Preoptic-Hypothalamo-Hypophyseal System of Sturgeons

The presence of numerous cells and fibers containing different hypophysiotrophic factors, such as corticoliberin (Belenky et al. 1985), corticotropin-releasing factor (CRF, González et al. 1992), gonadotropin-releasing hormone (GnRH, Leprêtre et al. 1993; Amiya et al. 2011), neuropeptide Y (NPY, Chiba and Honma 1994; Amiya et al. 2011), methionine-enkephalin (Met-enk, Rodríguez-Moldes et al. 1997), melanin-concentrating hormone (MCH, Baker and Bird 2002), serotonin (5-HT, Adrio et al. 1999; Piñuela and Northcutt 2007), tyrosine hydroxylase (TH, Adrio et al. 2002), neurophysin (NPH, Adrio et al. 2005), galanin (GAL, Adrio et al. 2005; Amiya et al. 2011), substance P, dopamine and leucine-enkephalin (SP, DA, Leu-enk, Piñuela and Northcutt 2007), and somatostatin (SOM, Adrio et al. 2008), has been observed in the preoptic region and the rostral and caudal hypothalamus of the sturgeons by the use of immunohistochemical techniques. All these neuromodulators showed high concentrations and had a similar distribution in the hypothalamus of all the sturgeon species studied. Although some neurons immunoreactive to those factors were observed in nuclei located away from the ventricle, most of them were located in the periventricular region, and they were of the CSF-C type. These CSF-C cells containing those factors were observed both in the preoptic area and in the hypothalamus of all the sturgeon species studied, and they were very abundant in the parvocellular preoptic nucleus; in the dorsal, lateral, and ventrolateral walls of the lateral recesses (lateral and dorsal periventricular hypothalamus of Rupp and Northcutt 1998); and in the tuberal and caudal hypothalamus where they were located in the ventromedial walls of the infundibulum, corresponding to the position of the anterior and lateral tuberal nuclei (ventral periventricular hypothalamus of Rupp and Northcutt 1998). CSF-C cells are considered phylogenetically ancient in type and may play an important role in the neurotransmission and/or neuromodulation of neuroendocrine pathways in fish (Vigh-Teichmann et al. 1983). In fact, these hypothalamic CSF-C cells showed an intense staining to the different hypophysiotrophic factors studied in both the subventricular perikarya and ventricular processes.

Furthermore, the basal processes of these CSF-C neurosecretory cells coursed ventrolaterally toward the external surface and along the hypothalamic floor toward the hypophysis forming a *preoptic-hypothalamo-hypophyseal tract*, where fibers immunoreactive to neuropeptides (corticoliberin (Belenky et al. 1985), CRF (González et al. 1992), GnRH (Leprêtre et al. 1993), NPY (Chiba and Honma 1994), Met-enk (Rodríguez-Moldes et al. 1997), MCH (Baker and Bird 2002), GAL and NPH (Adrio et al. 2005), SOM (Adrio et al. 2008), and Leu-enk (Adrio, unpublished results) were observed in different sturgeon species. Some of these fibers innervate the proximal neurosecretory contact region, and others coursed more caudally in the thin floor of the hypothalamus and the hypophyseal stalk toward the neurointermediate lobe of the neurohypophysis, where they appear close to glandular cells of the pars intermedia. Although the origin of the hypothalamus-hypophyseal projections in the sturgeon has not been studied experimentally, these fibers could arise, at least in part, from neurosecretory neurons observed in the preoptic and/or the hypothalamic region, such as those of the anterior and lateral tuberal nuclei, as previously reported in the hypothalamus of teleosts (Holmqvist and Ekström 1995). In fact, the preoptic nuclei and the lateral tuberal nucleus were described in *Acipenser fulvescens* as neurosecretory nuclei by Sathyanesan and Chavin (1967) who used Gomori's aldehyde fuchsin method for neurosecretion (a classical marker of the hypothalamo-hypophyseal neurosecretory system), and, more recently, cells immunoreactive to neurophysin (NPH-ir) were described in the preoptic (magnocellular preoptic nucleus, Figs. 13.3b, c and 13.4k, l) and hypothalamic (anterior and lateral tuberal nuclei, Figs. 13.3d–f and 13.5e) areas in *Acipenser baerii* suggesting that both regions contain neurosecretory cells related with the hypothalamo-hypophyseal system (Adrio et al. 2005). Therefore, the presence of different hypophysiotrophic factors in those nuclei appears general to fishes (see references in Adrio et al. 2005).

In addition, the presence of cells immunoreactive to hypophysiotrophic factors such as CRF (González et al. 1992), NPY (Chiba and Honma 1994), Met-enk (Rodríguez-Moldes et al. 1997), MCH (Baker and Bird 2002), and SOM (Adrio et al. 2008), in the pars distalis of the sturgeon, suggests that these factors are released to the blood in the adenohypophysis and have peripheral functions. Alternatively, some of these factors could have local actions in the pituitary.

Therefore, the hypothalamus of the sturgeons contains neurosecretory cells related to the hypophysis and most likely acting as hypophysiotropic factors modulating the release of adenohypophyseal hormones. However, the hypophysis-related CSF-C neuronal groups of the sturgeon hypothalamus are heterogeneous in terms of their neurochemical content, and, therefore, different types of CSF-C cells could play different roles in neurosecretion and neurotransmission (Adrio et al. 2005).

13.2.1 Galaninergic System

Galanin is a 29-amino acid peptide widely distributed in the central nervous system of vertebrates, and the hypothalamus is particularly rich in galanin-synthesizing neurons and nerve processes. Depending on the destination of galanin in nerve terminals, this neuropeptide may function as a neuromodulator/neurotransmitter when it innervates other neurons or as a hypophysiotropic messenger when it is released into the hypothalamo-hypophyseal portal circulation and reaches the anterior pituitary (Merchenthaler et al. 2013). The presence of galaninergic structures in the preoptic-hypothalamic regions seems to be highly conserved among vertebrates. Thus, the location of most of the galanin-immunoreactive (GAL-ir) cell bodies in the preoptic-hypothalamic area and the high galaninergic innervation observed in the hypophysis of all of the vertebrates studied so far (fish, amphibians, reptiles, birds, and mammals; see Adrio et al. (2005) and Mensah et al. (2010) for a review), and the observation that galanin may directly influence hormone release from the pituitary gland (Murakami et al. 1987; Maiter et al. 1990; López et al. 1991; Rao et al. 1996), has led to consider it as a hypophysiotropic peptide in both mammal and nonmammalian vertebrates, including humans (see Mechenthaler (2008), Merchenthaler et al. (2013) and Mensah et al. (2010) for a review). Moreover, galanin expression appears to be modulated by gonadal steroids both in mammals (Park et al. 1997; Rugarn et al. 1999; Shen et al. 1999; Scheffen et al. 2003; Splett et al. 2003) and in teleosts (Olivereau and Olivereau 1991b), which suggests that

Fig. 13.3 Schematic drawings of transverse sections (**a–h**) through the preoptic region and hypothalamus of *Acipenser baerii* (from rostral to caudal) showing the distribution of neurons (*solid circles*) and fibers (*dotted areas*) immunoreactive to galanin (GAL), serotonin (5-HT), tyrosine hydroxylase (TH), or somatostatin (SOM). On the left, neurons (*solid circles*) and fibers (*dotted areas*) immunoreactive to neurophysin (NPH) are shown, and the main anatomical regions are also indicated. The levels of the sections are indicated in a lateral view of the brain on the top. Correspondence with photomicrographs in other figures is indicated by *boxed areas*. Schematic drawings were modified from Adrio et al. (1999, 2002, 2005, 2008). Abbreviations: *CB*, corpus cerebelli; *CP*, choroid plexus; *Dd*, dorsal part of the dorsal telencephalon; *Dl*, lateral part of the dorsal telencephalon; *FR*, fasciculus retroflexus; *h*, habenula; *H*, hypophysis; *HL*, hypothalamic lobes; *i*, infundibulum; *III*, third ventricle; *IIIn*, oculomotor nucleus; *IV*, fourth ventricle; *IVn*, trochlear nucleus; *LR*, lateral hypothalamic recess; *ME*, median eminence; *MLF*, medial longitudinal fascicle; *MO*, medulla oblongata; *MPT*, medial nucleus of the posterior tubercle; *NAT*, anterior tuberal nucleus; *NIL*, neurointermediate lobe of the hypophysis; *NLT*, lateral tuberal nucleus; *NPOm*, magnocellular preoptic nucleus; *NPOp*, parvocellular preoptic nucleus; *NPT*, posterior tuberal nucleus; *NRP*, posterior recess nucleus; *OB*, olfactory bulb; *OC*, optic chiasm; *OT*, optic tectum; *OVLT*, organum vasculosum of the terminal lamina; *P*, pineal organ; *PC*, posterior commissure; *Pd*, pars distalis of the hypophysis; *Pi*, pars intermedia of the hypophysis; *POR*, posterior recess; *PR*, preoptic recess; *Pt*, pretectum; *PT*, posterior tubercle; *SV*, saccus vasculosus; *T*, telencephalon; *Td*, dorsal thalamus; *TG*, mesencephalic tegmentum; *tl*, torus longitudinalis; *TPp*, periventricular nucleus of the posterior tubercle; *vc*, valvula cerebelli; *VM*, ventromedial thalamic nucleus. Scale bars = 1 mm (lateral view), 500 μm (sections)

13 Chemical Neuroanatomy of the Hypothalamo-Hypophyseal System in Sturgeons

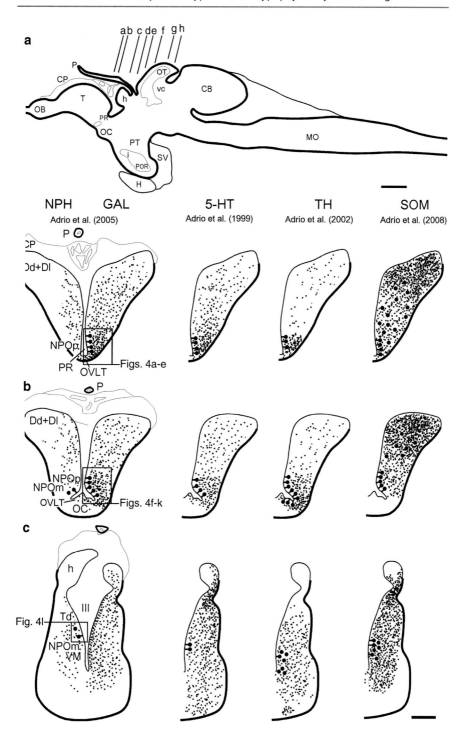

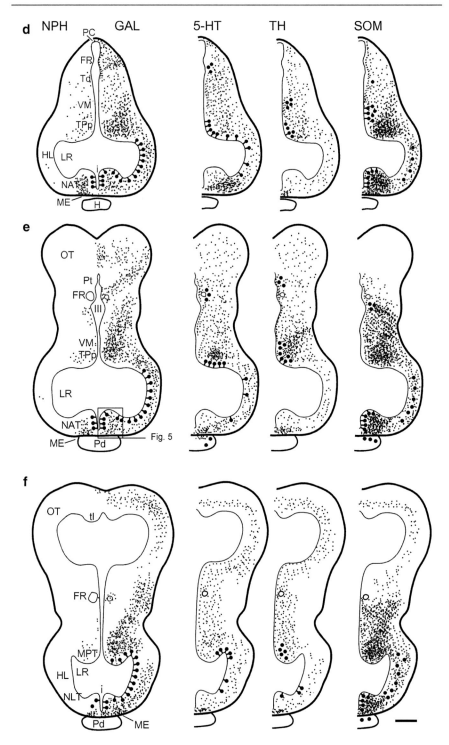

Fig. 13.3 (continued)

13 Chemical Neuroanatomy of the Hypothalamo-Hypophyseal System in Sturgeons

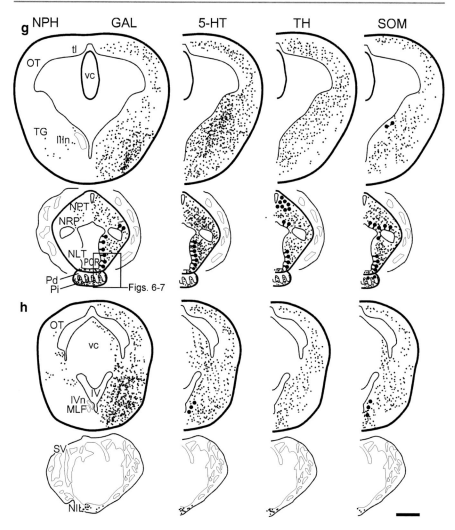

Fig. 13.3 (continued)

this peptide is involved in sexual and reproductive behavior. Therefore, fish galanin is implicated in the regulation of various physiological functions, such as feeding, growth, reproduction, or hormone release from the pituitary gland and gonads, and has an important role in the neuroendocrine integration of those functions in fishes (see Mensah et al. (2010) for a review).

Immunohistochemical studies on the presence of galanin in the brain of non-mammalian vertebrates mainly used fish as the model organism. In fact, the distribution of GAL-ir cells and fibers was studied in the central nervous system of cyclostomes, elasmobranchs, chondrosteans, and teleosts (see references in Mensah et al. 2010), and the majority of the galaninergic neurons were located in the preoptic-hypothalamic regions in all fish species studied. Thus, the highest density of GAL-ir neurons, most of them CSF-C cells, was observed in the preoptic area

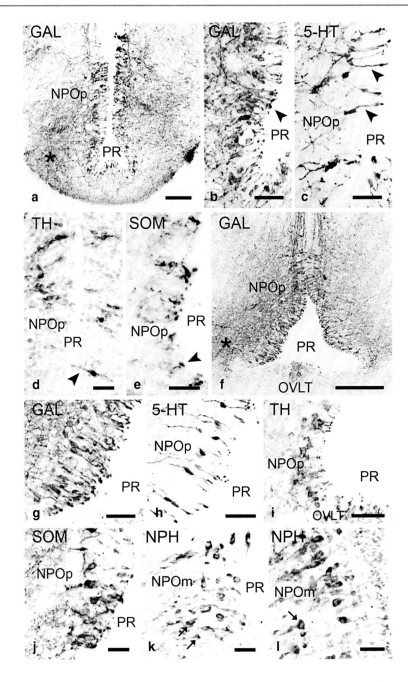

Fig. 13.4 Transverse sections through the preoptic region of *Acipenser*. (**a–e**) Sections at the rostral level of the preoptic region showing CSF-C cells immunoreactive to GAL (**a–b**), 5-HT (**c**), TH (**d**) or SOM (**e**) in the parvocellular preoptic nucleus (NPOp, note the typical apical dendrite of these CSF-C cells, *arrowheads* in **b–e**) and the dense GAL-ir innervation in the ventrolateral region (**a**, *asterisk*). (**f–l**) Sections at the caudal level of the preoptic region showing abundant CSF-C cells immunoreactive to GAL (**f–g**), 5-HT (**h**), TH (**i**), or SOM (**j**) in the parvocellular preoptic nucleus (NPOp) and to NPH in the rostral (**k**) and caudal (**l**) magnocellular preoptic nucleus (NPOm). In the NPOm not all NPH-ir cells were of the CSF-C type (*arrows* in **k–l**). Note GAL-ir fibers in the organum vasculosum of the lamina terminalis (**f**, OVLT) and the dense GAL-ir innervation in the ventrolateral region (**f**, *asterisk*). The levels of the sections correspond to those of Fig. 13.3a (**a–e**), 13.3b (**f–k**), and 13.3c ((**l**)). Abbreviations: *NPOm*, magnocellular preoptic nucleus; *NPOp*, parvocellular preoptic nucleus; *OVLT*, organum vasculosum of the terminal lamina; *PR*, preoptic recess. Scale bars = 500 μm (**f**), 250 μm (**a**), 100 μm (**b–c**, **g–i**, **k–l**), 50 μm (**d–e**, **j**)

←

and the hypothalamus of the sturgeon brain (Adrio et al. 2005). In the periventricular region, GAL-ir cells extended along the parvocellular preoptic nucleus bordering the preoptic recess (Figs. 13.3a, b and 13.4a, b, f, g); the lateral, ventrolateral, and dorsal walls of the lateral recesses (Fig. 13.3d–f); along the laterodorsal walls of the posterior recess (posterior recess nucleus, Fig. 13.3g); and in the ventromedial walls of the infundibulum corresponding to the position of the anterior (Figs. 13.3d, e and 13.5a, b) and lateral (Figs. 13.3f, g and 13.6a, b) tuberal nuclei (Adrio et al. 2005). The basal processes of these GAL-ir cells coursed ventrolaterally toward the external surface and along the ventral hypothalamus probably contributing to the dense plexus of GAL-ir fibers observed at the level of the median eminence (Figs. 13.3d–g, 13.5a, b and 13.6a–c) and more caudally in the neurointermediate lobe of the hypophysis (Figs. 13.3h and 13.7a, b) (Adrio et al. 2005). These GAL-ir fibers projecting from the preoptic-hypothalamic region onto the pituitary observed in the sturgeon are well characterized in fish (Cornbrooks and Parsons 1991a, b; Moons et al. 1991; Olivereau and Olivereau 1991a; Anglade et al. 1994; Power et al. 1996; Rodríguez et al. 2003; Rodríguez Díaz et al. 2011) and mammals (Ch'ng et al. 1985; Arai et al. 1990; Gai et al. 1990).

Numerous studies in mammals indicate that galanin is functionally related to other neuroactive substances, and the activity of GAL-ir neurons is regulated by afferents containing different neuropeptides, catecholamines, and indolamines and by hormones acting via their corresponding membrane or nuclear receptors (see Merchenthaler 2010 for a review).

Anatomical relations between galanin and other hypophysiotrophic factors, such as NPH, 5-HT, and catecholamines (Adrio et al. 2005) or NPY (Amiya et al. 2011), were studied in the sturgeon brain. GAL, NPH, 5-HT, and catecholamines (TH) are related in the hypophysiotrophic nuclei, such as preoptic (Figs. 13.3a, b and 13.4a–d, f–l) and tuberal nuclei (Figs. 13.3d–g, 13.5a–e and 13.6a–f) and hypophysis (Figs. 13.3g, h and 13.7a–e) (Adrio et al. 1999, 2002, 2005). These cell populations are clearly overlapped

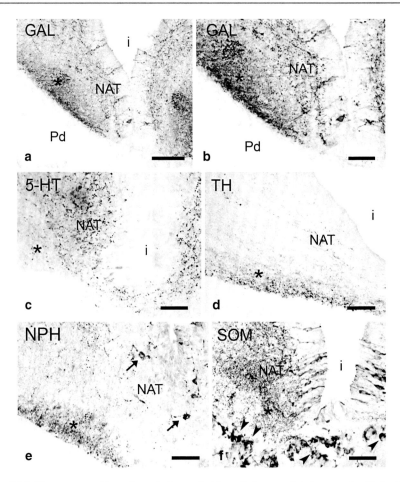

Fig. 13.5 Transverse sections through the rostral hypothalamus of *Acipenser*. (**a–b**) Sections showing GAL-ir CSF-C cells in the anterior tuberal nucleus (NAT). The basal process of these GAL-ir cells appeared to end in the median eminence (*asterisk*) close to the pars distalis of the hypophysis (Pd). (**c**) Section showing abundant 5-HT-ir fibers that innervate the NAT and scarce 5-HT-ir fibers in the median eminence (*asterisk*). (**d**) While TH-ir fibers are very abundant in the median eminence (*asterisk*), only a few innervate the NAT. (**e**) Detail of scarce NPH-ir CSF-C cells in the NAT (*arrows*) and abundant NPH-ir fibers in the median eminence (*asterisk*). (**f**) Detail of SOM-ir CSF-C cells in the NAT. Note the abundant SOM-ir fibers in the median eminence (*asterisk*) and the presence of SOM-ir cells in the Pd (*arrowheads*). The level of the sections corresponds to that of Fig. 13.3e. Abbreviations: *i*, infundibulum; *NAT*, anterior tuberal nucleus; *Pd*, pars distalis of the hypophysis. Scale bars = 200 μm (**a**), 100 μm (**b-f**)

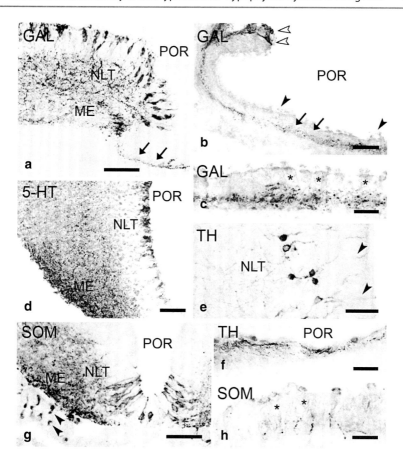

Fig. 13.6 Transverse sections through the caudal hypothalamus of *Acipenser*. (**a**) Section through the tuberal region showing abundant GAL-ir CSF-C cells in the lateral tuberal nucleus (NLT) and GAL-ir fibers coursing in the hypothalamic floor toward the neurohypophysis (*arrows*). (**b**) Section through the caudal hypothalamus showing some GAL-ir CSF-C cells in the NLT (*empty arrowheads*) and GAL-ir fibers coursing in the hypothalamic floor toward the neurohypophysis (*arrows*). Note also the typical thick apical process of coronet cells, which were not GAL-ir (*arrowheads*). (**c**) Detail of GAL-ir fibers coursing in the hypothalamic floor toward the neurohypophysis. Note that coronet cells were not GAL-ir (*asterisk*). (**d**) Detail of 5-HT-ir CSF-C cells in the NLT and dense serotoninergic innervation in the median eminence (ME). (**e**) Detail of TH-ir CSF-C cells in the NLT. Note the very long apical dendrites exhibited by these cells (*arrowheads*). (**f**) Detail of TH-ir fibers coursing in the hypothalamic floor toward the neurohypophysis. (**g**) SOM-ir CSF-C cells in the NLT and SOM-ir cells in the pars distalis of the hypophysis (*arrowheads*). Note the dense SOM-ir fibers in lateral regions and in the ME. (**h**) Detail of very scarce SOM-ir fibers coursing in the hypothalamic floor toward the neurohypophysis. Coronet cells were not SOM-ir (*asterisk*). The level of the sections corresponds to that of Fig. 13.3g. Abbreviations: *ME*, median eminence; *NLT*, lateral tuberal nucleus; *POR*, posterior recess. Scale bars = 100 μm (**a-b, d-f**), 50 μm (**c, h**)

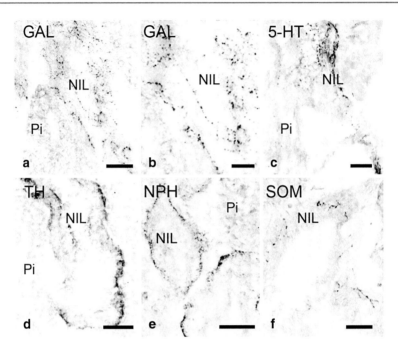

Fig. 13.7 Transverse sections through the caudal region of the hypophysis of *Acipenser* showing fibers immunoreactive to GAL (**a–b**), 5-HT (**c**), TH (**d**), NPH (**e**), or SOM (**f**) in the neurointermediate lobe (NIL) of the neurohypophysis near the glandular cells of the pars intermedia (Pi) of the adenohypophysis. The level of the sections corresponds to that of Fig. 13.3g. Scale bars = 100 μm (**a, d, e**), 50 μm (**c, f**), 25 μm (**b**)

in the sturgeon (Figs. 13.3, 13.4, 13.5, 13.6 and 13.7), but double immunolabeling experiments which have compared the location of GAL-ir neurons with that of these NPH-ir neurons have revealed no coexistence of GAL and NPH in preoptic and tuberal cells of *Acipenser*, although GAL-ir buttons are surrounding NPH-ir cells in the NPOp (see Fig. 9A–C already published in Adrio et al. 2005). However, the comparison of the location of GAL-ir neurons with that of these 5HT-ir or TH-ir neurons shows partial codistribution in some cells of the NPOp (see Figs. 9D–F and I already published in Adrio et al. 2005), but not in the hypothalamus, although in the rodent brain most of the GAL-ir neurons located in the hypothalamic arcuate nucleus also contained TH (Melander et al. 1986) or GAL-ir fibers made direct contact on TH-ir neuronal cell bodies of that nucleus (Kageyama et al. 2008; Merchenthaler et al. 2013). As far as we know, there are no studies that evidence a functional or anatomical relationship of galaninergic, serotoninergic, and catecholaminergic cells of the vertebrate preoptic/hypothalamic region, although it has been demonstrated that galanin modulates the metabolism of serotonin and dopamine in several brain areas of mammals (Fuxe et al. 1988; Jansson et al. 1989) and catecholaminergic fibers regulate the activity of GAL-ir hypothalamic neurons in rodents and humans (Merchenthaler et al. 2013). Therefore, the relationship between galaninergic, serotoninergic, and catecholaminergic cells observed in the preoptic and tuberal areas of the sturgeon could represent a primitive condition, but more studies in different vertebrate species will be necessary to prove this.

Galanin and NPY are among the most abundant neuropeptides in the hypothalamus of vertebrates, and their role in the regulation of the secretory activity of the anterior pituitary has been well established (see Merchenthaler et al. 2010 for a review). Although in the sturgeon hypothalamus both neuropeptides have a similar distribution (Chiba and Honma 1994; Adrio et al. 2005), the study of the interaction between galanin and NPY did not show neurons containing both neuropeptides, but NPY-ir fibers in close contact with GAL-ir neurons of the anterior tuberal nucleus were observed (Amiya et al. 2011). These results suggest the existence of reciprocal connections between the NPY-ir and GAL-ir neurons in the brain of the sturgeon which may mediate effects of NPY on neuronal systems innervated by galanin and therefore may play a pivotal role in the regulation of reproduction, growth, energy, and metabolism, as it was suggested for teleosts (Volkoff et al. 2005; Amano et al. 2009) and mammals (Horvath et al. 1996; Takenoya et al. 2002; Merchenthaler et al. 2010) where numerous NPY-ir nerve terminals also surrounded the majority of the GAL-ir neurons in the hypothalamus.

Finally, a possible coexistence of GAL and other neuroactive substances in nerve terminals in the median eminence and in the neurohypophysis must also be considered taking in account that GAL-ir, NPH-ir, 5-HT-ir, TH-ir, and NPY-ir cell populations are overlapped in hypophysiotrophic nuclei of the sturgeon and which basal processes coursed ventrolaterally and along the hypothalamic floor toward the pituitary (Figs. 13.3g, h, 13.4a, f, 13.5a–e, 13.6a–e, f and 13.7a–e) (Chiba and Honma 1994; Adrio et al. 1999, 2002, 2005; Amiya et al. 2011). There are no data about this possible colocalization which will indicate an interaction of those substances in the neuroendocrine system of chondrosteans. For instance, the presence of GAL-ir and NPY-ir fibers in the sturgeon hypothalamic floor and the median eminence (Chiba and Honma 1994; Adrio et al. 2005) could indicate that both neuropeptides play roles in the neuronal circuitry regulating the secretion of hormones and have a similar effect on the hypothalamo-pituitary-gonadal axis as it was reported in mammals (see Merchenthaler et al. 2010 for a review).

13.2.2 Somatostatinergic System

Somatostatin (SOM) is a neuropeptide that is widely distributed in the central nervous system of vertebrates. In mammals, two major forms of SOM (SOM-14 and SOM-28) are produced by tissue-dependent processing of the same precursor protein (Bohlen et al. 1980; Patzelt et al. 1980; Schindler et al. 1996). The primary structure of SOM-14 has been strongly conserved during evolution, although different molecular forms with similar biological activity have been found (Tostivint et al. 2004). Two isoforms of SOM-14 (SS1, which is identical to mammalian S14, and SS2, which is a variant with one amino acid changed ([Pro2] S14)) have been characterized in sturgeon (Nishii et al. 1995; Kim et al. 2000; Li et al. 2009). Moreover, two SOM precursors, which are encoded by two distinct genes, have been characterized in sturgeons: PSS1, which generates SOM-14, and PSS2, which gives rise to the [Pro2] SOM-14 variant (Trabucchi et al. 2002). In situ hybridization studies have demonstrated that the two SOM precursors (PSS1 and PSS2) are differentially expressed in numerous regions of the sturgeon brain (Trabucchi et al. 2002).

Numerous immunohistochemical studies have demonstrated the wide distribution of SOM in the central nervous system of many vertebrate taxa, including fishes (see references in Adrio et al. 2008 and Coveñas et al. 2011). SOM-immunoreactive (SOM-ir) cells and fibers are widely distributed throughout the central nervous system of the Siberian sturgeon (*Acipenser baerii*) where most SOM-ir cells were found in the preoptic area and hypothalamus as observed in the brain of other fish and tetrapods, mainly at the level of hypothalamic regions, where the majority of these SOM-ir cells in the sturgeon were CSF-C, as previously reported in the hypothalamus of several fish and amphibians (see references in Adrio et al. 2008). Thus, in the periventricular region, SOM-ir CSF-C cells were found in the parvocellular preoptic nucleus bordering the preoptic recess (Figs. 13.3a, b and 13.4e, j), the lateral walls of the lateral recesses (Figs. 13.3d–f), along the laterodorsal walls of the posterior recess (posterior recess nucleus, Fig. 13.3g), and in the anterior and lateral tuberal nuclei (Figs. 13.3d–h, 13.5f and 13.6g) (Adrio et al. 2008). The evolutionarily conserved expression pattern of SOM suggests that, at least in the preoptic area and hypothalamus, this peptide serves a basic function that was already present in ancestral vertebrates. Moreover, the abundance of SOM-ir CSF-C cells observed in the brain of *Acipenser* (Figs. 13.4e, j, 13.5f and 13.6g) (Adrio et al. 2008) suggests that SOM may act as a neurotransmitter of neuroendocrine pathways, as reported in fish (Vigh-Teichmann et al. 1983), or as a regulator of neural circuits related to cerebrospinal fluid homeostasis as observed in amphibians (Mathieu et al. 2004). SOM-ir adenohypophysial cells were also observed in the pars distalis of the hypophysis of *A. baerii*, which suggests that SOM is released to the blood in the adenohypophysis, and has peripheral functions (Figs. 13.3e, f, 13.5f and 13.6g) (Adrio et al. 2008).

In situ hybridization studies have demonstrated that mRNAs of the two SOM precursors (PSS1 and PSS2) are differentially expressed in neurons of the brain of the white sturgeon (Trabucchi et al. 2002). Thus, both precursors are expressed in cells of some hypothalamic regions, but only PSS1 mRNA is expressed in the preoptic region (Trabucchi et al. 2002), so observations in *A. baerii* with immunohistochemical techniques indicate that the used antiserum by Adrio et al. (2008) probably only reveals the SS1 isoform in the Siberian sturgeon brain.

Abundant SOM-ir fibers, presumably arising from preoptic and tuberal cells, coursed along the hypothalamic floor toward the median eminence, where SOM-ir innervation was very dense (Figs. 13.3d–g, 13.5f and 13.6g, h), and some of them reaching the neurointermediate lobe of the neurohypophysis (Figs. 13.3h and 13.7f) (Adrio et al. 2008). The presence of SOM in the hypothalamo-hypophyseal system of *Acipenser* (Adrio et al. 2008) indicates that this peptide has a role in the regulation of secretion of pituitary hormones, as reported in other vertebrates (Eigler and Ben-Shlomo 2014) including teleosts, in which the projection of SOM-ir cells to the pituitary was related to the control of prolactin (Grau et al. 1985), growth hormone

(Marchant et al. 1989; Lin et al. 1993), or adrenocorticotropic hormone (Langhorne 1986) secretion. In fact, the distribution of SOM-ir elements observed in the hypothalamus of *A. baerii* (Adrio et al. 2008) is similar to the distribution of hypophysiotrophic neurons containing other peptides in the hypothalamus of other sturgeon species (CRF: González et al. 1992; GnRH: Leprêtre et al. 1993; NPY: Chiba and Honma 1994; MCH: Baker and Bird 2002; GAL: Adrio et al. 2005), which again suggests a hypophysiotrophic role for this peptide.

Finally, and as it was indicated previously for galanin, several neuropeptides, such as somatostatin, often exert their actions together with catecholamines and other classical neurotransmitters. In fact, the overlapping between distributions of SOM-ir cells and fibers (Figs. 13.3d–h, 13.5f, 13.6g, h and 13.7f) (Adrio et al. 2008) and catecholaminergic cells and fibers (Figs. 13.3d–h, 13.5d, 13.6e and 13.7d) (Adrio et al. 2002) in the hypothalamus and the median eminence of *Acipenser* suggests interaction between these systems which could influence each other's functions, as has been shown in mammals (Ibata et al. 1982; Sakanaka et al. 1990) and amphibians (González et al. 2003).

13.2.3 Gonadotropin Releasing-Hormone System

Gonadotropin-releasing hormone (GnRH) plays a major role in growth regulation in fish and is one of the key regulators of the reproduction in all vertebrates and some invertebrates, since it acts in the brain-pituitary-gonad axis inducing synthesis and release of gonadotropins in the pituitary. Gonadotropins in turn stimulate synthesis of the steroid hormones in gonads, and some of these steroids, besides regulating steroidogenesis and gametogenesis, also exert feedback regulation upon the hypothalamus and/or pituitary gland in order to complete the reproduction cycle as it was reported in teleosts (Weltzien et al. 2004; Kim et al. 2006; Chang et al. 2009; Taranger et al. 2010; Hildahl et al. 2011).

Using a combination of HPLC and radioimmunoassay, two variants of gonadotropin-releasing hormone (GnRH) have been identified in the brain of the white sturgeon (*Acipenser transmontanus*) and other phylogenetically ancient bony fish (Sherwood et al. 1991). One of these forms corresponds to the mammalian decapeptide (mGnRH; pGlu-His-Trp-Ser-His-Gly-Trp-Tyr-Pro-Gly-NH2) and the other to chicken GnRH-II (cGnRH-II; pGlu-His-Trp-Ser-Tyr-Gly-Leu-Arg-Pro-Gly-NH2). The presence of these two forms was confirmed in other sturgeon species such as the Siberian (*Acipenser baerii*, Leprêtre et al. 1993), the Russian (*Acipenser gueldenstaedtii*, Lescheid et al. 1995), the Beluga (*Huso huso*, Gharaei et al. 2010), and the Chinese sturgeon (*Acipenser sinensis*, Yue et al. 2013). Only two studies examined the distribution of the GnRH neurons in the brain of sturgeons revealed by immunohistochemical techniques (Leprêtre et al. 1993; Amiya et al. 2011). Cell bodies mGnRH-immunoreactive (mGnRH-ir) were observed in the olfactory nerves and bulbs, the telencephalon, the preoptic region, and the mediobasal hypothalamus (Fig. 13.8) (Leprêtre et al. 1993; Amiya et al. 2011). In the preoptic area, mGnRH-ir cells were located in the parvocellular preoptic nucleus and the organum vasculosum of the terminal lamina (Fig. 13.8a, b) and in the anterior and lateral tuberal nuclei in the hypothalamus (Fig. 13.8d, e). Most of these mGnRH-ir neurons were of the CSF-C type and exhibited short processes reaching the ventricular lumen (Fig. 13.8a, b, d, e). All these cell bodies were observed along mGnRH-ir fiber tracts that could be followed from the olfactory nerve to the hypothalamo-pituitary interface. Thus, numerous mGnRH-ir fibers were observed in all brain regions, in particular in the anterior brain (Fig. 13.8a–e), although mGnRH-ir fibers were not observed in the anterior lobe of the pituitary, but a few were seen to enter the neurointermediate lobe (Fig. 13.8f), similar to the situation existing in teleosts (Kah et al. 1986, 1991). In contrast, cGnRH-II was more abundant in the posterior brain, although a few fibers could be detected in the preoptic region and the hypothalamus (Fig. 13.8d, e). The only cGnRH-II-positive cells, which were negative for mGnRH, were consistently observed in the midbrain located close to the nucleus of the medial longitudinal fasciculus (Fig. 13.8f), similar to what has been reported in many teleosts or even tetrapods (see references in Leprêtre et al. 1993). These observations suggest that in sturgeons, mGnRH has a hypophysiotrophic role regulating the release of gonadotropin and also functions as a neuromodulator, whereas cGnRH-II has only neuromodulatory functions (Leprêtre et al. 1993) as it was reported for salmon GnRH (sGnRH) and cGnRH-II in the masu salmon (Amano et al. 1991).

Despite the fact that the sturgeon specimens used in Leprêtre et al. (1993) and Amiya et al. (2011) studies were immature, both reported the presence of an overall distribution of mGnRH in the brain that was very close similar to that of the distribution of salmon GnRH in teleosts (Kah et al. 1986, 1991; Oka and Ichikawa 1990; Amano et al. 1991), which suggests that this GnRH system is established early during development as shown in other vertebrate species and is probably activated by increasing level of sex steroids at the time of puberty.

Fig. 13.8 Schematic drawings of transverse sections through the preoptic region (**a–c**) and hypothalamus (**d–f**) of *Acipenser baerii* (from rostral to caudal) showing the distribution of neurons (*solid circles*) and fibers (*dotted areas*) immunoreactive to mammalian (mGnRH) and chicken (cGnRH-II) gonadotropin-releasing hormone. At the *left*, anatomical regions are summarily indicated. The levels of the sections are indicated in a lateral view of the brain on the top. Schematic drawings were modified from Leprêtre et al. (1993). Abbreviations: *CP*, choroid plexus; *Dd*, dorsal part of the dorsal telencephalon; *Dl*, lateral part of the dorsal telencephalon; *FR*, fasciculus retroflexus; *h*, habenula; *H*, hypophysis; *HL*, hypothalamic lobes; *Hyp*, hypothalamus; *i*, infundibulum; *III*, third ventricle; *IV*, fourth ventricle; *IVn*, trochlear nucleus; *LR*, lateral hypothalamic recess; *ME*, median eminence; *MLF*, medial longitudinal fascicle; *MPT*, medial nucleus of the posterior tubercle; *NAT*, anterior tuberal nucleus; *NIL*, neurointermediate lobe of the hypophysis; *NLT*, lateral tuberal nucleus; *NPOp*, parvocellular preoptic nucleus; *OB*, olfactory bulb; *OC*, optic chiasm; *OT*, optic tectum; *OVLT*, organum vasculosum of the terminal lamina; *P*, pineal organ; *PC*, posterior commissure; *Pd*, pars distalis of the hypophysis; *POR*, posterior recess; *PR*, preoptic recess; *SV*, saccus vasculosus; *T*, telencephalon; *Td*, dorsal thalamus; *tl*, torus longitudinalis; *TPp*, periventricular nucleus of the posterior tubercle; *vc*, valvula cerebelli; *VM*, ventromedial thalamic nucleus. Scale bars = 1 mm (lateral view), 500 µm (sections)

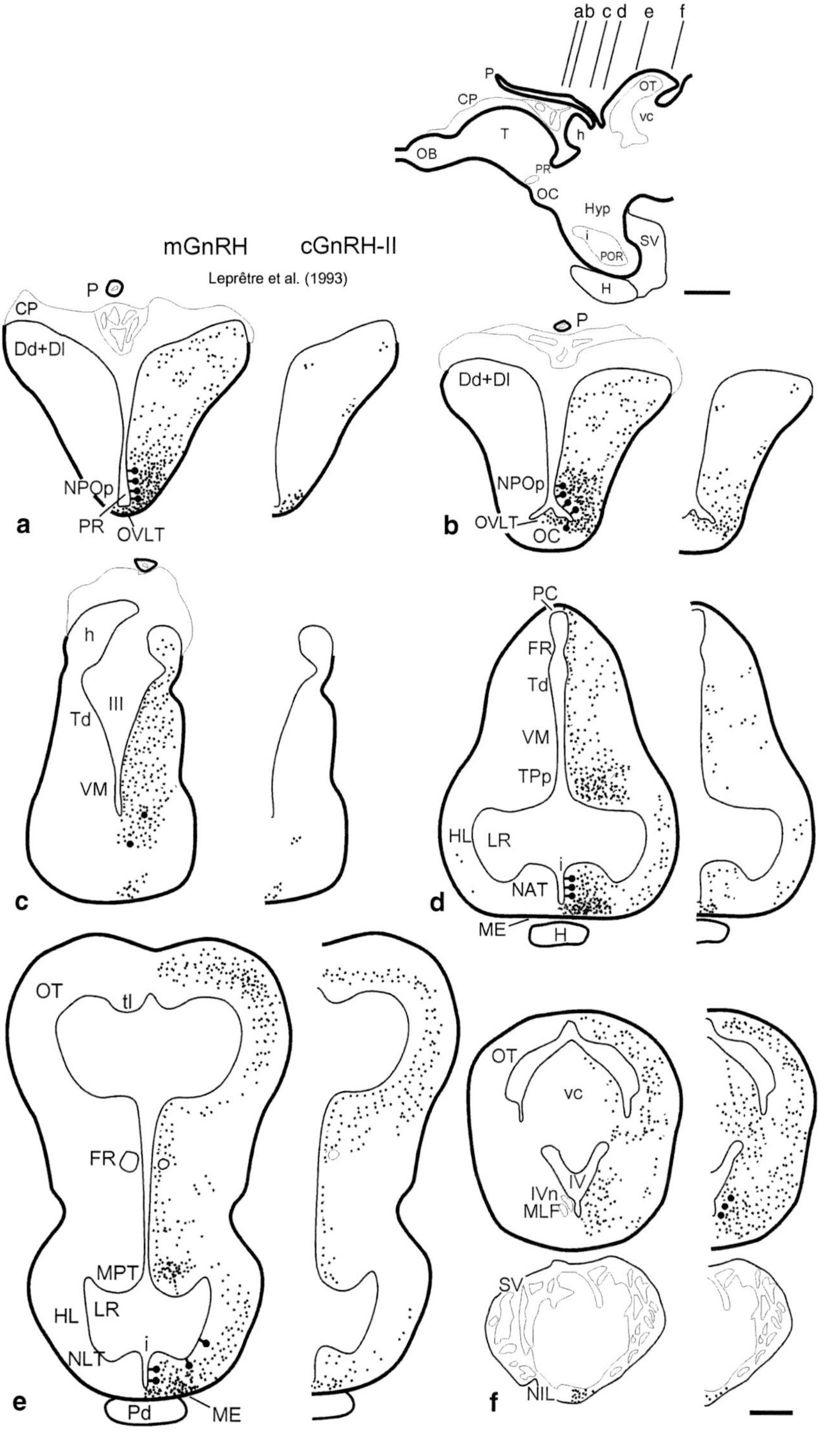

ab c d e f
P
CP
OT
vc
T
h
OB
PR
OC
Hyp
i
POR
SV
H
mGnRH
cGnRH-II
Leprêtre et al. (1993)
P
CP
Dd+Dl
NPOp
PR
OVLT
a
P
Dd+Dl
NPOp
OVLT
OC
b
h
III
Td
VM
c
PC
FR
Td
VM
TPp
HL
LR
i
NAT
ME
H
d
OT
tl
FR
MPT
HL
LR
i
NLT
Pd
ME
e
OT
vc
IV
IVn
MLF
SV
NIL
f

In the brain of the Siberian sturgeon, reciprocal connections were reported between the NPY and GnRH neurons (Amiya et al. 2011). NPY-ir profiles were observed in close contact with GnRH-ir cell bodies in the preoptic area, while NPY-ir cell bodies were contacted by GnRH-ir fibers. This suggests that NPY and GnRH neural activities are reciprocally regulated in the brain of sturgeons (Amiya et al. 2011).

Conclusions

In conclusion, the hypothalamus of sturgeons is characterized by the importance of a very large ventricle and the presence of a majority of CSF-C cells most likely representing a basal character that can be also observed in teleosts such as the eel. In early teleosts, such as the zebrafish, the hypothalamic ventricle is not so large, although CSF-C cells are also very abundant. On another hand, most of the neuropeptides and neurohormones found in tetrapods are present in sturgeons, suggesting that their common ancestors, before the split between sarcopterygians and actinopterygians, already possessed such regulatory systems. Unfortunately, because of the difficulty in approaching the physiology of sturgeons (size, cost, etc.), the number of experimental studies aiming at deciphering the roles of such neuropeptides and neurohormones is very limited, although we can speculate that part of the functions supported by these neurohormones would be similar.

References

Adrio F, Anadón R, Rodríguez-Moldes I (1999) Distribution of serotonin (5HT)-immunoreactive structures in the central nervous system of two chondrostean species (*Acipenser baeri* and *Huso huso*). J Comp Neurol 407:333–348

Adrio F, Anadón R, Rodríguez-Moldes I (2002) Distribution of tyrosine hydroxylase (TH) and dopamine beta-hydroxylase (DBH) immunoreactivity in the central nervous system of two chondrostean fishes (*Acipenser baeri* and *Huso huso*). J Comp Neurol 448:280–297

Adrio F, Anadón R, Rodríguez-Moldes I (2008) Distribution of somatostatin immunoreactive neurons and fibres in the central nervous system of a chondrostean, the Siberian sturgeon (*Acipenser baeri*). Brain Res 1209:92–104

Adrio F, Rodríguez MA, Rodríguez-Moldes I (2005) Distribution of galanin-like immunoreactivity in the brain of the Siberian sturgeon (*Acipenser baeri*). J Comp Neurol 487:54–74

Amano M, Amiya N, Hiramatsu M, Tomioka T, Oka Y (2009) Interaction between neuropeptide Y immunoreactive neurons and galanin immunoreactive neurons in the brain of the masu salmon, *Oncorhynchus masou*. Neurosci Lett 462:33–38

Amano M, Oka Y, Aida K, Okumoto N, Kawashima S, Hasegawa Y (1991) Immunocytochemical demonstration of salmon GnRH and chicken GnRH-II in the brain of the Masu salmon. J Comp Neurol 314:587–597

Amemiya Y, Sogabe Y, Nozaki M, Takahashi A, Kawauchi H (1999) Somatolactin in the white sturgeon and African lungfish and its evolutionary significance. Gen Comp Endocrinol 114:181–190

Amiya N, Amano M, Tabuchi A, Oka Y (2011) Anatomical relations between neuropeptide Y, galanin, and gonadotropin-releasing hormone in the brain of chondrostean, the Siberian sturgeon *Acipenser baeri*. Neurosci Lett 503:87–92

13 Chemical Neuroanatomy of the Hypothalamo-Hypophyseal System in Sturgeons 273

Anglade I, Wang Y, Jensen J, Tramu G, Kah O, Conlon JM (1994) Characterization of trout galanin and its distribution in trout brain and pituitary. J Comp Neurol 350:63–74

Arai R, Onteniente B, Trembleau A, Landry M, Calas A (1990) Hypothalamic galanin-immunoreactive neurons projecting to the posterior lobe of the rat pituitary: a combined retrograde tracing and immunohistochemical study. J Comp Neurol 299:405–420

Arochena M, Anadón R, Díaz-Regueira SM (2004) Development of vimentin and glial fibrillary acidic protein immunoreactivities in the brain of gray mullet (*Chelon labrosus*), an advanced teleost. J Comp Neurol 469:413–436

Baker BI, Bird DJ (2002) Neuronal organization of the melanin-concentrating hormone system in primitive actinopterygians: evolutionary changes leading to teleosts. J Comp Neurol 442:99–114

Belenky MA, Kuzik VV, Chernigovskaya EV, Polenov AL (1985) The hypothalamo-hypophysial system in Acipenseriade. X. Corticoliberin-like immunoreacticity in the hypothalamus and hypophysis of *Acipenser ruthenus* L. Gen Comp Endocrinol 60:20–26

Bohlen P, Brazeau P, Benoit R, Ling N, Esch F, Guillemin R (1980) Isolation and amino acid composition of two somatostatin-like peptides from ovine hypothalamus: somatostatin-28 and somatostatin-25. Biochem Biophys Res Commun 96:725–734

Braford MR Jr, Northcutt RG (1983) Organization of the diencephalon and pretectum of the ray-finned fishes. In: Davis RE, Northcutt RG (eds) Neurobiology. Higher brain areas and functions, vol 2. University of Michigan Press, Ann Arbor, pp 117–164

Chang JP, Johnson JD, Sawisky GR, Grey CL, Mitchell G, Booth M, Volk MM, Parks SK, Thompson E, Goss GG, Klausen C, Habibi HR (2009) Signal transduction in multifactorial neuroendocrine control of gonadotropin secretion and synthesis in teleosts-studies on the goldfish model. Gen Comp Endocrinol 161:42–52

Chiba A, Honma Y (1994) Neuropeptide Y-immunoreactive structures in the telencephalon and diencephalon of the white sturgeon, *Acipenser transmontanus*, with special regard to the hypothalamo-hypophyseal system. Arch Histol Cytol 57:77–86

Ch'ng JL, Christofides ND, Anand P, Gibson SJ, Allen YS, Su HC, Tatemoto K, Morrison JF, Polak JM, Bloom SR (1985) Distribution of galanin immunoreactivity in the central nervous system and the responses of galanin-containing neuronal pathways to injury. Neuroscience 16:343–354

Cornbrooks EB, Parsons RL (1991a) Sexually dimorphic distribution of a galanin-like peptide in the central nervous system of the teleost fish *Poecilia latipinna*. J Comp Neurol 304:639–657

Cornbrooks EB, Parsons RL (1991b) Source of sexually dimorphic galanin-like immunoreactive projections in the teleost fish *Poecilia latipinna*. J Comp Neurol 304:658–665

Coveñas R, Mangas A, Medina LE, Sánchez ML, Aguilar LA, Díaz-Cabiale Z, Narváez JA (2011) Mapping of somatostatin-28 (1-12) in the alpaca diencephalon. J Chem Neuroanat 42:89–98

Eigler T, Ben-Shlomo A (2014) Somatostatin system: molecular mechanisms regulating anterior pituitary hormones. J Mol Endocrinol 53:R1–R19

Fuxe K, Ogren SO, Jansson A, Cintra A, Harfstrand A, Agnati LF (1988) Intraventricular injections of galanin reduces 5-HT metabolism in the ventral limbic cortex, the hippocampal formation and the frontoparietal cortex of the male rat. Acta Physiol Scand 133:579–581

Gai WP, Geffen LB, Blessing WW (1990) Galanin immunoreactive neurons in the human hypothalamus: colocalization with vasopressin-containing neurons. J Comp Neurol 298:265–280

Gharaei A, Mahboudi F, Esmaili-Sari A, Edalat R, Adeli A, Keyvanshokooh S (2010) Molecular cloning of cDNA of mammalian and chicken II gonadotropin-releasing hormones (mGnRH and cGnRH-II) in the beluga (*Huso huso*) and the disruptive effect of methylmercury on gene expression. Fish Physiol Biochem 36:803–817

Gómez A, Durán E, Ocaña FM, Jiménez-Moya F, Broglio C, Domezain A, Salas C, Rodríguez F (2009) Observations on the brain development of the sturgeon *Acipenser naccarii*. In: Carmona R, Domezain A, García-Gallego M, Hernando JA, Rodríguez F, Ruiz-Rejón M (eds) Biology, conservation and sustainable development of sturgeons, Fish & fisheries series, vol 29. Springer, Netherlands, pp 155–174

González GC, Belenky MA, Polenov AL, Lederis K (1992) Comparative localization of corticotropin and corticotropin releasing factor-like peptides in the brain and hypophysis of a primitive vertebrate, the sturgeon *Acipenser ruthenus* L. J Neurocytol 21:885–896

González A, Moreno N, Morona R, López JM (2003) Somatostatin-like immunoreactivity in the brain of the urodele amphibian *Pleurodeles waltl*. Colocalization with catecholamines and nitric oxide. Brain Res 965:246–258

Grandi G, Chicca M (2004) Early development of the pituitary gland in *Acipenser nacarii* (Chondrostei, Acipenseriformes): an immunocytochemical study. Anat Embryol 208:311–321

Grau EG, Nishioka RS, Young G, Bern HA (1985) Somatostatin-like immunoreactivity in the pituitary and brain of three teleosts fish species: somatostatin as a potential regulator of prolactin cell function. Gen Comp Endocrinol 59:350–357

Hansen GH (1971) On the structure and vascularization of the pituitary gland in some primitive actinopterygians (*Acipenser*, *Polyodon*, *Calamoichthys*, *Polypterus*, *Lepisosteus* and *Amia*). Biol Skr 18:1–64

Hansen GH, Hansen BL (1975) Inmmunohistochemical localization of growth hormone and prolactin in the pituitary gland of *Acipenser güldenstaedti* Brandt (Chondrostei). Acta Zool 56:29–41

Herrick CJ (1910) The morphology of the forebrain in amphibia and reptilia. J Comp Neurol 20:413–547

Herrick CJ (1933) Morphogenesis of the brain. J Morphol 54:233–258

Hildahl J, Sandvik GK, Edvardsen RB, Fagernes C, Norberg B, Haug TM, Weltzien FA (2011) Identification and gene expression analysis of three GnRH genes in female Atlantic cod during puberty provides insight into GnRH variant gene loss in fish. Gen Comp Endocrinol 172:458–467

Holmqvist BI, Ekström P (1995) Hypophysiotropic systems in the brain of the Atlantic salmon. Neuronal innervation of the pituitary and the origin of pituitary dopamine and nonapeptides identified by means of combined carbocyanine tract tracing and immunocytochemistry. J Chem Neuroanat 8:125–145

Horvath TL, Naftolin F, Leranth C, Sahu A, Kalra SP (1996) Morphological and pharmacological evidence for neuropeptide Y-galanin interaction in the rat hypothalamus. Endocrinology 137:3069–3078

Ibata Y, Fukui K, Obata HL, Tanaka M, Hisa Y, Sano Y, Ishigami T, Imagawa K, Sin S (1982) Postnatal ontogeny of catecholamine and somatostatin neuron systems in the median eminence of the rat as revealed by a colocalization technique. Brain Res Bull 9:407–415

Jansson A, Fuxe K, Eneroth P, Agnati L (1989) Centrally administered galanin reduces dopamine utilization in the median eminence and increases dopamine utilization in the medial neostriatum of the male rat. Acta Physiol Scand 135:199–200

Johnston JB (1901) The brain of *Acipenser*. A contribution to the morphology of the vertebrate brain. Zool Jahrb Abt Anat Ontog 15:59–260

Joss JMP, Dores RM, Crim JW, Beshaw M (1990) Immunocytochemical location of pituitary cells containing ACTH, α-MSH, and β-endorphin in *Acipenser transmontanus*, *Lepisosteus spatula*, and *Amia calva*. Gen Comp Endocrinol 78:459–468

Kageyama H, Takenoya F, Hori Y, Yoshida T, Shioda S (2008) Morphological interaction between galanin-like peptide- and dopamine-containing neurons in the rat arcuate nucleus. Regul Pept 145:165–168

Kah O, Breton B, Dulka JG, Nunez-Rodriguez J, Peter RE, Corigan A, Rivier JJ, Vale WW (1986) A reinvestigation of the Gn-RH (gonadotropin-releasing hormone) systems in the goldfish brain using antibodies to salmon Gn-RH. Cell Tissue Res 244:327–337

Kah O, Zanuy S, Mañanós E, Anglade I, Carrillo M (1991) Distribution of salmon gonadotrophin releasing-hormone in the brain and pituitary of the sea bass (*Dicentrarchus labrax*). An immunocytochemical and immunoenzymoassay study. Cell Tissue Res 266:129–136

Kim JB, Gadsboll V, Whittaker J, Barton BA, Conlon JM (2000) Gastroenteropancreatic hormones (insulin, glucagon, somatostatin, and multiple forms of PYY) from the pallid sturgeon, *Scaphirhynchus albus* (Acipenseriformes). Gen Comp Endocrinol 120:353–363

Kim J, Hayton WL, Schultz IR (2006) Modeling the brain–pituitary–gonad axis in salmon. Mar Environ Res 62(Suppl):S426–S432

Kotrschal K, Krautgartner WD, Adam H (1983) Crown cells in the diencephalon of *Acipenser ruthenus* (Acipenseridae, Chondrostei). J Hirnforsch 24:655–657

Kotrschal K, Krautgartner WD, Adam H (1985) Distribution of aminergic neurons in the brain of the sterlet, *Acipenser ruthenus* (Chondrostei, Actinopterygii). J Hirnforsch 26:65–72

Langhorne P (1986) Somatostatin stimulates ACTH release in brown trout (*Salmo trutta* L.) Gen Comp Endocrinol 61:71–75

Leprêtre E, Anglade I, Williot P, Vandesande F, Tramu G, Kah O (1993) Comparative distribution of mammalian GnRH (gonadotrophin-releasing hormone) and chicken GnRH-II in the brain of the immature Siberian sturgeon (*Acipenser baeri*). J Comp Neurol 337:568–583

Lescheid DW, Powell JF, Fischer WH, Park M, Craig A, Bukovskaya O, Barannikova IA, Sherwood NM (1995) Mammalian gonadotropin-releasing hormone (GnRH) identified by primary structure in Russian sturgeon, *Acipenser gueldenstaedti*. Regul Pept 55:299–309

Li CJ, Wei QW, Zhou L, Cao H, Zhang Y, Gui JF (2009) Molecular and expression characterization of two somatostatin genes in the Chinese sturgeon, *Acipenser sinensis*. Comp Biochem Physiol A Mol Integr Physiol 154:127–134

Lin XW, Lin HR, Meter RE (1993) Growth hormone and gonadotropin secretion in the common carp (*Cyprinus carpio* L.): in vitro interactions of gonadotropin-releasing hormone, somatostatin, and the dopamine agonist apomorphine. Gen Comp Endocrinol 89:62–71

López FJ, Merchenthaler I, Ching M, Wisniewski MG, Negro-Vilar A (1991) Galanin: a hypothalamic-hypophysiotropic hormone modulating reproductive functions. Proc Natl Acad Sci U S A 88:4508–4512

Maiter DM, Hooi SC, Koenig JI, Martin JB (1990) Galanin is a physiological regulator of spontaneous pulsatile secretion of growth hormone in the male rat. Endocrinology 126:1216–1222

Marchant TA, Dulka JG, Peter RE (1989) Relationship between serum growth hormone levels and the brain and pituitary content of immunoreactive somatostatin in the goldfish, *Carassius auratus* L. Gen Comp Endocrinol 73:458–468

Mathieu M, Bruzzone F, Chartrel N, Serra GP, Spiga S, Vallarino M, Vaudry H (2004) Somatostatin in the brain of the cave salamander, *Hydromantes genei* (Amphibia, Plethodontidae): immunohistochemical localization and biochemical characterization. J Comp Neurol 475:163–176

Mechenthaler I (2008) Galanin and the neuroendocrine axes. Cell Mol Life Sci 65:1826–1835

Melander T, Hökfelt T, Rökaeus A (1986) Distribution of galaninlike immunoreactivity in the rat central nervous system. J Comp Neurol 248:475–517

Mensah ET, Volkoff H, Unniappan S (2010) Galanin systems in non-mammalian vertebrates with special focus on fishes. EXS 102:243–262

Menuet A, Anglade I, Le Guevel R, Pellegrini E, Pakdel F, Kah O (2003) Distribution of aromatase mRNA and protein in the brain and pituitary of female rainbow trout: comparison with estrogen receptor alpha. J Comp Neurol 462:180–193

Merchenthaler I (2010) Galanin and the neuroendocrine axes. In: Hökfelt T (ed) Galanin. Springer, Basel, pp 71–86

Merchenthaler I, Rotoli G, Grignol G, Dudas B (2010) Intimate associations between the neuropeptide Y system and the galanin-immunoreactive neurons in the human diencephalon. Neuroscience 170:839–845

Merchenthaler I, Rotoli G, Peroski M, Grignol G, Dudas B (2013) Catecholaminergic system innervates galanin-immunoreactive neurons in the human diencephalon. Neuroscience 238:327–334

Moons L, Batten TF, Vandesande F (1991) Autoradiographic distribution of galanin binding sites in the brain and pituitary of the sea bass (*Dicentrarchus labrax*). Neurosci Lett 123:49–52

Murakami Y, Kato Y, Koshiyama H, Inoue T, Yanaihara N, Imura H (1987) Galanin stimulates growth hormone (GH) secretion via GH-releasing factor (GRF) in conscious rats. Eur J Pharmacol 136:415–418

Nakane Y, Ikegami K, Iigo M, Ono H, Takeda K, Takahashi D, Uesaka M, Kimijima M, Hashimoto R, Arai N, Suga T, Kosuge K, Abe T, Maeda R, Senga T, Amiya N, Azuma T, Amano M, Abe H, Yamamoto N, Yoshimura T (2013) The saccus vasculosus of fish is a sensor of seasonal changes in day length. Nat Commun 4:2108

Nieuwenhuys R (1998) Chondrostean fishes. In: Nieuwenhuys R, Ten Donkelaar HJ, Nicholson C (eds) The central nervous system of vertebrates, vol 1. Springer, Berlin, pp 701–757

Nishii M, Movérus B, Bukovskaya OS, Takahashi A, Kawauchi H (1995) Isolation and characterization of (Pro2)somatostatin-14 and melanotropins from Russian sturgeon, *Acipenser gueldenstaedti* Brandt. Gen Comp Endocrinol 99:6–12

Northcutt RG (1995) The forebrain of gnathostomes: in search of a morphotype. Brain Behav Evol 46:275–318

Oka Y, Ichikawa M (1990) Gonadotropin-releasing hormone (GnRH) immunoreactive system in the brain of the dwarf gourami (*Colisa lalia*) as revealed by light microscopic immunocytochemistry using a monoclonal antibody to common amino acid sequence of GnRH. J Comp Neurol 300:511–522

Olivereau M, Olivereau JM (1991a) Immunocytochemical localization of a galanin-like peptidergic system in the brain and pituitary of some teleost fish. Histochemistry 96:343–354

Olivereau M, Olivereau JM (1991b) Galanin-like immunoreactivity is increased in the brain of estradiol- and methyltestosterone-treated eels. Histochemistry 96:487–497

Park JJ, Baum MJ, Tobet SA (1997) Sex difference and steroidal stimulation of galanin immunoreactivity in the ferret's dorsal preoptic area/anterior hypothalamus. J Comp Neurol 389:277–288

Patzelt C, Tager HS, Carroll RJ, Steiner DE (1980) Identification of prosomatostatin in pancreatic islets. Proc Natl Acad Sci U S A 77:2410–2414

Pelissero C, Núñez-Rodríguez J, Le Menn F, Kah O (1988) Immunohistochemical investigation of the pituitary of the sturgeon (*Acipenser baeri*, Chondrostei). Fish Phisiol Biochem 5:109–119

Piñuela C, Northcutt RG (2007) Immunohistochemical organization of the forebrain in the white sturgeon, *Acipenser transmontanus*. Brain Behav Evol 69:229–253

Polenov AL, Belenky MA, Garlov PE, Konstantinova MS (1976) The hypothalamo-hypophysial system in Acipenseriade. VI. The proximal neurosecretory contact region. Cell Tissue Res 170:129–144

Polenov AL, Efimova NA, Konstantinova MS, Senchik YI, Yakovleva IV (1983) The hypothalamo-hypophysial system in Acipenseriade. IX. Formation of monoaminergic neurosecretory cells in the preoptic nucleus region during early ontogeny. Cell Tissue Res 232:651–667

Polenov AL, Garlov PE (1971) The hypothalamo-hypophysial system in Acipenseriade. I. Ultrastructural organization of large neurosecretory terminals (herring bodies) and axoventricular contacts. Z Zellforsch 116:349–374

Polenov AL, Garlov PE (1973) The hypothalamo-hypophysial system in Acipenseriade. III. The neurohypophysis of *Acipenser güldenstädti* Brandt and *Acipenser stellatus* Pallas. Z Zellforsch 136:461–477

Polenov AL, Garlov PE, Konstantinova MS, Belenky MA (1972) The hypothalamo-hypophysial system in Acipenseriade. II. Adrenergic structures of the hypophysial neurointermediate complex. Z Zellforsch 128:470–481

Polenov AL, Kuzik VV, Danilova OA (1997) The hypothalamo-hypophysial system in Acipenseriade. XI. Morphological and immunohistochemical analysis of nonapeptidergic and cortociliberin-immunoreactive elements in hypophysectomized starlet (*Acipenser ruthenus* L.) Gen Comp Endocrinol 105:314–322

Polenov AL, Pavlovic M (1978) The hypothalamo-hypophysial system in Acipenseriade. VII. The functional morphology of the peptidergic neurosecretory cells in the preoptic nucleus of the sturgeon, *Acipenser güldenstädti* Brandt. A quantitative study. Cell Tissue Res 186:559–570

Power DM, Canario AV, Ingleton PM (1996) Somatotropin release-inhibiting factor and galanin innervation in the hypothalamus and pituitary of seabream (*Sparus aurata*). Gen Comp Endocrinol 101:264–274

Puelles L, Rubenstein JLR (1993) Expression patterns of homeobox and other putative regulatory genes in the embryonic mouse forebrain suggest a neuromeric organization. TINS 16:472–479

Rao PD, Murthy CK, Cook H, Peter RE (1996) Sexual dimorphism of galanin-like immunoreactivity in the brain and pituitary of goldfish, *Carassius auratus*. J Chem Neuroanat 10:119–135

Rodríguez MA, Anadón R, Rodríguez-Moldes I (2003) Development of galanin-like immunoreactivity in the brain of the brown trout (*Salmo trutta fario*), with some observations on sexual dimorphism. J Comp Neurol 465:263–285

Rodríguez Díaz MA, Candal E, Santos-Durán GN, Adrio F, Rodríguez-Moldes I (2011) Comparative analysis of met-enkephalin, galanin and GABA immunoreactivity in the developing trout preoptic-hypophyseal system. Gen Comp Endocrinol 173:148–158

Rodríguez-Moldes I, Candal E, Huesa G, Adrio F, Anadón R (1997) Distribución de neuronas inmunorreactivas a la Met-encefalina en el SNC del esturión. Rev Neurol 25:1800

Rugarn O, Theodorsson A, Hammar M, Theodorsson E (1999) Effects of estradiol, progesterone, and norethisterone on regional concentrations of galanin in the rat brain. Peptides 20:743–748

Rupp B, Northcutt RG (1998) The diencephalon and pretectum of the white sturgeon (*Acipenser transmontanus*): a cytoarchitectonic study. Brain Behav Evol 51:239–262

Rustamov EK (2006a) Organization of diencephalon of the sturgeons. Preoptic area. J Evol Biochem Physiol 42:195–207

Rustamov EK (2006b) Organization of hypothalamic area of diencephalon in sturgeons. J Evol Biochem Physiol 42:342–353

Sakanaka M, Magari S, Inoue N (1990) Somatostatin co-localizes with tyrosine hydroxylase in the nerve cells of discrete hypothalamic regions in rats. Brain Res 516:313–317

Sathyanesan AG, Chavin W (1967) Hypothalamo-hypophyseal neurosecretory system in the primitive actinopterygian fishes (Holostei and Chondrostei). Acta Anat (Basel) 68:284–299

Scheffen JR, Splett CL, Desotelle JA, Bauer-Dantoin AC (2003) Testosterone-dependent effects of galanin on pituitary luteinizing hormone secretion in male rats. Biol Reprod 68:363–369

Schindler M, Humphrey PP, Emson PC (1996) Somatostatin receptors in the central nervous system. Prog Neurobiol 50:9–47

Shen ES, Hardenburg JL, Meade EH, Arey BJ, Merchenthaler I, López FJ (1999) Estradiol induces galanin gene expression in the pituitary of the mouse in an estrogen receptor alpha-dependent manner. Endocrinology 140:2628–2631

Sherwood NM, Doroshov S, Lance V (1991) Gonadotropin-releasing hormone (GnRH) in bony fish that are phylogenetically ancient: reedfish (*Calamoichthys calabaricus*), sturgeon (*Acipenser transmontanus*), and alligator gar (*Lepisosteus spatula*). Gen Comp Endocrinol 84:44–57

Smeets WJAJ, Nieuwenhuys R, Roberts BL (1983) The central nervous system of cartilaginous fishes. Structure and functional correlations. Springer Verlag, New York

Splett CL, Scheffen JR, Desotelle JA, Plamann V, Bauer-Dantoin AC (2003) Galanin enhancement of gonadotropin-releasing hormone-stimulated luteinizing hormone secretion in female rats is estrogen dependent. Endocrinology 144:484–490

Sueiro C, Carrera I, Ferreiro S, Molist P, Adrio F, Anadón R, Rodríguez-Moldes I (2007) New insights on Saccus vasculosus evolution: a developmental and immunohistochemical study in elasmobranchs. Brain Behav Evol 70:187–204

Takenoya F, Funahashi H, Matsumoto H, Ohtaki T, Katoh S, Kageyama H, Suzuki R, Takeuchi M, Shioda S (2002) Galanin-like peptide is co-localized with alpha-melanocyte stimulating hormone but not with neuropeptide Y in the rat brain. Neurosci Lett 331:119–122

Taranger GL, Carrillo M, Schulz RW, Fontaine P, Zanuy S, Felip A, Weltzien FA, Dufour S, Karlsen O, Norberg B, Andersson E, Hansen T (2010) Control of puberty in farmed fish. Gen Comp Endocrinol 165:483–515

Tostivint H, Trabucchi M, Vallarino M, Conlon JM, Lihrmann I, Vaudry H (2004) Molecular evolution of somatostatin genes. In: Patel YC (ed) Somatostatin endocrine updates. Kluwer Academic, Dordrecht

Trabucchi M, Tostivint H, Lihrmann I, Sollars C, Vallarino M, Dores RM, Vaudry H (2002) Polygenic expression of somatostatin in the sturgeon *Acipenser transmontanus*: molecular cloning and distribution of the mRNAs encoding two somatostatin precursors. J Comp Neurol 443:332–345

Vázquez M, Rodríguez F, Domezain A, Salas C (2002) Development of the brain of the sturgeon *Acipenser nacarii*. J Appl Ichthyol 18:275–279

Vigh-Teichmann I, Vigh B, Korf HW, Oksche A (1983) CSF-contacting and other somatostatin-immunoreactive neurons in the brains of *Anguilla anguilla, Phoxinus phoxinus* and *Salmo gairdneri* (Teleostei). Cell Tissue Res 233:319–334

Volkoff H, Canosa LF, Unniappan S, Cerdá-Reverter JM, Bernier NJ, Kelly SP, Peter RE (2005) Neuropeptides and the control of food intake in fish. Gen Comp Endocrinol 142:3–19

Weltzien FA, Andersson E, Andersen O, Shalchian-Tabrizi K, Norberg B (2004) The brain–pituitary–gonad axis in male teleosts, with special emphasis on flatfish (Pleuronectiformes). Comp Biochem Physiol A Mol Integr Physiol 137:447–477

Yáñez J, Rodríguez M, Pérez S, Adrio F, Rodríguez-Moldes I, Manso MJ, Anadón R (1997) The neuronal system of the saccus vasculosus of trout (*Salmo trutta fario* and *Oncorhynchus mykiss*): an immunocytochemical and nerve tracing study. Cell Tissue Res 288:497–507

Yue H, Ye H, Chen X, Cao H, Li C (2013) Molecular cloning of cDNA of gonadotropin-releasing hormones in the Chinese sturgeon (*Acipenser sinensis*) and the effect of 17β-estradiol on gene expression. Comp Biochem Physiol A Mol Integr Physiol 166:529–537

An Updated Version of Histological and Ultrastructural Studies of Oogenesis in the Siberian Sturgeon *Acipenser baerii*

14

Françoise Le Menn, Catherine Benneteau-Pelissero, and René Le Menn

Abstract

Oogenesis of Siberian sturgeon *Acipenser baerii* is studied on farm fish using light and electron microscopy. We have identified five stages correlated with physiological state of the ovarian follicle constituted by the oocyte surrounded by its cellular (theca and granulosa) and a-cellular (zona radiata) layers: Stages I and II before vitellogenesis, Stages III and IV during vitellogenesis, and Stage V during maturational processes.

Following the oogonial stage, Stage I presents an elevated nucleoplasmic index, and the nucleus contains only one nucleolus. In this early stage, lipid globules were identified. Stage II is characterized by nucleoli multiplication and their migration toward the nuclear periphery. During this stage, the number of lipid globules increases in the oocyte cytoplasm. At the end of this stage, cortical alveoli get synthesized. The beginning of Stage III, called sub-Stage IIIa, corresponds to the first features of vitellogenin incorporation simultaneously with the elaboration of the *zona radiata externa*. During sub-Stage IIIb the *zona radiata interna* 1 is built. Stage IV is characterized by the first apparition of pigment granules and the elaboration of the *zona radiata interna* 2. Yolk accumulation increases and the oocyte volume grows considerably. Stage V corresponds to maturation with the beginning of nucleus migration toward the oocyte membrane. The *zona radiata* is completely synthesized, and a "jelly coat" is deposited on the outer surface of the *zona radiata externa* by synthesis of granulosa cells. Yolk accumulation ends at this stage.

In conclusion, ultrastructural data allow accurate determination of oogenesis stages.

F. Le Menn (✉) • R. Le Menn
26 rue Gustave Flaubert, 33600 Pessac, France
e-mail: francoiselemenn@orange.fr

C. Benneteau-Pelissero
University of Bordeaux, 351 Cours de la Libération, 33405 Talence Cedex, France

U862 Inserm, Magendie Neurocentre, 146, Rue Léo Saignat, 33077 Bordeaux Cedex, France

Bordeaux Sciences Agro, Cours du Général de Gaulle, 33175 Gradignan, France

© Springer International Publishing AG, part of Springer Nature 2018
P. Williot et al. (eds.), *The Siberian Sturgeon (Acipenser baerii*, Brandt, 1869)
Volume 1 - Biology, https://doi.org/10.1007/978-3-319-61664-3_14

Keywords

Ovarian follicle • Electron microscopy • Light microscopy • Oogenesis • *Acipenser baerii*

Introduction

Most studies on oogenesis in sturgeon species have been performed using light microscopy (Ginzburg and Detlaf 1969; Caloianu-Yordachel 1971; Kornienko 1975; Trusov 1975; Fedorova 1976; Kondrat'yev 1977; Akimova et al. 1979; Badenko et al. 1981; Veshchev 1982; Kijima and Maruyama 1985; Flynn and Benfey 2007). A few investigations used electron microscopy (Markov 1975; Raïkova 1976; Cherr and Clark 1982, 1984; Grandi and Chicca 2008; Rzepkowska and Ostaszewska 2013). Nevertheless we have not found studies dealing with *Acipenser baerii* gametogenesis of fish living in natural conditions in Siberian rivers. This species, which arrived in France in 1976 and later in 1982, is reared in fish farms to produce caviar and in experimental fish farms as a biological model to help restore the European sturgeon (*Acipenser sturio*) (Williot and Brun 1982a, b).

This paper deals with the description of the oogenesis of the Siberian sturgeon reared in French fish farm conditions until the end of vitellogenesis. This study was done to identify the different stages of the reproduction process comparing the data provided by samplings of the same tissue treated with the two photonic and ultrastructural technologies.

Oogenesis was compared in teleostean and chondrichtian, with first a rapid review of initial events from evolution of primordial germ cells (PGC) toward differentiated oogonia. Meiosis of oogonia gave rise to a primary oocyte surrounded by its cellular and a-cellular layer, constituting a primary follicle. This first meiotic division is stopped at the end of the prophase in the diplotene stage for a longer time in sturgeon compared with teleostean. During this period the development of primary follicle was divided into four stages: Stage I and Stage II before vitellogenesis and Stage III and Stage IV during vitellogenesis. In these stages we presented the results dissociating the events occurring in the oocyte from those occurring in its surrounding layers. The Stage V corresponded to maturational events when metaphase of the first meiotic division occurred.

14.1 Material and Methods

Ovarian samples were taken from female sturgeon born in the former USSR in 1982, which arrived into France the same year, and reared at the INRA fish farm-CEMAGREF (now IRSTEA) hatchery, Donzacq (southwest of France). The fish farm was supplied with underground water at a constant temperature of 17 °C. Fish was fed with trout compound diets.

Biopsies were performed for light microscopy. Fixations were made with Bouin's fluid for 72 h and tissue embedded in cytoparaffin. Serial sections 7 μm thick were stained with Groat eosin-hematoxylin and periodic acid-Schiff. Immunocytochemical staining using a specific antibody raised against Siberian sturgeon vitellogenin was achieved using peroxidase-antiperoxidase complex. This showed the time of incorporation of exogenous yolk into oocytes at light microscopy level (Pelissero 1988).

Very few females taken from the same 4- to 8-year-old age batch were sacrificed for electron microscopy every 3 months for 4 years. Gonads of stunned but live fish were perfused for 30 min with 3% glutaraldehyde in sodium cacodylate buffer 0.04 M, 0.05% CaCl$_2$, with an osmotic pressure of 300 mosm at a pH of 7.4. Gonads were then cut in small pieces and immersed for 90 min in the same fixative and washed three times for 15 min in a washing buffer of the same composition as the fixative fluid except that glutaraldehyde was replaced by sucrose to obtain a similar osmolarity. Osmic postfixation by 2% OsO$_4$ in sodium cacodylate buffer 0.04 M for 30 min was followed by washing in graded ethanols for dehydration and immersion in propylene oxide before embedding in Epon 812. Semi-thin sections were stained with 0.1% toluidine blue; lipids were revealed by azure blue 2 (Parry 1973) and slides submitted to photonic observations. Ultrathin sections were contrasted by uranyl acetate and lead citrate (Reynolds 1963) and observed on a 100S JEOL microscope (Département de Microscopie Electronique, Bordeaux University, France).

14.2 Oogenesis Description

We chose a classification of oocyte stages based upon the distinction of the successive physiological steps in the process of the oocyte development (Fig. 14.1).

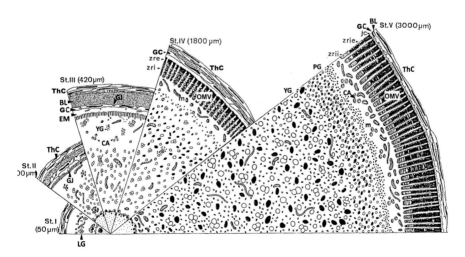

Fig. 14.1 Schematic representation of the main development stages of the Siberian sturgeon oocytes reared in French conditions (Credit: F Le Menn). Abbreviations: *BL*, basement lamina; *CA*, cortical alveoli; *EM*, extracellular matrix; *GC*, granulosa cell; *GJ*, gap junction; *LG*, lipid globule; *OMV*, oocyte microvilli; *PG*, pigment granule; *ThC*, thecal cell; *YG*, yolk globule; *zre*, zona radiata externa; *zri*, zona radiata interna; *zrie*, zona radiata interna externa; *zrii*, zona radiata interna interna

14.2.1 Short Review of Initial Events

During embryogenesis, primordial diploid germ cells (PGC) of the germinal line are singled out very early and migrate toward the genital crests from which ovaries or testes derive. As soon as the PGC colonize the ovary, they connect with somatic cells. In fish as in other vertebrates, the development of oocytes in the ovary is inevitably linked with that of the surrounding somatic cells. An oocyte and its specialized somatic cell layers constitute an ovarian follicle (Wallace and Selman 1981; Guraya 1986; Selman and Wallace 1989; Selman et al. 1993; Matova and Cooley 2001).

Undifferentiated PGC give rise to differentiated oogonia. After a number of mitotic divisions, diploid oogonia undergo their first meiotic division differentiating into primary oocytes (Fig. 14.2). This first meiotic division stops at the end of the prophase in the diplotene stage. This arrest lasts for a few days or months in most Teleostei but for years in Chondrostei as for sturgeons. During this period, the female germinal cells accumulate informational components and nutritional reserves needed for embryo development in a process called vitellogenesis and complete differentiation of its cellular and a-cellular envelopes.

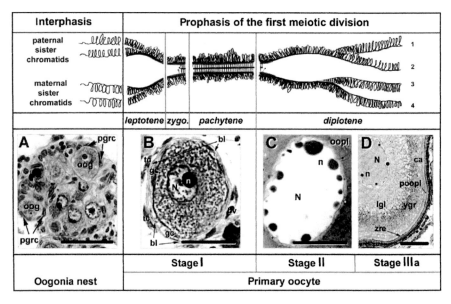

Fig. 14.2 Nuclear stages correlated with primary oocyte development during the prophase of the first meiotic division. (Credit: F Le Menn). Synaptonemal complex structure schemas during nuclear stages are indicated in the upper figure, while corresponding primary follicle development stages are indicated in the lower par. (**a**) Oogonial nest in *Acipenser baerii* (Chondrostei, Acipenseridae). (**b**) Stage I ovarian follicle of *A. baerii*. (**c**) nucleus detail of *Gobius niger* Stage II ovarian follicle (Teleostei, Gobiidae). (**d**) *Oreochromis niloticus* Stage III ovarian follicle (Teleostei, Cichlidae). Abbreviations: *bl*, basement lamina; *bv*, blood vessel; *ca*, cortical alveoli; *gc*, granulosa cell; *lgl*, lipid globule; *N*, nucleus; *n*, nucleolus; *oog*, oogonium; *oopl*, ooplasm; *poopl*, peripheral ooplasm; *pgrc*, pre-granulosa cell; *ygr*, yolk granule; *zre*, zona radiata externa; *tc*, thecal cell. Scale bars = 30 μm in (**a**), 50 μm in (**b**), 100 μm in (**c**, **d**). Data (**a**–**c**) from F. Le Menn, University Bordeaux, France, and (**d**) from P. Ndiaye, IFAN, Dakar, Senegal, respectively

This first meiotic division gives rise to two cells differing greatly in size: a very small first polar body which degenerates and a large secondary oocyte.

Ovulation occurs once maturation is completed. The secondary oocyte separates from its envelopes (Fig. 14.3) and drops into the lumen of the ovary. In fish, as in most vertebrates, the second meiotic division proceeds to metaphase and pauses until fertilization, which activates the end of the division. This leads to the formation

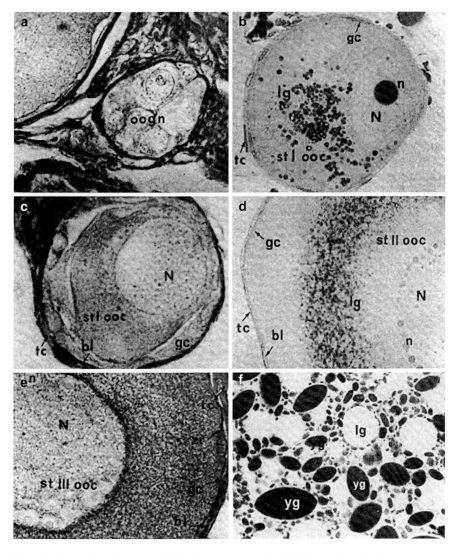

Fig. 14.3 Photonic microscopy of Siberian sturgeon ovarian follicle at various development stages (Credit: F Le Menn). (**a**) Oogonia nest. (**b, c**) Early Stage I ovarian follicle. In French fish farm conditions this Stage I exhibits lipid globules. (**d**) Stage II ovarian follicle. (**e**): Stage III ovarian follicle. (**f**) Stage IIIb ovarian follicle yolk. Abbreviations: *bl*, basement lamina; *gc*, granulosa cell; *lg*, lipid globule; *N*, nucleus; *n*, nucleoli; *ooc*, oocyte; *oogn*, oogonia nest; *tc*, thecal cell; *yg*, yolk globule

of a transient haploid female gamete, called ovum, and of a second polar body which degenerates like the first one. Fertilization occurs immediately by fusion of the haploid ovum nucleus with the haploid spermatozoon nucleus to form a diploid egg.

14.2.2 From PGC to Oogonia

In several fish species including the sturgeon *Acipenser baerii,* the gradual transformation of PGC into oogonia has been observed using electron microscopy (Bruslé and Bruslé 1978; Le Menn et al. 2007; Grandi and Chicca 2008; Zelazowska et al. 2015). PGC are larger than somatic cells. They exhibit heavy electron density and a high nucleus-to-cytoplasm ratio. Nuclei are round and eccentrically located. They contain generally only one nucleolus and have few nuclear pores. The cytoplasm contains few organelles and some relatively large mitochondria located near the nucleus. They are intimately linked by dense material termed "ciment" (Clérot 1976) or germinal dense bodies (Hamagushi 1985). A dense material of the same kind, termed "nuages," appears scattered through the perinuclear cytoplasm. These electron-dense "ciment" and "nuages" come from the nucleus (Mazabraud et al. 1975) and are often observed passing through the nuclear pores. They constitute excellent germinal cell markers on electron micrographs (Hogan 1978; Bruslé 1980).

In sturgeon, oogonia vary in size between 8 and 15 μm. Their nucleoplasmatic ratio is greater than one. They occur in clusters of 10–20 cells surrounded by adipose tissue. Each cluster is enclosed by a common single layer of pre-granulosa cells with a basal lamina (Figs. 14.2a and 14.3a).

Ultrathin sections show numerous lipid globules and mitochondrial aggregates intermingled with a granular endoplasmic reticulum. The oogonium exhibits a large, centrally located ovoid nucleus with a prominent spherical single nucleolus. The perinuclear ooplasm shows some "nuages" near the outer nuclear membrane and a lot of mitochondria aggregates around the "ciment."

The oogonia multiply rapidly by successive divisions before the start of meiosis. Formed oogonia remain linked together with few surrounding somatic cells constituting the pre-granulosa cells. Each oogonia nest (Figs. 14.2a and 14.3a) is separated from the ovarian stroma by a basal lamina, extracellular matrix secreted by the granulosa epithelium at its base.

14.2.3 From Oogonium to Primary Oocyte and Ovarian Folliculogenesis

Normally, in diploid fish each oogonium is a diploid cell, with a nucleus containing two copies of each chromosome, called homolog, one from the father and one from the mother (Fig. 14.2). During the interphase following the last mitotic division of a diploid oogonium and preceding the first meiotic division of the primary oocyte, each homolog is duplicated by DNA replication, giving rise to a pair of identical chromatids, known as a sister chromatid pair. For each chromosome, a sister chromatid pair from the mother and one from the father are present in the nucleus of the primary oocyte. As a result, the development of the primary oocyte occurs in the presence of tetraploid chromatids, i.e., a double amount of DNA for RNA synthesis. This feature is unique to primary oocytes. Oogonia begin meiosis in the oogonial nest (Grier 2000) before separating from it. The karyotype evolution of germinal cells follows the same principle in sturgeons although the evolution of Acipenseriformes (sturgeon and paddlefish) is linked with polyploidization events. Indeed, *Acipenser baerii* with 249 ± 5 chromosomes is considered to be tetraploid (Fontana et al. 2008) even though hexaploid *A. baerii* were reported recently in fish farm conditions (Havelka et al. 2014).

When each oogonium gave rise to a primary oocyte, the granulosa epithelium and the basal lamina invaginated the cluster and surrounded each cells. Thecal cells were present outside the basal lamina (Figs. 14.2b and 14.3c). The oocyte and its surrounding layers formed a follicle (Fig. 14.3b, c). Thecal cells, as granulosa cells, are involved in sexual steroidogenesis during oocyte growth and maturation (Nagahama et al. 1995; Nagahama 1997). We advise using the term "granulosa cells" as opposed to "follicle cells," because thecal cells and oocyte are also part of the follicle.

In sturgeon, as in all species studied, the development of the primary follicle is divided into five stages, which reflect the characteristic physiological steps in the development of the oocyte and the development of its surrounding somatic cell layers (Fig. 14.1).

The Stage I follicle is characterized by the nuclear genetic recombination of chromosomal chromatids at synaptonemal complexes. During Stage II, or pre-vitellogenesis, organelles and molecules used during the later stages are synthesized. Stage III, or vitellogenesis, is characterized by yolk deposition in the oocyte and the formation of the *zona radiata* (ZR), a protective structure surrounding the future egg. Stage IV is characterized by the rapid growth phase of the oocyte preceding Stage V: the maturation stage.

14.2.4 Ovarian Follicle Development Before Vitellogenesis

During the prophase of the first meiotic division, the nucleus of the oocyte passes through five successive stages: leptotene, zygotene, pachytene, diplotene, and diacinesis (Fig. 14.2). The first three nucleus stages occur during Stage I of primary oocyte development.

14.2.4.1 Stage I Follicle

Oocyte Events
We observed this stage for the first time in the gonad of *Acipenser baerii* taken in a 1-year-old batch with a mean size of 40 cm and a mean of weight of 800 g.

On semi-thin sections, a large centrally round nucleus contained a few nucleoli close to the smooth nuclear envelope (Figs. 14.2b and 14.3b). The cell diameter varied between 20 and 80 μm. In the vicinity of the nucleus, a crown of lipid granules was observed (Fig. 14.3b).

At the ultrastructural level, numerous mitochondrial aggregates (Fig. 14.4a), intimately associated with "ciment," are spread in the peripheral ooplasm. These structures correspond to an intense multiplication of mitochondria (Clérot and Wegnez 1977). In the vicinity of the outer nuclear membrane, a crown of spotted highly electron-dense "nuages" seems to emerge from the nuclear envelope. The oocyte membrane, or oolemma, is smooth.

In the leptotene phase (Fig. 14.2), both maternal and paternal sister chromatid pairs attach each of their ends to the inner membrane of the nucleus envelope, and, recognizing each other at a distance, they move closer together. In the following zygotene stage, the maternal and paternal homologous sister chromatids join together. A like-with-like recognition is mediated by a long central proteinous ladderlike core that gradually forms a synaptonemal complex with a maternal sister chromatid pair on one side and a paternal homologous sister chromatid pair on the other side. This tetrad chromatid structure named bivalent is observable by electron microscopy. During the pachytene stage, chromosomal exchanges occur between opposite non-sister chromatids with the help of recombination modules located along the central protein core of the synaptonemal complex.

Surrounding Events

The somatic granulosa cells, associated with the germinal cells since the PGC stage, multiply forming a regular epithelium of flattened cells. They complete the secretion of the basement lamina from their basal membrane. Outside the basement lamina, multilayers of unorganized meso-epithelial cells constitute the thecal layer, or theca, irrigated by numerous blood vessels.

Granulosa epithelium, closely adjacent to the oolemma, exhibit ultrastructural features of metabolic activity (mitochondria, rough endoplasmic reticulum, Golgi complexes, and coated vesicles) (Fig. 14.3c). Coated pits are present on the outer basal granulosa membrane against the basement lamina.

14.2.4.2   Stage II Follicle

In this stage, the oocyte diameters vary between 80 and 120 µm. The Stage II appeared for the first time in gonad of 3-year-old fish with an average weight of 2 kg.

Oocyte Events

This is the pre-vitellogenesis stage, i.e., preceding the entry of yolk precursors into the oocyte (Le Menn and Burzawa-Gerard 1985). Semi-thin observations showed, by azure blue coloration, an increase of the number of lipid globules (Fig. 14.3d).

The nucleus exhibited a new feature, characteristic of the diplotene stage of the prophase of the first meiotic division. Two kinds of processes occurred. The first was the de-structuring of the synaptonemal organization by disassembling the lateral protein axes of the ladderlike central core (Fig. 14.2 higher part). Compared to somatic cells, each sister chromatid pair was highly active in RNA synthesis, due to the double amount of DNA. The newly transcribed RNAs, packed into dense RNA/protein complexes, were visible on electron micrographs on large chromatin loops emanating from the linear chromatid axis. At this pre-vitellogenic stage, the so-called lampbrush chromosome structure appeared and persisted throughout the

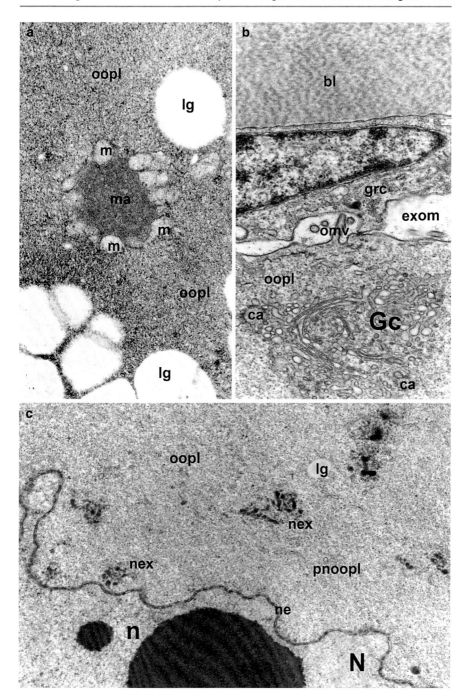

Fig. 14.4 Stage II ovarian follicle (Credit: F Le Menn). (**a**) Mitochondrial aggregate, (**b**) peripheral ooplasm of early Stage II, (**c**) perinuclear area. Abbreviations: *bl*, basement lamina; *ca*, cortical alveoli; *exom*, extra-oocyte matrix; *Gc*, Golgi complex; *gcr*, granulosa cell; *lg*, lipid globule; *m*, mitochondrium; *N*, nucleus; *n*, nucleoli; *ne*, nuclear envelope; *nex*, nuclear extrusions; *omv*, oocyte microvilli; *oopl*, ooplasm; *pnoopl*, perinuclear ooplasm

diplotene phase, i.e., during vitellogenic Stages IIIa and IIIb. The second process was an increase in nucleoli number due to a huge amplification of the nucleolar organizer genes. They were located in the concavities of the wrinkles of the nuclear envelope near the inner nuclear membrane (Figs. 14.2c and 14.4c). At the same time, both ribosomal RNA and mRNAs were transported toward the ooplasm, appearing in the perinuclear ooplasm as electron-dense nuclear extrusions (Fig. 14.4c). For example, large quantities of vitellogenin receptor mRNA were accumulated in the ooplasm during this Stage II (Davail et al. 1998; Perazzolo et al. 1999; Agulleiro et al. 2007).

The ooplasm was very electron dense, due to numerous tightly packed ribosomes. It exhibited a scattered smooth endoplasmic reticulum. Mitochondria migrated toward the peripheral ooplasm where few Golgi complexes were observed. The remaining part of the cytoplasm also contained Golgi complexes with numerous Golgian vesicles.

The oolemma separated from some areas of the apical granulosa cell surface. The intervening space was filled with a flocculent, electron-clear material, corresponding to an extra-oocyte matrix (Fig. 14.4b). In this matrix, oocyte microvilli protruded from the oocyte surface toward the granulosa epithelium (Fig. 14.4b). In numerous species, microvilli are grouped together in batches. In *Acipenser baerii*, circular oocyte microvilli batches are under umbrellalike structures formed by the apex of the granulosa epithelium (Fig. 14.5b). This apical granulosa membrane adhered to the oolemma as gap junctions around each batch of microvilli (Fig. 14.5c). Then, at the end of this Stage II, as soon as these junctions disappeared, oocyte microvilli appeared in the extra-oocyte matrix in a parallel radial position toward the granulosa cells (Fig. 14.6a), as observed on cryofracture preparations (Le Menn, personal observations).

Surrounding Events

Granulosa cells multiplied to form an epithelial coating of cubic cells, joined together by gap junctions (Fig. 14.5d). In species such as the Siberian sturgeon, the basement lamina might be exceptionally thick up to five times thicker than as the granulosa epithelium (Fig. 14.5a). The thecal layer consisted of a few cells close to blood vessels just outside the basement lamina (Fig. 14.5a).

At the end of pre-vitellogenic Stage II, the oocyte contained all the molecules and organelles needed for its subsequent endocytotic and exocytotic activities during oocyte vitellogenesis. It was surrounded in a centripetal pattern by cellular and a-cellular layers, i.e., theca, basement lamina, granulosa, and extra-oocyte matrix. The oocyte started forming microvilli which became, at the next stage, the essential means for managing exchanges with the general blood flow.

Fig. 14.5 Stage II ovarian follicle (Credit: F Le Menn). (**a**) Ovarian follicle envelopes, (**b**) relationship granulosa cell/oocyte, (**c**) gap junction granulosa cell/oocyte, (**d**) gap junction granulosa cell/granulosa cell. Abbeviations: *bl*, basement lamina; *cp*, endocytotic coated pit; *erc*, erythrocyte in thecal blood vessel; *exom*, extra-oocyte matrix; *gc*, granulosa cell; *gp*, gap junction; *m*, mitochondrium; *N*, nucleus; *omv*, oocyte microvilli; *oopl*, ooplasm; *tc*, thecal cell; *arrows*, gap junction

14 Histological and Ultrastructural Description of Oogenesis of the Siberian Sturgeon

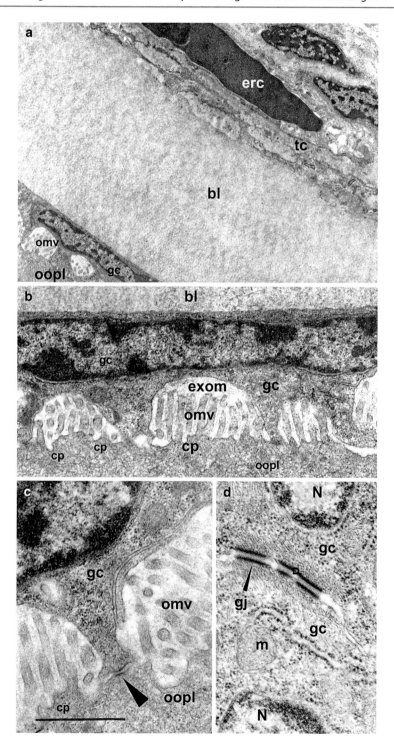

14.2.5 Ovarian Follicle Development During Vitellogenesis

Ovarian follicle enlargement occurred in Stage III, while the nucleus remained in the diplotene stage (Fig. 14.2). During this stage the oocyte accumulated, from the blood stream, the yolk containing nutritional reserves needed for embryo development. It also completed the differentiation of its cellular and a-cellular envelopes. In sturgeon as in other oviparous vertebrates, the sexual cycle and ovarian follicle development are controlled by environmental factors. In response to these factors, the central nervous system induces a cascade of neurohormones leading to the secretion of gonadotropin-releasing hormone (GnRH) by specialized hypothalamic neurons. The GnRH acts on pituitary cells, which secretes the undifferentiated gonadotrophin hormones (GTH). The GTH signal, mediated by specific ovarian receptors located on thecal and granulosa cells, leads to the synthesis of sexual steroid hormones, such as 17ß-estradiol (E2). E2 is secreted into the theca blood vessels to reach the blood stream. In response specific receptors in hepatocytes mediate the synthesis and release into the blood of vitellogenins (Vtgs), the main yolk precursors in plasma. Vtgs are specifically incorporated by the oocytes via receptor mediated endocytosis (Stifani et al. 1990; Chan et al. 1991; Davail et al. 1998; Hiramatsu et al. 2004). The term vitellogenesis has a double meaning: strictly speaking the hepatic synthesis of Vtgs but also the incorporation of Vtgs by the oocyte and further processing into yolk proteins.

14.2.5.1 Stage IIIa Follicle

This substage was characterized by the first discrete entry of Vtgs into the oocyte ooplasm, only detectable in electron microscopy (Figs. 14.3e and 14.6b). As in many fish species, a huge accumulation of lipid globules was seen in the Siberian sturgeon Stage IIIa oocytes. This was previously described as Type I vitellogenesis (Breton et al. 1983).

Oocyte diameter varied from 120 to 600 µm. This sub-Stage IIIa appeared for the first time in the gonad of 3.5-year-old females with a mean weight of 4 kg.

Oocyte Events

The Vtgs reached the oolemma by passing first between thecal cells, through the basement lamina, between the granulosa cells, and, finally, along the oocyte microvilli in the surrounding oocyte matrix (Selman and Wallace 1982; Abraham et al. 1984; Le Menn et al. 2007) (Fig. 14.7a).

It is now well established that Vtgs is selectively sequestered by growing ovarian follicles via specific receptors in the oolemma, which become effective at this early stage of vitellogenesis. Numerous endocytotic clathrin-coated pits (Pearse 1976) developed in the oolemma leading to the formation of coated vesicles that moved into the peripheral ooplasm (Fig. 14.7a). These Vtg-containing coated vesicles are fused with lysosomes, as multivesicular bodies (MVB), originating from the Golgi apparatus. The MVB contain lysosomal enzymes, including cathepsin D, that possibly cleave the Vtgs into smaller yolk proteins (Sire et al. 1994; Carnaveli et al. 1999).

14 Histological and Ultrastructural Description of Oogenesis of the Siberian Sturgeon

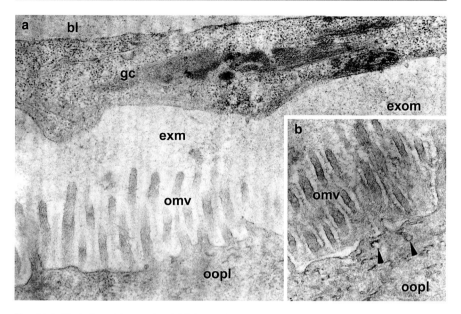

Fig. 14.6 Early Stage IIIa ovarian follicle (credit: F Le Menn). (**a**) Disappearance of gap junction granulosa cells/oocyte. (**b**) Detail of VG endocytosis. Abbreviations: *bl*, basement lamina; *exom*, extra-oocyte matrix; *gc*, granulosa cell; *omv*, oocyte microvilli; *oopl*, ooplasm; *arrow*, coated pit and coated vesicles

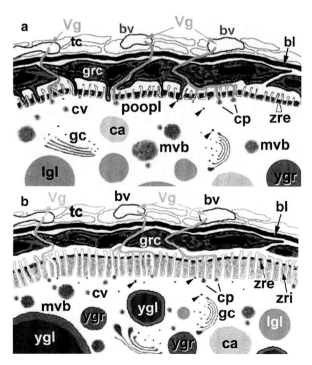

Fig. 14.7 Schematic representation of Stage III ovarian follicle (Credit: F Le Menn). (**a**) Stage IIIa, (**b**) Stage IIIb. Abbreviations: *bl*, basement lamina; *bv*, blood vessel; *ca*, cortical alveoli; *cp*, endocytotic coated pit; *cv*, coated vesicle; *gc*, Golgi complex; *grc*, granulosa cell; *lgl*, lipid globule; *mvb*, multivesicular body; *poopl*, peripheral ooplasm; *tc*, theca cell; *Vg*, vitellogenin; *ygl*, yolk globule; *ygr*, yolk granule; *zre*, zona radiata externa; *zri*, zona radiata interna; *arrow*, Golgian exocytotic vesicle

In the Siberian sturgeon oocytes, the MVB increased in size and were gradually transformed into small yolk granules, then in larger yolk globules (Fig. 14.7a). Spots of electron-dense "nuages" were still located near nuclear pores, but the patches of "ciment" had disappeared. Mitochondria were scattered throughout the ooplasm. It contained an abundant smooth endoplasmic reticulum. In the nucleus, nucleoli nested in crenelated nuclear envelope concavities and exhibited a granulated feature. Exocytosis events occurred in the peripheral ooplasm and oolemma, associated with a first outer deposit of a protective layer (Fig. 14.7a).

Surrounding Events

This deposit occurred in the oocyte matrix, between oocyte microvilli. In ultrathin tangential sections, this zona radiata externa (ZRE) appeared as an amorphous structure perforated by channels containing the oocyte microvilli (Fig. 14.8a–c). This was correlated with the first entry of Vtgs into the oocyte and indicated the beginning of Type I vitellogenesis (Fig. 14.6b) (Le Menn et al. 1999). In all species studied, the ZRE was visible using high-magnification light microscopy. It appeared as a thin, neutral glycoprotein, periodic acid-Schiff positive line located at the oolemma surface. In electron microscopy, some cells of the thecal layer had irregularly shaped nuclei, well-developed smooth endoplasmic reticulum, and characteristic crested mitochondria. These steroidogenic cells were involved in androgen synthesis induced by GTH. The androgens were converted into estrogens in the underlying granulosa cells. E2 passing through basement lamina and thecal cells was discharged into the blood stream via thecal vessels (Kagawa et al. 1981; Kagawa 1985; Nakamura and Naghama 1885; Nakamura et al. 1993).

14.2.5.2 Stage IIIb Follicle

At this substage, the acting molecules and organelles were identical as in the previous sub-Stage IIIa. However, the proportion of lipid and yolk globules was inverted, and the follicles increased rapidly in size (Fig. 14.7b). This stage was previously described as Type II vitellogenesis (Breton et al. 1983).

Oocyte Events

The Vtgs endocytosis was very active at this stage, inducing a rapid growth of the follicle from 600 to 900 µm in sturgeon. All yolk globules and lipid globules were intermingled in the ooplasm (Fig. 14.3f).

In light microscopy, a clear crown of lipid globules was observed in the medium cytoplasm. With PAS treatment, a very thin glucidic layer was observed. This was the first visualization of the ZRE located between the oolemma and the granulosa epithelium. Small granules located in the peripheral ooplasm were PAS positive and corresponded to the cortical alveoli.

Cortical alveoli appeared clustered in groups of three or four near a Golgi complex and mitochondria on electron micrographs (Fig. 14.9a–c). They contained fibrillar material. These organelles were not used by the embryo and cannot be considered as part of the yolk. At the end of vitellogenesis, they were distributed in a single layer underlying the oolemma. At fertilization, they fused with the oolemma.

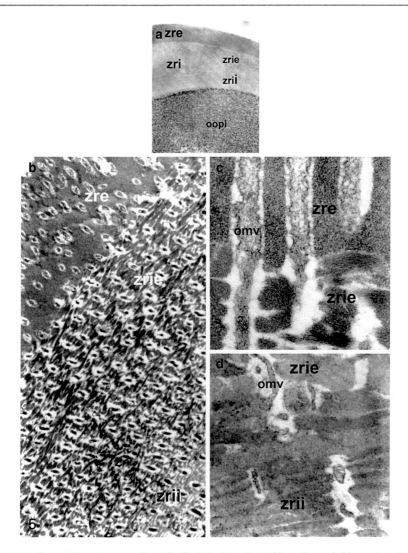

Fig. 14.8 Stage IIIb and zona radiate (Credit: F Le Menn). (**a**) Photonic semi-thin section, (**b–d**) ultrastructural micrographs. Abbreviations: *omv*, oocyte microvilli; *oopl*, ooplasm; *zre*, zona radiata externa; *zri*, zona radiata interna; *zrie*, zona radiata interna externa; *zrii*, zona radiata interna interna

They discharged their glycoprotein content at the oocyte surface by exocytosis during the cortical reaction of the female gamete.

In the peripheral ooplasm, a huge amount of endocytotic vesicles of Vtgs moved centripetally toward the MVB. In the meantime, numerous dense-cored vesicles originating from the Golgi apparatus reached the oolemma to deposit the zona radiata interna *(ZRI)* at the oocyte surface by exocytosis (Fig. 14.7b). These transfers were probably helped by the presence of a cytoskeleton network underlying the oolemma. This network, known as the terminal web, was formed by actin filaments,

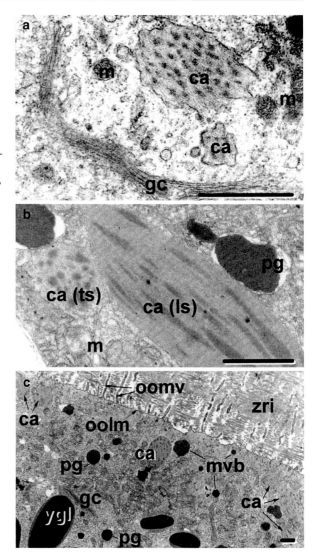

Fig. 14.9 Cortical alveoli formation (Credit: F Le Menn). (**a**) Transversal section, (**b**) longitudinal section, (**c**) peripheral ooplasm in Stage IV ovarian follicle. Abbreviations: *ca*, cortical alveolus; *ca (ls)*, in longitudinal section; *ca (ts)*, in transversal section; *gc*, Golgi complex; *m*, mitochondrium; *mvb*, multivesicular body; *ooml*, oolemma; *oomv*, oocyte microvilli; *pg*, pigment granule; *ygl*, yolk globule; *zri*, zona radiata interna

where the internal actin skeleton of each microvillus was anchored. A great number of endocytotic vesicles budded off from the oolemma to the peripheric ooplasm near the base of the oocyte microvilli, in correlation with the enormous amount of yolk stored. These coated pits and coated vesicles have not previously been observed in the oolemma of the microvilli, probably due to bundles of axial actin filaments forming an internal skeleton preventing any distortion of the oolemma (Fig. 14.6a, b).

In sturgeon as in numerous other species, the formation of crystals in yolk globules (yolk platelets) seems to enhance the capacity of the oocyte to store yolk (review in Wallace and Selman 1981). Yolk platelets appeared as regularly or irregularly shaped crystalline cores distributed in an apparently amorphous matrix (Fig. 14.10) (Karasaki 1967).

Fig. 14.10 Late Stage IV ovarian follicle (Credit: F Le Menn). Abbreviations: *ca*, cortical alveolus; *omv*, oocyte microvilli; *m*, mitochondrium; *pg*, pigment granule; *yg*, yolk granule; *ypl*, yolk platelet

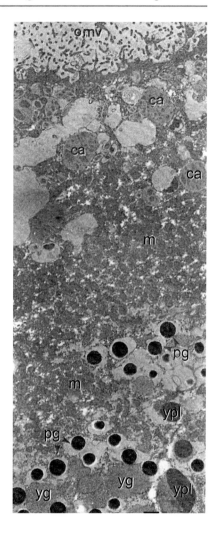

Surrounding Events

The ZR interna was formed in the extra-oocyte matrix by successive deposits of inner electron-clear reticulated layers, gradually displacing the ZR externa toward the granulosa cells (Fig. 14.8c, d). Unlike the amorphous structure of the ZR externa, the ZR interna had reticulated deposits with twisted arrangements, giving it an arched appearance of polymerized fibrillar secretion in a matrix. The parabolic lamellar pattern appeared clearly on oblique ultrathin sections (Fig. 14.8b). The ZR interna structure was common to all oviparous Teleostei and Chondrostei species studied. Many other biological polymers produce these twisted fibrous arrangements by deposition of microfibrils in regular helicoidal arrays, providing pliant structures, having generally a protective function (Bouligand 1972; Giraud-Guille 1996). These formations showed a pattern similar to the well-defined molecular state known as the cholesteric phase, as first described in microscopic preparations

Fig. 14.11 Stage V ovarian follicle (Credit: F Le Menn). (**a**) Migration of the germinal vesicle of a fully grown ovarian follicle after boiling and medial cutting, (**b**) one micropyle of *A. baerii,* observed with scanning microscope after ovulation. Abbreviations: *ap*, animal pole; *jc*, jelly coat; *N*, germinal vesicle; *pg*, pigment granules accumulated beneath the oolemma; *zr*, zona(s) radiata(s)

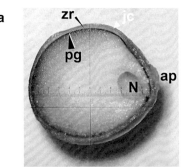

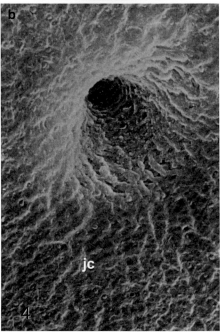

of cholesterol derivatives (Fridel 1922). These supramolecular arrangements appeared in a viscous state with some fluidity, as molecules are free to move with respect to each other (Besseau and Giraud-Guille 1995). In ray-finned fish, this particular structure might gradually stretch, due to very great increase of oocyte volume during its rapid growth phase in this Stage IIIb. The basement lamina very thick (over 10 µm) at the previous stage decreased to 1 µm.

The micropyles were fully formed during sub-Stage IIIb (Fig. 14.11b). The micropylar cells originating from the granulosa epithelium induced a local reorganization of the zona radiata into one or more funnels, depending on the species. In the Siberian sturgeon, the number of micropyles was reported to be 8.2 ± 3.2 (Ginzburg and Detlaf 1969). Micropyles constitute the only entrance into the egg for the spermatozoa (Dumont and Brummet 1980; Cherr and Clark 1982; Amanze and Lyengar 1990). The location of the micropyles on the oocyte surface delineates the future animal pole (Fig. 14.11a). ZR glycoproteins were reported to have an affinity for spermatozoa and to guide sperm into the micropyle (Amanze and Lyengar 1990; Iwamatsu et al. 1997).

14.2.6 From Primary Oocyte Undergoing Vitellogenesis to Secondary Oocyte: Stage IV and Maturational Processes

14.2.6.1 Stage IV Follicle

During Stage IV, the primary oocyte left the diplotene stage and underwent the first meiotic division. Unfortunately, the term "maturation" has very often been used improperly to describe primary oocyte size enlargement during arrest in the prophase of the first meiotic division. This has led to misunderstandings between scientists and fish farmers. Maturation processes were induced by gonadotropic hormones and occurred in the ooplasm and nucleus of primary oocytes. The maturation signal consisted of a new pulse of gonadotrophic hormones, stimulating ovarian somatic follicle cells and triggering synthesis of the maturation-inducing steroid (MIS). MIS acted on specific receptors located on the oolemma, which, in turn, triggered the synthesis and activation of maturation-promoting factor (MPF) in the ooplasm (Yamashita 1998; Senthilkumaran et al. 2004).

This stage corresponded to the rapid vitellogenic growth phase leading to a considerable increase in oocyte diameter (900–2800 µm). It was found for the first time in a 5-year-old fish with a mean weight of 5 kg.

In light microscopy, yolk globules were first observed in the outer part of the cytoplasm, then in the whole cytoplasm intermingled with lipid globules (Fig. 14.3f). In the surrounding layers, the basement lamina became too thin to be observable.

On semi-thin sections, yolk globules were ovoid and lipid globules were spheric (Fig. 14.3f). The zona radiata interna exhibited two layers, numerous narrow stratifications, and a new zona radiata interna interna (ZRII) deposited under the zona radiata interna externa (ZRIE) (Fig. 14.8a–c). Electron micrographs showed that the number of yolk globules increased considerably compared to those of the cortical alveoli and lipid globules. The yolk globules showed yolk platelets (Fig. 14.10) inside a granular matrix. Mitochondria and rough endoplasmic reticulum were scattered throughout these inclusions. Pigment granules appeared underneath the peripheral ooplasm and a very thick layer of mitochondria. These pigments showed a very electron-dense matrix inside a vacuole with dark granulations (Fig. 14.10).

14.2.6.2 Stage V Follicle or Maturational Events

This stage corresponds to the maturation process. We observed these features in gonads of 6-year-old females with an average weight of 5.5 kg, and the size of oocyte can increase to 3200 µm. These fish were maintained in the INRA fish farm (Donzacq) at a constant temperature of 17 °C during the whole year.

On semi-thin sections, the nucleus migrated to the periphery of the oocyte toward the animal pole, associated with the micropyles in the granulosa layer (Fig. 14.11a).

Fig. 14.12 Jelly-coat synthesis during Stage V ovarian follicle (Credit: F Le Menn). Abbreviations: *ch*, zona radiata externa channel; *exom*, extra-oocyte matrix; *gc*, granulosa cell; *jc*, jelly coat; *zre*, zona radiata externa; *arrow*, granulosa exocytotic coated vesicles

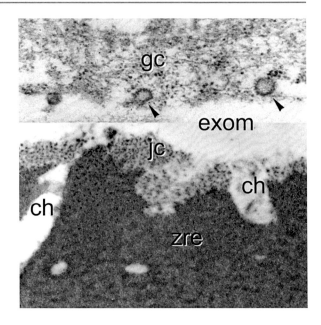

The morphological events during maturation generally followed a similar sequence in all Teleostei and Chondrichthyes. However, in sturgeon it can take several months. For *Acipenser baerii* the phenomena took up to 6 months under the conditions of the Donzacq fish farm.

On ultrathin sections, at the end of this stage, just before the germinal vesicle break down (GVBD), the microvilli had retracted. The structure of the zona radiata was denser (Fig. 14.12). Yolk granules could fuse with lipid globules to form a liquid yolk mass located at the vegetative pole of the oocyte. Above the ZRE, a new electron-clear spongy layer, the "jelly coat," was deposited by granulosa cells (Fig. 14.12). This layer partially entered the empty pores of the ZRE. The thickness of the jelly coat varied from 0.5 to 0.9 μm (Le Menn and Pelissero 1991).

MPF acted on the lamina covering the inside of the internal nuclear membrane, phosphorylating lamina serines, and leading to de-structuring of the nuclear envelope, setting in motion the GVBD.

Metaphase of the first meiotic division occurred. It produced two cells, a large secondary oocyte and a small first polar body which degenerated. The secondary oocyte separated from its somatic surrounding layers (Fig. 14.13) during ovulation. In the wild, spawning occurred spontaneously. In fish farm, sturgeon's spawning required carp or mammalian LH injections as well as manual stripping. This was a gentle pressure on the female's abdomen. In all cases, the spawn cells were matured secondary oocytes or ova. Fusion with a spermatozoon produced fertilized egg, i.e., the starting point of embryo development.

Fig. 14.13 Dissection after boiling of a Stage V ovarian follicle (Credit: P Williot). In the upper left part is a heap which corresponds to the two granulosa and theca externa layers irrigated by blood vessels. Underneath and fold to the left is the jelly coat. The most complete and thicker layer is the zona radiata externa. Inside and fold to the right is the zona radiata interna

14.3 Discussion

Oogenesis in *Acipenser baerii* resembles oogenesis in numerous fresh water teleosts (Nagahama 1983). Nevertheless some synthetic and structural particularities were worth noting.

14.3.1 *Acipenser baerii* Synthetic and Endocytotic Features

14.3.1.1 Type 1 and Type 2 Vitellogenesis

Type 1 vitellogenesis corresponds to the formation of cortical alveoli and lipid globules in the oocyte. It is usually called endogenous vitellogenesis. Type 2 vitellogenesis corresponds to the concomitant incorporation of vitellogenin, a hepatically secreted phosphoglycoprotein which is the plasma yolk precursor (review of De Vlaming et al. 1980; Wallace et al. 1983). In Type 2 vitellogenesis, endocytotic events predominate greatly compared with synthetic events. Pre-vitellogenesis corresponds to Type 1 vitellogenesis (Jalabert, Le Bail and Le Menn, unpublished data).

In *Acipenser baerii* lipid globules are identified by histochemical staining from the oogonial stage onward (Pelissero et al. 1985). This has not been observed in other species (Wallace and Selman 1981), even in freshwater species where lipid synthesis occurs for a very long time before Type 2 vitellogenesis (Droller and Roth 1966, for *Anguilla anguilla*; Ulrich 1969, for *Brachydanio rerio* or Busson-Mabillot 1969, for *Lebistes reticulatus*). However, we should recall that *Acipenser baerii* is reared in France under specific temperature and feeding conditions, and no firm conclusion can be reached until gametogenesis in wild populations living in natural conditions in the Russia has been studied.

14.3.1.2 Cortical Alveoli Synthesis

In the oocyte at the end of Stage 2, at the beginning of oocyte microvillosities protrusion, we observed a very considerable Golgian synthetic activity in the medium and peripheral ooplasm with numerous Golgi complexes (Fig. 14.4b). In the vicinity of each Golgi complex, several spherical aggregates of Golgian vesicles led to the formation of small vesicles by deposition of the Golgian secretions in central smooth endoplasmic reticulum vesicles. These vesicles were filled with granular material resembling material contained by cortical alveoli in the next stage (Figs. 14.9 and 14.10). We suppose that these features are the first visualization of synthesis of cortical alveoli. Further investigations are necessary to confirm this hypothesis, by identifying the chemical nature of their content on ultrathin sections.

14.3.2 Specific Structural Features

14.3.2.1 Basement Lamina

This extracellular matrix was located between the granulosa epithelium and the thecal cells. The basement lamina is generally secreted by epithelium (Hay 1981). In this case, a granulosa source was indicated by observations of exocytotic features on the plasmic granulosa membrane related to the matrix. Its secretion occurred until the follicle reached Stage III where the thickness could be 10 μm. During the rapid growth phase, the stretching of the follicle caused a decrease of the thickness to 1 μm making it invisible in light microscopy.

This matrix was identified in *Huso huso*, *Acipenser gueldenstaedtii*, *Acipenser ruthenus,* and *Acipenser stellatus* (Caloianu-Yordachel 1971; Kornienko 1975). In *Acipenser baerii*, it is a polysaccharidic layer stained with PAS. In this species electron, micrographies showed a simple organization of small circular electron-clear conglomerates in an amorphous matrix, unlike the usual three-layered structure of a true basement lamina (Vrackq 1974) with a lamina lucida between two lamina densa. The literature leads us to expect a biochemical composition of glycosaminoglycans, elastin, and collagen and the function of a semipermeable filter (Hay 1981). This function would be interesting to observe in an investigation of the pathway of plasma vitellogenin from thecal capillaries to the oocyte (Selman and Wallace 1982; Abraham et al. 1984). Another explanation could be the protection of early oocyte stages from shocks. As most cartilaginous fish, the sturgeon species are known to be fragile to manipulation and therefore to any shock which could occur in rapid water in the wild. Because the early vitellogenic stage lasts for a long time in this species at a time when the fish is still of modest size, this protection may help in ensuring the preservation of future progeny.

14.3.2.2 Extra-Oocyte Matrix

This matrix deposited between the oolemma and the granulosa epithelium is particularly emphasized in *Acipenser baerii*. Few authors mention its presence in other fish: its electron-clear fibrillar material mentioned without specification by Caporiccio and Connes (1977) has been interpreted as a cell coat by Busson-Mabillot (1973), as the first deposited zone of the primary envelope (Anderson 1967; Wourms 1976), as a follicular matrix by Abraham et al. (1984), and as an extra-oocyte matrix by Le Menn (1984) and Nunez Rodriguez (1985). In all fish we studied, ultrastructural features gave no indication about its cellular origin.

This matrix was deposited as soon as the granulosa epithelium separated from oolemma at the end of Stage 1 follicle. It is in this environment that the zona radiata externa is deposited. The chemical composition of this matrix is not defined, but it probably contains proteoglycans (Hascall and Hascall 1981) well known for their ability to capture and release free ions. This structure could influence the pH in the vicinity of the oocyte and act on the incorporation of exogenous vitellogenin (VTG). Indeed this molecule reaches the oolemma by passing through this matrix along the oocyte microvilli. It has been demonstrated in birds that the binding of the VTG to its receptors is pH dependent with the best binding for acid pH values (Woods and Roth 1980; Yusko et al. 1981). So, such a structure may interfere in the process of internalization by providing an adapted microenvironment. As for the basement lamina, a protective function of this extracellular matrix toward the future progeny cannot be simply ruled out.

14.3.3 Zona Radiata

This proteoglycan matrix is built up between the oocyte and the granulosa epithelium synthesized apparently by the oocyte. The first part deposited is then in an outer position compared with the later parts deposited. At the end of vitellogenesis, at Stage 4, just before process of maturation, the zona radiata is formed by three layers: we observed in order of deposition the zona radiata externa (ZRE), the zona radiata interna externa (ZRIE), and the zona radiata interna interna (ZRII). During the formation of this matrix, we observed numerous dense-cored Golgian vesicles associated with Golgi complexes located in the peripheral ooplasm just under the oolemma. These features seem to confirm as in other fish species the oocyte origin of this matrix (Tesoriero 1977; Le Menn 1984; Nunez Rodriguez 1985). These three layers have been observed in other sturgeon species (Magnin 1967; Ginzburg 1968; Caloianu-Yordachel 1971; Markov 1975; Cherr and Clark 1982), but the interpretations of Russian authors are not in agreement with ours. They consider the ZRE as the outer layer which allows the eggs to stick on the substrate after laying, and they call it the "jelly coat." In fact at the end of vitellogenesis during maturation, this new spongy layer is deposited on the zona radiata externa (Pelissero et al. 1985) causing the fixation of the egg when it is spawned in water. This layer is described in *Acipenser transmontanus* (Cherr and Clark 1982, 1984). Such specific secretions involved in the fixation of eggs have been described in teleosts: *Cichlasoma nigrofasciatum* (Busson-Mabillot 1977), *Cynolebias melanotoenia* and *Cynolebias ladigesi* (Wourms and Sheldon 1976), *Pomatoschistus minutus* (Riehl 1978), and *Gobius niger* (Le Menn 1984). In all these species, as in *Acipenser baerii*, ultrastructural features indicated that the sticking material was secreted by the granulosa cells.

Until the end of vitellogenesis, only one microvillus is observed in each channel of the zona radiata. Granulosa cells never put out microvilli toward the oocyte. From the beginning of Stage III onward, the follicular epithelium is completely separated from the surface of the oocyte as soon as the junctions with oolemma disappeared. We have never observed on this species junctions between the apical part of oocyte microvilli and the apical plasmic membrane of the granulosa epithelium as shown in *Plecoglossus altivelis* (Toshimori and Yasuzuni 1979).

References

Abraham M et al (1984) The cellular envelope of oocytes in Teleosts. Cell Tissue Res 235:403–410

Agulleiro M et al. (2007) High transcript level of fatty acid-protein but not very low-density lipoprotein receptor is correlated to ovarian follicle in teleost fish (*Solea senegalensis*). Biol Reprod PMID: 17554079

Akimova NB et al. (1979) Growth and gametogenesis of the Siberian sturgeon (*Acipenser baerii* B.) under experimental and natural conditions. Proc. 7th Japan Soviet joint Symp Aquacult Tokyo: 179

Amanze D, Lyengar A (1990) The micropyle: a sperm guidance system in teleost fertilization. Development 109:495–500

Anderson E (1967) The formation of the primary envelope during oocyte differentiation in teleosts. J Cell Biol 35:193–212

Badenko LV et al (1981) Method for evaluating the quality of sturgeon spawners as exemplified in the sevryuga (*Acipenser stellatus*) from the Kuban river. J Ichthyol 21:96–103

Besseau L, Giraud-Guille MM (1995) Stabilization of fluid cholesteric phases of collagen to ordered gelateed matrices. J Mol Biol 251:197–202

Bouligand Y (1972) Twisted fibrous arrangements in biological materials and cholesteric mesophases. Tissue Cell 4:189–217

Breton B et al (1983) Maturational glycoprotein gonadotropin and estradiol-17ß during the reproductive cycle of the female brown tour (*Salmo trutta*). Gen Comp Endocrinol 49:220–231

Bruslé S (1980) Fine structure of early previtellogenic oocytes in *Mugil* (*Lisa*) *auratus* Risso, 1810 (Teleostei, Mugilidae). Cell Tissue Res 207:121–134

Bruslé S, Bruslé J (1978) An ultrastructural study of early germs cells in *Mugil* (*Liza*) *auratus* Risso, 1810 (Teleostei, Mugilidae). Ann Biol Anim Biophys 18:1141–1153

Busson-Mabillot S (1969) Données récentes sur la vitellogénèse. Ann Biol 8:199–227

Busson-Mabillot S (1973) Evolution des enveloppes de l'ovocyte et de l'œuf chez un poisson téléostéen. J Microscopie 18:23–44

Busson-Mabillot S (1977) Un type particulier de sécrétion exocrine, celui de l'appareil adhésif de l'oeuf d'un poisson téléostéen. Biol Cell 30:233–244

Caloianu-Yordachel M (1971) L'ovogénèse chez les poissons Acipenséridés, la morphogénèse et la constitution histochimique des membranes externes. Rev Roum Biol Zool 16(2):113–120

Caporiccio B, Connes R (1977) Etude ultrastructurale des enveloppes périovocytaires et périovulaires de *Dicentrarchus labrax* L. (Poisson téléostéen). Ann Sei Nat Zool Paris 19(12ème série):351–368

Carnaveli O et al (1999) Yolk formation and degradation during oocyte maturation in bream, *Sparus aurata*: involvement of two lysosomal proteinases. Biol Reprod 60:140–146

Chan L et al (1991) Vitellogenin purification and development of assay for vitellogenin receptor in oocyte membranes of the tilapia (*Oreochromis niloticus* Linnaeus) 1766. J Exp Zool 257:96–109

Cherr GN, Clark WH Jr (1982) Fine structure of the envelope and the micropyles of the eggs of the white sturgeon *Acipenser transmontanus*. Develop Growth Differ 24:341–352

Cherr GN, Clark WH Jr (1984) Jelly release in the eggs of the white sturgeon *Acipenser transmontanus*: an enzymatically mediated event. J Exp Zool 230:145–149

Clérot JC (1976) Les groupements mitochondriaux des cellules germinales de poissons Téléostéens Cyprinidés. 1. Etude ultrastructurale. J Ultrastruct Res 54:461–475

Clérot JC, Wegnez M (1977) Etude ultrastructurale et biochimique de l'ovocyte en prévitellogénèse de vertébrés inférieurs. Biol Cell 29:23a

Davail B et al (1998) Evolution of oogenesis: the receptor fot vitellogenin from the rainbow trout. J Lipid Res 39:1929–1937

De Vlaming VL et al (1980) Golsfish *Carassius auratus* vitellogenin: induction, isolation, properties and relationship to yolk proteins. Comp Biochem Physiol 67 b:613–623

Droller MJ, Roth TF (1966) An electron microscope study of yolk formation during oogenesis in *Lebistes reticulatus* Guppyi. J Cell Biol 28:209–232

Dumont JN, Brummet AR (1980) The viteline envelope, chorion and micropyle of *Fundulus heteroclitus* eggs. Gamete Res 3:25–44

Fedorova LS (1976) Physiological and biochemical characteristics of the reproductive products and larvae of sturgeons during artificial rearing. J Ichthyol 16:427–436

Flynn SR, Benfey TJ (2007) Sex differentiation and aspects of gametogenesis in shortnose sturgeon *Acipenser brevirostrum* Lesueur. J Fish Biol 70:1027–1044

Fontana F et al (2008) Evidence of hexaploid karyotype in shortnose sturgeon. Genome 51(2):113–119

Fridel G (1922) Les états mésomorphes de la matière. Ann Phys 18:273–474

Ginzburg AS (1968) Fertilisation in fishes and the problem of polyspermy. N O A A and National Scientific Foundation Translation, Silver Spring, p 290

Ginzburg AS, Detlaf TA (1969) Razvitie Osetrovykh Ryb. Sozrevanie yats, oplodotvorenie i embriogenez. Izdatel'stvo "NAUKA", Moskva, p 134

Giraud-Guille MM (1996) Twisted liquid crystalline supramolecular arrangements in morphogenesis. Int Rev Cytol 166:59–101

Grandi G, Chicca M (2008) Histological and ultrastructural investigation of early gonad development and sex differentiation in Adriatic sturgeon *Acipenser nacarii*, (Acipenseriformes, Chondrostei). J Morphol 269:1238–1262

Grier H (2000) Ovarian germinal epithelium and folliculogenesis in the common snook *Centropomus undecimalis* (Teleostei, Centropomidae). J Morphol 243:265–281

Guraya SS (1986) The cell and molecular biology of fish oogenesis. In: Sauer HW (ed) Monographs in developmental biology, vol 18. Karger, Basel, pp 1–223

Hamagushi S (1985) Changes in the morphology of the germinal dense bodies in primordial germ cells of the teleost *Oryzias latipes*. Cell Tissue Res 240:669–673

Hascall VC, Hascall GK (1981) Proteoglycans. In: Hay ED (ed) Cell biology of extracellular matrix. Plenum Press, New York, pp 39–60

Havelka M et al (2014) Fertility of a spontaneous hexaploid male Siberian sturgeon, *Acipenser baerii*. BMC Genet 15:5

Hay ED (1981) Cell biology of extracellular matrix. In: Hay ED (ed) . Plenum Press, New York, p 417

Hiramatsu N et al (2004) Molecular characterisation and expression of vitellogenin receptor from the white perch (*Morone Americana*). Biol Reprod 70:1720–1730

Hogan JC (1978) An ultrastructural analysis of "cytoplasmic maker" in germ cells of *Oryzias latipes*. J Ulrastruct Res 62:237–250

Iwamatsu T et al (1997) Effect of micropylar morphology and size on rapid sperm entry into the eggs of medaka. Zool Sci 14:626–628

Kagawa H (1985) Ultrastructural and histochemical observations regarding the ovarian follicles of the amago salmon (*Oncorhynchus rhodurus*). J UOEH 7:27–35

Kagawa H et al (1981) Correlation of plasma estradiol-17-β and progesterone levels with ultrastructure and histochemistry of ovarian follicles in the spoted char, *Salvenilus leucomaenis*. Cell Tissue Res 218:315–329

Karasaki S (1967) An electron microscope study on the crystalline structure of the yolk platelets of the lamprey egg. J Ultrastruct Res 18:377–390

Kijima T, Maruyama T (1985) Histological research for the development of the gonad of the hybrid sturgeon bester (*Acipenser ruthenus* male x *Huso huso* female). Bull Nati Res Inst Aquacult 8:23–29

Kondrat'yev AK (1977) The functional morphology of oocytes in the period of previtellogenesis in the Siberian sterlet *Acipenser ruthenus* M. At different time of its annual biological cycle. J Ichthyol 17:769–778

Kornienko GG (1975) Early degenerative changes in the oocyte of the Kuban sevryuga *Acipenser stellatus*. J Ichthyol 15:503–507

Le Menn F (1984) Aspects ultrastructuraux, biochimiques et endocriniens de la vitellogénèse d'un poisson téléostéen *Gobius niger* L. Thèse de Doctorat d'Etat, Univ. Bordeaux I, n° 814

Le Menn F, Burzawa-Gerard E (1985) Effect of carp gonadotrophin (cGTH) and a fraction unadsorbed on concavalin A-sepharose obtained from c-GTH on vitellogenesis in the hypophysectomized marine teleost *Gobius niger*. Gen Comp Endocrinol 57:23–36

Le Menn F, Pelissero C (1991) Histological and ultrastructural studies of oogenesis of the Siberian sturgeon (*Acipenser baerii*). In: Williot P (ed) , vol 57. CEMAGREF, Antony, France, pp 23–36

Le Menn F et al. (1999) A new approach to fish oocyte vitellogenesis. Proc 6th Int. Symp Reprod Physiol Fish, Bergen, Norway: 281–284

Le Menn F et al. (2007) Ultrastructural aspects of the ontogeny and differentiation of ray-finned fish ovarian follicles. In: Babin PJ et al (eds) The fish oocyte: from basic studies to biochemical applications, p 1–37

Magnin E (1967) Recherches sur les cycles de reproduction des esturgeons *Acipenser fulvescens* Raf. de la rivière Nottaway tributaire de la Baie James. Verh Int Ver Limnol 16:1018–1024

Markov KP (1975) Scanning electron microscope study of the microstructure of the egg membrane in the Russian sturgeon *Acipenser gueldenstaedtii* B. J Ichthyol 15:739–749

Matova N, Cooley L (2001) Comparative aspects of animal oogenesis. Dev Biol 231:291–320

Mazabraud A et al (1975) Biochemical research on oogenesis. RNA accumulation in the oocytes of teleosts. Dev Biol 44:326–332

Nagahama Y (1983) The functional morphology of teleost gonad. In: Hoar WS, Randall DJ, Donaldson EM (eds) Fish physiology. Acad Press, NY IX A, pp 223–275

Nagahama Y (1997) 17-20ß-dihydroxy-pregnen-3-one, a maturation inducing hormone in fish oocytes: mechanisms of synthesis and action. Steroids 62:190–196

Nagahama Y et al (1995) Regulation of oocyte growth and maturation in fish. Curr Yop Dev Biol 30:103–145

Nakamura M, Nagahama Y (1985) Steroid producing cells during ovarian differentiation of the tilapia, *Sarotherodon niloticus*. Dev Growth Diffr 27:701–708

Nakamura M et al (1993) Ultrastructural analysis of the developing follicle during early vitellogenesis in tilapia, *Oreochromis niloticus*, with special references in the steroid-producing cells. Cell Tissue Res 272:33–39

Nunez Rodriguez J (1985) Contribution à l'étude de la biologie de la reproduction de la sole (*Solea vulgaris* Quensel 1806). Approche ultrastructurale et physiologique. Thèse 3ème Cycle, Université de Bordeaux I, n° 2061

Parry EW (1973) Methylene blue and azure-2 as stains for lipid in osmium-fixed tissues embeded in araldite. J Clin Pathol 16:546–548

Pearse (1976) Clathrin: a unique protein associated with intracellular transfer of membrane by coated vesicles. Proc Natl Acad Sc USA 73:1255–1259

Pelissero C (1988) Mise en place des bases méthodologiques pour l'étude de la reproduction chez l'esturgeon *Acipenser baerii* femelle. Thèse de 3ème cycle, Université de Bordeaux I, n° 2229

Pelissero C et al (1985) Ultrastructural characteristic features of the oocyte of the sturgeon *Acipenser baerii* B, 7th Conf Fish Culture Europ Soc Comp Physiol Biochem Barcelona, vol A3, Promociones Publicationes Universitarias, Barcelona, p 8

Perazzolo LM et al (1999) Expression and localization of messenger ribonucleic acid for the vitellogenin receptor in ovarian follicles throughout oogenesis in the rainbow trout, *Oncorhynchus mykiss*. Biol Reprod 60:1057–1068

Raïkova EV (1976) Evolution of the nuclear apparatus during oogenesis in Acipenseridae. J Embryol Exp Morphol 35(8):667–687

Reynolds ES (1963) The use of lead citrate at hight pH as an electron opaque stain in clectron microscopy. J Cell Biol 17:208

Riehl R (1978) Electronen mikroskopische und autoradiographische Untersuchungen an den Dotterkernen in den Oocyten von *Noemacheilus barbatulus* L. und *Phoxinus phoxinus* L. (Pisces, Teleostei). Cytobiologie 17:137–145

Rzepkowska M, Ostaszewska T (2013) Proliferating cell nuclear antigen and Vasa protein expression during gonadal development and sexual differentiation in cultured Siberian (*Acipenser baerii* Brandt, 1869) and Russian (*Acipenser gueldenstaedtii* Brandt, Ratzeburg, 1833) sturgeon. Rev Aquac 5:1–14

Selman K, Wallace RA (1982) The inter- and intracellular passage of protein through the ovarian follicle in teleosts. In: Proc. Int Symp Reprod Physiol Fish. Wageningen, The Netherlands, p 57

Selman K, Wallace RA (1989) Cellular aspects of oocyte growth in teleost. Zool Sci 6:211–231

Selman K et al (1993) Stages of oocyte development in the zebrafish, *Brachidanio rerio*. J Morphol 218:203–224

Senthilkumaran B et al (2004) A shift un steroidogenesis occurring in ovarian follicles prior to oocyte maturation. Mol Cell Endocrinol 215:11–18

Sire MF et al (1994) Involvement of the lysosomal system in yolk protein deposit and degradation during vitellogenesis and embryonic development in trout. J Exp Zool 269:69–83

Stifani S et al (1990) Regulation of oogenesis: the piscine receptor for vitellognin. Biochem Biophys Acta 1045:271–279

Tesoriero JV (1977) Formation of the chorion (zona pellucida) in the teleost *Oryzias latipes*. I. Morphology of early oogenesis. J Ultrastruct Res 59:282–291

Toshimori K, Yasuzuni F (1979) Gap junctions between microvillosities of the oocyte and follicular cells in the teleost *Plecoglossus altivelis*. Z Mikrosk Anat Forsch 93:458–164

Trusov VZ (1975) Maturation in the gonads of the female sevryuga *Acipenser stellatus* during its life in the ocean. J Ichthyol 15:61–72

Ulrich E (1969) Etude des ultrastructures au cours de l'ovogenèse d'un poisson téléostéen le Danio *Brachydanio rerio*. J Microsc 8:447–473

Veshchev PV (1982) Reproduction of sterlet *Acipenser ruthenus* (Acipenseridae) in the lower volga. J Ichthyol 22:40–46

Vrackq R (1974) Basal lamina scaffold: anatomy and significance for maintenance of orderly tissue structure. Am J Pathol 77:314–346

Wallace RA, Selman K (1981) Cellular and dynamic aspects of oocyte growth in teleosts. Am Zool 21:325–343

Wallace RA et al. (1983) The oocyte as an endocytotic cell. In: Molecular biology of egg maturation. Ciba Found Symp Eds Pitman Books London: p 228–248

Williot P, Brun R (1982a) Résultats sur la reproduction d'*Acipenser baerii* en 1982. Bull Fr Piscic 287:19–22

Williot P, Rouault T (1982b) Compte rendu d'une première reproduction en France de l'esturgeon sibérien *Acipenser baerii*. Bull Fr Piscic 286:255–261

Woods JW, Roth TF (1980) Selective protein transport: identity of the solubilized phosvitin receptor from chicken oocyte. J Supramol Struct 14:473–480

Wourms JP (1976) Annual fish oogenesis. I. Differentiation of the mature oocyte and formation of the primary envelope. Dev Biol 50:338–354

Wourms JP, Sheldon H (1976) Annual fish oogenesis. II. Formation of the secondary egg envelope. Dev Biol 50:355–366

Yamashita M (1998) Molecular mechanisms of meiotic maturation and arrest in fish and amphibian oocytes. Semin Cell Dev Biol 9:569–579

Yusko S et al (1981) Receptor-mediated vitellogenin binding to chicken oocytes. Biochem J 200:43–50

Zelazowska M et al. (2015) Ovarian nests in immature and mature sturgeons (*Acipenser gueldenstaedtii*) and paddlefish (*Polyodon spathula*) (Chondrostei, Acipenseriformes) comprise early previtellogenic oocytes. Tissue Cell

Sperm and Spermatozoa Characteristics in the Siberian Sturgeon

Martin Pšenička and Andrzej Ciereszko

Abstract

Sperm and spermatozoa in Siberian sturgeon are very interesting and specific from several points of view. Siberian sturgeon usually produces high volume of semen with relatively low sperm and protein concentration, which is partially explained by the atypical testicular morphology where spermatozoa are mixed with urine during passage through the kidneys to the Wolffian ducts. Sodium and chloride ions contribute most to the osmolality of the seminal fluid. Potassium ions are critical for immobilization of spermatozoa, while its antagonist is calcium ion, which triggers spermatozoa motility. The motility period is relatively long (2–3 min) with flagellum beat frequency about 50 Hz. The main characteristics of sturgeon spermatozoa are an elongated head with an acrosome containing acrosomal proteins. The flagellum is equipped with a fin for more efficient movement. During penetration into the egg micropyle, the acrosome undergoes acrosomal reactions, which include formation of fertilization filament and opening of posterolateral projections. The fertilization filament activates the egg and causes the formation of a perivitelline space, while the posterolateral projections serve as an anchor against release from the micropyle. The acrosomal reaction has been recognized to be important for fertilization and development.

M. Pšenička (✉)
Faculty of Fisheries and Protection of Waters, Laboratory of Germ Cells, Research Institute of Fish Culture and Hydrobiology, South Bohemian Research Center of Aquaculture and Biodiversity of Hydrocenoses, University of South Bohemia in Ceske Budejovice, Vodnany, Czech Republic
e-mail: psenicka@frov.jcu.cz

A. Ciereszko
Institute of Animal Reproduction and Food Research, Polish Academy of Sciences, Tuwima 10, 10-747 Olsztyn, Poland
e-mail: a.ciereszko@pan.olsztyn.pl

> **Keywords**
> Semen composition • Spermatozoon ultrastructure • Acrosomal reaction • Siberian sturgeon

Introduction

Siberian sturgeon semen is characterized by high volume and low numbers of spermatozoa and ions as well as low protein concentration (Tables 15.1 and 15.2). Significant variation in these values is observed and most likely relates to age, time after hormonal stimulation, number of sperm collections, and maintenance. The lowest volume (8 mL) was reported from a 34-month-old male, whereas older males of 7+ years produced ejaculates >200 mL (Glogowski et al. 2002). Sperm concentration can be as low as 0.2 × 109 spermatozoa, but values ten times higher have been recorded. Such variation can be partially explained by dilution of semen with urine in the Wolffian ducts (Gallis et al. 1991). Some variation of seminal plasma osmolality and ionic composition are observed as well (Table 15.2).

The mode of spermatozoa-ova gamete interaction is unique in sturgeons. The spermatozoa possess an acrosome, which is generally used to penetrate through the egg envelopes after entering one of the eggs several micropyles so as to provide increased access for spermatozoa.

15.1 Ions of Seminal Plasma

Basic parameters of Siberian sturgeon seminal plasma are shown in Table 15.2. Sodium and chloride ions are the most significant components that contribute to the seminal osmolality, which is generally lower than in teleostean species. Potassium

Table 15.1 Semen volume and sperm concentration of Siberian sturgeon milt

Feature	Mean ± sd	Reference
Semen volume (mL)	25.9 ± 14.3	Li et al. (2011)
	73 ± 58	Sieczyński et al. (2012)
	8 ($n = 1$)	Glogowski et al. (2002)
	267.5 ± 38.9	Glogowski et al. (2002)
	85–300	Piros et al. (2002)
	356 ± 135	Tsvetkova et al. (1996)
Sperm concentration ($\times 10^9$)	0.20 ± 0.17	Li et al. (2011)
	0.22 ($n = 1$)	Glogowski et al. (2002)
	0.58 ± 0.37	Sieczyński et al. (2012)
	0.61 ± 0.37	Pšenička et al. (2008a)
	0.64 ± 0.42	Sarosiek et al. (2004)
	1.21 ± 0.45	Judycka et al. (2015a)
	1.21 ± 0.41	Glogowski et al. (2002)
	1.68 ± 0.33 (1)[a]	Piros et al. (2002)
	2.42 ± 0.78 (2)[a]	Piros et al. (2002)

[a]First and second semen collections as described in Piros et al. 2002. All values are mean ± SD with the exception of sperm concentration data of Piros et al. 2002 which are expressed as mean ± SEM

15 Sperm and Spermatozoa Characteristics in the Siberian Sturgeon

Table 15.2 Basic parameters of Siberian sturgeon seminal plasma

Feature	Mean ± sd	Reference
pH	7.97 ± 0.42	Li et al. (2011)
	8.16 ± 0.18	Pšenička et al. (2008a)
	8.70 ± 0.13	Judycka et al. (2015a)
Osmolality	38 ± 3	Gallis et al. (1991)
	46.2 ± 11.6	Li et al. (2011)
	77.2 ± 52.8	Pšenička et al. (2008a)
	93.6 ± 7.3 (1)[a]	Piros et al. (2002)
	95.7 ± 5.4 (2)[a]	Piros et al. (2002)
Na^+ (mM)	14.6 ± 6.1	Li et al. (2011)
	28 ± 0.7	Gallis et al. (1991)
	31.4 ± 10.2	Pšenička et al. (2008a)
K^+ (mM)	2.5 ± 0.3	Gallis et al. (1991)
	3.5 ± 1.1	Pšenička et al. (2008a)
	4.5 ± 1.1	Li et al. (2011)
	0.24 ± 0.06	Pšenička et al. (2008a)
Ca^{2+} (mM)	0.27 ± 0.09	Li et al. (2011)
Mg^{2+} (mM)	0.48 ± 0.22	Li et al. (2011)
Cl^- (mM)	6.2 ± 1.2	Li et al. (2011)
	14.0 ± 4.3	Pšenička et al. (2008a)
Protein conc. (mg mL^{-1})	0.38 ± 0.16	Judycka et al. (2015a)
	0.39 ± 0.19	Sarosiek et al. (2004)
	0.58 ± 0.11 (1)[a]	Piros et al. (2002)
	0.57 ± 0.06 (2)[a]	Piros et al. (2002)

[a]First and second semen collections as described in Piros et al. 2002. All values are mean ± SD with the exception of osmolality data of Piros et al. 2002 which are expressed as mean ± SEM

ions are critical for immobilization of spermatozoa in Wolffian duct (Gallis et al. 1991; Toth et al. 1997). However, inhibition of sperm motility is not always complete because some movement is often observed in non-diluted semen (Glogowski et al. 2002). Inhibition of motility by K^+ ions can be modulated by Ca^{2+} ions which are also present in seminal plasma (Alavi et al. 2012a). Values of pH seem to be less variable and are alkaline (8.0 and higher) which is a characteristic for sperm maturation in fishes (Morisawa and Morisawa 1988). Lower values of pH have been attributed to contamination with urine (Li et al. 2011).

15.2 Proteins of Seminal Plasma

15.2.1 Electrophoretic Pattern

Protein concentration in seminal plasma of Siberian sturgeon is low. Generally, in sturgeon it is not higher than 1 mg mL^{-1} (Table 15.2). Two-dimensional electrophoresis of seminal plasma has not been performed yet, and at present only a one-dimensional SDS-PAGE protein profile is available (Li et al. 2011). The latter distinguished five mutual protein bands with molecular weight ranging from 29 to

71.3 kDa; a band of 71.3 kDa is speculated to be β-N-acetylglucosaminidase which was isolated and characterized by Sarosiek et al. (2008).

15.2.2 Enzymes and Inhibitors

Data on biochemical composition of Siberian sturgeon seminal plasma were described by Piros et al. (2002). Activities of lactic dehydrogenase (LDH), arylsulfatase, acid phosphatase, and β-N-acetylglucosaminidase were found; the presence of the latter had been postulated by Li et al. (2011; see above). These activities seem to be low, because values several times higher were found in spermatozoa (Piros et al. 2002). This seems to reflect a low concentration of protein in seminal plasma, and it is possible that the presence of these enzymes in seminal plasma is related to their release from damaged spermatozoa. It has been postulated that similar to mammals, Siberian sturgeon arylsulfatase may modify the spermatozoan surface charge by removal of sulfate groups of galactosyl conjugates (Piros et al. 2002). Superoxide dismutase, glutathione reductase, and glutathione peroxidase were detected both in seminal plasma and spermatozoa of Siberian sturgeon (Shaliutina et al. 2013). This indicates the existence of protection against oxidative stress.

Despite the presence of proteolytic enzymes in spermatozoa, anti-proteinase activity (APA) in seminal plasma is also low, but seems to be exclusively present in seminal plasma, because APA does not change after freezing-thawing of semen (Piros et al. 2002). Activity of APA is clearly affected by the season (Słowińska et al. 2015), being highest in December (87.3 ± 17.0 U mL^{-1}) compared to February (12.6 ± 1.4 U mL^{-1}) and April (12.4 ± 2.7 U mL^{-1}). Before stimulation, fish were transferred from ponds (water temperatures of 1.2 and 5.3 °C, for December and April, respectively) to tanks. The water temperature in the tanks was gradually increased (1 °C day 1) to 16 °C and then maintained at that temperature for 7 days. This suggests that protease inhibitors are especially important during spermatogenesis and sperm maturation. Target protease for APA are unknown at present, but it is likely that some of several proteases recently identified in seminal plasma, including serine proteases and metalloproteases, can be controlled by APA (Słowińska et al. 2015).

15.3 Proteins of Spermatozoa

15.3.1 Electrophoretic Pattern

Contrary to information on seminal plasma, two-dimensional electrophoretograms of spermatozoa extracts are available (Li et al. 2010, 2011). The latter identified 95 protein spots (Li et al. 2011) to over 100 (Li et al. 2010); these can be divided into those common for sturgeons, presumably highly conserved proteins, and protein spots which seems to be species-specific. At present, a lack of data on cDNA and gene sequence hampers identification of protein spots, but several isoforms of enolase B and lactate dehydrogenase have been identified, which seem to be highly conserved in sturgeon sperm.

15.3.2 Enzymes

The main proteinase in the acrosome of sturgeons is acrosin which was first discovered in white sturgeon (Ciereszko et al. 1994, 1996), and recently, it was reported from sterlet by immunohistochemistry (Pšenička et al. 2009, see Sect. 15.5.2.) and Siberian sturgeon using Western blotting (Słowińska et al. 2015). The latter have demonstrated the presence of four bands of acrosin in sperm extracts which likely represent a proacrosin-acrosin system. Proacrosin is present in an inactive zymogen form which is converted to mature enzyme through a series of proteolytic cleavages. The role of acrosin is likely involved in the acrosomal reaction. The role of the acrosome reaction in sturgeon fertilization is paradoxical because of the presence of micropyles in the eggs.

The presence of enzymes other than acrosin-acrosomal in Siberian sturgeon acrosome is unclear. It is possible that arylsulfatase and β-N-acetylglucosaminidase are acrosomal enzymes (Piros et al. 2002); however, direct evidence is not yet available. There are some indirect indications for the presence of both enzymes in the acrosome. Both enzymes are localized in mammalian acrosomes (Nikolajczyk and O'Rand 1992; Brandon et al. 1997). Moreover, arylsulfatase was not found in anacrosomal spermatozoa of teleost fishes. On the other hand, β-N-acetylglucosaminidase is present in teleostean sperm (Sarosiek et al. 2012), which complicates the understanding of its localization and role. Recent data suggest that arylsulfatase, β-N-acetylglucosaminidase, and acid phosphatase are related to sperm fertilizing ability in Siberian sturgeon (Sarosiek et al. 2014; Sarosiek et al. 2015).

Other enzymes identified in spermatozoa (enolase, LDH) are related to metabolism (glycolysis). Their role is likely related to provide energy (ATP) for sperm movement. This can be demanding task for sturgeon sperm, because duration of motility is much longer, compared to teleostean fish spermatozoa (see below).

15.4 Spermatozoa Morphology

Generally, the spermatozoon function in the transport of the male haploid chromosome sets into the oocyte. For this purpose, it is composed of four different compartments—the acrosome, the head, the midpiece, and the flagellum (Callard and Callard 1999; Knobil and Neill 1999). The acrosome has lytic activities so as to enable the entrance of the spermatozoon into the oocyte through egg envelope; the head contains the nucleus and therefore the DNA material. The mitochondria are located in the midpiece, which delivers the energy for flagellar beating. The flagellum itself is the motor of the spermatozoon. The centriolar complex consists of the proximal and the distal centriole whereby the latter is the basal body of the flagellum. This centriolar complex anchors the flagellum at the sperm cell and is normally located in close proximity to the nucleus (Figs. 15.1 and 15.4a).

Fish gametes are diverse in morphology and ultrastructure, including the number and location of different organelles (Baccetti et al. 1984; Baccetti 1986; Jones and Butler 1988). Species-specific morphological and physiological features have been shown in spermatozoa and eggs of several fish species, which reflect

Fig. 15.1 Shows basic structure of surgeon spermatozoa. *A*, acrosome; *PPs*, posterolateral projections; *EC*, endonuclear canals; *IF*, implantation fossa; *N*, nucleus; *PC*, proximal centriole; *DC*, distal centriole; *M*, mitochondria; *CC*, cytoplasmatic canal; *F*, flagellum; *Fi*, fin

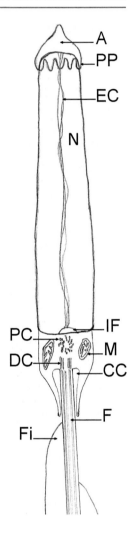

differences in functional capabilities and adaptations for spawning strategy. Morphology and fine structure are considered to be the major sources of information in comparative spermatology (Baccetti 1986; Jamieson 1991; Jamieson 1999; Lahnsteiner and Patzner 1997).

Spermatozoa of sturgeons have characteristic differences compared to those of teleost fishes in terms of morphology (Jamieson 1991), by the presence of acrosomal structure elongate nucleus and a flagellum with fins (Cherr and Clark 1984; Dettlaff et al. 1993; Pšenička et al. 2007; Wei et al. 2007). Moreover, there are also significant differences in parameters of spermatozoa among sturgeon species. Morphological parameters of Siberian sturgeon spermatozoa are compared with other sturgeon species (Table 15.3). It follows that Siberian sturgeon spermatozoa belong to middle-size sturgeon spermatozoa. Although Pšenička et al. (2010a)

Table 15.3 Comparison of morphological parameters of spermatozoa among sturgeon species

Sturgeon species	AL	AW	HL	ANW	PNW	ML	MW	FL	TL	n	References
Beluga	1.12 (0.14)	0.87 (0.10)	7.14 (0.47)	0.69 (0.06)	0.98 (0.08)	2.10 (0.42)	0.61 (0.09)	42.21 (3.82)	51.27 (4.71)	4	Linhartova et al. (2013)
Russian sturgeon	1.18 (0.01)	1.05 (0.01)	6.84 (0.04)	1.10 (0.01)	1.48 (0.01)	1.64 (0.03)	0.92 (0.01)	49.42 (0.37)	57.08 (0.79)	5	Hatef et al. (2012)
Persian sturgeon	1.15 (0.15)	1.16 (0.15)	7.05 (0.51)	1.24 (0.15)	1.54 (0.17)	1.81 (0.46)	0.90 (0.12)	50.31 (5.87)	59.18 (6.23)	5	Hatef et al. (2011)
Sterlet	0.79 (0.07)	0.75 (0.07)	3.30 (0.31)	0.67 (0.07)	0.85 (0.08)	0.97 (0.23)	0.64 (0.12)	42.47 (1.89)	47.61 (1.89)	3	Pšenička et al. (2009)
Chinese sturgeon	0.54 (0.15)	0.68 (0.06)		0.59 (0.05)	1.84 (0.45)	2.17 (0.36)	1.57 (0.27)	33.26 (2.74)	38.7		Wei et al. (2007)
Siberian sturgeon	**0.95 (0.17)**	**0.93 (0.12)**	**4.98 (0.83)**	**0.87 (0.13)**	**1.14 (0.18)**	**1.09 (0.43)**	**0.80 (0.25)**	**44.75 (4.93)**	**51.76**	**8**	Pšenička et al. (2007)
Pallid sturgeon	1.07 (0.10)	0.82 (0.06)	3.78 (0.33)	0.68 (0.04)	0.89 (0.06)	1.23 (0.16)	0.67 (0.08)	37.16	43.23	16	DiLauro et al. (2001)
Lake sturgeon	0.73 (0.14)	0.81 (0.07)	5.69 (0.43)	0.68 (0.07)	1.04 (0.08)	2.68 (0.43)	0.70 (0.08)	47.53	56.63	14	DiLauro et al. (2000)
Shortnose sturgeon	0.78 (0.08)	0.91 (0.06)	6.99 (0.83)	0.75 (0.11)	1.21 (0.12)	1.91 (0.35)	0.81 (0.09)	36.7	46.41	15	DiLauro et al. (1999)
Atlantic sturgeon	0.83 (0.11)	1.00 (0.07)	3.15 (0.36)	0.92 (0.06)	0.55 (0.08)	1.37 (0.16)	0.51 (0.07)	37.08	42.74	12	DiLauro et al. (1998)
White sturgeon	1.31	1.34	9.21	1.25	1.44	2.13	1.08	30-40	41.82-51.82	1	Cherr and Clark (1985)
Stellate sturgeon	0.97	1.22	6.66	0.98	1.49	3.43	1.38	40-70	51.05-81.05	1	Ginsburg (1977)

Measurements are shown as means ± SD (in parentheses) for n sperm. Data are measured in µm. AL, acrosome length; ANW, anterior nucleus width; AW, acrosome width; FL, flagellum length; HL, head length; ML, midpiece length; MW, midpiece width; PNW, posterior nucleus width; TL, total length (head with acrosome, midpiece, and flagella)

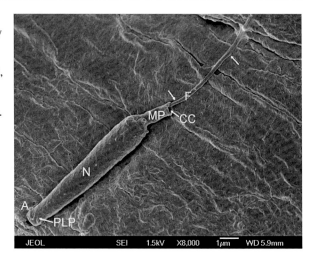

Fig. 15.2 Micrograph from scanning electron microscopy shows acrosome (A), cytoplasmic channel (CC), flagellum (F), midpiece (MP), and nucleus (N). The *arrows* indicate the development of fins along the flagellar length. Scale bar is 1 μm. (Pšenička et al. 2007)

showed highly significant differences, with deviation up to 30%, in size determination of all measured values for one specimen (sterlet spermatozoa) using different electron microscopic methods, they suggested that for correct comparative studies, the samples should be prepared in the same way and at the same time, which unfortunately cannot be done (a posteriori). A scheme of common structure of Siberian sturgeon spermatozoa is illustrated (Fig. 15.1).

15.4.1 The Sturgeon Sperm Head

The shape of Siberian sturgeon sperm head has an elongated trapezoidal shape, tapering from the anterior end to the posterior end, similar to other sturgeon species spermatozoa (Fig. 15.2). The head is composed of the acrosome, endonuclear canals (ECs), an implantation fossa, and a nucleus with electron dense and slightly granular material surrounded by a nuclear membrane.

The acrosomal components include the acrosome, a subacrosome, and posterolateral projections (PPs); the latter are located at the posterior part of acrosome. Numbers and sizes of PPs vary between sturgeon species; the acrosome of Siberian sturgeon possesses a high number (10) and size (940 nm) of PPs (Fig. 15.3a, b) (Pšenička et al. 2007). For comparison, Chinese sturgeon *A. sinensis* has 10 (370 nm) (Wei et al. 2007), and there are 9–10 (295 nm) in sterlet *A. ruthenus* (Pšenička et al. 2009), 8 (760 nm) in pallid sturgeon *Scaphirhynchus albus* (DiLauro et al. 2001), and 7–9 (490 nm) in beluga *Huso huso* (Linhartova et al. 2013).

Ciereszko et al. (1996) found that the trypsin-like activity in sturgeon spermatozoa shares many properties with mammalian acrosin. Acrosin appears to be a widely distributed and conserved protein (Baccetti et al. 1989), which developed a billion years ago during the early period of eukaryote evolution (Klemm et al.

Fig. 15.3 Longitudinal sagittal section (**a**) and cross section (**b**) captured by transmission electron microscopy shows the acrosome (A) and the subacrosome (SA), with the ten posterolateral projections (PLPs) and the endonuclear canals (ENC) traversing the nucleus (N). Scale bar, 500 nm. (Pšenička et al. 2007)

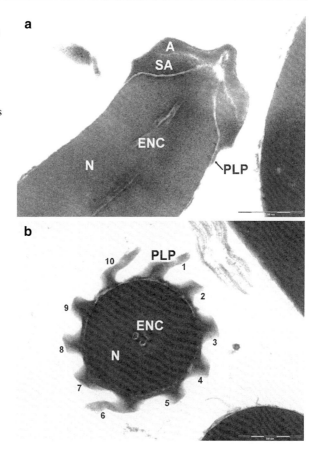

1991). For this reason, the presence of acrosin in sturgeon sperm was expected (see Sect. 15.4.2). Immunolabeling shows that acrosin is present in sturgeon (sterlet) spermatozoa and is localized in the acrosome and the implantation fossa. In addition, scanning electron microscopy on cryofracture shows evidence of the opening of ECs to the acrosome and implantation fossa as well. That means that the acrosome is connected to the implantation fossa by the ECs. Therefore it is suggested that the ECs and implantation fossa are components of the acrosome (Pšenička et al. 2009).

Except for the Atlantic sturgeon (DiLauro et al. 1998), which has two ECs, all other sturgeon species have three ECs in their spermatozoa (Ginsburg 1977; Cherr and Clark 1984; DiLauro et al. 2000, 2001; Pšenička et al. 2007; Wei et al. 2007; Linhartova et al. 2013). Pšenička et al. (2008a) also found that some spermatozoa of Siberian sturgeon had two to four ECs. There also are differences in the canal diameters between sturgeon species, with values of 35, 97, 49, 57, 40, and 44 nm in Atlantic, shortnose, lake, pallid, sterlet, and Siberian sturgeons, respectively.

Fig. 15.4 Longitudinal sagittal section (**a**) and cross section (**b**) of the midpiece captured by transmission electron microscopy shows the distal (DC) and proximal (PC) centrioles in the implantation fossa (IF) with triplets of microtubules (TM). Numerous mitochondria (M) and vesicles (V) were irregularly dispersed in the cytoplasm. The nucleus (N) is surrounded by the nuclear membrane (NM). The flagellum (F) is separated by the cytoplasmic channel (CC). Scale bars are 500 nm (**a**) and 200 mm (**b**). (Pšenička et al. 2007)

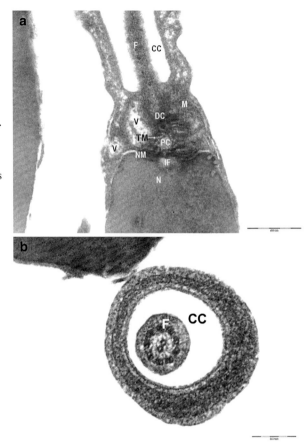

15.4.2 The Sturgeon Sperm Midpiece

The midpiece has a cylindrical shape and an elongated caudal base. Siberian sturgeon spermatozoa contain from three to six mitochondria in a peripheral section of the midpiece. The mitochondria provide energy in the form of ATP for sperm movement. The proximal centriole (185.28 nm × 147.42 nm) is posteriorly the implantation fossa in nucleus. Both centrioles are composed of nine peripheral triplets of microtubules in a cylindrical shape. The axoneme is formed as an extension of the distal centriole. Around the flagellum, there is a plasma membrane with cytoplasmic channel, which is formed by an invagination of the membrane. There is an extracellular space between the cytoplasmic sheath and the flagellum called cytoplasmic channel (Fig. 15.4a, b) (Pšenička et al. 2007).

Fig. 15.5 Cross section of the flagellum (**a**) and axoneme (**b**) captured by transmission electron microscopy shows the peripheral doublets of microtubules (PDM) and central doublets of microtubules (CDM) with radial joins (RJ) and a bridge (B). The propelling machinery is the internal (IDA) and external (EDA) dynein arms. The fins (F) make the flagellum more effective. Scale bars are 200 nm (**a**) and 50 nm (**b**). (Pšenička et al. 2007)

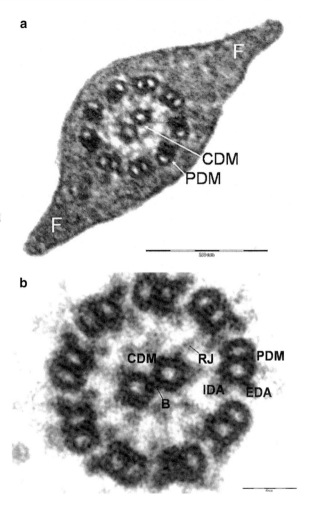

15.4.3 The Sturgeon Flagellum

In sturgeon, the fibrillar part of the flagellum, called axoneme, consists of nine peripheral doublets and a central pair of single microtubules. The pair of central microtubules are linked by bridges and they are encased in a central sheath. Nine radial spokes, or joints, connect the two central and peripheral doublets. From the end of the cytoplasmatic sheath of the midpiece to the terminal region of the flagellum, two independent lateral extensions of the flagellar plasma membrane gradually taper to form the fins (Figs. 15.2 and 15.5a, b). These fin structures are oriented along the horizontal plane, parallel to the central microtubules (Billard 1970), and probably help to increase the efficiency of wave propagation (Cosson et al. 2000). In silver salmon *Oncorhynchus kisutch*, the spermatozoon tail has the membrane in central part formed into a spiral with 12–15 coils (Lowman 1953), but Pšenička et al. (2007) described this fin as (straight) in Siberian sturgeon spermatozoa, indicating no spiral (rotation), but exactly parallel to the plane of the two central microtubules of the flagellum. This means that the central microtubules also are orientated in one plane. The fins are not formed the same on either side of the flagellum. The first and second fins start 0.7 and 5.3 μm post-midpiece, respectively, and continue along the flagellum. The first and second fins end 3.4 and 5.1 μm from the end of the flagellum, respectively. They extend distally up to 705.87 ± 220.04 nm in the middle part of flagellum. The sperm flagellar membrane in fishes with this unusual fin-shaped structure also significantly increases the membrane surface area, thus, favoring increased water exchange. This has considerable implications for water exchange/osmotic regulation at activation (Cosson et al. 2000; Pšenička et al. 2007).

The length of the flagellum and content of mitochondria are positively correlated with the swimming velocity in tench spermatozoa (Pšenička et al. 2010b). Pšenička et al. (2008a) compared sperm characteristics in Siberian and sterlet sturgeons. There also was a correlation between the flagellum length and velocity, and therefore motility, especially at the end of motility.

15.5 Sperm Motility Characteristics

Sperm motility of sturgeons differs markedly from that of teleosts. Duration of sturgeon sperm movement is longer, for example, about 2–3 min for Siberian sturgeon, and also the number of active spermatozoa regularly declines to 5–10% at 2–3 min post-activation (Billard et al. 1999). During this period, the flagellum beat frequency is stable at about 50 Hz for 30 s and then drops to about 30 Hz after 60 s. Beat frequency is affected by cryopreservation; flagellar waves are asymmetrical in post-thawed spermatozoa (Billard et al. 2000). During the motile period, a constant decline in ATP concentration takes place, but it seems that the efficiency of swimming performance of Siberian sturgeon spermatozoa is also improved by the fins. Swimming performance of Siberian sturgeon spermatozoa is facilitated by the presence of the paddle-like fins that extend along most of flagella length (Gillies et al. 2013).

Initiation of sperm motility consists of a cascade of events, starting from a decrease in osmolality and potassium ion concentrations, which result in an efflux of these ions from spermatozoa (Alavi et al. 2012b). Next, membrane hyperpolarization occurs, followed by Ca^{2+} ion influx and membrane depolarization. A rise in Ca^{2+} and calmodulin induces initiation of sperm movement through the activation of phosphodiesterase.

Due to low osmolality of sturgeon semen (see Table 15.2), activating solutions are characterized by low osmolality as well; for example, Tsvetkova et al. (1996) used 50 mM Tris-HCl, pH 8.0, and Billard et al. (1999) 30 mM Tris-HCl, pH 8.0,

15 Sperm and Spermatozoa Characteristics in the Siberian Sturgeon

Table 15.4 Sperm motility parameters of Siberian sturgeon semen

Feature	Mean ± sd	Reference
Motility (%)		
Before activation	0.47 ± 1.1	Glogowski et al. (2002)
	2.4 ± 2.5	Sieczyński et al. (2012)
After activation	41.3 ± 14.1	Sieczyński et al. (2012)
	61.7 ± 17.6	Glogowski et al. (2002)
	70	Urbanyi et al. 2004
	76	Sarosiek et al. (2004)
	80	Judycka et al. (2015b)
	88 ± 4.4	Tsvetkova et al. (1996)
	90	Billard et al. (1999)
	95–100	Shaliutina et al. 2013
	95.2	Pšenička et al. (2011)
	100	Pšenička et al. (2008a)
	100	Judycka et al. (2015a)
Duration of motility	2 min	Shaliutina et al. (2013)
	2 min.	Pšenička et al. (2008a)
Velocity (μm s^{-1})	123.7 ± 42.6	Li et al. (2011)
	165 ± 14	Shaliutina et al. (2013)
	181.0 ± 3	Pšenička et al. (2011)
VCL (μm s^{-1})	121.9 ± 17.3	Sieczyński et al. (2012)
	311.5 ± 12.9	Judycka et al. (2015a)
VAP (μm s^{-1})	97.6 ± 14.0	Sieczyński et al. (2012)
	292.6 ± 16.1	Judycka et al. (2015a)
VSL (μm s^{-1})	82.5 ± 14.3	Sieczyński et al. (2012)
	237.5 ± 17.9	Judycka et al. (2015a)
LIN (%)	58.56 ± 11.0	Sieczyński et al. (2012)
	72.9 ± 4.8	Judycka et al. (2015a)
ALH (μm)	10.4 ± 2.7	Sieczyński et al. (2012)
	2.3 ± 0.2	Judycka et al. (2015a)

for motility analysis of Siberian sturgeon spermatozoa. Values of pH for optimal movement are 8.0 and higher; these values resemble pH of seminal plasma. Sperm motility parameters are shown in Table 15.4. Although spermatozoa remain quiescent under inhibitory action of K^+ ions, some motility in undiluted semen has been observed, possibly due to mixture with urine. Recently, Judycka et al. (2015a) established that spontaneous motility of undiluted semen may be related to season. These authors observed a 40% spontaneous sperm motility (estimated subjectively by one experimenter, as in Williot et al. 2000) in samples collected in December. In this context, CASA analysis (computer-assisted sperm analysis) showed that 1/3 of the sperm population obtained in December was motile upon activation in seminal plasma compared to only single spermatozoon in semen from April.

Parametric data for sperm motility obtained with CASA are now available for Siberian sturgeon. This system is capable of measuring several parameters related to speed and trajectory of movement. The most popular parameters for characterizing spermatozoa are straight-line velocity (VSL), curvilinear velocity (VCL), average path velocity (VAP), linearity (LIN = 100 × VSL/VCL), amplitude of lateral head displacement (ALH) and percentage of motile sperm. Values of sperm motility parameters clearly indicate, that the percentage of motility can reach 100%, which translates into elevated values of sperm velocity (up to 300 μm s^{-1}) and high linearity of movement (LIN values 70% and higher). These values are lower for poor-quality semen as can be demonstrated by comparison of the data of Judycka et al. (2015a) with data of Sieczyński et al. (2012); they recorded 100% sperm motility after activation versus 41% of sperm motility, respectively.

15.6 The Acrosome Reaction

In general, the typical organelle of many vertebrate spermatozoon, the acrosome, is considered to be responsible for enabling the spermatozoon to traverse the investment coats surrounding an egg. Most spermatozoa undergo an acrosomal reaction in response to an egg component; this acrosomal reaction exposes the contents of the acrosome, which includes enzymes and binding proteins (Dan 1967; Shapiro and Eddy 1980; Lopo 1983). Tunicate, lampreys, hagfish (*Eptatretus burgeri* and *E. stouti*), and sturgeons are reported to possess sperm that form fertilization filaments during the acrosomal reaction (Cherr and Clark 1984; Dettlaff et al. 1993; Morisawa 1995, 1999a, 1999b; Pšenička et al. 2009). Lamprey eggs are covered by an impenetrable envelope with no micropyles (Kille 1960). Morisawa and Cherr (2002) described the acrosomal reaction in hagfish, where the spermatozoa must penetrate a U-shape layer that fills the bottom of the micropyle (Morisawa 1999b). However, sturgeon eggs possess an impenetrable envelope, which is perforated by numerous micropyles (3–16 in Siberian sturgeon, Debus et al. 2002). The latter provide spermatozoon direct access to the oolemma. The multiplicity of micropyles could be advantageous to facilitate egg location by spermatozoon in fast-flowing water; however, at the same time, it also increases the risk of polyspermy. Therefore, the presence of acrosomes and long acrosomal processes in sturgeon spermatozoa is inconclusive, and the contradictions to others do not agree with models of acrosomal evolution (Baccetti and Afzelius 1976; Baccetti 1979). The micropyles of sturgeon eggs are filled with spermatozoa within about 15 s postfertilization. The spermatozoon, which reached the bottom vicinity of the micropyle, merges with a cytoplasmic projection of the egg that is located in the micropyle. The membrane of the acrosome decomposes, starting from the apical part. During the acrosomal reaction, material in the three ECs of the sperm head is ejected forward through the acrosome like a fine harpoon-like acrosomal filament (Fig. 15.6a–c; Pšenička et al. 2010c). With reference to multiplicity of micropyles, it has been speculated that the filament serves as a quick signal transducer, conveying information to the egg about the presence of spermatozoon in close proximity, and induces a cytoplasmic projection from the egg for fusion with the spermatozoon and blocking polyspermy through the formation of a perivitelline space in all other micropyles (Pšenička et al. 2010c).

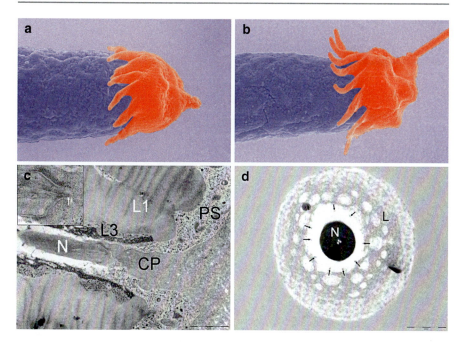

Fig. 15.6 Micrographs show acrosome before (**a**) and after (**b**) acrosomal process (scale bar, 100 nm); spermatozoon into the egg 60 s after fertilization showing the fusion of spermatozoon acrosome and nucleus (N) with egg cytoplasmic projection (CP) and the formation of perivitelline space (PS) under egg layers (L1 and L3). Inset: formation of fertilization filament (*arrow*) (**c**, scale bar, 2 μm); cross section at the level of ten extended posterolateral projections (*arrows*) of spermatozoon in egg micropyle with nucleus (N) and egg layer (L) (**d**, scale bar, 1 μm) (Pšenička et al. 2010c)

15.6.1 Function of Posterolateral Projections

The PPs of the acrosomes of sturgeon spermatozoa are radially distributed lobes. The diameter of the micropylar base in Siberian sturgeon eggs (1.95 μm) is only slightly larger than the diameter of the spermatozoon's head (1.13 μm). As a result, no more than one sperm can enter the micropyle. When a spermatozoon of Siberian sturgeon penetrates through the micropyle into the egg, the ten PPs open and function as an anchor; this is a unique feature of sturgeon spermatozoa. To confirm the opening of the PPs, the distance of PPs from the nucleus was measured in cross sections in the PP region of in vitro activated spermatozoa (0.19 ± 0.12 μm) and nonactivated spermatozoa (0.14 ± 0.04 μm). The difference was highly significant ($p < 0.01$) as well as the distance of PPs of spermatozoa in micropyles (0.20 ± 0.16 μm) (Fig. 15.6a, b, d; Pšenička et al. 2010c).

15.6.2 Prevention of Polyspermy

Immediately after fusion, when the first spermatozoon penetrates the oocyte plasma membrane, the entire egg envelope separates from the oocyte plasma membrane by

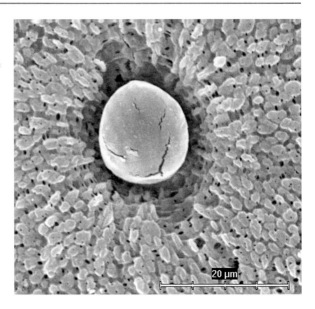

Fig. 15.7 Scanning electron micrograph shows fertilization cone in the micropyle 180 s after fertilization. Scale bar, 20 μm

the circumferential spreading of a perivitelline space which is filled with the material from cortical granules; this agglutinates other spermatozoa. The perivitelline space expands around the oocyte from the site of sperm attachment to the vegetal part of the oocyte (Fig. 15.6c) (Cherr and Clark 1985; Dettlaff et al. 1993; Pšenička et al. 2010c).

Mature eggs of several fish species respond to sperm entry by the formation of fertilization cone at the sperm entry site (Kudo 1980; Kobayashi and Yamamoto 1981; Iwamatsu et al. 1991; Linhart and Kudo 1997). In Siberian sturgeon, approximately 60 s postfertilization, the fertilization cone extends to about 20–30 μm in width, with the ball-like enlarged apex of the cone which reaches the micropylar vestibule. Spermatozoa are never able to fuse with the cone membrane (Fig. 15.7, Pšenička et al. 2010c).

15.6.3 Monitoring and Importance of the Acrosomal Reaction

Pšenička et al. (2008b) tested a soybean trypsin inhibitor conjugated with Alexa Fluor® 488 for screening of acrosomal reaction in sturgeon spermatozoa. This method seems to be the most objective for acrosomal integrity evaluation (see chapter Siberian sturgeon sperm cryoconservation). There were differences in the integrity of acrosome in sperm within males. Moreover, while using this staining, suitable acrosome activation medium containing 2.5 mM Ca^{2+} and 15 mM tris with pH 10 was selected for sturgeon spermatozoa. Sperm was treated with the medium and used for fertilization. The hatching percentage rapidly decreased and in addition was negatively correlated within males with percentage of disrupted acrosomes; nevertheless, the ability of movement was sustained. These results strongly suggest a usefulness of acrosome for fertilization and development in sturgeon (Pšenička et al. 2008b).

Conclusions

The characteristics of Siberian sturgeon semen are considerably affected by dilution with urine in the Wolffian ducts. The concentrations of spermatozoa, proteins, ions, and, therefore, osmolality are lower than in teleostean species, and semen volume is higher. Moreover, the structure of spermatozoa as well as fertilization process is unique among fishes. Sturgeon spermatozoa possess an acrosome which contains typical acrosomal proteins, and while sperm penetration is facilitated by a number of egg micropyles which disagrees with evolutionary biology in fish gametes. It is suggested that the acrosome with the posterolateral projections has a more likely function of an anchor within micropyle instead of a mechanism for penetration.

Acknowledgments The study was financially supported by COST Office (Food and Agriculture COST Action FA1205: AQUAGAMETE); by the Ministry of Education, Youth and Sports of the Czech Republic, projects "CENAKVA" (No. CZ.1.05/2.1.00/01.0024) and "CENAKVA II" (No. LO1205 under the NPU I program); and by the Czech Science Foundation (No. P502/13/26952S), the National Science Centre granted for research project (No. 2011/01/D/NZ9/03738) and funds appropriated to Institute of Animal Reproduction and Food Research. Authors also express thanks to prof. MSc. William L. Shelton, PhD, for English corrections.

References

Alavi SMH, Hatef A, Pšenička M et al (2012b) Sperm biology and control of reproduction in sturgeon: (II) sperm morphology, acrosome reaction, motility and cryopreservation. Rev Fish Biol Fisher 22:861–886

Alavi SMH, Rodina M, Gela D et al (2012a) Sperm biology and control of reproduction in sturgeon: (I) testicular development, sperm maturation and seminal plasma characteristics. Rev Fish Biol Fisher 22:695–717

Baccetti B (1979) The evolution of the acrosomal complex. In: Fawcett DW, Bedford JM (eds) The spermatozoon. Urban and Schwarzenberg, Baltimore-Munich, pp 305–329

Baccetti B (1986) Evolutionary trends in sperm structure. Comp Biochem Physiol A 85:29–36

Baccetti B, Afzelius BA (1976) The biology of the sperm cell. Monogr Dev Biol 10:1–4

Baccetti B, Burrini AG, Callaini G et al (1984) Fish germinal cell. I. Comparative spermatology of seven cyprinid species. Gamete Res 10:373–396

Baccetti B, Burrini AG, Collodel G et al (1989) Localization of acrosomal enzymes in arthropoda, echinodermata and vertebrata. J Submicrosc Cytol Pathol 21(2):385–389

Billard R (1970) Ultrastructure comparée de spermatozoides de quelques poissons téléostéens. In: Baccetti B (ed) Comparative spermatology. Academic Press, New York, pp 71–79

Billard R, Cosson J, Fierville F et al (1999) Motility analysis and energetics of the Siberian sturgeon *Acipenser baerii* spermatozoa. J Appl Ichthyol 15:199–203

Billard R, Cosson J, Linhart O (2000) Changes in the flagellum morphology of intact and frozen/thawed Siberian sturgeon *Acipenser baerii* (Brandt) sperm during motility. Aquac Res 31:283–287

Brandon CI, Srivastava PN, Heusner GL et al (1997) Extraction and quantification of acrosin, β-Nacetylglucosaminidase, and arylsulfatase-a from equine ejaculated spermatozoa. J Exp Zoo 279:301–308

Callard GV, Callard IP (1999) Spermatogenesis in nonmammals. In: Knobil E, Neill JD (eds) Encyclopedia of reproduction IV. Academic Press, New York, pp 563–570

Cherr GN, Clark WH (1984) An acrosome reactions in sperm from the white sturgeon, *Acipenser transmontanus*. J Exp Zool 232:129–139

Cherr GN, Clark WH (1985) Gamete interaction in the white sturgeon *Acipenser transmontanus*: a morphological and physiological review. Environ Biol Fish 14:11–22

Ciereszko A, Dabrowski K, Lin F et al (1994) Identification of trypsin-like activity in sturgeon spermatozoa. J Exp Zool 268:486–491

Ciereszko A, Dabrowski K, Ochkur SI (1996) Characterization of acrosin-like activity of lake sturgeon (*Acipenser fulvescens*) spermatozoa. Mol Reprod Dev 45:72–77

Cosson J, Linhart O, Mims S et al (2000) Analysis of motility parameters from paddlefish (*Polyodon spathula*) and shovelnose sturgeon (*Scaphirhynchus platorynchus*) spermatozoa. J Fish Biol 56:1348–1367

Dan JC (1967) Acrosome reaction and lysins. In: Metz CB, Monroy A (eds) Fertilization, comparative morphology, biochemistry and immunology. Academic Press, New York, pp 237–288

Debus L, Winkler M, Billard R (2002) Structure of micropyle surface on oocytes and caviar grains in sturgeons. Internat Rev Hydrobiol 87(5-6):585–603

Dettlaff TA, Ginsburg AS, Schmalhausen OI (1993) Sturgeon fishes. Developmental biology and aquaculture. Springer-Verlag, Berlin

DiLauro MN, Kaboord W, Walsh RA (1998) Sperm-cell ultrastructure of north American sturgeons. I. The Atlantic sturgeon (*Acipenser oxyrhynchus*). Can J Zool-Rev Can Zool 76:1822–1836

DiLauro MN, Kaboord WS, Walsh RA (1999) Sperm-cell ultrastructure of north American sturgeons. II. The shortnose sturgeon (*Acipenser brevirostrum*, Lesueur, 1818). Can J Zool-Rev Can Zool 77:321–330

DiLauro MN, Kaboord WS, Walsh RA (2000) Sperm-cell ultrastructure of north American sturgeon. III. The Lake sturgeon (*Acipenser fulvescens* Rafinesque, 1817). Can J Zool 78:438–447

DiLauro MN, Walsh RA, Peiffer M (2001) Sperm-cell ultrastructure of north american sturgeons. IV. The pallid sturgeon (*Scaphirhynchus albus* Forbes and Richardson, 1905). Can J Zool-Rev Can Zool 79:802–808

Gallis JL, Fedrigo E, Jatteau P et al (1991) Siberian sturgeon spermatozoa: effects of dilution, pH, osmotic pressure, sodium and potassium ions on motility. In: Williot P (ed) Acipenser. Cemagref, Antony, France, pp 143–151

Gillies EA, Bondarenko V, Cosson J et al (2013) Fins improve the swimming performance of fish sperm: a hydrodynamic analysis of the Siberian sturgeon *Acipenser baerii*. Cytoskeleton 70:85–100

Ginsburg AS (1977) Fine structure of the spermatozoon and acrosome reaction in *Acipenser stellatus*. In: Beljaev DK (ed) Problemy eksperimentalnoj biologii. Nauka, Moscow, pp 246–256

Glogowski J, Kolman R, Szczepkowski M et al (2002) Fertilization rate of Siberian sturgeon (*Acipenser baerii*, Brandt) milt cryopreserved with methanol. Aquaculture 211:367–373

Hatef A, Alavi SMH, Noveiri SH et al (2011) Morphology and fine structure of *Acipenser percius* (*Acipenseridae, Chondrostei*) spermatozoon: interspecies comparison in *Acipensiformes*. Anim Reprod Sci 123:81–88

Hatef A, Alavi SMH, Rodina M et al (2012) Morphology and fine structure of the Russian sturgeon, *Acipenser gueldenstaedtii* (*Acipenseridae, Chondrostei*) spermatozoa. J Appl Ichthyol 28:978–983

Iwamatsu T, Onitake K, Yoshimoto Y, Hiramoto Y (1991) Time sequence of early events in fertilization in the medaka egg. Develop Growth Differ 33:479–490

Jamieson BGM (1991) Fish evolution and systematics: evidence from spermatozoa. Cambridge University Press, Cambridge

Jamieson BGM (1999) Spermatozoal phylogeny of the vertebrata. In: Gagnon C (ed) The male gamete: from basic science to clinical applications. Cache River Press, Vienna, pp 303–331

Jones PR, Butler RD (1988) Spermatozoon ultrastructure of *Platichthys flesus*. J Ultra Mol Struct R 98:71–82

Judycka S, Szczepkowski M, Ciereszko A et al (2015a) Characterization of Siberian sturgeon (*Acipenser baerii* Brandt) sperm obtained out of season. J Appl Ichthyol 31:34–40

Judycka S, Szczepkowski M, Ciereszko A et al (2015b) New extender for cryopreservation of Siberian sturgeon (*Acipenser baerii*) semen. Cryobiology 70:184–189

Kille RA (1960) Fertilization of lamprey egg. Exp Cell Res 20:12–27

Klemm U, Mulleresterl W, Engel W (1991) Acrosin, the peculiar sperm-specific serine protease. Hum Genet 87(6):635–641

Knobil E, Neill D (1999) Encyclopaedia of reproduction. Academic Press, New York

Kobayashi W, Yamamoto TS (1981) Fine structure of micropylar apparatus of the chum salmon egg, with a discussion of the mechanism for blocking polyspermy. J Exp Biol 217:265–275

Kudo S (1980) Sperm penetration and the formation of a fertilization cone in the common carp egg. Develop Growth Differ 22:403–414

Lahnsteiner F, Patzner RA (1997) Fine structure of spermatozoa of four littoral teleosts Symphodus ocellatus, *Coris julis, Thalassoma pavo* and *Chromis chromis*. J Submicr Cytol Path 29:477–485

Li P, Hulak M, Rodina M et al (2010) Comparative protein profiles: potential molecular markers from spermatozoa of *Acipenseriformes (Chondrostei, Pisces)*. Comp Biochem Physiol D 5:302–307

Li P, Rodina M, Hulák M et al (2011) Physico-chemical properties and protein profiles of sperm from three freshwater chondrostean species: a comparative study among Siberian sturgeon (*Acipenser baerii*), sterlet (*Acipenser ruthenus*) and paddlefish (*Polyodon spathula*). J Appl Ichthyol 27:673–677

Linhart O, Kudo S (1997) Surface ultrastructure of paddlefish eggs before and after fertilization. J Fish Biol 51:573–582

Linhartova Z, Rodina M, Nebesarova J et al (2013) Morphology and ultrastructure of beluga (*Huso huso*) spermatozoa and a comparison with related sturgeons. Anim Reprod Sci 137:220–229

Lopo AC (1983) Sperm-egg interaction in invertebrates. In: Hartman JF (ed) Mechanisms and control fertilization. Academic Press, New York, pp 269–324

Lowman FG (1953) Electron microscope studies of silver salmon spermatozoa (*Oncorhynchus kisutch* W.) Exp Cell Res 5:335–360

Morisawa S (1995) Fine structure of spermatozoa of the hagfish *Eptatretus burgeri* (Agnatha). Biol Bull 189:6–12

Morisawa S (1999a) Acrosome reaction in spermatozoa of the hagfish *Eptatretus burgeri* (Agnatha). Develop Growth Differ 41:109–112

Morisawa S (1999b) Fine structure of micropylar region during late oogenesis in eggs of the hagfish *Eptatretus burgeri* (Agnatha). Develop Growth Differ 41:611–618

Morisawa S, Cherr GN (2002) Acrosome reaction in spermatozoa from hagfish (Agnatha) *Eptatretus burgeri* and *Eptatretus stouti*: acrosomal exocytosis and identification of filamentous actin. Develop Growth Differ 44:337–344

Morisawa S, Morisawa M (1988) Induction of potential for sperm motility by bicarbonate and pH in rainbow trout and chum salmon. J Exp Biol 136:13–22

Nikolajczyk BS, O'Rand MG (1992) Characterization of rabbit testis β-galactosidase and arylsulfatase-a – purification and localization in spermatozoa during the acrosome reaction. Biol Reprod 46:366–378

Piros B, Glogowski J, Kolman R et al (2002) Biochemical characterization of Siberian sturgeon *Acipenser baerii* and sterlet *Acipenser ruthenus* milt plasma and spermatozoa. Fish Physiol Biochem 26:289–295

Pšenička M, Alavi SMH, Rodina M et al (2007) Morphology and ultrastructure of Siberian sturgeon, *Acipenser baerii*, spermatozoa using scanning and transmission electron microscopy. Biol Cell 99(2):103–115

Pšenička M, Alavi SMH, Rodina M et al (2008a) Morphology, chemical contents and physiology of chondrostean fish sperm: a comparative study between Siberian sturgeon (*Acipenser baerii*) and sterlet (*Acipenser ruthenus*). J Appl Ichthyol 24:371–377

Pšenička M, Cosson J, Alavi SMH et al (2008b) Staining of sturgeon spermatozoa with trypsin inhibitor from soybean, Alexa Fluor® 488 conjugate for visualization of sturgeon acrosome. J Appl Ichthyol 24:514–516

Pšenička M, Flajšhans M, Hulák M et al (2010b) The influence of ploidy level on ultrastructure and motility of tench (*Tinca tinca* L.) spermatozoa. Rev Fish Biol Fish 20(3):331–338

Pšenička M, Kašpar V, Rodina M et al (2011) Comparative study on ultrastructure and motility parameters of spermatozoa of tetraploid and hexaploid Siberian sturgeon *Acipenser baerii*. J Appl Ichthyol 27:683–686

Pšenička M, Rodina M, Linhart O (2010c) Ultrastructural study on fertilization process in sturgeon (*Acipenser*), function of acrosome and prevention of polyspermy. Anim Reprod Sci 117:147–154

Pšenička M, Těšitel J, Tesařová M et al (2010a) Size determination of *Acipenser ruthenus* spermatozoa in different types of electron microscopy. Micron 41:455–460

Pšenička M, Vancová M, Koubek P et al (2009) Fine structure and morphology of sterlet (*Acipenser ruthenus* L. 1758) spermatozoa and acrosin localization. Anim Reprod Sci 111:3–16

Sarosiek B, Ciereszko A, Rzemieniecki A et al (2004) The influence of semen cryopreservation on the release of some enzymes from Siberian sturgeon (*Acipenser baerii*) and sterlet (*Acipenser ruthenus*) spermatozoa. Arch Pol Fish 12:13–21

Sarosiek B, Dryl K, Judycka S et al (2015) Influence of acid phosphatase and arylsulfatase inhibitor additions on fertility rate of Siberian sturgeon (*Acipenser baerii* Brandt, 1869). J Appl Ichthyol 31:154–148

Sarosiek B, Glogowski J, Cejko BI et al (2014) Inhibition of β-N-acetylglucosaminidase by acetamide affects sperm motility and fertilization success of rainbow trout (*Oncorhynchus mykiss*) and Siberian sturgeon (*Acipenser baerii*). Theriogenology 81:723–732

Sarosiek B, Kowalski RK, Dryl K et al (2012) Isolation and characteristics of beta-N-acetylglucosaminidase present in rainbow trout (*Oncorhynchus mykiss*) and Siberian sturgeon (*Acipenser baerii*) milt. J Appl Ichthyol 28:984–989

Sarosiek B, Kowalski R, Glogowski J (2008) Isolation and preliminary characteristics of beta-N-acetylglucosaminidase in the sperm of Siberian sturgeon (*Acipenser baerii*) and rainbow trout (*Oncorhynchus mykiss*). J Appl Ichthyol 24:492–496

Shaliutina A, Hulak M, Gazo I et al (2013) Effect of short-term storage on quality parameters, DNA integrity, and oxidative stress in Russian (*Acipenser gueldenstaedtii*) and Siberian (*Acipenser baerii*) sturgeon sperm. Anim Reprod Sci 139(1-4):127–135

Shapiro BM, Eddy EM (1980) When sperm meets egg: biochemical mechanisms of gamete interaction. Int Rev Cytol 66:257–302

Sieczyński P, Glogowski J, Cejko B et al (2012) Characteristics of Siberian sturgeon and sterlet sperm motility parameters compared using CASA. Arch Pol Fish 20:137–143

Słowińska M, Liszewska E, Dietrich GJ et al (2015) Effect of season on proteases and serine protease inhibitors of Siberian sturgeon (*Acipenser baerii* Brandt, 1869) semen. J Appl Ichthyol 31:125–131

Toth GP, Ciereszko A, Christ SA et al (1997) Objective analysis of sperm motility in the lake sturgeon, *Acipenser fulvescens*: activation and inhibition conditions. Aquaculture 154:337–348

Tsvetkova LI, Cosson J, Linhart O et al (1996) Motility and fertilizing capacity of fresh and frozen-thawed spermatozoa in sturgeons *Acipenser baerii* and *A. ruthenus*. J Appl Ichthyol 12:107–112

Urbanyi B, Horvath A, Kovacs B (2004) Successful hybridization of *Acipenser* species using cryopreserved sperm. Aquacult Int 12:47–56

Wei Q, Li P, Pšenička M et al (2007) Ultrastructure and morphology of spermatozoa in Chinese sturgeon (*Acipenser sinensis* gray 1835) using scanning and transmission electron microscopy. Theriogenology 67:1269–1278

Williot P, Kopeika EF, Goncharov BF (2000) Influence of testis state, temperature and delay in semen collection on spermatozoa motility in the cultured Siberian sturgeon (*Acipenser baerii* Brandt). Aquaculture 189:53–61

Gonadal Steroids: Synthesis, Plasmatic Levels and Biological Activities in Sturgeons

16

Denise Vizziano-Cantonnet

Abstract

The information on sex steroids in sturgeons is fragmented and comes from different species. In females, circulating oestradiol-17β increases during vitellogenesis and resulted a good marker of puberty in some species. However, researchers failed to induce oocyte growth and vitellogenesis using treatments with oestradiol-17β. Androgens also increase during this period but remained elevated during ovulation and post-ovulation suggesting a physiological role at peri-ovulatory period. 17,20,21P seems to be a good candidate as mediator of gonadotropin to induce follicle maturation since it is produced *in vitro* by sturgeons ovaries, it has a high potency to induce follicle maturation *in vitro*, and its plasmatic concentration increases after hormonal induction of ovulation; however, other C21 steroids as 17,21-dihydroxy-4-pregnene-3,20-dione (17,21P), 17,20β-dihydroxy-4-pregnen-3-one (17,20βP) and 17,20β-dihydroxy-4-pregnen-3-one (17,20βP) need to be further investigated as possible maturation-inducing steroids. Studies on potency of C21 and C19 steroids to induce resumption of meiosis revealed that C21 steroids are more potent than testosterone and that 11-oxygenated androgens do not induce maturation. Aromatase expression in immature males and plasmatic changes of testosterone and oestradiol-17β suggest their participation in early testicular development. 11-Ketotestosterone increased significantly during spermatogenesis both *in vitro* and *in vivo* as it has been shown in teleosts. The C21 steroid (17,20βP and 17,20,21P) increases in blood plasma after hormonal induction of spermiation suggesting their participation in the control of sperm maturation and release. Biological activity of steroids and *in vitro* steroid production by gonads are almost none studied in sturgeons.

Keywords

Sturgeons · Sex steroids · Steroidogenesis · Plasma concentrations · Gonads
Acipenser baerii

D. Vizziano-Cantonnet
Facultad de Ciencias, Laboratorio de Fisiología de la Reproducción y Ecología de Peces, Iguá 4225, Montevideo 11400, Uruguay
e-mail: vizziano@gmail.com

Introduction

Sex steroid hormones play important roles in the regulation of different processes as sex differentiation, metabolism, immune responses, circadian rhythms, stress response and reproduction of vertebrates. This chapter is focused on gonadal steroids involved in the control of reproduction in sturgeons. A general introduction using the information on teleost fish will be done in order to have a reference to review information reported in sturgeons. The meaning of common abbreviations of steroids used in text is shown in the Table 16.1. The name of enzymes is abbreviated, and the corresponding name of gene is given in the text, and meaning is shown in Table 16.2.

In teleost fish steroid hormones, i.e. oestradiol-17β, testosterone, 11-ketotestosterone (11KT) and 17,20β-dihydroxy-4-pregnen-3-one (17,20βP), are highly produced by ovaries and testis under control of gonadotropins and are key factors regulating gametogenesis (see reviews from Billard et al. 1982; Fostier et al. 1983; Scott and Sumpter 1983; Scott and Canario 1987; Jalabert et al. 1991; Nagahama 1994; Borg 1994; Schulz and Miura 2002; Vizziano et al. 2007; Scott et al. 2010; Schulz et al. 2010).

During early gonad development, Yamamoto (1969) proposed that oestrogens are the natural inducers of female differentiation and androgens the natural inducers of male differentiation. More than 45 years of research suggest that oestrogens play an important endogenous role in the control of female teleost sex differentiation (Baroiller et al. 1999; Piferrer and Guiguen 2008; Guiguen et al. 2010). Nevertheless androgens seem not to be the universal endogenous regulators of male teleost sex

Table 16.1 Meaning of common abbreviations of steroids used in text following Kime (1995)

Abbreviation	Full name	Other abbreviations or names
P5	3β-Hydroxy-5-pregnen-20-one	Pregnenolone
P	4-Pregnene-3,20-dione	Progesterone
17P	17-Hydroxy-4-pregnene-3,20-dione	17-Hydroxyprogesterone
17P5	3β,17-Dihydroxy-5-pregnen-20-one	17-Hydroxypregnenolone
17,20ßP	17,20ß-Dihydroxy-4-pregnen-3-one	DHP
17,20αP	17,20α-Dihydroxy-4-pregnen-3-one	
17,21P	17,21-Dihydroxy-4-pregnene-3,20-dione	11-Deoxycortisol
17,20,21P	17,20β,21-Trihydroxy-4-pregnen-3-one	20β-S
cortisol	11β,17,21-Trihydroxy-4-pregnene-3,20-dione	Hydrocortisone
A	4-Androstene-3,17-dione	Androstenedione
T	17ß-Hydroxy-4-androsten-3-one	Testosterone
11βOHT		11β-Hydroxytestosterone
11KT	17ß-Hydroxy-4-androstene-3,11-dione	11-Ketotestosterone
11βOHA	11ß-Hydroxy-4-androstene-3,17-dione	11-Beta-hydroxyandrostenedione
Ad	4-Androstene-3,11,17-trione	Adrenosterone, androstenetrione
E1	3-Hydroxy-1,3,5(10)-estratrien-17-one	Oestrone
E2	1,3,5(10)-Estratriene-3,17β-diol	Oestradiol-17β
E17α	1,3,5(10)-Estratriene-3,17α-diol	Oestradiol-17α

16 Gonadal Steroids in Sturgeons

Table 16.2 Meaning of common abbreviations of steroidogenic enzymes used in text and figures following zebrafish nomenclature

Abbreviation	Full name	Gene name
P450 scc	Side-chain cleavage enzyme	*cyp11a1*
P450-17H	17 alpha-hydroxylase/17,20 lyase	*cyp17a1*
P450-11H	11 beta-hydroxylase	*cyp11c1*
P450-arom	aromatase	*cyp19l a*
P450-21H	21-Hydroxylase	*cyp21a2*
3β-HSDH	3 beta-hydroxysteroid dehydrogenase	*hsd3b1*
11β-HSD	Hydroxysteroid (11-beta) deshydrogenase 2	*hsd11b2*
17β HSD-type3	Hydroxysteroid (17-beta) dehydrogenase 3	*hsd17b3*
17β-HSD-type1	Hydroxysteroid (17-beta) dehydrogenase 1, or	*hsd17b1*
20αHSD	Hydroxysteroid (20-alpha) dehydrogenase	
20β-HSD	Hydroxysteroid (20-beta) dehydrogenase2	*hsd20b2*

differentiation in fish because they are produced only in some species previously instead to sex differentiation (Nakamura et al. 1998; Vizziano et al. 2007; Ijiri et al. 2008; Hattori et al. 2009; Blasco et al. 2013).

Contrasting with the high number of publications on steroid production and release and its effects in the regulation of gametogenesis reported in teleost fish, the number of studies published in sturgeons is very limited. This review provides an overview about steroid hormone biosynthesis at gonadal level and its release in blood plasma in sturgeons with special emphasis in Siberian sturgeon. It also contributes with original unpublished data on steroid profiles *in vivo* during puberty. There are few data on functional roles of steroids in sturgeons, and they will be introduced in sections of steroid synthesis and its plasmatic level along the biological cycle.

16.1 Steroidogenesis and Biological Activity Along the Course of the Biological Cycle

Steroidogenesis occurs in fish primarily in gonads, interrenal gland (Fostier et al. 1983) and brain (Diotel et al. 2011). All classes of steroid hormones are synthesized *de novo* from the common precursor cholesterol (Gower and Fotherby 1975). A key step is the conversion of cholesterol (C27 molecule) into pregnenolone (C21 steroid) by the action of the enzyme p450 side-chain cleavage enzyme (p450 scc) that removes the side chain of cholesterol (Gower and Fotherby 1975) (Fig. 16.1). This step is under the control of steroidogenic acute regulatory protein (star) that transfers cholesterol between the outer and inner mitochondrial membrane. The cytochrome p450 side-chain cleavage is the only enzyme able to convert the cholesterol into the first steroid (pregnenolone) of the steroidogenic route (Gower and Fotherby 1975); thus, its presence defines the steroidogenic capacity of one tissue. Downstream several enzymes modify pregnenolone to reach to the major steroids identified in fish. The pathway proposed for teleost fish is shown in Fig. 16.1. One of the key enzymes involved in C21 steroid biosynthesis is 3β-hydroxysteroid dehydrogenase (3β-HSD, *hsd3b*) that uses different substrates and is able to convert pregnenolone into progesterone and 17-hydroxypregnenolone (17P5) into 17-hydroxyprogesterone (17P) (Fig. 16.1). This enzyme transforms delta 5 steroids (steroids with double bound between carbon 5 and 6 and a hydroxyl group in the carbon 3) in delta 4 steroids (steroids with double bound between carbon 4 and 5 and a ketone group in the carbon 3). 17P is a metabolic crossroad point because it can be converted in other C21 steroids as 17,20βP, 17,20α-dihydroxy-4-pregnen-3-one (17,20αP), 17,21-dihydroxy-4-pregnene-3,20-dione (17,21P) and 17,20β,21-trihydroxy-4-pregnen-3-one (17,20,21P) or in androstenedione, the first step in the androgen (C19 steroid) synthetic route (Fig. 16.1). Depending on tissue and stage of development 17P can be converted in androgens (C19 steroids) or in other C21 steroids (Fig. 16.1).

In tissues in which 20β-hydroxysteroid dehydrogenase (20βHSD) is activated, the substrate 17P is converted into 17,20βP, while the activation of 20α-hydroxysteroid dehydrogenase (20αHSD) is conduced to the synthesis of 17,20αP. When the enzyme 21-hydroxylase (21-OH or P450-21H) is activated, 17,20βP can also be converted into 17,20,21P (Fig. 16.1).

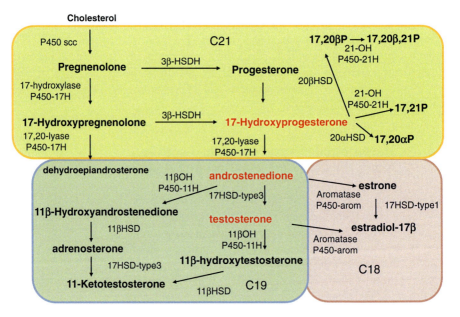

Fig. 16.1 Pathways of steroid synthesis in teleost fishes based on Fostier et al. (1983) and Borg (1994). Steroids are grouped following the number of carbons. C21 is highlighted in *yellow*, C19 or androgens in *blue* and C18 or oestrogens in *pink*. In red it is shown two crossroad points of gonad steroid metabolism: 17-hydroxyprogesterone and aromatizable androgens. The meaning of common name of steroid mentioned is given in Table 16.1 and of enzymes in Table 16.2

16 Gonadal Steroids in Sturgeons

When 17-hydroxylase/lyase (P450-17H) is active and the substrate 17P is produced, the conversion of 17P (C21 steroid) into androstenedione (C19 steroid) can occur. Androstenedione is the first steroid of delta 4 androgen synthetic route and can be converted in testosterone (C19 steroid) by the enzyme 17-hydroxysteroid dehydrogenase type 3 (17HSD-type3). Testosterone can be converted in 11β-hydroxytestosterone (11βOHT) by the action of the enzyme 11β-hydroxylase (11βOH or P450-11H) and 11βOHT into 11-ketotestosterone (11KT) by the action of 11β-hydroxysteroid dehydrogenase (11βHSD) (Fig. 16.1, Fostier et al. 1983; Borg 1994).

Androstenedione and testosterone (C19 steroids) are other metabolic crossroad points, because they can be converted into 11-oxygenated androgens by the action of the enzyme 11βOH or in oestrogens (C18 steroids) by the enzyme aromatase (P450-arom, *cyp19a1*). Aromatase is a key enzyme for oestrogen synthesis and converts androstenedione into oestrone and testosterone into oestradiol-17β (Fig. 16.1, Gower and Fotherby 1975; Fostier et al. 1983).

16.1.1 Ovarian Steroidogenesis

16.1.1.1 Pre-vitellogenesis and Vitellogenesis

In females of teleost fish, aromatase is active during pre-vitellogenesis, and its activity is highly increased during vitellogenic ovaries converting testosterone in oestradiol-17β which is released to the blood mainly during exogenous phase of vitellogenesis (Fig. 16.1, Fostier et al. 1983). Oestradiol-17β was detected in almost all fish studied at the time of vitellogenesis and decreases during follicle maturation period (Billard et al. 1978; Fostier et al. 1983; Scott and Sumpter 1983; Nagahama 1987, 1994). This oestrogen is the main regulator of vitellogenin production by the liver which constitutes the precursor of exogenous vitellus integrated in oocytes during the growing period of oocytes and ovaries (Lubzens et al. 2010).

Studies on steroid metabolism in female sturgeons are scarce. For pre-vitellogenic and vitellogenic females of Russian sturgeon, major metabolites of pregnenolone were 17-hydroxyprogesterone, androstenedione and testosterone, while those of androstenedione were testosterone and 5α- and 5β-androstanediols. No evidence was found for the gonadal production of oestrogens studied by TLC/HPLC/reduction/oxidation and crystallization with authentic standard (Bukovskaya et al. 1997). The precursors androstenedione and testosterone are produced by pre-vitellogenic and vitellogenic tissues, but oestrogens were not identified indicating that aromatase were not active. The absence of oestrogen production by vitellogenic follicles of Russian sturgeons is unexpected since plasma concentrations of oestradiol-17β have been measured in vitellogenic females of the species (see Sect. 16.3) opening the question of the primary source of oestrogens released to blood plasma in the species.

Siberian sturgeon ovaries in pre-vitellogenesis express *cyp17a1*—the gene coding for the enzyme P450-17H—that converts C21 into C19 steroids (Table 16.2, Fig. 16.1). At this stage ovaries also express aromatase (*cyp19a1a*, see Table 16.2), the gene coding for enzyme that converts androgens (C19) to oestrogens (C18) (Vizziano-Cantonnet et al. 2016). These results suggest that ovaries at pre-vitellogenesis are able to produce oestrogens.

In bester (*Huso huso* L. females x *Acipenser ruthenus* L. males), gonads at different stages incubated with precursors as pregnenolone, 17P and testosterone produced oestradiol-17β measured by ELISA (Amiri et al. 1999). The conversion of 17P and testosterone into oestradiol-17β involves the activation of several enzymes of steroid synthesis (P450-17H, 17β HSD-type 3, P450-11H, Fig. 16.1). Oestradiol-17β showed high concentrations in incubation media from primary yolk stage to tertiary yolk stages (vitellogenic ovaries) with respect to oil droplet stage (presence of oil droplets in the periphery of the nucleus) and decreased sharply at migratory nucleus stage, suggesting changes in aromatase activity as reported in teleosts (Kagawa et al. 1983; Young et al. 1983a, b).

Estrogen production has been poorly investigated in sturgeons during pre-vitellogenesis and vitellogenesis, but seems to be produced at least in Siberian sturgeons and in bester. The absence of oestrogen production in Russian sturgeons is unsual when considered the fish literature.

16.1.1.2 Follicle Maturation

In teleost fish, during last phases of ovogenesis when follicles starts its maturation, 20β-hydroxysteroid dehydrogenase (20βHSD, *hsd20b*) is activated, and the substrate 17P is converted into 17,20βP (C21, steroid) considered as the most potent inducer of follicle maturation in almost all teleost fish studied (Fig. 16.1, Fostier et al. 1973; Fostier et al. 1983; Scott and Canario 1987; Nagahama 1994). 17,20βP is produced by ovaries and released to plasma mainly during the maturation period of several teleost fish under control of gonadotropin (Nagahama 1987; Jalabert et al. 1991). However, in female Sciaenidae 17,20βP can be converted into 17,20β,21-trihydroxy-4-pregnen-3-one (17,20,21P) by the activity of the enzyme 21-hydroxylase (21-OH or P450-21H) (Fig. 16.1, García-Alonso et al. 2003, 2004). This tri-hydroxylated steroid is largely produced during maturation of the follicles of Atlantic croaker (*Micropogonias undulatus*), and it has been established as the maturation-inducing steroid in Sciaenidae fish (Trant et al. 1986; Thomas et al. 1997; Thomas 2012). In sum, two C21 steroids act as main mediators of follicle maturation in teleost fish: 17,20βP and 17,20,21P.

Few studies have been made in sturgeons during follicle maturation. In bester, 17,20βP has been measured by ELISA in culture media of ovaries at different stages (primary yolk stage, secondary yolk granule stage, tertiary yolk stage, migratory nucleus stage) with an increase at tertiary and migratory nucleus stages, the latter corresponding to the process of follicle maturation or meiosis resumption (Amiri et al. 1999). However, other potential C21 steroids were not measured to conclude about which is the potential mediator of follicle maturation in bester. The major steroids produced *in vitro* by follicles during maturation of Siberian sturgeon incubated with tritiated 17P and analysed by TLC/HPLC/oxidation were androstenedione, 17,20,21P, 17,20αP and 17,21P (Alberro et al. 2008). 17,20βP was not confirmed by HPLC in the zone of TLC co-migrating with 17,20βP/17,21P, and 17,20,21P resulted the more prominent C21 steroid produced *in vitro* (Alberro 2009). In white sturgeons among C21 steroids produced by maturing follicles, 17,21P, 17,20βP and 17,20,21P had the major activity as maturation-inducing steroid (MIS) (Webb et al. 2000).

Interestingly, in other sturgeons as sterlet, 17,20,21P increased ten times in culture media of follicles 16 h after gonadotropic stimulated maturation (Semenkova et al. 2006).

In sum, di- and tri-hydroxylated progestins (C21 steroids) are produced by sturgeon ovarian follicles during maturation. They were 17,20βP, 17,21P and 17,20,21P for white sturgeon and bester; 17,21P, 17,20αP and 17,20,21P for Siberian sturgeon; and 17,20,21P and 17,20βP for sterlet. 17,20P was not produced *in vitro* by Siberian sturgeon during the period of follicle maturation. Collectively, the results suggest that the main mediator of follicle maturation could be 17,20,21P in some sturgeon species, including Siberian sturgeon. Other C21 steroid produced could also contribute to regulate the follicle to reach the stage needed to be ovulated and fertilized as it was previously proposed (Webb et al. 2000). The results suggest that some sturgeon species share with Sciaenidae fish 17,20,21P as mediator of follicle maturation.

Another steroid studied during ovarian development of sturgeons was 11KT (Siberian sturgeon, Cuisset et al. 1995; Russian sturgeon, Bukovskaya et al. 1997; white sturgeon Webb et al. 2002). Experiments made *in vitro* showed that the yield of 11KT production was very low in ovarian follicles of Siberian sturgeons at maturation time (Cuisset et al. 1995), while no 11KT was detected using similar stages of ovarian follicle development for the species (Alberro et al. 2008). Moreover, 11KT was not identified in follicles of white sturgeon during maturation (Webb et al. 2002) or in pre-vitellogenic and vitellogenic ovarian follicles of Russian sturgeon (Bukovskaya et al. 1997). The results suggest that 11KT is not a mediator of follicle maturation in sturgeons.

16.1.2 Testicular Steroidogenesis

16.1.2.1 Spermatogenetic Period

Spermatogenesis comprises three major phases: mitotic proliferation of spermatogonia, meiosis of spermatocytes and spermiogenesis (Schulz and Miura 2002). When spermatozoa are morphologically completed after spermiogenesis, sperm maturation and hydration occurs under regulation of gonadotropins and steroids; see review (Vizziano et al. 2007).

Spermatogonial renewal is stimulated by low doses of oestradiol-17β in eels (Miura et al. 1999) at early phases of testis development when testis were composed of spermatogonias. Physiological role of oestrogens in males has been poorly studied in fish. In Salmonidae, evidence that oestradiol-17β is produced in testes suggest that oestrogens participates in paracrine control of testicular function (see review Vizziano et al. 2007). In sturgeons the oestrogen production by testes has not been reported. However, aromatase—the enzyme that converts androgens to oestrogens—(Fig. 16.1) is expressed in testis of immature (only spermatogonias) male

Siberian sturgeons suggesting that oestrogens are produced very early and could be involved in regulating early testicular development of this species (Vizziano-Cantonnet et al. 2016).

Contrasting with oestrogens, androgens have been well studied in fish. In adult males of teleosts, testis in spermatogenesis converts testosterone in 11β-hydroxytestosterone (11βOHT) by the action of the enzyme 11β-hydroxylase (11βOH or P450-11H, *cyp11b/cyp11c1*) and 11βOHT into 11-ketotestosterone (11KT) by the action of 11β-hydroxysteroid dehydrogenase (11βHSD, *hsd11b*) (Fostier et al. 1983; Borg 1994). This seems to be the predominant way of synthesis. As consequence spermatogenetic period in fish is characterized by the production of high level of 11-oxygenated androgens (11βOHT, 11KT, 11ßOHA, Ad) especially 11KT considered as the most potent androgen in fish (Fostier et al. 1983; Borg 1994). Androgens are effective in supporting spermatogenesis (Billard et al. 1982; Fostier et al. 1983; Borg 1994; Nagahama et al. 1994), some early steps as spermatogonial proliferation (Schulz and Miura 2002) and were stimulated by gonadotropins along all testicular cycle (see review Vizziano et al. 2007). It is known that gonadotropin and 11KT induces spermatogonial proliferation in teleosts (Schulz and Miura 2002), and can induce spermiation in some species. But they were clearly less effective than progestins to induce sperm release (Ueda et al. 1985; Schulz et al. 2010).

There are few studies on steroid metabolism of sturgeon testes. Testosterone and 5α- and 5β-androstanediols, but no 11KT were produced after incubation of pre-spermiating testes of Russian sturgeons with radiolabeled androstenedione (Bukovskaya et al. 1997). Contrasting with that in bester, 11KT has been measured by ELISA in culture media of testis at different developmental stages incubated with inert pregnenolone, 17P and testosterone (Amiri et al. 1999). Levels of 11KT resulted higher during late spermatogenesis and pre-spermiation to decrease at degeneration stage (post-spawning).

The few reports in male sturgeons prevent any general conclusion, but at least in bester, 11KT is high during spermatogenetic period as expected.

16.1.2.2 Sperm Maturation and Release

In Salmonidae, 17,20βP is produced from immature to spermiating testes and is under gonadotropin regulation both *in vivo* and *in vitro* (Scott and Sumpter 1983; Schulz et al. 1992, 2010; Vizziano et al. 2007; Schulz and Miura 2002; Scott et al. 2010). 17,20βP is considered an essential factor for the initiation of meiosis at least in eels (Miura et al. 2006), but the effect of this C21 steroid on the initiation of meiosis remains to be better studied in other species. A physiological role of 17,20βP has been shown in Salmonidae and Cyprinidae (Ueda et al. 1985), in the amplification of milt volume production (Baynes and Scott 1985; Yueh and Chang 1997) and in acquisition of the potential motility of male gametes (Miura et al. 1992; Miura and Miura 2003). In male Sciaenidae, the main metabolite produced by testes from spermiating fish is 17,20βP (Vizziano-Cantonnet et al. 2015), but its participation as mediator of sperm maturation and release remains to be studied.

16 Gonadal Steroids in Sturgeons

Steroids produced by testes of sturgeons are unknown during spermiation, and very scarce data are published on steroid metabolism of testes at different stages of adult cycle. In Russian sturgeons, pregnenolone was added to incubation media of pre-spermiant testes and metabolites analysed by TLC/HPLC/reduction/oxidation and crystallization with authentic standard. The metabolites produced were progesterone, 17P and androgens, but no 17,20βP were reported (Bukovskaya et al. 1997). Contrasting with that, in bester the addition of precursors as pregnenolone and 17P to culture media of testes from males at different testicular stages resulted in detectable levels of 17,20βP measured by ELISA, reaching. High levels in late spermatogenesis and pre-spermiation (Amiri et al. 1999). The trends observed in male bester are in agreement with a physiological role of 17,20βP from early spermatogenesis to pre-spermiation as it was observed for Salmonidae. The work made in bester (Amiri et al. 1999) suggest a change in synthesis from 11KT during spermatogenesis to 17,20βP during pre-spermiation, i.e. from C19 to C21, similarly to changes described for teleost fish (Schulz and Miura 2002; Vizziano et al. 2007). However the lack of production of 17,20βP by pre-spermiating testis of Russian sturgeons (Bukovskaya et al. 1997) open the question about the mediators used by this species.

In Siberian sturgeon, *in vivo* studies after hormonal stimulation of spermiation show a high increase in plasma concentrations of 17,20βP together with an increase in the volume of sperm released (Vizziano et al. 2006) suggesting that 17,20βP is a main mediator of sperm maturation and release as it was reported in teleost fish; see review (Vizziano et al. 2007). In Stellate sturgeon males, plasmatic 17,20,21P showed an increase after LHRH analogue stimulation of spermiation. However, other C21 steroids as progesterone and 17,20βP were not stimulated. Testosterone and cortisol were also stimulated during induced spermiation (Bayunova et al. 2006). Other studies in stellate sturgeon showed that only testosterone was increased significantly after hormonal stimulation of spermiation, while progestins 17,20,21P and 17,21P increased significantly only at post-spawning period (Bayunova et al. 2008). Finally, in another report on stellated sturgeon, progesterones were clearly increased in blood plasma after injection of sturgeon pituitary extracts, but 17,20,21P was not measured (Semenkova et al. 2002). For Persian sturgeon testosterone, 11KT and progesterone resulted more elevated in spermiating males stimulated by LHRH analogue when compared to non-spermiating males (Aramli et al. 2014). Similar results were obtained for testosterone and 11KT in Russian sturgeon males (Artyukhin et al. 2006).

The *in vivo* studies after hormonal stimulation of spermiation show that C21 steroids as 17,20βP and 17,20,21P can be increased, while in other cases androgens were stimulated.

Results on C21 steroid production in sturgeons females and males at the end of gametogenetic cycle show that two major steroids seem to be MIS in females (17,20βP and 17,20,21P), while 17,20βP could be the progestin involved in the regulation of sperm maturation and release of Siberian sturgeons; however, other steroids as 17,20,21P remain to be studied for the species.

A physiological role of androgens during spermiation cannot be discarded.

16.2 Circulating Steroids During Puberty of Females

Puberty is a critical endocrine period that comprises the transition between immature juvenile to a mature adult state of the reproductive system; see review of Taranger et al. (2010). The fish became capable to reproduce for the first time, and this implies that hypothalamus-pituitary-gonad axis became functional (Taranger et al. 2010). During the entry in puberty of females—considered as the transition to first batch of oocytes accumulating cortical alveoli—oestradiol-17β increases in some salmons and yellowtail together with circulating fish (Taranger et al. 2010).

In farmed fish it is interesting to control puberty to advance or delay it. In natural conditions Siberian sturgeon reach puberty at different ages depending on the river or lake in which it lives, ranging from 9 to 34 years for females and 8 to 29 years for males (see Chap. 1). In captivity, this species reaches puberty at 5 years old for males and 7 years old for females in France (Pelissero and Le Menn 1991). In Uruguayan culture conditions, males are spawning at 3 years old, while females reach the puberty at 5 years old (Ryncowski, personal communication). In Russian sturgeon sexual maturity is reached at 8–13 years for males and 10–16 years for females under natural conditions (Vlassenko et al. 1989). Puberty occurs in farms of Israel at 6 years old, and oestradiol-17β was a very good indicator of ovarian development for the species (Hurvitz et al. 2008).

Studies made during puberty in Siberian sturgeon maintained in captivity in France (females and males at 4, 5 and 10 years old) showed that plasmatic concentration of oestradiol-17β was not different between males and females (Pelissero and Le Menn 1988) as expected considering the trends observed in teleosts. Moreover, oestradiol-17β showed high concentrations in pre-vitellogenetic fish and do not show a sharp increase during vitellogenesis (see Table 16.5). In addition the maximum levels of oestradiol-17β in plasma were reached at pre-ovulating time and not in vitellogenic females as expected. Oestrone increased between pre-vitellogenic and vitellogenic females, but remained high during pre-ovulation period. Oestrogens do not follow the trends observed in other fish, and the authors measured oestrogens in food to better understand their results. They suggested that oestrogens measured in males and females could be the result of an alteration of the natural production altered by the exogenous supply coming from food.

Studies made by our laboratory in cooperation with sturgeon farmers of Uruguay showed that Russian sturgeon reaches puberty at 5 years old as indicated by the presence of females at stage 3 (follicle diameter ~2 mm) and stage 4 (~2.5 mm). Fish data and maturity stages are given in Table 16.3. Serum anti-oestradiol-17β was very specific as it is shown in Table 16.4. Plasmatic concentrations of oestradiol-17β increased significantly ($p < 0.05$) between immature females (stage 1) and females at stage 2 (small white oocytes) and stage 3 (follicle diameter of 2 mm) (Fig. 16.2). At stage 4 (follicle diameter of 2.5 mm) oestradiol-17β decreased significantly ($p < 0.05$). Oestradiol-17β resulted in a good indicator of female puberty for Russian sturgeon. We used oestradiol-17β for the diagnosis of a delayed cohort of 5-year-old fish that were not well developed and for which oestradiol-17β

16 Gonadal Steroids in Sturgeons

Table 16.3 Plasma concentrations of E2 in female *A. gueldenstaedtii* at immature and first maturity stages in fish farmed in Uruguay

Species/country	Puberty in captivity	Maturity stage	Oocyte diameter (mm)	E2 (ng mL^{-1})
A. gueldenstaedtii/Uruguay	5 years old	Stage 1—undistinguishable oocytes ($n = 10$)	nd	0.074 ± 0.09
		Stage 2—white tissue, undistinguishable oocytes ($n = 13$)	nd	1.53 ± 3.25 ($p < 0.05$)
		Stage 3—grey oocytes ($n = 8$)	2.05 ± 0.23 ($n = 100$)	4.37 ± 2.03 ($p < 0.05$)
		Stage 4—grey oocytes ($n = 11$)	2.47 ± 0.11 ($n = 159$)	0.64 ± 0.86 ($p < 0.05$)

nd = not determined

Table 16.4 Cross-reactivity of serum anti-oestradiol-17β used for radioimmunoassay

Steroids	Anti E2
17,20,21-P	nd
17P	nd
P4	nd
17,21P	nd
cortisol	nd
T	<0.01
E2	100%
A	nd
11ßOHA	nd
17,20ßP	<0.01
E3	<0.01
E17α	<0.01
11KT	<0.01

nd: assay not done

concentrations were undetectable. We were able to answer to farmers that this cohort was unable to produce caviar in next months.

In order to control vitellogenesis, experiments with oestradiol-17β implants during immature period of females have been made in *Huso huso*, but growing of oocytes was not observed (Akhavan et al. 2015), while oestradiol-17β causes suppression of growth in stellate sturgeon (Khara et al. 2013). A more ample knowledge on the axis of hypothalamus-pituitary-gonads during puberty remains to be further studied in sturgeons.

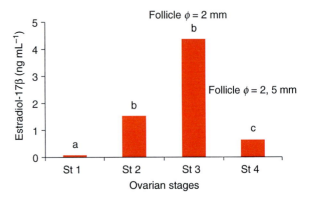

Fig. 16.2 Concentrations of plasmatic oestradiol-17β (ng mL^{-1}) during ovarian development at puberty of Russian sturgeons of 5 years old in farmed conditions in Uruguay. Statistical differences come from an ANOVA

16.3 Circulating Steroids During Adult Phase

16.3.1 Females

During gonadal cycle, oestrogens (C18 steroids) are produced by vitellogenic ovaries, and a shift towards progestins (C21 steroids) occurs when follicles enter to the process of maturation (resumption of meiosis). The shift from the gonadal production of oestradiol-17β to 17,20βP has been first observed in teleosts (Kagawa et al. 1983; Young et al. 1983a, b; Yaron 1995). Steroids are released to the blood, and measurements of plasmatic steroids reflect in some extent the gonadal steroid production of ovaries. 17,20β P is produced in some teleosts during ovarian maturation induced by gonadotropins and is released and measured in plasma during maturation and ovulation (Fostier et al. 1981; Young et al. 1982, 1983b; Ueda et al. 1984). In this section the steroid changes in blood plasma are reviewed for female sturgeons.

16.3.1.1 Pre-vitellogenesis and Vitellogenesis

In Siberian sturgeon (Hamlin et al. 2011) and in bester (Amiri et al. 1996a), oestradiol-17β showed a significant increase of 6–15 times from pre-vitellogenic to vitellogenic fish reaching 4–7.5 ng mL^{-1} with a significant decline at post-vitellogenic and maturation (migratory nucleus) stages (Table 16.5). Similar results have been observed for Russian sturgeons during puberty (see Sect. 16.3). These are the expected profiles of oestradiol-17β taking account the information coming from teleost fish and are coherent with a functional role of oestradiol-17β as stimulator of exogenous vitellogenesis. However, the primary source of oestradiol-17β needs to be further investigated at least for Russian sturgeons, for which ovaries at vitellogenesis are not able to produce oestradiol-17β when precursors are given.

Testosterone concentrations in blood plasma of sturgeons can be one or two orders higher than oestrogens for a given species (Table 16.5). A significant increase

16 Gonadal Steroids in Sturgeons   339

in testosterone concentrations occurs from pre-vitellogenic to vitellogenic sturgeons (bester, Amiri et al. 1996a, Siberian sturgeon, Hamlin et al. 2011, Table 16.5) as it was observed for oestradiol-17β. Oestrogens decreases significantly at the end of oogenesis during migratory nucleus stage in sturgeons (Hamlin et al. 2011, Table 16.5) as it is the case in teleosts. Contrasting with that Siberian sturgeons, testosterone can increase significantly at follicle maturation stage (Pelissero and Le Menn 1988; Hamlin et al. 2011) and remain at high concentrations during post-ovulated period in Stellate sturgeon (Bayunova et al. 2011) or during post-vitellogenesis in bester (Amiri et al. 1996a, Table 16.5). The high testosterone concentration around the time of ovulation is discussed below. C21 steroids as progesterone, 17,20βP and 17,20,21P remain low (0.05–0.75 ng mL^{-1}) and stable from pre-vitellogenesis to post-vitellogenesis (Table 16.5) suggesting that this C21 steroids are not involved in the regulation of this process.

16.3.1.2 Follicle Maturation

As previously mentioned progestins measured in blood plasma (P and 17,20βP) are very low (order of pg mL^{-1}) and do not change between immature and post-vitellogenic fish (Table 16.5). Progesterone and 17,20βP remain very low (≤ 0.3 ng mL^{-1}) except for European sturgeon (*Acipenser sturio*) that shows ~1 ng mL^{-1} during maturing stages (Table 16.5). The progestins measured in sturgeons without hormonal stimulation were in general low, but 17,20,21P is highly increased in blood plasma after hormonal stimulation of follicle maturation (Sect. 16.2.1.2). In fact in sterlet females 17,20,21P, but not 17,20β P was clearly stimulated after hormonal treatments to induce follicle maturation (Bayunova 2016). In addition, in Stellate sturgeon a surge of 17,20,21P has been observed after hormonal stimulation (GnRH) of follicle maturation (Semenkova et al. 2002; Bayunova et al. 2006; Semenkova et al. 2006; Bayunova 2016). In accordance with these data 17,20β P became detectable but very low ($0,09$ ng ml^{-1}) after hormonal stimulation (GnRH) of follicle maturation (100% of GVBD) of Siberian sturgeon females (Vizziano et al. 2006). These *in vivo* data together with *in vitro* studies (see 16.2.1.2) suggest that 17,20,21P is the main mediator of follicle maturation in several sturgeon species including Siberian sturgeon

Siberian sturgeon females show high plasmatic levels of 11KT during the whole period of ovarian development and reached its maximum concentrations at early and mid vitellogenesis. Moreover, a huge and significantly high amounts of testosterone (120 ng mL^{-1}) were described during follicle maturation of this species (Table 16.5, Williot et al, Chap. 17 of this book, Hamlin et al. 2011). A possible explanation is that testosterone acts in sturgeons as follicle maturation mediator of LH. However, studies on potency of C21 and C19 steroids to induce resumption of meiosis (follicle maturation or GVBD) in sturgeons revealed that C21 steroids (17,21P, P, 17P, 17,20P, 17,20,21P and cortisol) are more potent than testosterone and that 11-oxygenated androgens do not induce maturation (Webb et al. 2002). This result discards androgens as follicle maturation inducers.

Table 16.5 Plasma steroid concentrations (ng mL^{-1}) measured in female sturgeons

Species	Ovarian stage/month	E2	E3	T	11KT	17,20P	17,20,21P	P	Source
A. baerii	Pre-vitellogenic	11.6 ± 1.2	5.1 ± 3.2	8.2 ± 0.3	–	–	–	–	Pelissero and Le Menn (1988)
	Vitellogenic	16.6 ± 1.34	10.8 ± 7.18	5.8 ± 0.54	–	–	–	–	
	Pre-ovulation	19.3 ± 2.68	10.3 ± 1.53	109 ± 25.5	–	–	–	–	
A. baerii	Migratory nucleus Follicle Φ = 3.3–4.7 mm	0.15–0.41	–	16–133	–	nd	–	–	Vizziano et al. (2006)
A. baerii	Pre-vitellogenic (St2) Follicle Φ = 0.6 mm	~0.5	–	~10	~5	–	–	0.15	Hamlin et al. (2011)
	Early-vitellogenic (St3) Follicle Φ = 0.6–1.8 mm	~3.2 ($p < 0.05$)	–	~40 ($p < 0.05$)	~20 ($p < 0.05$)	–	–	0.1	
	Mid-vitellogenic (St4) Follicle Φ = 1.8–2.2 mm	~7.5 ($p < 0.05$)	–	~65	~20	–	–	0.12	
	Migratory nucleus (St5) Follicle Φ = 1.8–2.2 mm	~6 ($p < 0.05$)	–	~120 ($p < 0.05$)	~10 ($p < 0.05$)	–	–	0.13	

16 Gonadal Steroids in Sturgeons

A. baerii	Migratory nucleus Follicle $\Phi = 3.3$–4.7 mm	~0.7–2.7	–	~40—160	~58–90	~1.8–3.8	–	Williot et al see Chap. 17
A. sturio	Mature female	~33	–	12–15	<2	<2	–	Davail-Cuisset et al. (2008)
A. sturio	Maturing (IP = 0.1–0.3)	4.5	–	9.4	1.7	0.05	–	Davail-Cuisset et al. (2011)
	90% GVBD/ovulation	7.9	–	9.6	2.2	1		
Huso huso L. Females x Acipenser ruthenus L. males	Immature	<0.6	–	5–15	–	0.05–0.3	–	Amiri et al. (1996a)
	Vitellogenic	2–4 ($p < 0.01$)	–	20–80 ($p < 0.05$)	–	0.05–0.3	–	
	Post-vitellogenic	1–2 ($p < 0.01$)	–	20–80	–	0.05–0.3	–	
A. gueldenstaedtii	Pre-vitellogenic oocytes	0.2–0.48	–	–	–	–	–	Hurvitz et al. (2008)
A. gueldenstaedtii	Stage III	0.68 ± 0.93	–	46 ± 67	13.7 ± 3.4	–	–	Barannikova et al. (2006)
	Stage IV		–	105 ± 30	–	–	–	
	Stage V	−0.11 ± 0.22	–	19 ± 7	28.8 ± 6.7	–	–	

(continued)

Table 16.5 (continued)

Species	Ovarian stage/month	E2	E3	T	11KT	17,20P	17,20,21P	P	Source
A. stellatus	Spring	~3	–	~55	~70	~0.025	–	–	Ceapa et al. (2002)
	Early May	<2.5	–	~35	~18	~0.25	–	–	
	Late May	~6	–	~30	~39	~0.2	–	–	
	Autumn	~2.5	–	~48	~78	~1	–	–	
A. stellatus[a]	Stage IV (April–May)	0.9 ± 0.09	–	172 ± 13	26 ± 1.3	–	–	0.9 ± 0.3	Barannikova et al. (2002)
	Stage III–IV (May–June)	0.4	–	98 ± 7	35	–	–	0.6 ± 0.1	
	Stage IV (June)	0.38 ± 0.2	–	99 ± 9	28 ± 5.3	–	–	0.9 ± 0.1	
	Stage IV (July)	–	–	29 ± 8	–	–	–	0.3 ± 0.1	
A. stellatus	Post-ovulated	–	–	111 ± 43	9.5 ± 2			0.4 ± 0.12	Bayunova et al. (2011)
A. stellatus	Pre-spawning	~1.4	–	~180	~25	~0.1	~0.25	~0.8	Bayunova et al. (2006)
A. stellatus	Captured in wild and delivered to the farm	1.8 ± 1	–	94 ± 14	25 ± 4	–	<0.5	0.15 ± 0.1	Semenkova et al. (2002)
	14 days holding	1.6 ± 0.8	–	167 ± 21	40 ± 4	–	<0.5	0.33 ± 0.05	
A. ruthenus	Pre-spawning	–	–	~18	~10	~0.5	~0.75	–	Semenkova et al. (2006)

[a]Stages III–IV, incomplete gametogenesis; Stage IV, complete gametogenesis; Stage V, mature gonads (Barannikova et al. 2002)

The high concentrations of testosterone in female before and after ovulation (Vizziano et al. 2006, Hamlin et al. 2011, see Chap. 17, Bayunova et al. 2011) and in LH-stimulated culture media of maturing follicles (Semenkova et al. 2006) suggest an important function during peri-ovulatory stage (Table 16.5). The presence of androgens in coelomic fluid leads instead to propose that androgens are functional to maintain the viability of ovules in the abdominal cavity (Bayunova et al. 2003).

A shift from oestrogens (C18 steroids) to progestins (C21 steroids) is observed at circulating levels of sturgeons as it is the case in some teleost fish.

16.3.2 Males

Among steroids measured in plasma of males (oestradiol-17β, T, 11KT, 17,20βP, 17,20,21P, 17,21P, P), the major concentrations were observed for androgens as T (13–257 ng mL^{-1}) and 11KT (1–82 ng mL^{-1}), while low and stable concentrations were measured for estradiol-17β (<0.7 ng mL^{-1}) in most of the cases studied (Table 16.6). Contrasting with that, oestrogens measured in males were very high in Siberian sturgeons farmed in France, and these unusual high levels could come from the oestrogens given in food (Pelissero and Le Menn 1988). In the case of Siberian sturgeon cultured in Florida (USA) amounts of testosterone and oestradiol-17β increased significantly between pre-meiotic (only spermatogonias) and meiotic (spermatogenesis) fish (Hamlin et al. 2011, Table 16.5), while in bester 11KT increased significantly at pre-spermiation testicular stage (Amiri et al. 1996b). During first reproductive cycle of Siberian sturgeon, high concentrations of 11KT were detected in blood plasma of males during spermatogenesis (Cuisset et al. 1994), while plasmatic testosterone was high both during spermatogenesis and spermiation (Pelissero and Le Menn 1988). The high level of androgens observed in sturgeons during spermatogenesis plaids in favour of a role in the control of male gametogenesis as it was stated for teleost fish (Schulz and Miura 2002; Vizziano et al. 2007; Schulz et al. 2010). Plasma levels of steroids observed in sturgeons are coherent with a participation of C18 and C19 steroids in the control of fish spermatogenesis as it was proposed for teleosts (see Sect. 16.2.2.1).

C21 steroids were poorly studied in sturgeon blood plasma and are generally low. 17,20βP were not detectable or low (0.5 ng mL^{-1}) during pre-spermiation (Vizziano et al. 2006; Bayunova et al. 2006, 2008) and changes with seasons in Stellate sturgeons reaching 1 ng mL^{-1} in Autumn (Ceapa et al. 2002, Table 16.6). After hormonal induction of spermiation, 17,20βP increases in blood plasma of Siberian sturgeons with the volume of sperm released (Vizziano et al. 2006), and 17,20,21P increased in Stellate sturgeons (Bayunova et al. 2006) suggesting a physiological role in the control of sperm maturation and release.

Table 16.6 Plasma steroids concentrations (ng mL^{-1}) measured in male sturgeons

Species	Maturity stage/month	E2	T	11KT	17,20P	17,20,21P	17,21P	P	Source
A. baerii	Spermatogenesis	14 ± 0.95	104 ± 47	–	–	–	–	–	Pelissero and Le Menn (1988)
	Spermiation	27 ± 4.7	178 ± 6.5	–	–	–	–	–	
A. baerii	Pre-spermiating	0.05–0.23	70–116	–	nd	–	–	–	Vizziano et al. (2006)
A. baerii	Pre-meiotic	~0.1	~18	~18	–	–	–	~0.14	Hamlin et al. (2011)
	Meiotic	~0.2 ($p < 0.05$)	~80 ($p < 0.05$)	~80	–	–	–	~0.15	
A. sturio	Mature male	<1	~40	~6.6	–	–	–	–	Davail-Cuisset et al. (2008)
A. gueldenstaedtii	Stage A—primary spermatogonias	–	–	<1	–	–	–	–	Hurvitz et al. (2008)
	Stage B—small number of germ cells	–	–	~4	–	–	–	–	
	Stage C—whole spermatogenesis	–	–	~8	–	–	–	–	
A. gueldenstaedtii	May	0.018 ± 0.01	75.5 ± 27	15.6 ± 2.8	–	–	–	–	Barannikova et al. (2006)
	May	–	185 ± 23	–	–	–	–	–	
	May	–	49 ± 13	26.9 ± 4.6	–	–	–	–	

A. stellatus	Spring	~<0.5	~55	~70	~0.025	–	–	–	Ceapa et al. (2002)
	Early May	<0.5	~35	~18	~0.25	–	–	–	
	Late May	~0.7	~30	~39	~0.2	–	–	–	
	Autumn	~0.7	~48	~78	~1	–	–	–	
A. stellatus[a]	Stage IV (April–May)	0.042 ± 0.009	257 ± 11	–	–	–	–	0.37 ± 0.04	Barannikova et al. (2002)
	Stages III–IV (May–June)	0.015	117 ± 10	32.9	–	–	–	0.6 ± 0.06	
	Stage IV (June)	0.034 ± 0.004	157 ± 18	82 ± 11	–	–	–	1 ± 0.14	
	Stage IV (July)	–	22 ± 16	–	–	–	–	0.24 ± 0.08	
A. stellatus	Pre-spermiating	–	46 ± 12	54 ± 17	0.3 ± 0.09	0.9 ± 0.19	1.1 ± 0.24	–	Bayunova et al. (2008)
A. gueldenstaedtii	In spermatogenesis	–	~50	15–20	–	–	–	–	Artyukhin et al. (2006)
A. stellatus	Pre-spermiating	–	~20	~15	~0.5	~3	–	<0.5	Bayunova et al. (2006)
Huso huso L. Females x *Acipenser ruthenus* L. males	Spermatogonial proliferation	–	13	12	–	–	–	–	Amiri et al. (1996b)
	Early spermatogenesis	–	28 ($p < 0.01$)	55 ($p < 0.01$)	–	–	–	–	
	Pre-spermiation	–	28	72 ($p < 0.01$)	–	–	–	–	
	Post-spawning	–	28	10 <10	–	–	–	–	

nd = not detected; –: not determined

[a]Stage III, incomplete gametogenesis; Stage IV, fully completed spermatogenesis (Barannikova et al. 2006)

Conclusions

Some general patterns emerge for sturgeons. In females androgens and oestrogens are the main steroids produced by ovaries and released to blood plasma during gonadal growth, while di- and tri-hydroxylated progestins (C21 steroids) have a maturational activity. Androgens are the precursors of oestrogen steroid production, but the higher levels detected at peri-ovulatory period suggest a physiological role at the last phases of biological cycle of females. In males androgens are the main steroids during the spermatogenesis, while C21 could have a maturational activity. Very poor efforts have been made to understand the biological activity of steroids in sturgeons. The role of other gonadal steroids in sturgeon reproduction remains to be studied. Many gaps need to be filled into the current knowledge of gonad regulation by steroids.

Acknowledgments Special thanks are due to Dr. Patrick Williot to receive the author in the experimental installations of the CEMAGREF (Bordeaux, France) to develop the *in vitro* work made using ovarian samples of adult female Siberian sturgeons. Andrés Alberro, Valeria Camarero and Florencia Barrios helped in technical assistance. Many thanks are due to Dr. Alexis Fostier (INRA, Rennes, France) for the generous donation of serum anti-steroids.

References

Akhavan SR, Falahatkar B, Gilani MHT, Lokman PM (2015) Effects of estradiol-17β implantation on ovarian growth, sex steroid levels and vitellogenin proxies in previtellogenic sturgeon *Huso huso*. Anim Reprod Sci 157:1–10

Alberro A (2009) Síntesis de esteroides durante la maduración ovocitaria en el esturión siberiano *Acipenser baerii*. Trabajo especial II de la Licenciatura en Bioquímica. Universidad de la República Oriental del Uruguay, Montevideo

Alberro A, Williot P, Vizziano D (2008) Steroid synthesis during oocyte maturation in the Siberian sturgeon *Acipenser baerii*. Cybium 32(2):255–255

Amiri B, Maebayashi M, Adachi S, Moberg G, Doroshov S, Yamauchi K (1999) *In vitro* steroidogenesis by testicular fragments and ovarian follicles in a hybrid sturgeon, Bester. Fish Physiol Biochem 21(1):1–14

Amiri B, Maebayashi M, Adachi S, Yamauchi K (1996a) Testicular development and serum sex steroid profiles during the annual sexual cycle of the male sturgeon hybrid the bester. J Fish Biol 48(6):1039–1050

Amiri B, Maebayashi M, Hara A, Adachi S, Yamauchi K (1996b) Ovarian development and serum sex steroid and vitellogenin profiles in the female cultured sturgeon hybrid, the bester. J Fish Biol 48(6):1164–1178

Aramli M, Kalbassi M, Nazari R (2014) Sex steroid levels of Persian sturgeon, *Acipenser persicus*, Borodin, 1897, males in negative and positive responding to LH-RH-analogue. J Appl Ichthyol 30(1):18–19

Artyukhin EN, Semenkova T, Bayunova L, Lunev G, Barannikova I (2006) Histological assessment of the testes coupled with determinations of sex steroid levels in *Acipenser gueldenstaedtii* males responding negatively to pituitary treatment. J Appl Ichthyol 22:361

Barannikova I, Bayunova L, Semenkova T (2006) Serum sex steroids and their specific cytosol binding in the pituitary and gonads of Russian sturgeon (*Acipenser gueldenstaedtii Brandt*) during final maturation. J Appl Ichthyol 22:331

Barannikova I, Dyubin V, Bayunova L, Semenkova T (2002) Steroids in the control of reproductive function in fish. Neurosci Behav Physiol 32(2):141–148

Baroiller JF, Guiguen Y, Fostier A (1999) Endocrine and environmental aspects of sex differentiation in fish. CMLS 55(6–7):910–931

Baynes S, Scott A (1985) Seasonal variations in parameters of milt production and in plasma concentration of sex steroids of male rainbow trout (*Salmo gairdneri*). Gen Comp Endocrinol 57(1):150–160

Bayunova L (2016) The effect of hormonal stimulation on steroid levels in tissue incubates of the sterlet (*Acipenser ruthenus L.*) J Evol Biochem Physiol 52(1):17–27

Bayunova L, Barannikova I, Dyubin V, Gruslova A, Semenkova T, Trenkler I (2003) Sex steroids concentrations in Russian sturgeon (*Acipenser gueldenstaedtii Br.*) serum and coelomic fluid at final oocyte maturation. Fish Physiol Biochem 28(1–4):325–326

Bayunova L, Canario AV, Semenkova T, Couto E, Gerasimov A, Barannikova I (2008) Free androgens and progestins and their conjugated forms in serum and urine of stellate sturgeon (*Acipenser stellatus* Pallas) males. Cybium 32(2):273–274

Bayunova L, Canario AV, Semenkova T, Dyubin V, Sverdlova O, Trenkler I, Barannikova I (2006) Sex steroids and cortisol levels in the blood of stellate sturgeon (*Acipenser stellatus* Pallas) during final maturation induced by LH-RH-analogue. J Appl Ichthyol 22(s1):334–339

Bayunova L, Semenkova T, Canario AV, Gerasimov A, Barannikova I (2011) Free and conjugated androgen and progestin levels in the serum of stellate sturgeon (*Acipenser stellatus* Pallas) males treated with female coelomic fluid. J Appl Ichthyol 27(2):655–659

Billard R, Breton B, Fostier A, Jalabert B, Weil C (1978) Endocrine control of the teleost reproductive cycle and its relation to external factors: salmonid and cyprinid models. In: Gaillard PJ, Boer HH (eds) Comparative endocrinology. Elsevcier/North Holland Biomedical Press, Amsterdam, pp 37–48

Billard R, Fostier A, Weil C, Breton B (1982) Endocrine control of spermatogenesis in teleost fish. Can J Fish Aquat Sci 39(1):65–79

Blasco M, Somoza GM, Vizziano-Cantonnet D (2013) Presence of 11-ketotestosterone in pre-differentiated male gonads of *Odontesthes bonariensis*. Fish Physiol Biochem 39(1): 71–74

Borg B (1994) Androgens in teleost fishes. Comp Biochem Physiol Part C: Pharmacology, Toxicology and Endocrinology 109(3):219–245

Bukovskaya O, Lambert J, Kime D (1997) *In vitro* steroidogenesis by gonads of the Russian sturgeon, *Acipenser gueldenstaedtii* Brandt. Fish Physiol Biochem 16(4):345–353

Ceapa C, Williot P, Le Menn F, Davail-Cuisset B (2002) Plasma sex steroids and vitellogenin levels in stellate sturgeon (*Acipenser stellatus* Pallas) during spawning migration in the Danube River. J Appl Ichthyol 18(4–6):391–396

Cuisset B, Fostier A, Williot P, Bennetau-Pelissero C, Le Menn F (1995) Occurrence and in vitro biosynthesis of 11-ketotestosterone in Siberian sturgeon, *Acipenser baerii* Brandt maturing females. Fish Physiol Biochem 14(4):313–322

Cuisset B, Pradelles P, Kime DE, Kühn ER, Babin P, Davail S, Le Menn F (1994) Enzyme immunoassay for 11-ketotestosterone using acetylcholinesterase as label: application to the measurement of 11-ketotestosterone in plasma of Siberian sturgeon. Comp Biochem Physiol 108C(2):229–241

Davail-Cuisset B, Lacomme S, Viaene E, Williot P, Lepage M, Gonthier P, Davail S, Rouault T (2008) Hormonal profile in adult European sturgeon, *Acipenser sturio*, adapted to hatchery conditions in France. Cybium 32:169–170

Davail-Cuisset B, Rouault T, Williot P (2011) Estradiol, testosterone, 11-ketotestosterone, 17, 20β-dihydroxy-4-pregnen-3-one and vitellogenin plasma levels in females of captive European sturgeon, *Acipenser sturio*. J Appl Ichthyol 27(2):666–672

Diotel N, Do Rego JL, Anglade I, Vaillant C, Pellegrini E, Gueguen MM, Mironov S, Vaudry H, Kah O (2011) Activity and expression of steroidogenic enzymes in the brain of adult zebrafish. Eur J Neurosci 34(1):45–56

Fostier A, Breton B, Jalabert B, Marcuzzi O (1981) Evolution of plasma levels of glycoproteic gonadotropins and of 17 alpha hydroxy-20 beta dihydroprogesterone during maturation and ovulation of rainbow trout, *Salmo gairdneri*. CR Acad Sci Serie III, Sciences de la vie 293(15):817–820

Fostier A, Jalabert B, Billard R, Breton B, Zohar Y (1983) The gonadal steroids. In: Hoar WS, Randall DJ (eds) Fish physiology, vol 9. Academic Press, New York, pp 277–372

Fostier A, Jalabert B, Terqui M (1973) Predominant action of a hydroxylated derivative of progesterone on the *in vitro* maturation of ovocytes of the rainbow trout (*Salmo gairdneri*). CR Acad Sci, Serie D: Sciences naturelles 277(4):421–424

García-Alonso J, Nappa A, Somoza G, Rey A, Vizziano D (2003) *In vitro* steroid metabolism during final oocyte maturation in white croaker *Micropogonias furnieri* (Sciaenidae). Fish Physiol Biochem 28(1–4):337–338

García-Alonso J, Nappa A, Somoza G, Rey A, Vizziano D (2004) Steroid metabolism *in vitro* during final oocyte maturation in white croaker *Micropogonias furnieri* (Pisces: Scianidae). Braz J Biol 64(2):211–220

Gower D, Fotherby K (1975) Biosynthesis of the androgens and oestrogens. In: HLJ M (ed) Biochemistry of steroid hormones. Blackwell, Oxford, pp 77–104

Guiguen Y, Fostier A, Piferrer F, Chang CF (2010) Ovarian aromatase and estrogens: a pivotal role for gonadal sex differentiation and sex change in fish. Gen Comp Endocrinol 165(3):352–366

Hamlin HJ, Milnes MR, Beaulaton CM, Albergotti LC, Guillette LJ (2011) Gonadal stage and sex steroid correlations in Siberian sturgeon, *Acipenser baerii*, habituated to a semitropical environment. J World Aquacul Soc 42(3):313–320

Hattori RS, Fernandino JI, Kishii A, Kimura H, Kinno T, Oura M, Somoza GM, Yokota M, Strussmann CA, Watanabe S (2009) Cortisol-induced masculinization: does thermal stress affect gonadal fate in pejerrey, a teleost fish with temperature-dependent sex determination? PLoS One 4(8):e6548

Hurvitz A, Jackson K, Yom-Din S, Degani G, Levavi-Sivan B (2008) Sexual development in Russian sturgeon (*Acipenser gueldenstaedtii*) grown in aquaculture. Cybium 32:283–285

Ijiri S, Kaneko H, Kobayashi T, Wang DS, Sakai F, Paul-Prasanth B, Nakamura M, Nagahama Y (2008) Sexual dimorphic expression of genes in gonads during early differentiation of a teleost fish, the Nile tilapia *Oreochromis niloticus*. Biol Reprod 78(2):333–341

Jalabert B, Fostier A, Breton B, Weil C (1991) Chapter 2: Oocyte maturation in vertebrates. In: PKT P, Schreibman MP (eds) Vertebrate endocrinology: fundamentals and biomedical implications, vol 4A. Academic Press, New York, pp 23–90

Kagawa H, Young G, Nagahama Y (1983) Relationship between seasonal plasma estradiol-17 beta and testosterone levels and in vitro production by ovarian follicles of amago salmon (*Oncorhynchus rhodurus*). Biol Reprod 29(2):301–309

Khara H, Falahatkar B, Meknatkhah D, Ahmadnezhad M, Efatpanah I, Poursaeid S, Rahbar M (2013) Effect of dietary estradiol 17 on growth, hematology and biochemistry of stellate sturgeon *Acipenser stellatus*. WORLD 5(2):113–120

Kime D (1995) Steroid Nomenclature. Gen Comp Endocrinol 98(2):119–120

Lubzens E, Young G, Bobe J, Cerdà J (2010) Oogenesis in teleosts: how fish eggs are formed. Gen Comp Endocrinol 165(3):367–389

Miura T, Higuchi M, Ozaki Y, Ohta T, Miura C (2006) Progestin is an essential factor for the initiation of the meiosis in spermatogenetic cells of the eel. PNAS 103(19):7333–7338

Miura T, Miura CI (2003) Molecular control mechanisms of fish spermatogenesis. Fish Physiol Biochem 28(1–4):181–186

Miura T, Miura C, Ohta T, Nader MR, Todo T, Yamauchi K (1999) Estradiol-17β stimulates the renewal of spermatogonial stem cells in males. Bioch Biophys Res Comm 264(1):230–234

Miura T, Yamauchi K, Takahashi H, Nagahama Y (1992) The role of hormones in the acquisition of sperm motility in salmonid fish. J Exp Zool 261(3):359–363

Nagahama Y (1987) 17α, 20β-dihydroxy-4-pregnen-3-one: a teleost maturation-inducing hormone. Dev Growth Diff 29(1):1–12

Nagahama Y (1994) Endocrine regulation of gametogenesis in fish. Int J Dev Biol 38:217–217

16 Gonadal Steroids in Sturgeons

Nagahama Y, Hirose K, Young G, Adachi S, Suzuki K, Tamaoki B-i (1983) Relative *in vitro* effectiveness of 17α, 20β-dihydroxy-4-pregnen-3-one and other pregnene derivatives on germinal vesicle breakdown in oocytes of ayu (*Plecoglossus altivelis*), amago salmon (*Oncorhynchus rhodurus*), rainbow trout (*Salmo gairdneri*), and goldfish (*Carassius auratus*). Gen Comp Endocrinol 51(1):15–23

Nagahama Y, Yoshikuni M, Yamashita M, Tanaka M (1994) Regulation of oocyte maturation in fish. Fish Physiol 13:393–439

Nakamura M, Kobayashi T, Chang XT, Nagahama Y (1998) Gonadal sex differentiation in teleost fish. J Exp Zool 281(5):362–372

Pelissero C, Le Menn F (1988) Détermination des taux plasmatiques de stéroides sexuels et de la vitellogénine chez l'esturgeon sibérien *Acipenser baeri* élevé en pisciculture. CR Acad Sci Paris 3007(Série III):749–754

Pelissero C, Le Menn F (1991) Evolution of sex steroid levels in males and first time maturing females of the Siberian sturgeon (*Acipenser baerii*) reared in a French fish farm. In: Williot P (ed) Acipenser. Cemagref Publ, Antony, pp 87–97

Piferrer F, Guiguen Y (2008) Fish gonadogenesis. Part II: molecular biology and genomics of sex differentiation. Rev Fish Sci 16(S1):35–55

Schulz R, Andriske M, Lembke P, Blüm V (1992) Effect of salmon gonadotropic hormone on sex steroids in male rainbow trout: plasma levels and testicular secretion *in vitro*. J Comp Physiol B 162(3):224–230

Schulz RW, de Franca LR, Lareyre JJ, Le Gac F, Chiarini-Garcia H, Nobrega RH, Miura T (2010) Spermatogenesis in fish. Gen Comp Endocrinol 165(3):390–411

Schulz RW, Miura T (2002) Spermatogenesis and its endocrine regulation. Fish Physiol Biochem 26(1):43–56

Scott A, Canario A (1987) Status of oocyte maturation-inducing steroids in teleosts. In: Idler DR, Crim LW, Walsh JM (eds) Proceedings of the third international symposium on reproductive physiology of fish. Memorial University of Newfoundland St. John's, Newfoundland, pp 224–234

Scott A, Sumpter J (1983) The control of trout reproduction: basic and applied research on hormones. In: Rankin JC, Pitcher TJ, Duggan RT (eds) Control processes in fish physiology. Croom Helm, London, pp 200–220

Scott A, Sumpter J, Stacey N (2010) The role of the maturation-inducing steroid, 17, 20β-dihydroxypregn-4-en-3-one, in male fishes: a review. J Fish Biol 76(1):183–224

Semenkova T, Barannikova I, Kime D, McAllister B, Bayunova L, Dyubin V, Kolmakov N (2002) Sex steroid profiles in female and male stellate sturgeon (*Acipenser stellatus* Pallas) during final maturation induced by hormonal treatment. J Appl Ichthyol 18(4–6):375–381

Semenkova TB, Canário AV, Bayunova LV, Couto E, Kolmakov NN, Barannikova IA (2006) Sex steroids and oocyte maturation in the sterlet (*Acipenser ruthenus* L.) J Appl Ichthyol 22(s1): 340–345

Taranger GL, Carrillo M, Schulz R, Fontaine P, Zanuy S, Felip A, Weltzien FA, Dufour S, Karlsen O, Norberg B, Andersson HT (2010) Control of puberty in farmed fish. Gen Comp Endocrinol 165:483–515

Thomas P (2012) Rapid steroid hormone actions initiated at the cell surface and the receptors that mediate them with an emphasis on recent progress in fish models. Gen Comp Endocrinol 175(3):367–383

Thomas P, Breckenridge-Miller D, Detweiler C (1997) Binding characteristics and regulation of the 17, 20β, 21-trihydroxy-4-pregnen-3-one (20β-S) receptor on testicular and sperm plasma membranes of spotted seatrout (*Cynoscion nebulosus*). Fish Physiol Biochem 17(1–6):109–116

Trant JM, Thomas P, Shackleton CH (1986) Identification of 17α, 20β, 21-trihydroxy-4-pregnen-3-one as the major ovarian steroid produced by the teleost *Micropogonias undulatus* during final oocyte maturation. Steroids 47(2):89–99

Ueda H, Hiroi O, Hara A, Yamauchi K, Nagahama Y (1984) Changes in serum concentrations of steroid hormones, thyroxine, and vitellogenin during spawning migration of the chum salmon, *Oncorhynchus keta*. Gen Comp Endocrinol 53(2):203–211

Ueda H, Kambegawa A, Nagahama Y (1985) Involvement of gonadotrophin and steroid hormones in spermiation in the amago salmon, *Oncorhynchus rhodurus*, and goldfish, *Carassius auratus*. Gen Comp Endocrinol 59(1):24–30

Vizziano D, Barrios F, Astigarraga I, Breton B, Williot P (2006) Unusual conditions for Siberian sturgeon (*Acipenser baerii* Brandt) spawning. J Appl Ichthyol 22(s1):325–330

Vizziano D, Fostier A, Le Gac F, Loir M (1996) 20 beta-hydroxysteroid dehydrogenase activity in nonflagellated germ cells of rainbow trout testis. Biol Reprod 54(1):1–7

Vizziano D, Randuineau G, Baron D, Cauty C, Guiguen Y (2007) Characterization of early molecular sex differentiation in rainbow trout, *Oncorhynchus mykiss*. Dev Dyn 236(8):2198–2206

Vizziano-Cantonnet D, Di Landro S, Lasalle A, Martínez A, Mazzoni TS, Quagio-Grassiotto I (2016) Identification of the molecular sex-differentiation period in the siberian sturgeon. Mol Reprod Dev 83(1):19–36

Vizziano-Cantonnet D, Mateo M, Alberro A, Barrios F, Fostier A (2015) 17, 20β-P and cortisol are the main in vitro metabolites of 17-hydroxy-progesterone produced by spermiating testes of *Micropogonias furnieri* (Desmarest, 1823) (Perciformes: Sciaenidae). Neotrop Ichthyol 13(3):613–624

Vlassenko AD, Pavlov AV, Sokolov LI, Vasil'ev VP (1989) *Acipenser gueldenstaedtii* Brandt, 1833. In: Holcik L (ed) The freshwater fishes of Europe. General introduction to fishes acipenseriformes. AULA-Verlag, Wiesbaden, pp 294–344

Webb MA, Feist GW, Trant JM, Van Eenennaam JP, Fitzpatrick MS, Schreck CB, Doroshov SI (2002) Ovarian steroidogenesis in white sturgeon (*Acipenser transmontanus*) during oocyte maturation and induced ovulation. Gen Comp Endocrinol 129(1):27–38

Webb M, Van Eenennaam J, Doroshov S (2000) Effects of steroid hormones on *in vitro* oocyte maturation in white sturgeon (*Acipenser transmontanus*). Fish Physiol Biochem 23(4):317–325

Yamamoto TO (1969) Sex differentiation. In: Hoar WS, Randall DJ (eds) Fish physiology, vol 3. Academic Press, New York, pp 117–175

Yaron Z (1995) Endocrine control of gametogenesis and spawning induction in the carp. Aquaculture 129(1):49–73

Young G, Crim LW, Kagawa H, Kambegawa A, Nagahama Y (1983a) Plasma 17α, 20β-dihydroxy-4-pregnen-3-one levels during sexual maturation of amago salmon (*Oncorhynchus rhodurus*): correlation with plasma gonadotropin and *in vitro* production by ovarian follicles. Gen Comp Endocrinol 51(1):96–105

Young G, Kagawa H, Nagahama Y (1982) Oocyte maturation in the amago salmon (*Oncorhynchus rhodurus*): In vitro effects of salmon gonadotropin, steroids, and cyanoketone (an inhibitor of 3β-hydroxy-Δ5-steroid dehydrogenase). J Exp Zool 224(2):265–275

Young G, Ueda H, Nagahama Y (1983b) Estradiol-17β and 17α, 20β-dihydroxy-4-pregnen-3-one production by isolated ovarian follicles of amago salmon *(Oncorhynchus rhodurus)* in response to mammalian pituitary and placental hormones and salmon gonadotropin. Gen Comp Endocrinol 52(2):329–335

Yueh W, Chang C (1997) 17α, 20β, 21-trihydroxy-4-pregnen-3-one and 17α, 20β-dihydroxy-4-pregnen-3-one stimulated spermiation in protandrous black porgy, *Acanthopagrus schlegelii*. Fish Physiol Biochem 17(1–6):187–193

Steroid Profiles Throughout the Hormonal Stimulation in Females Siberian Sturgeon *Acipenser baerii* Brandt

17

Patrick Williot, Sylvain Comte, Françoise Le Menn, and Blandine Davail-Cuisset

Abstract

By setting up a cannulation on farmed Siberian sturgeon spawners, patterns of testosterone (T), 11-ketotestosterone (11-KT), 17,20β-dihydroxy-4-pregnen-3-one (17,20βP) and oestradiol (E_2) are established in non-stressed females throughout complete two alternatives of spawning procedures. Treatment 1 consisted in (a) setting up the cannula together with the ovarian follicles sample; (b) later on, injecting the fish with hormones without any handling; and (c) collecting the ovulated eggs. Treatment 2 contracted the setting up of the cannula, the sampling of the ovarian follicles and the hormonal injection in one handling; collection of ovulated eggs underwent later on. Oestradiol remained at a very low level from 0.3 to 1.2 ng/mL throughout the experiments. Figures for T show a surge 6–12 h post-hormonal injection from 150–210 ng/mL up to 260–320 ng/mL for treatment 1 and 2, respectively. Peaks of T declined rapidly in treatment 1 and last around 20 h in treatment 2 before declining to low initial values. 11-KT increased slowly postinjection in a lower extent than for T with a maximum

---

P. Williot (✉)
4, Rue du pas de madame, 33980 Audenge, France

Cemagref (Now Irstea), 50 avenue de Verdun, 33612 Cestas, France
e-mail: williot.patrick@neuf.fr

S. Comte
Cemagref (Now Irstea), 50 avenue de Verdun, 33612 Cestas, France

F. Le Menn
Former Laboratory of Biology of the Reproduction of Fish, University of Bordeaux I, Avenue des Facultés, F-33405 Talence Cedex, France

B. Davail-Cuisset
Former Laboratory of Biology of the Reproduction of Fish, University of Bordeaux I, Avenue des Facultés, F-33405 Talence Cedex, France

Laboratory of Physical and Toxical Chemistry, Environmental Toxicology, University of Bordeaux 1, 33405 Talence Cedex, France

20–25 h post-hormonal injection with about 60 and 80 ng/mL for treatments 1 and 2, respectively. Figures for treatment 2 are pulsation-like. 17,20βP increased postinjection more rapidly in treatment 2 and decreased before ovulation. Altogether, treatment 2, characterised by a two-step spawning procedure instead of three in treatment 1, showed higher values and more durable elevated levels of steroids than for treatment 1. We discuss the above results in the light of sturgeon-related literature.

Keywords

Acipenser baerii • Farming • Cannulation • Induced spawning • Steroids profiles

Introduction

Sex steroid hormones play a major role in fish reproductive functions in teleosts; see Fostier et al. (1983) for a review. With regard to sturgeons, testosterone (T) was identified for in mature plasma females of *Acipenser oxyrhynchus* (Sangalang et al. 1971). Later on, T was shown on wild stellate sturgeon (*Acipenser stellatus*) during the spawning induction (Bukovskaya 1991), on wild Russian sturgeon (*Acipenser gueldenstaedtii*) in spring form just before spawning (Barannikova 1991) and on bester (*H. huso* × *A. ruthenus*) at migrating nucleus stage (Mojazi Amiri et al. 1996). Concerning the Siberian sturgeon (*Acipenser baerii*), high plasmatic contents in testosterone were shown for females from 5 years old and older (Pelissero and Le Menn 1988), in androgens of females of which the vitellogenesis was achieved (Pelissero and Le Menn 1991). Further investigations in post-vitellogenic females of Siberian sturgeon (*Acipenser baerii*) (i.e. females of which the migration of the germinal vesicle towards the animal pole of the ovarian follicles was achieved) revealed the presence of 11-ketotestosterone (11-KT) in the plasma of females (Cuisset et al. 1995). At a time of the present study (mid-1990s), no other result was published in the field.

The most potent maturation-inducing steroids (MIS) were the 17-hydroxy-20β-dihydro-progesterone (17,20β-dihydroxy-4-pregnen-3-one[1] abbreviated as 17,20βP) as it was shown to be present in responsive sturgeon of *Acipenser transmontanus* and *Acipenser oxyrinchus* females, respectively (Lutes et al. 1987; Van Eenenaam et al. 1996), and the tri-hydroxylated progestin (17,20β,21-trihydroxy-4-pregnen-3-one abbreviated as 20βS) previously described as the MIS in sciaenid fish (Thomas 1994).

However, most of published data from in vivo studies were obtained from punctually blood sampling. Therefore, to limit the impacts of handling, the fish were cannulated as already described by Williot (1997) and later by Williot et al. (2011) (see Chap. 24 for details) which allows determining the steroid profiles in non-stressed females. Moreover, as in the ovary of teleosts, 11-KT and oestradiol have the same precursor, the testosterone; the oestradiol was also assayed. In addition, we

---

[1] The correct description of the molecule is 17-hydroxy-20β-dihydro-progesterone, while the international nomenclature is 17,20β-dihydroxy-4-pregnen-3-one after Dr. A. Fostier.

17 Steroid Profiles During Controlled Spawning

took the opportunity of this work to compare the profiles obtained in two different ways of reproduction procedures of the females, that is assessing firstly the in vitro maturation competence of the ovarian follicles prior to hormonal stimulation or not. The same experiment was used to compare the two procedures with regard to stress indicators and reproductive potential (Williot et al. 2011).

The objective of the present study was to establish the patterns of testosterone (T), 11-ketotestosterone (11-KT), 17,20βP and oestradiol (E_2) throughout the spawning of non-stressed Siberian sturgeon submitted to two managing spawning procedures. At a time of the present study, we unfortunately were unable to measure the 20βS.

17.1 Material and Methods

17.1.1 Fish and Preliminary Handling

All tagged fish were already-spawned 11-year-old females. They were randomly chosen in spring (February to April) from preselected fish in the previous November according to findings of Williot and Brun (1998). Three days before the beginning of each experiment of the study, about 15 randomly chosen females were isolated at the head of the raceway to make them easily fishable. One day before first fish handling in rearing raceways, the fish were starved. In order to establish the basic levels of the plasma characteristics, the fish were prepared as follows: at the beginning of the experiment, females were immediately blood-sampled in situ (in the raceways) in the caudal vasculature. This procedure lasts no more than 2 min and then is presumed to be representative of the basic level. The fish were then carried each one in a well-oxygenated 2-m-diameter tank with a water depth of 40–50 cm. All the fish exhibited fully developed ovarian follicles (3.3–3.4 mm) and almost completely achieved migration of germinal vesicle (polarisation index = 3–6%) (Williot 1997; Williot et al. 2011).

17.1.1.1 Experimental Design

Two spawning procedures were applied: treatment 1 had three steps, sampling of ovarian follicles, hormonal injection and spawning; treatment 2 had two steps, sampling ovarian follicles and hormonal injection simultaneously and further spawning (Table 17.1A). The cannulation of the fish was performed at a time of the first respective step. It is worthy to note that hormonal injection in treatment 1 do not need handling as it was practised directly in fish moving in 2-m-diameter tanks. Hormonal injection consisted in IM injection of aqueous suspension of carp pituitary powder (Argent Chemical Laboratories, USA) at a rate of 5 and 2 mg/kg of body weight for females and males, respectively (Williot et al. 1991). From that time, temperature was maintained at 15 °C onwards in agreement with our previous experimental works. Two trials were carried out, the first in February and the second in April (Table 17.1B). Due to overload timetable, control with sham-injected females was organised soon after the second trial (Table 17.1B). In each trial, eight fish were randomly chosen and separated into two groups of four females. The treatment 2 females corresponded to our shortened management procedure (the two-step

Table 17.1 Protocol

A) Scheme of handling and blood sample schedule depending on spawning procedure in a given experiment (modified after Williot et al. 2011)

Spawning procedure	Operation and blood sample timetable		
Treatment 1 (3 steps)	Ovarian follicle sample & cannulation (handling)	Hormonal injection (no handling)	Ovulation, Spawning, (handling)
Blood sampling schedule (h)	0	30/32 33 35 38 44 50 56 62 68 72 79	
Treatment 2 (2 steps)		Ovarian follicle sample, cannulation, hormonal injection (handling)	Ovulation, Spawning (handling)
Blood sampling schedule (h)		30/32 33 35 38 44 50 56 62 68 72 79	

B) Number of females and time table

	10/02	10/04	18/04
	First trial	Second trial	control
Treatment 1	4 (females 1–4)	4 (females 43–46)	
Treatment 2	4 (females 5–6)	4 (females 47–50)	
Control			6 (sham injected)

one) that is with close successive follicle sampling and hormonal injection. Both of these interventions were preceded by cannula installation. Females were blood-sampled according to the timetable given in Table 17.1A.

The treatment 1 females (three-step one) followed the normal procedure with firstly cannula installation immediately succeeded by follicle sampling and 30–32 h later by hormonal injection on dorsal muscle as usual, carried out directly in the tank without handling the fish. After each operation, the time schedule for blood sampling was the same as for treatment 2 females.

17.1.2 Fish Cannulation (for Details See Williot et al. (2011) and Chap. 24)

To carry out the operation, females were placed ventral side up on an appropriate table with a continuous supply of water in the mouth as described by Doroshov et al. (1983). Fish were tranquilised for 5 min in a bath with clove oil at a concentration of 40 ppm. First, we carefully placed the cannula in the dorsal artery behind the genital papilla. The operation together with further ovarian follicles sample (very short indeed) lasts half an hour maximum (Williot et al. 2011).

17.1.3 Blood Samples, Preparation and Analysis

For each time sample, the first 0.5 mL blood-collected fraction was rejected. After collecting 2–2.5 mL, the rest of the blood was carefully pushed back with an adapted Ringer solution (Salin 1992) composed of NaCl (65 mmol/L), KCl (1.35 mmol/L), CaCl (0.95 mmol/L), $MgCl_2$ (0.45 mmol/L), $NaHCO_3$ (2.5 mmol/L), glucose (0.5 mmol/L) and Li-heparin (50 UI/mL), and then the cannula was again closed with a flame. The blood was collected in microtubes previously supplied with Li-heparin at a ratio of 20 µL of a solution containing 100 UI/mL for 1.25 mL of blood and then immediately centrifuged for 7 min at 7000 rpm. Plasma was then frozen (-20 °C) for further analysis. 11-KT was measured out by an enzyme immunoassay (EIA) according to the method described by Cuisset et al. (1994). Testosterone was determined by radioimmunoassay (RIA) modified from Fostier et al. (1982). The antibody to testosterone raised in rabbits (product AB-1030 from BioClin, UK) has been diluted (1/600) which gave the maximum binding of 40–50%. The comparative cross-reactions of this antibody from the supplier are: 16% (5α-dihydrotestosterone), 5.8% (5α-androstane-3α,17β-diol), 3.7% (5α-androstane-3β,17β-diol) and 2.1% (androstenedione). The tritiated T was purified by chromatography (Sephadex LH20 column) and then tested in benzene ethanol. E_2 were determined by the same procedure as for 11-KT with an antibody anti-E_2. The 17,20β-dihydroxy-4-pregnen-3-one (17,20βP) was determined by RIA. The characteristics of T and E2 methods are similar to those described by Pelissero and Le Menn (1988). With regard to 11-KT, assay sensitivity and specificity are given in Cuisset et al. (1994). The 17,20βP antibody was diluted (1/10), and 17,20βP radioactive tracer contained 2 µCi/mL. Characteristics of the dosage of 17,20βP are those described by Fostier et al. (1981).

17.2 Results and Discussion

Due to mistakes, we were unable to use properly the data of the females 43 and 44 in the second experiment. As there were no significant differences between the two experiments, results were pooled. Indeed, all the cannulated females exhibited a fairly good reproductive potential with about 70% (Williot et al. 2011).

17.2.1 Oestradiol

Sham-injected females (controls) exhibited a somewhat more variable concentration in E_2 with a declining trend over the experimental period. Taking into account all the females and time samples, E_2 remained in the range of 0.05–3 ng/mL (Fig. 17.1) which is in accordance with the findings already reported in fully vitellogenic females (Pelissero and Le Menn 1991).

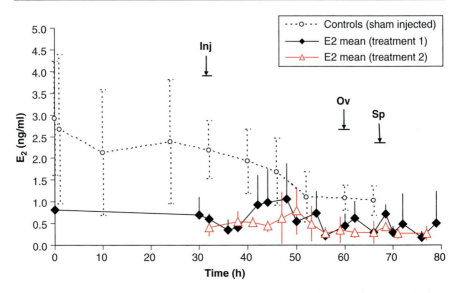

Fig. 17.1 Profiles of plasmatic oestradiol depending on reproductive procedure. *Inj*, injection; *Ov*, ovulation; *Sp*, spawning

17.2.2 Testosterone

The results are shown (Fig. 17.2) for both treatments. The controls exhibited a low content with a small variation around 20–30 ng/mL along the experiment. The general figures are similar whatever the group of females with a rise about 10–12 h after the hormonal injection. However, in the treatment 2 females (two steps), the high level of T (310–340 ng/mL) is shown to be maintained for about 20 h that is up closed to the ovulation time in contrast to rapid declining trend observed in the treatment 1 females up to about 70 ng/mL. Later on the ovulation time, the T content in treatment 2 females declined as well. The only available data on sturgeon were those obtained on wild Stellate sturgeon breeders (*Acipenser stellatus*) by Bukovskaya (1991) who reported a peak ($\leq$ 30 ng/mL) when fish were LH-RH stimulated as compared with an aci-GTH stimulation which lead to a decrease of T. Immature Siberian sturgeon females stimulated with carp pituitary homogenates did not show significant increase in plasmatic T (Cuisset 1993) and then, compared with the present finding, provide putative involvement of T in the reproduction of females. Preovulating Siberian sturgeon (without further precision on gonadal stage) exhibited T content closed to 110 ng/mL (Pelissero and Le Menn 1988).

17.2.3 11-Ketotestosterone

The 11-KT values of controls show a large variation together with a declining trend along the experimental period (Fig. 17.3). Whatever the treatment, a slight increase in plasmatic 11-KT content from 50–70 ng/mL up to 70–90 ng/mL is shown

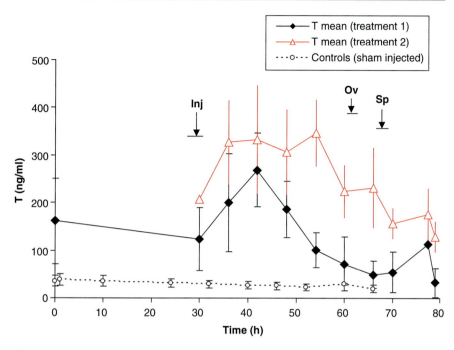

Fig. 17.2 Profiles of plasmatic testosterone depending on reproductive procedure. *Inj*, injection; *Ov*, ovulation; *Sp*, spawning

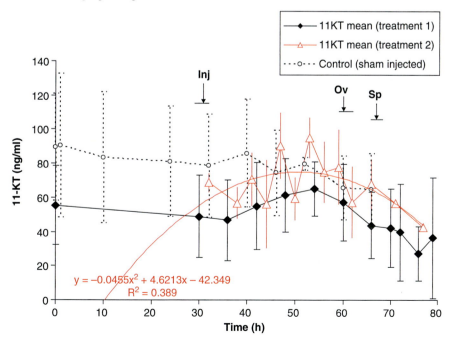

Fig. 17.3 Profiles of plasmatic 11-ketotestosterone depending on reproductive procedure. *Inj*, injection; *Ov*, ovulation; *Sp*, spawning

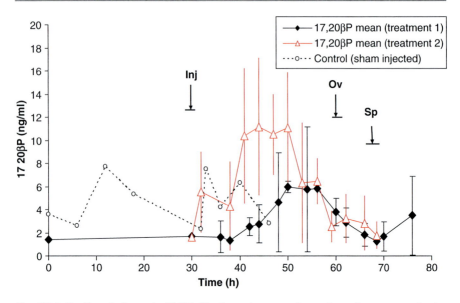

Fig. 17.4 Profiles of plasmatic 17,20β-dihydroxy-4-pregnen-3-one depending on reproductive procedure. *Inj*, injection; *Ov*, ovulation; *Sp*, spawning

20–25 h post-hormonal stimulation. Later on, a general decreasing trend is observed. The figures of treatment 2 females show a pulsative content in 11-KT of which the best regression line (parabola, Fig. 17.3) is about 20 ng/mL above the figure of that of treatment 1 females. These data are in good agreement with those already reported on Siberian sturgeon in an advanced maturational stage (Cuisset et al. 1995).

17.2.4 17,20βP

The profiles for both treatments show an elevation post-hormonal injection of the plasmatic content of 17,20βP from 2 to 6 ng/mL (treatment 1 females) and from 2 to 6 to about 11 ng/mL (treatment 2 females) (Fig. 17.4). Peaks last from about 6–10 h and underwent 10–15 h post-hormonal injection. Afterwards, the contents in 17,20βP decline prior to the initiation of ovulation. The present 17,20βP peaks are higher than those reported on other ovulating sturgeon (Lutes et al. 1987; Van Eenenaam et al. 1996; Webb et al. 2002); therefore we could be incline to suggest involvement of the steroid in the normal reproductive course of the Siberian sturgeon. But, more recently, it has been demonstrated that very low concentrations of the 17,20βP were present among the metabolites of Siberian sturgeon ovarian follicles incubated in vitro, in contrast to the 20βS (21-trihydroxy-4-pregnen-3-one) (Alberro et al. 2008; Chap. 16).

As a result, the question of which molecule is the MIS of the Siberian sturgeon remains on the table.

Since the time the present study was carried out, steroids were investigated in some sturgeon species with two objectives: i) early gender discrimination and ii) establishing of steroid profiles for females prior to or near the spawning time. The first group is exampled by Cuisset et al. (1994) in the farmed Siberian sturgeon with the use of 11-KT, by Ceapa et al. (2002) in the wild Stellate sturgeon (*Acipenser stellatus*) using ratios E2/T or E2/11-KT, by Feist et al. (2004) in farmed white sturgeon (*Acipenser transmontanus*) using 11-KT and T, by Qu et al. (2010) in the farmed Chinese sturgeon (*Acipenser sinensis*) using E2 and T and by Mola and Hovannisyan (2014) in the farmed beluga (*Huso huso*) using E2, 11-KT and T. The second group of studies is focused on steroid dosages and/or profiles for fish closed to or during spawning induction (Table 17.2).

Table 17.2 Synthesis of main steroid dosage in female sturgeon species closed to end of oogenesis or during induced spawning

Species (wild or farmed)	Water temperature (°C)	Stage (age) hormonal stimulation	Steroids levels (ng/mL)	Observation	Source
Acipenser baerii (farmed)	17	Preovulating (10y)	[T] ~ 110 ± 25 [E2] ~ 19 ± 3	Lack of precision in staging	(1)
		Vitellogenic (5y)	[T] ~ 6 ± 0.5 [E2] ~ 17 ± 1	Punctually sampled fish	
Acipenser gueldenstaedtii (wild, Volga)	?	Just before spawning	[T] ~ 14 ± 2.5	Lack of precision in staging	(2)
		Spawning next spring	[T] ~ 4 ± 0.5	Punctually sampled fish	
Bester (*H. huso*♀ × *A. ruthenus* ♂) (farmed)	3–18 or 7–14	Migratory nucleus stage, oocyte diameter > 2.6 mm	[T] around 45–50 [E2] around 1.5–2 vs. 4 for vitellogenic oocyte (1.6–2.6 mm) [17,20βP] < 0.2	Punctually sampled Uncertain staging	(3)
Acipenser gueldenstaedtii (wild, Volga)	14–16	Stage V LH-RH stimulated	[T] ~ 19 ± 7	Sample time uncertain (after induced final maturation)	(4)
			[11-KT] ~ 29 ± 7	Punctually sampled fish	
			[E2] ~ 0.1	Uncertain staging	

(continued)

Table 17.2 (continued)

Species (wild or farmed)	Water temperature (°C)	Stage (age) hormonal stimulation	Steroids levels (ng/mL)	Observation	Source
Acipenser stellatus (wild, Volga)	17–19.7	LH-RH stimulated	[T] shows a peak (250) and then decreased up to 100 at ovulation time	Absence of precise staging	(5)
			[11-KT] shows a decreasing trend from 35 to 25 at ovulation time	Punctually sampled fish (five samples)	
			[20βS] increased from 0.3 to 3.3 at ovulation time		
			[17,20βP] closed to 0.1	No data for samples four and five (before and after ovulation time)	
Acipenser ruthenus (wild, Dvina)	13–14	Sturgeon pituitary (2001) or LH-RH stimulated (2002)	[T] shows a peak around 95 after hormonal stimulation and then lower to about ten at ovulation time	Absence of precise staging	(6)
			[11-KT] shows an increase from about 8–35 after hormonal stimulation and then lower under ten at ovulation time	Fish sacrificed before sampling	
			[17,20βP] shows a surge up to six and lowers to less than two at ovulation time	Likely pooled results of 2 years	
			[20βS] shows a decline from 0.6 to about 0.2 at ovulation time		

Table 17.2 (continued)

Species (wild or farmed)	Water temperature (°C)	Stage (age) hormonal stimulation	Steroids levels (ng/mL)	Observation	Source
Acipenser baerii (farmed)	~15 (France)	GnRH stimulated (10y, Fr and 6y, Ur)	[T] shows an increase after hormonal stimulation (70–165 and 133–159) in Uruguay, whereas it remains at a rather low level (25–35) in France	The two females in Ur exhibited 100% GVBD in vitro and ovulated	(7)
	19 (Uruguay)		[E2] remains at a low levels in every females (0.05–0.5)	The four females in Fr exhibited a poor GVBD and did not ovulated	
Acipenser sturio (farmed)	?	Reputed mature female Hormonal stimulation	[T] shows peak about 12–15	No ovulation	(8)
			[11-KT] < 2	Staging uncertain	
			[17,20βP] < 2	Sampling time uncertain with regard to staging	
Acipenser baerii (farmed)	18–27	Stage V migratory nucleus Oocyte diameter > 2.2 mm	[T] shows a peak at 140 vs. 70 for stage IV	Punctually sampled fish	(9)
			[11-KT] shows a peak at 120 vs. 30 for lower stages	Staging uncertain	
			[E2] shows a decrease to 0.6 vs. 0. 7 at stage IV		

(continued)

Table 17.2 (continued)

Species (wild or farmed)	Water temperature (°C)	Stage (age) hormonal stimulation	Steroids levels (ng/mL)	Observation	Source
Acipenser sturio (farmed)	11–18	GnRH stimulated (12y)	[T] closed to 1 vs. 0.7 2 months earlier	Punctually sampled fish	(10)
			[11-KT] closed to 2.2 vs. 1.7 1 month earlier	One sample at ovulation time	
			[E2] closed to 8 vs. 4.5 1 month earlier	Progenies were obtained	
			[17,20βP] closed to 1 vs. 0.05 1 month earlier		
Acipenser baerii (farmed)		Carp pituitary stimulated Two different reproductive procedures	[T] shows a peak (260–350) about 10 h post-hormonal injection and then lowers at ovulation and later on	Cannulated fish allowing unstressed fish sample	(11)
			[11-KT] shows a less pronounced peaks (70–95) about 25 h post injection and further lower	Fertilisation rate closed to 70%	
			[E2] steady state in a range of 0.5–1.5		
			[17,20βP] shows peaks (6–10) 10–20 h post-hormonal injection and further lower		

(*1*) Pelissero and Le Menn (1988), (*2*) Barannikova (1991), (*3*) Mojazi Amiri et al. (1996), (*4*) Barannikova et al. (2006), (*5*) Bayunova et al. (2006), (*6*) Semenkova et al. (2006), (*7*) Vizziano et al. (2006), (*8*) Davail-Cuisset et al. (2008), (*9*) Hamlin et al. (2011), (*10*) Davail-Cuisset et al. (2011), (*11*) Present study

Conclusions

In conclusion, our design allowed us to show for the first time in non-stressed reproductive Siberian sturgeon females the simultaneous profiles of oestradiol, testosterone, 11-ketotestosterone and 17,20β-dihydroxy-4-pregnen-3-one. The main findings are as follows. Oestradiol remains at a very low level in accordance with the complete development of the ovaries of the experimental females. A peak of T occurred the first similarly to Pelissero and Le Menn (1988) around about 10 h post-hormonal injections and decreased to its lowest values after spawning. Later on, less pronounced 11-KT peaks appeared before ovulation. It has been demonstrated that 11-KT could be produced by interrenal tissue under cortisol precursor (Cuisset et al. 1995). And, high peaks of cortisol were shown in non-stressed females (treatment 1 females) post-hormonal injection (Williot et al. 2011). However, the involvement of androgens in Siberian sturgeon reproduction is still to be explained. Even increase content in 17,20βP is shown between hormonal stimulation and spawning, its role as MIS is questionable in the light of recent in vitro findings (Chap. 16). Altogether, it is worth noting that the treatment 2 (two-step procedure) reproductive procedure leads to an increase in all three steroids investigated in the present study as compared with the treatment 2 (the three-step one).

Acknowledgments Authors are grateful to Drs. Zanui and Carillo for the gift of E_2 antibody and to Dr. Fostier for the gifts of 17,20βP as well as 17,20βP radioactive tracer.

References

Alberro A, Williot P, Vizzano D (2008) Steroid synthesis during ovarian follicle maturation in the Siberian sturgeon *Acipenser baerii*. Cybium 32(2):255

Barannikova IA (1991) Peculiarities of intrapopulational differentiation of sturgeon (*Acipenser gueldenstaedtii*) under present-day conditions. In: Williot P (ed) Acipenser. Cemagref Publ, Antony, pp 137–142

Barannikova IA, Bayunova LV, Semenkova TB (2006) Serum sex steroids and their specific cytosol binding in the pituitary and gonads of Russian sturgeon (*Acipenser gueldenstaedtii* Brandt) during final maturation. J Appl Ichthyol 22:331–333

Bayunova L, Canario AVM, Semenkova T et al (2006) Sex steroids and cortisol levels in the blood of stellate sturgeon (*Acipenser stellatus* Pallas) during final maturation induced by LH-RH analogue. J Appl Ichthyol 22:334–339

Bukovskaya OS (1991) Serum sex steroid levels in stellate sturgeon (*Acipenser stellatus* Pallas) during induced maturation and ovulation. In: Scott AP, Sumpter JP, Kime DE, Rolfe MS (eds) Proceedings of the fourth international symposium on the reproductive physiology of fish. Fish Symp 91, Sheffield, p 98

Ceapa C, Williot P, Le Menn F et al (2002) Plasma steroids and vitellogenin levels in stellate sturgeon (*Acipenser stellatus* Pallas) during spawning migration in the Danube River. J Appl Ichthyol 18:391–396

Cuisset B (1993) Etude endocrinologique de la fonction de reproduction chez l'esturgeon sibérien *Acipenser baeri* Brandt: application au sexage des populations sauvages ou élevées en acipensériculture. Thèse No. 1039, Université Bordeaux 1, p 215

Cuisset B, Fostier A, Williot P et al (1995) Occurrence and in vitro biosynthesis of 11-ketotestosterone in Siberian sturgeon, *Acipenser baerii* Brandt maturing females. Fish Physiol Biochem 14(4):313–322

Cuisset B, Pradelles P, Kime DE et al (1994) Enzyme immunoassay for 11-ketotestosterone using acetylcholinesterase as label: application to the measurement of 11-ketotestosterone in plasma of Siberian sturgeon. Comp Biochem Physiol 108c(2):229–241

Davail-Cuisset B, Lacomme S, Viaene E et al (2008) Hormonal profile in adults of Atlantic European sturgeon, *Acipenser sturio*, adapted to hatchery in France. Cybium 32((2) suppl):169–170

Davail-Cuisset B, Rouault T, Williot P (2011) Hormones and vitellogenin levels in females of Atlantic sturgeon, *Acipenser sturio* changes during maturation process. J Appl Ichthyol 27:666–672

Doroshov SI, Clark WH, Lutes PB et al (1983) Artificial propagation of the white sturgeon, *Acipenser transmontanus* Richardson. Aquaculture 32:93–104

Feist G, Van Eenennaam JP, Doroshov SI et al (2004) Early identification of sex in cultured white sturgeon, *Acipenser transmontanus*, using plasma steroid levels. Aquaculture 232:581–590

Fostier A, Billard R, Breton B et al (1982) Plasma 11-oxotestosterone and gonadotrophin during the beginning of spermiation in rainbow trout (*Salmo gairdnerii*). Gen Comp Endocrinol 46:428–434

Fostier A, Jalabert B, Billard R et al (1983) The gonadal steroids. In: Hoar WS, Randall DJ, Donaldson EM (eds) Fish physiology, vol 9. Academic Press, New York, pp 277–372

Fostier A, Jalabert B, Campbell C et al (1981) Cinétique de libération in vitro de 17α-hydroxy-20β-dihydroprogesterone par des follicules de Truite arc-en-ciel *Salmo gairdnerii*. C R Acad Sci Paris 292:777–780

Hamlin HJ, Milnes MR, Beaulaton CM et al (2011) Gonadal stage and sex steroid correlations in Siberian sturgeon, *Acipenser baerii*, habituated to a semitropical environment. J World Aquacult Soc 42(3):313–320

Lutes PB, Doroshov SI, Chapman F et al (1987) Morpho-physiological predictors of ovulatory success in white sturgeon, *Acipenser transmontanus* Richardson. Aquaculture 66:43–52

Mojazi Amiri B, Maebayashi M, Hara A et al (1996) Ovarian development and serum sex steroid and vitellogenin profiles in the female cultured sturgeon hybrid, the bester. J Fish Biol 48:1164–1178

Mola AE, Hovannisyan HG (2014) Measurements of serum steroid hormones (testosterone, 11-ketotestosterone, and 17 –estradiol) in farmed great sturgeon. Comp Clin Pathol. doi:10.1007/s00580-014-1931-9

Pelissero C, Le Menn F (1988) Détermination des taux plasmatiques de stéroïdes sexuels et de la vitellogénine chez l'esturgeon sibérien élevé en pisciculture. C R Acad Sci Paris 307(série III):749–754

Pelissero C, Le Menn F (1991) Evolution of sex steroid levels in males and first time maturing females of the Siberian sturgeon (*Acipenser baerii*) reared in a French fish farm. In: Williot P (ed) Acipenser. Cemagref Publ, Antony, pp 87–97

Qu QZ, Sun DJ, Wan BQ et al (2010) The relationships between gonad development and sex steroid levels at different ages in *Acipenser schrenckii*. J Appl Ichthyol 26:1–5

Salin D (1992) La toxicité de l'ammoniaque chez l'esturgeon sibérien, *Acipenser baeri*: effets morphologiques, physiologiques et métaboliques d'une exposition à des doses sublétales et létales. Thèse N° 749, Université Bordeaux I, p 134

Sangalang GB, Weisbart M, Idler DR (1971) Steroids of a chondrostean: corticosteroids and testosterone in the plasma of the American Atlantic sturgeon, *Acipenser oxyrhynchus* Mitchill. J Endocrinol 50:413–421

Semenkova TB, Canario AVM, Bayunova LV et al (2006) Sex steroids and oocyte maturation in the sterlet (*Acipenser ruthenus* L.) J Appl Ichthyol 22:340–345

Thomas P (1994) Hormonal control of final oocyte maturation in sciaenid fishes. In: Davey K, Peter R, Tobe S (eds) . S S National Research Council of Canada, Ottawa, pp 619–625

Van Eenenaam JP, Doroshov SI, Moberg GP et al (1996) Reproductive conditions of the Atlantic sturgeon (*Acipenser oxyrinchus*) in the Hudson River. Estuaries 19(4):769–777

Vizziano D, Barrios F, Astirraga I et al (2006) Unusual conditions for Siberian sturgeon (*Acipenser baerii* Brandt) spawning. J Appl Ichthyol 22:325–330

Webb MAH, Feist GW, Trant J et al (2002) Ovarian steroidogenesis in white sturgeon (*Acipenser transmontanus*) during oocyte maturation and induced ovulation. Gen Comp Endocrinol 129:27–38

Williot P (1997) Reprodeuction de l'esturgeon sibérien (*Acipenser baeri* Brandt) en élevage: gestion des génitrices, compétence à la maturation in vitro de follicules ovariens et caractéristiques plasmatiques durant l'induction de la ponte. Thèse n°1822, Université Bordeaux I, p 227.

Williot P, Brun R (1998) Ovarian development and cycles in cultured Siberian sturgeon, *Acipenser baerii*. Aquat Liv Ressour 11(2):111–118

Williot P, Brun R, Rouault T et al (1991) Management of female spawners of the Siberian sturgeon, *Acipenser baerii* BRANDT: first results. In: Williot P (ed) Acipenser. Cemagref Publ, Antony, pp 365–379

Williot P, Comte S, Le Menn F (2011) Stress indicators throughout the reproduction of farmed Siberian sturgeon, *Acipenser baerii* (Brandt) females. Int Aquat Res 3:31–43

Part III

Ecophysiology: Adaptation to Environment

Respiratory and Circulatory Responses to Hypoxia in the Sturgeon, *Acipenser baerii*

Guy Nonnotte, Patrick Williot, and Valérie Maxime

Abstract

Siberian sturgeon, *Acipenser baerii*, when exposed to progressive hypoxia, was able to maintain standard oxygen consumption until a low critical level of ambient PO_2 ($PwO_2 < 40$ mm Hg). During the post-hypoxic period, an O_2 debt was repaid by an elevated oxygen consumption indicating that a shift to anaerobic metabolism had occurred during the exposure to severe hypoxia. Gradually increasing ambient hypoxia initially induced a respiratory alkalosis. Below the critical level of PwO_2 and during normoxic recovery, a flush of lactate into the blood was associated with a metabolic acidosis which was totally compensated 3.5 h after return to normoxia. Respiratory responses of the sturgeon to progressive hypoxia reveal a typical O_2 regulatory behavior.

An acute severe hypoxia ($PwO_2 = 10$ mmHg) followed by a rapid return to normoxia caused a significant stress to the fish, as revealed by high levels of plasma catecholamines and cortisol. The moderate rise in heart rate and in dorsal aortic blood pressure observed during the first phase of hypoxia represented typical results of increased plasma catecholamines. These effects were then masked by a vagal reflex resulting in bradycardia. Deep hypoxia induced a hyperventilatory response followed by a marked ventilatory depression at the lowest level of PwO_2. The initial ventilatory alkalosis was combined with a moderate metabolic acidosis. The latter was amplified during the first 2 h of the recovery period in

G. Nonnotte (✉)
12 rue Marcel Pagnol, 33260 La Teste de Buch, France
e-mail: guy.nonnotte@wanadoo.fr

P. Williot
4 rue du Pas de Madame, 33980 Audenge, France

V. Maxime
Département Sciences de la Matière et de la Vie, University of South Britain, BP92116, F-56321 Lorient-cedex, France

normoxia, concomitantly with a flush of lactate into the blood and an increase in plasma sodium concentration. During normoxic recovery, hyperventilation resumed, consistent with the repayment of an oxygen debt.

Keywords
Siberian sturgeon • Hypoxia • Acid-base status • O_2 regulator • Ventilation Circulation • Heart rate • Catecholamines • Cortisol • Fish

Abbreviations

MO_2	Standard oxygen consumption
PaO_2	Oxygen partial pressure in arterial blood
PCO_2	Carbon dioxide partial pressure
PO_2	Oxygen partial pressure
PwO_2	Oxygen partial pressure in water
αwO_2	O_2 solubility in water

Introduction

With respect to the relationship between oxygen consumption and ambient oxygen tension, aerobic organisms have been classically described as oxyconformers, when their O_2 consumption varies directly with ambient PO_2, or oxyregulators if O_2 consumption is independent of this factor, at least within a certain range above a critical O_2 pressure, Pc. Oxyregulation is made possible by a number of compensatory responses increasing the various O_2 conductances in the gas exchange system and allowing maintenance of an unchanged tissue O_2 supply, in spite of a reduced ambient O_2 availability (see Dejours 1981). Oxyconformity, throughout the whole PO_2 range, or failure of oxyregulation below the P_c, results from either the absence or the limitation of such mechanisms, so that the oxygen supply cannot match the oxygen demand anymore. In these cases, tissue energy expenditure may either be depressed or roughly maintained by shifting to anaerobic metabolism, at the price of a so-called O_2 debt which must be repaid upon return to normoxic conditions.

Oxyconformity has rarely been described unambiguously in vertebrates (see Prosser 1973; Ultsch et al. 1981). One of the best documented examples is the sturgeon *Acipenser transmontanus*, in which both O_2 consumption and gill water flow rate were reported to be reduced steadily with declining ambient oxygen tension (Burggren and Randall 1978). Furthermore, the absence of any O_2 debt repayment and of any decrease of blood pH indicated that a reduced total energy expenditure took place during hypoxic exposure, rather than a shift to anaerobic metabolism maintaining energy production.

Although based on clear and apparently consistent findings and fitting well the ancientness of the Chondrostei among vertebrate groups, this notion of oxyconformity being a characteristic of sturgeons was not supported by a large amount of previous data from Russian workers (see Vinberg 1956; Klyashtorin 1981) and has even been challenged by a more recent investigation on the same species (Ruer et al. 1987). However, these criticisms were only based on measurements of O_2 consumption as a function of O_2 tension. Therefore, in order to settle this problem, we have investigated as completely as possible the respiratory and acid-base responses to declining oxygen tension in the Siberian sturgeon *Acipenser baerii*.

The frequent occurrence of hypoxia in aquatic environments has entailed many studies carried out in fish, primarily rainbow trout, dealing with various aspects of respiratory and circulatory responses to decreased ambient PwO_2 (Tetens and Lykkeboe 1985; Boutilier et al. 1988). Environmental hypoxia initiates immediate ventilatory and cardiovascular reflexes, allowing the maintenance of arterial oxygen saturation at its optimum level despite a reduced transbranchial oxygen gradient (for review, see Randall and Perry 1992). The typical response of fish to hypoxia consists in increased ventilation rate and amplitude and bradycardia. The nature of the mechanisms that regulate these processes has received large interest. Circulatory and ventilatory control by catecholamines released into the plasma under adverse conditions such as environmental hypoxia has been more specifically and extensively studied (see Fritsche and Nilsson 1993). Moreover, in extreme hypoxia, changing levels of circulating catecholamines have other numerous physiological effects, both direct and indirect, all of which lead to either increases in or maintenance of energy turnover and oxygen supply (Randall and Perry 1992).

When compared with the reply to hypoxia of another species of sturgeons, *Acipenser naccarii* (Randall et al. 1992) and of other studied fishes, the sturgeons seem to present a high ability to cope with the lack of ambient oxygen. Considering that the time course and severity of hypoxia greatly influence the degree of physiological responses of fish (Boutilier et al. 1988), it was also pertinent to attempt to determine the limits of this ability by measuring respiratory, acid-base, and circulatory responses in the Siberian sturgeon when subjected to acute hypoxia followed by a rapid return to normoxic conditions. In order to test if environmental hypoxia is also a potent stimulus for catecholamines and cortisol release into the blood plasma in the Siberian sturgeon as well as in the teleosts, this study was completed by measurements of plasma catecholamines and cortisol concentrations to evaluate the magnitude of the stress generated by hypoxia.

18.1 Materials and Methods

18.1.1 Fish and Water

A lot of 3-year-old sturgeons (*Acipenser baerii*), weighing about 1.8 kg, were obtained from the experimental hatchery of IRSTEA, formerly CEMAGREF (St Seurin sur l'Isle, Gironde, France). Prior to experimentation, they were maintained for 3 weeks in a large outdoor circular tank supplied with well-aerated running tap water at seasonal temperature. They were fed once a day (1% of live weight) with a commercial diet (Aqualim, France) until 48 h before experiments.

Fish were anesthetized by immersion in 2-phenoxyethanol (1/200) in an air-saturated solution for approximately 10 min. After weighing, they were transferred to an operating table where the gills were continuously irrigated with aerated water containing a maintenance dose of 2-phenoxyethanol (1/600). Each fish was placed in a moist tray and chronically fitted with a dorsal aortic cannula (PE50) implanted by blind puncture at the caudal level (Salin 1992; Williot et al. 2011; Williot and Le Menn, see Chap. 24). A polyethylene PE190 catheter was then inserted through the cleithrum bone into the branchial cavity. The dorsal aorta cannula was flushed daily with a saline containing 10 i u mL^{-1} of heparin (lithium salt, Sigma) in a Ringer solution adapted for sturgeon.

Following the operation, the animals were placed on a perforated Plexiglas platform and transferred into a cylindrical darkened respirometer (37 L) in which aerated tap water was circulated along an open circuit at a flow rate of 1 L min^{-1} at 15 °C. They were allowed to recover there for 24–48 h. An immersed pump was placed into the respirometer to ensure homogenization of the water. The size of the respirometer was such that the fish remained at rest and hence could be regarded as being in a state of standard metabolism. During experimental periods, water was recirculated in a closed circuit by a centrifugal pump (constant flow rate: 1 L min^{-1}) through the respirometer from a large thermostated (15 °C) tank (120 L).

The acid-base balance of this water was automatically regulated with a pH-PCO$_2$ stat. As CO$_2$-free air was bubbled constantly, this device intermittently injected small quantities of pure CO$_2$ into the water through a solenoid valve set to open when water pH rose above a preset value. The titratable alkalinity of the water (TAw) was measured each day using a modified Gran titration procedure (for details see Chap. 25) and adjusted to a constant value. This allowed water pH and carbon dioxide tension to be regulated at the same value in normoxic and hypoxic conditions (pH = 8.050; TAw = 1.90 meq L^{-1}; PCO$_2$ = 0.75 mmHg).

18.1.2 Respiratory Responses to Progressive Ambient Hypoxia

- *Experimental protocols.* Fish ($n = 7$) were sequentially exposed to various stages of environmental hypoxia (PwO$_2$ = 60, 40, and 20 mmHg) obtained by bubbling pure nitrogen in the thermostated tank in which immersed marbles allowed quick O$_2$ depletion (within 10–15 min). Blood sampling and simultaneous measurements of respiratory and acid-base variables at each stage of imposed hypoxia were performed after a 1-h period of steady state in the water circuit. During the return to normoxic conditions, the same variables were then measured at 10, 20, and 30 min and 1, 2, 3, and 5 h.
- *Measurements and calculation of respiratory variables.* Standard oxygen consumption (MO$_2$) was determined at steady state by measuring the O$_2$ partial pressure difference between water flowing into (PO$_2$ inlet) and out of (PO$_2$ outlet) the respirometer. A peristaltic pump allowed alternate sampling of the inlet and outlet water which passed at a constant flow rate (5 mL min^{-1}) through a thermo-

stated PO_2 measuring cell (Radiometer E 5046). Oxygen consumption was calculated according to the relation:

$$MO_2 = (PO_2\,inlet - PO_2\,outlet)Q_w \cdot \alpha wO_2 / Body\,weight.$$

where Q_w is the constant water flow through the respirometer and αwO_2 is the O_2 solubility in freshwater = 2.01 μmol L^{-1} mmHg^{-1} at 15 °C.Frequency (VF) and amplitude (VA) of ventilatory movements were obtained from continuous recording of hydrostatic pressure changes in the branchial cavity, measured by connecting the catheter inserted through the cleithrum bone to a Honeywell 156PC pressure transducer. The maximal change of pressure observed during each breathing cycle was used as an estimation of ventilatory amplitude. An average value of this variable measured for each fish in steady state during a 10-min period before the exposure to hypoxia was used as an arbitrary unit to estimate ventilatory amplitude changes which occurred during hypoxia.Partial pressure of oxygen in arterial blood (PaO$_2$) was determined on a 200 μL blood sample using a thermostated PO_2 measuring cell (Radiometer E 5046). After each measurement, the blood was returned to the fish.

Measurements of blood acid-base characteristics and plasma ion concentrations. A blood sample (about 240 μL) was drawn from the arterial cannula into three heparinized glass capillaries. Blood pH was measured immediately from one glass capillary using a Radiometer microelectrode G 222A calibrated with Radiometer precision buffer solutions type S1500 and S1510 and connected to a Radiometer PHM72 pH meter. The blood partial pressure of carbon dioxide (PaCO$_2$) was determined by the Astrup method (Astrup 1956) using the two other capillaries. The bicarbonate concentration in arterial blood was calculated by the Henderson-Hasselbalch equation, using a CO_2 solubility coefficient and an operational pK' from Boutilier et al. (1985).

The measurement of plasma ionic concentrations needed an additional arterial blood sample of about 200 μL. Chloride concentration [Cl$^-$] was measured with a chloride titrator Radiometer CMT10. Sodium [Na$^+$] and potassium [K$^+$] concentrations were obtained with a flame photometer (Instrumentation Laboratory 243). Lactate concentration ([Lact$^-$]) was measured by an enzymatic method (Boehringer-Mannheim, kit 139 084).

18.1.3 Respiratory Responses to Acute Exposure to Deep Hypoxia

The fish was placed in a respirometer that could be supplied in closed circuit with normoxic or hypoxic water from two thermostated (15 °C) tanks (120 L each). By operating a three-way tap, normoxic or hypoxic water was pumped at a constant flow rate (about 3 L min^{-1}) to reduce, as much as possible, the complete replacement time of experimental medium in the respirometer.

- *Experimental protocol*

By using the setup described above, the imposed deep hypoxia level (10 mmHg) was reached in 30'. Blood sampling and recordings of ventilatory and circulatory variables were performed at 10' (PwO$_2$ = 30 mmHg) and 30' during the PwO$_2$ decrease and then at 2 h and 5 h30 during the return to normoxic conditions. In order to avoid possible disturbances created by blood sampling, recordings were always performed first. Averaged values were obtained from continuous recordings during 10-min periods.

Blood samples (about 1.7 mL) were withdrawn slowly from the dorsal aortic cannula into pre-heparinized syringes. Oxygen partial pressure in arterial blood (PaO$_2$), extracellular acid-base characteristics, and hematocrit were immediately determined from aliquots of whole blood of 200, 240, and 25 µL, respectively. The remaining blood was centrifuged (3000 rpm for 15 min) for subsequent measurements of adrenaline and noradrenaline (700 µL), cortisol (150 µL), main ions (150 µL), lactate (50 µL), and glucose (10 µL) concentrations in the plasma.

- *Measurements of plasma catecholamines and cortisol*

Plasma adrenaline and noradrenaline were extracted with boric acid gel (Affigel 601; Biorad) activated by the Maruta method (Maruta et al. 1984) slightly modified. The efficiency of this extraction method, determined by an internal standard procedure, was about 90%. The catecholamines eluated from the gel by 0.75 M acetic acid were separated by high-performance liquid chromatography and detected by electrochemical device (Bioanalytical Systems).

Cortisol assay was performed by radioimmunoassay (CIS Bio International, ref.: SB-CORT).

- *Measurements and calculation of circulatory variables*

Dorsal aortic pressure was recorded by connecting the dorsal aortic cannula to a pressure gauge (Honeywell 156PC). Mean pressure (P$_{DA}$) and differential pressure (P$_{diff}$) were calculated according to the following relations:

P_{DA} = 1/3 [systolic pressure + (2. diastolic pressure)]

P_{diff} = systolic pressure−diastolic pressure

The heart rate (HR) was determined from the pulse rate on the P$_{DA}$ record.

- *Measurements of respiratory variables, blood acid-base characteristic, plasma ionic concentrations, and blood glucose concentrations*

Frequency (VF) and amplitude (VA) of ventilatory movements and PaO$_2$ were obtained as previously described for blood acid-base characteristics and plasma

18 Respiratory Responses to Hypoxia in the Sturgeon, *Acipenser baerii*

ionic concentrations. Glucose concentrations were measured by enzymatic methods, Sigma Diagnostics, Procedure no 16-U.

18.1.4 Statistical Analysis

The results have been statistically analyzed using unpaired Student's *t*-test for differences from pre-hypoxic values.

18.2 Results

The nonsignificant decrease in hematocrit throughout the experimental period (first sampling, 24.0 ± 1.8; last sampling, 19.4 ± 1.5), indicated that the impact of the numerous and large blood samplings (representing 3–5% of total blood volume) was limited, owing to the large size of the fish.

- Progressive Ambient Hypoxia

Hypoxia. When acclimated to the respirometer in normoxia, the sturgeon showed a spontaneously interrupted ventilatory pattern with ventilatory periods consisting of 6–8 breathing cycles lasting 10–15 s, separated by apneic periods of about the same duration. Usually, a progressive decrease of the depth of ventilation occurs during the ventilatory periods, as illustrated by the typical sample of ventilation recording shown in Fig. 18.1a. When PO_2 into the respirometer (PwO_2) began to decrease, the periodic breathing pattern was quickly altered, with reduced duration of apneas and lengthening of the ventilatory periods, until complete disappearance of apneas (Fig. 18.1b). This effect was associated with a highly significant increase in overall as well as intraburst ventilatory frequency. The amplitude of the breathing movements also increased significantly.

During the gradually imposed hypoxia, the standard oxygen consumption did not significantly differ from normoxia until the stage of 40 mmHg PwO_2. Then, it markedly decreased, so that the value measured at 20 mmHg was about half of control. Thus, the critical oxygen tension below which aerobic metabolism became dependent on ambient PO_2 was lower than 40 mmHg and higher than 20 mmHg. Oxygen tension in arterial blood (PaO_2) decreased progressively until 40 mmHg, then more abruptly below the critical oxygen tension (Fig. 18.2).

Carbon dioxide tension in arterial blood ($PaCO_2$) initially decreased, while pHa rose significantly (Fig. 18.3). At the same time, both ventilatory frequency and amplitude increased. Then, at the lowest PwO_2 value, pHa returned to a level not significantly different from control, while plasma lactate increased sharply (Fig. 18.4). Plasma Na^+, K^+, and Cl^- concentrations remained unchanged throughout hypoxia exposure.

Back to normoxia. When PwO_2 was returned to normoxia, the ventilatory frequency and amplitude of pressure changes in the branchial cavity remained at their highest level for about 30 min, then fell back progressively to control values.

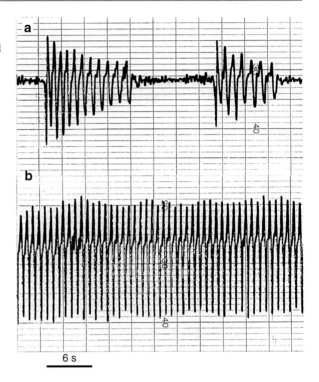

Fig. 18.1 Typical examples of ventilatory patterns recorded in normoxia (**a**) and in hypoxia (**b**) at $PwO_2 = 60$ mmHg

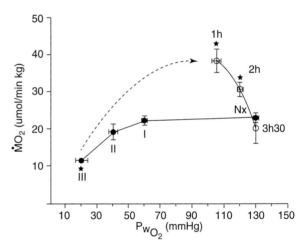

Fig. 18.2 Changes in standard oxygen consumption (MO_2) resulting from progressive hypoxia (*filled circle*) and return to normoxic conditions (*empty circle*)

Measurements of standard oxygen consumption could be performed only 1 h after the end of hypoxic exposure, because during the initial period of restoration of normoxia, the time course of variations of PwO_2 was too rapid to get a steady state of outlet PO_2 during measurement time, and consequently reliable values of MO_2 could not be obtained. The mean value of MO_2 observed at 1 h was about double the control value, while PwO_2 was only a little lower than the normoxic

18 Respiratory Responses to Hypoxia in the Sturgeon, *Acipenser baerii*

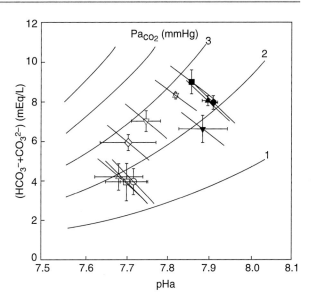

Fig. 18.3 Diagram [HCO_3^- + CO_3^{2-}] vs. pH illustrating acid-base changes in arterial blood resulting from progressive hypoxia and return to normoxic conditions. Filled symbols: normoxia (*filled square*) and hypoxia stages I (*filled triangle*), II (*filled diamond*), and III (*filled downward triangle*). Hollow symbols: back to normoxia, 10 min (*open circle*), 20 min (*open square*), 30 min (*open triangle*), 1 h (*open diamond*), 2 h (*open downward triangle*), and 3.5 h (*open star*)

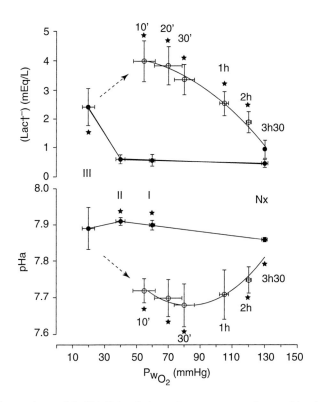

Fig. 18.4 Changes in arterial pH (pHa) and plasma lactate concentration resulting from progressive hypoxia (*filled circle*) and return to normoxic conditions (*open circle*)

level (Fig. 18.2). Then, MO_2 slowly returned within 3.5 h to control levels. Arterial O_2 partial pressure gradually increased toward pre-hypoxic levels in parallel with the progressive increase of PwO_2. However, as observed for ventilatory frequency and amplitude, there was a difference between pre- and post-hypoxic PaO_2 at the same PwO_2, with lower values during recovery than during progressive hypoxia at the same PwO_2.

During the initial 30 min of post-hypoxic period, pHa fell sharply to its lowest value (7.68 ± 0.06), then gradually increased toward the control pre-hypoxic level (Figs. 18.3 and 18.4). During these first 30 min into normoxic recovery, $PaCO_2$ remained almost steady at the lowest level observed during the deepest hypoxic stage. This coincided with the maximal hyperventilatory response. Then, ventilation decreased, while $PaCO_2$ returned to its normoxic level. Sodium, potassium, and chloride plasma concentrations did not change significantly from control values ($[Na^+] = 119.2 \pm 1.2$ mEq L^{-1}; $[K^+] = 2.32 \pm 0.07$ mEq L^{-1}; $[Cl^-] = 103.6 \pm 2.3$ mEq L^{-1}) during all the hypoxic and post-hypoxic periods. By contrast, plasma lactate concentration continued to rise at the beginning of the post-hypoxic period to reach a value about eight times higher than control at 30 min. Later on, $[Lact^-]$ steadily returned to the normoxic reference value, which was attained by 3.5 h (Fig. 18.4).

- Exposure to Deep Hypoxia

By the end of the 30-min exposure to deep hypoxia, the mean values of circulating adrenaline and noradrenaline reached, respectively, 300.88 ± 49.46 and 195.95 ± 32.29 pmol mL^{-1} (Fig. 18.5). Both catecholamines, especially adrenaline, rapidly increased at the onset of hypoxia, but the highest plasma levels corresponded to the lowest value of PwO_2. At the same time, the abrupt decrease in PwO_2 induced a sixfold increase in plasma cortisol concentration (Fig. 18.6) and a 54% rise in blood glucose (Fig. 18.7).

Deep hypoxia induced a moderate increase in heart rate at 10 min, followed by a marked bradycardia (from 52 beats min^{-1} at 10 min to 17 beats min^{-1} at 30 min). Meanwhile, the average dorsal aortic pressure increased approximately 1.5-fold during the first 10 min. This hypertensive response was followed by an acute fall to a value corresponding to two thirds of the control. Mean dorsal aortic pressure changes were associated with an approximate twofold increase in differential pressure between systolic and diastolic levels.

The time course of changes in both ventilatory amplitude and frequency presented a biphasic aspect during the hypoxic test. The first 10 min of hypoxia leads to a shift from a spontaneously interrupted ventilatory pattern to a continuous one and a marked hyperventilation resulting from an increase in amplitude (+ 260%) and frequency (+ 124%). In a second stage (below $PwO_2 = 30$ mmHg), sturgeons presented a ventilation depression characterized by the reappearance of frequent short apneas. The overall ventilatory frequency was therefore lower than the control value by 38%, while amplitude did not differ from normoxic value. The oxygen tension in arterial blood (PaO_2) decreased dramatically throughout the hypoxic test, reaching 2.2 ± 0.41 mmHg within 30 min.

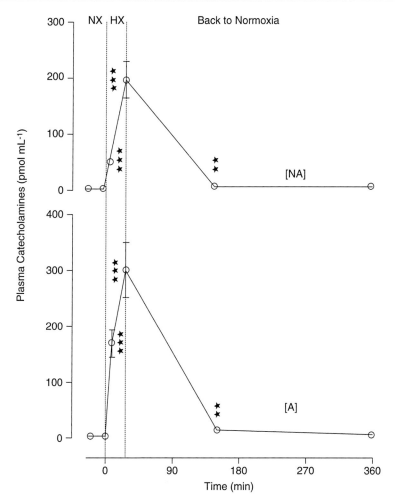

Fig. 18.5 Effects of acute deep hypoxia (HX) and rapid return to normoxia (NX) on plasma adrenaline (A) and noradrenaline (NA) concentrations. Stars indicate significant differences from normoxic value (**$P < 0.01$, ***$P < 0.005$, unpaired t-test)

The carbon dioxide washout caused by the initial hyperventilation resulted in a marked hypocapnia which was coupled with a moderate metabolic acidosis (Fig. 18.8) due to a small increase in the plasma lactate concentration (from 0.57 ± 0.10 in normoxia to 1.44 ± 0.19 mEq L^{-1} at PwO$_2$ = 30 mmHg and 3.10 ± 0.21 mEq L^{-1} at PwO$_2$ = 10 mmHg). Therefore, the arterial pH (7.85) was not different from control value during the first 10 min of hypoxia. The initiation of anaerobic metabolism then induced an important decrease in pHa (from 7.85 ± 0.01 in normoxia to 7.55 ± 0.04 at PwO$_2$ = 10 mmHg). At the same time, the carbon dioxide tension was paradoxically maintained at a low level despite ventilatory depression (Fig. 18.8). Plasma Na$^+$, K$^+$, and Cl$^-$ concentrations remained unchanged throughout hypoxia exposure.

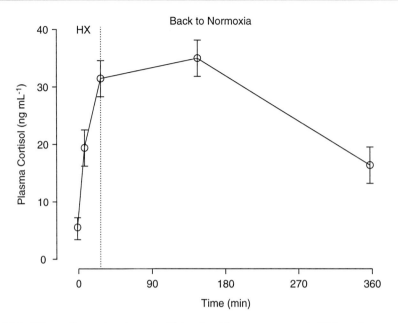

Fig. 18.6 Effects of acute deep hypoxia (Hx) and rapid return to normoxia (Nx) on plasma cortisol concentration. Stars indicate significant differences from normoxic value (*$P < 0.05$, **$P < 0.01$, ***$P < 0.005$, unpaired t-test)

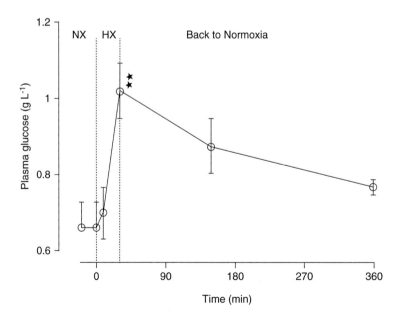

Fig. 18.7 Effects of acute deep hypoxia (HX) and rapid return to normoxia (NX) on plasma glucose concentration. *Stars* indicate significant differences from normoxic value (**$P < 0.01$, unpaired t-test)

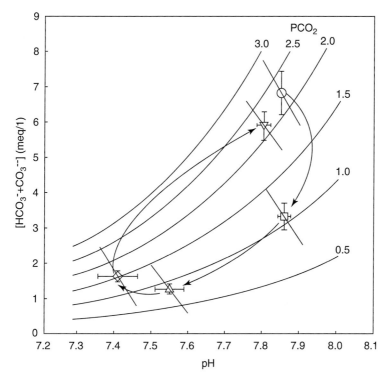

Fig. 18.8 Davenport diagram illustrating acid-base changes in arterial blood resulting from acute deep hypoxia and rapid return to normoxia. Normoxia (*circle*). Hypoxia: 10 min (*square*), 30 min (*triangle*). Back to normoxia: 2 h (*diamond*), 5 h30 (*downward triangle*)

18.2.1 Back to Normoxia

Plasma levels of catecholamines declined rapidly after cessation of the stress and returned to control value (Fig. 18.5). By contrast, plasma cortisol level remained at its high level during the first 2 h of return to normoxia, before tending to decrease (Fig. 18.6). 5h30 after the end of hypoxia, its concentration was still three times higher than the control value. Plasma glucose decreased progressively throughout the post-hypoxic period (Fig. 18.7).

The marked bradycardia observed at the lowest level of PwO_2 disappeared quickly during normoxic recovery. Moreover, 5h30 after the end of hypoxia, the heart rate value was significantly higher than that of the control. A rapid return to normoxic conditions induced a progressive increase in the mean dorsal aortic pressure from its lowest hypoxic level: no significant difference could be noted between control and post-hypoxic values by 5h30. In the same way, differential pressure between systolic and diastolic levels decreased swiftly to the normoxic value.

During rapid return to normoxia, the apneas, provoked by the acute hypoxia exposure, disappeared quickly. The frequency and amplitude of pressure changes in the branchial cavity increased quickly and reached upper values 2 h after the end of hypoxia. Then, these post-hypoxic hyperventilatory effects, affecting mostly amplitude, began to disappear. Finally, by 5h30 into recovery, post-hypoxic VA and VF values were not significantly different from the normoxic reference values. Moreover, recovery from hypoxia provoked an immediate increase in PaO$_2$.

Concerning the extracellular acid-base balance, the metabolic acidosis, having appeared during hypoxia, was emphasized during the first 2 h of the post-hypoxic period. However, this increase was moderate (ΔpHa = 0.14) in spite of an important release of lactate into the blood stream amounting to 12.98 $\pm$ 0.81 mEq L^{-1}. Then, the carbon dioxide tension and pHa progressively tended to recover their reference values (Fig. 18.8). Plasma Cl$^-$ level remained unchanged throughout the post-hypoxic period. At the same time, Na$^+$ and K$^+$ concentrations rose markedly to their maximum levels, respectively, 7 and 32% higher than normoxic levels, before tending to decrease.

18.3 Discussion

Our results clearly demonstrate that the sturgeon is able to maintain a standard O$_2$ uptake during progressive hypoxia down to a critical ambient PO$_2$ of 20–40 mmHg. This is made possible thanks to a marked hyperventilation, as shown by increases in both frequency and amplitude of gill respiratory movements, which lead to alkalosis and hypocapnia. These hyperventilatory responses result probably from a stimulation of branchial chemoreceptors by lack of oxygen in water. Burleson and Smatresk (1990) in the catfish and McKenzie et al. (1995) in the Adriatic sturgeon, *Acipenser naccarii,* suggested that two separate populations of chemoreceptors monitor the external (PwO$_2$) and internal (PO$_2$ blood) environments and are responsible for the cardioventilatory responses to hypoxia (Randall 1982).

Below the P$_c$, anaerobic metabolism is initiated, as indicated by lactate release to the blood, metabolic acidosis, and repayment of an O$_2$ debt after return to normoxia. These findings are in contrast with those reported by Burggren and Randall (1978) for similarly sized specimens of another sturgeon species, *Acipenser transmontanus.* These authors found a steady decrease not only of oxygen uptake but also of gill ventilatory flow rate during progressive hypoxia and concluded that *Acipenser transmontanus* behaves as an oxyconformer.

A more recent study by McKenzie et al. (2007) in the Adriatic sturgeon, *Acipenser naccarii*, proposes another interpretation. The Adriatic sturgeon behaves as an oxyconformer when exposed to progressive hypoxia under static conditions but as an oxyregulator when it was allowed to swim at a low sustained speed. In this case, the critical partial pressure of oxygen in water was evaluated to 36.75 $\pm$ 3.75 mmHg, a result which is close near to the one obtained in Siberian sturgeon (our study).

When corrected for the differences in weight by using a value of intraspecific mass exponent of 0.67 (Maxime et al. 1990), it appeared that, at a temperature of 15 °C, the standard MO$_2$ of *Acipenser baerii* was lower than that of *Acipenser transmontanus* and ranged between values reported for active fish (trout, salmon) and

Table 18.1 Standard oxygen consumption (μmol min^{-1} kg^{-1}) after correction for differences of weight (1 kg) and temperature (15 °C) by using an intraspecific mass exponent of 0.67 and a Q_{10} of 2

Species	MO$_2$	References
Acipenser baerii	29.09	Present study
Acipenser transmontanus	42.06	Burggren and Randall (1978)
Salmo salar	44.08	Maxime et al. (1990)
Oncorhynchus mykiss	37.93	Maxime et al. (1991)
Platichthys flesus	16.25	Duthie (1982)
Anguilla anguilla	15.67	Le Moigne et al. (1986)
Limanda limanda	5.73	Duthie (1982)

sluggish fish (eel) (Table 18.1). This observation can be related to the growth intensity of sturgeon in spite of its sluggish behavior.

A periodic nature of breathing rhythm, similar to that recorded in *Acipenser baerii*, was described in other teleosts in resting condition: *Tinca tinca* (Shelton and Randall 1962), *Cyprinus carpio* (Peyraud and Serfaty 1964; Hughes 1973), *Anguilla anguilla* (Peyraud-Waïtzenegger 1979), and *Anguilla australis* (Forster 1981; Hipkins and Smith 1983). Such a spontaneous interrupted ventilation in normoxia, only observed in bottom-dwelling species, known for their reduced motor activity and their ability to withstand hypoxic conditions, can be regarded as a typical eupneic ventilation.

An increased ventilation has been commonly reported in fish submitted to a lowering of ambient PO$_2$. Generally, the increase of ventilatory flow has been observed to result from a much more important rise of ventilated stroke volume than of frequency of breathing movements as shown in trout (Holeton and Randall 1967). This pattern of response occurred in the more active species. Showing roughly the same increase in both ventilatory frequency and amplitude, the sturgeon presented a pattern of ventilatory adjustments rather similar to those reported in other bottom-dwelling species such as carp (Peyraud 1965; Lomholt and Johansen 1979).

When submitted to gradual hypoxia, *Acipenser baerii* maintained standard oxygen consumption at about the mean normoxic level in a large range of decreased ambient PO$_2$. Thus, this species of sturgeon appeared like a typical O$_2$ regulator as able as the most fish to achieve the necessary ventilatory and possible cardiovascular adjustments to maintain standard MO$_2$ in the face of reduced O$_2$ availability. Moreover, the slight increase in pHa during the initial stage of hypoxic exposure allowed to take in account a left shift of the O$_2$ dissociation curve resulting in a steepening of the PO$_2$ gradient across the gills and an increase in arterial O$_2$ saturation.

The critical level of ambient PO$_2$ (Pc) for *Acipenser baerii* appeared lower, at similar water temperature, than the one reported for active fish such as trout (Randall and Shelton 1963). It was however of a same order of magnitude as the one determined in *Acipenser gueldenstaedtii* (Klyashtorin 1981) as well as in fish which commonly encounter environmental hypoxia in nature such as goldfish (Fry and Hart 1948). The low value of Pc observed in sturgeon may be related to the characteristics of Hb saturation curve. Indeed, according to data reported in *Acipenser transmontanus* (Burggren and Randall 1978) and obtained in *Acipenser baerii* by Nonnotte et al. (see Chap. 20), it appears that for a mean $PaO_2 = 23$ mmHg, measured at the second stage of imposed hypoxia ($PwO_2 = 40$ mmHg), about 65% of Hb was still O_2 saturated thereby allowing sufficient O_2 supply to tissues for aerobic metabolism.

Effects of gradual hypoxia. Increased gill ventilation has been commonly reported in fish subjected to a lowering of ambient PO_2. In active species such as trout (Holeton and Randall 1967) and Atlantic salmon (Maxime et al. unpublished), the increase in ventilatory flow rate results from a much more important rise of stroke volume than of frequency of breathing movements. The more active the species, the more marked this pattern of response appears.

Hyperventilation of the gills in hypoxic sturgeon led to a typical respiratory alkalosis with a decreased blood PCO_2 and an increased pH. This helped maintain tissue O_2 supply by a left shift of the O_2 dissociation curve that kept arterial saturation high. According to data reported by Burggren and Randall (1978) for *Acipenser transmontanus* and by Nonnotte et al. (see Chap. 20) for *Acipenser baerii*, blood O_2 affinity is relatively high in sturgeon, so that the mean PaO_2 of 23 mmHg recorded at the second stage of imposed hypoxia ($PwO_2 = 40$ mmHg) should correspond to a HbO_2 percent saturation of about 65%. This is probably a reason for a relatively low critical level of ambient PO_2 in *Acipenser baerii*.

Several aspects of the changes of respiratory variables during normoxic recovery deserve comment. Despite a short exposure time to the deepest hypoxic stage below P_c (1 h), the sturgeon obviously repaid an O_2 debt with an elevated MO_2 lasting about 3.5 h. This O_2 debt repayment was closely correlated with both the clearance of blood lactate and the recovery of blood pH (Fig. 18.7). It is additionally worth noting that blood lactate and the associated metabolic acidosis peaked together not during hypoxia but 10–30 min into normoxic recovery, suggesting a delayed but simultaneous release of both lactate ions and metabolic acid generated anaerobically during the hypoxic exposure. This result contrasts with the asynchronous release of lactate and metabolic acid found in most teleosts during recovery from exhausting exercise (see Truchot 1987). This is also in keeping with the absence of any significant change of plasma ion concentrations during either progressive hypoxia or normoxic recovery, which suggests that the generation and subsequent dissipation of the metabolic acidosis mainly involve internal metabolic events but no or minimal ionic exchanges.

At the same ambient PO_2, the values of PaO_2 measured during normoxic recovery were lower than those found during progressive hypoxia. This probably resulted from the increased tissue O_2 uptake which must have substantially reduced oxygenation of the venous blood. As shown by the recorded ventilatory frequencies and amplitudes, the gill water flow rate also exhibited a hysteresis during normoxic recovery, but with values higher than during gradual hypoxia, at the same ambient PO_2. This enhanced hyperventilatory response might be explained by a stronger hypoxic drive due to lower arterial PO_2 values, perhaps reinforced by a persistent acidosis which could disappear only after metabolic processing of released lactic acid. This relatively long-lasting post-hypoxic hyperventilation also delayed the restoration of normoxic $PaCO_2$ values.

Effects of exposure to deep hypoxia. An almost invariable neuroendocrine response of teleost to environmental stress is the release of catecholamines and corticosteroids into the blood stream (for review, see Mazeaud and Mazeaud 1981 and Donaldson 1981). In contrast to chromaffin tissue, interrenal tissue has been identified in sturgeon (Matty 1985). However, the chromaffin cells are often closely associated with interrenal cells in fish (Matty 1985), and they might also be present there in the sturgeon.

In *Acipenser naccarii*, Randall et al. (1992) reported resting normoxic levels of adrenaline and noradrenaline of 5.2 and 4.3 pmol mL^{-1}, respectively. Subsequently to a moderate acute hypoxia (PwO_2 = 80 mmHg), these levels increased slightly up to 29.9 and 45.1 pmol mL^{-1}, respectively. Resting levels obtained in the present study are a little lower than those of Randall et al. (1992). However, exposure to deep hypoxia (PwO_2 = 10 mmHg) promotes a considerable elevation in both adrenaline (x77) and noradrenaline (x159) levels in *Acipenser baerii* (Fig. 18.5). The discrepancy between our results and those of Randall et al. (1992) could be explained by the fact that the chromaffin cells release variable quantities of adrenaline and noradrenaline depending on the nature and duration of the stimulus (Randall and Perry 1992). Levels of catecholamines as high as those reported in the present study have been observed by Metcalfe and Butler (1989) in *Scyliorhinus canicula* submitted to moderate hypoxia (55 mmHg). In the same way, Thomas et al. (1991) have shown an increase in plasma adrenaline concentration in rainbow trout to about 300 pmol mL^{-1} at the midpoint of chronic (48 h) exposure to moderate hypoxia (50–70 mmHg). More recently Van Raij et al. (1996) have measured a noradrenaline concentration of about 300 pmol mL^{-1} in the carp plasma exposed to an acute hypoxia.

A useful index of affinity of hemoglobin is the partial pressure for which 50% of the pigment is oxygenated; this index is called P50. In the present study, the release of catecholamines is already large at arterial PO_2 value (about 13 mmHg) which allows us to predict a release threshold upper than an arterial PO_2 value about 13 mmHg, roughly equivalent to the hemoglobin P_{50} value as determined from oxygen dissociation curves (see Chap. 20). Moreover, because of the rapid clearance of catecholamines from the blood by the combined effects of tissue accumulation and metabolism, peak levels decrease within a few minutes after cessation of the stress.

Resting levels of cortisol observed for the first time for sturgeon in the present study (5.43 ± 1.63 ng mL^{-1}) are of the same order of magnitude as those reported for goldfish (Venkatesh et al. 1989) and rainbow trout (Pickering et al. 1991). There is a lack of information on the effects of hypoxic stress on plasma cortisol in fish. Effects of hypoxia initiated by deep anesthesia are the only ones that can be mentioned. Mazeaud et al. (1977) reported a moderate elevation of corticosteroids in coho salmon males but not in females. Surprisingly, Iwama et al. (1989) observed a progressive decline of cortisol concentrations in rainbow trout during hypoxia and recovery. Corticosteroid stress response in sturgeon is stronger and more persistent when compared with this data.

An analogy with known effects of catecholamines and corticosteroids in teleosts submitted to similar stress could be made to consider the role of adrenaline, noradrenaline, and cortisol in sturgeon submitted to extreme hypoxia. The main sites of action of catecholamines are the cardiovascular and the respiratory systems (see further discussion). The catecholamines also act directly on the liver to stimulate glycogenolysis (ß effect). Cortisol, like catecholamines, is responsible for the mobilization of energy reserves by activating liver glycogenolysis and inhibiting glycolysis (reviewed by Donaldson 1981). This results in a moderate increase of the plasma glucose concentration. So, during hypoxia, cortisol appears to reinforce the immediate catecholamine response to stress. Then, during the post-hypoxic period, plasma concentrations remaining transitionally high, cortisol prolongs the effects of catecholamines. However, the relative contributions of corticosteroid and adrenergic responses to the increased plasma glucose concentrations usually observed in fish are still unclear (Mazeaud and Mazeaud 1981).

Blood glucose levels vary considerably among species, within a species and over time, even within an individual (McDonald and Milligan 1992). The relatively low resting level measured in sturgeon and the moderate maximum increase observed at the lowest level of PwO_2 could in part be considered as an effect of the starvation during the experiments.

Early circulatory changes observed during the first 10 min of the abrupt decrease in PwO_2 could be considered as typical effects of increased catecholamines. Indeed, the initial moderate rise in HR associated with a relatively more acute increase in P_{diff} represents positive chronotropic and inotropic ß effects on the heart (Wood and Shelton 1980). In addition, catecholamines cause an α-adrenoceptor-mediated vasoconstriction of the systemic vasculature (Wood and Shelton 1975) which increases systemic resistance and results in a moderate elevation in P_{DA}. However, the disproportion between the small magnitude of these initial circulatory effects and the high corresponding levels of circulatory catecholamines must be underlined. This could indicate a low sensitivity of sturgeon to catecholamines resulting either from a density of adrenoceptors smaller than in teleosts or from fundamental differences with teleosts in the manner by which target tissue is stimulated consequently to receptors activation (Perry and Reid 1992). Thus, large differences in sensitivity of target tissues to catecholamines may exist in fish.

At the deepest level of PwO_2, the effects of catecholamines are masked by specific effects of hypoxia. The important bradycardia, occurring in spite of the highest plasma catecholamines concentration, can be regarded as a classical vagal reflex response in fish to severe hypoxia. This results in a pronounced decrease in P_{DA}. Indeed, *Acipenser baerii*, submitted to moderate progressive hypoxia ($PwO_2 = 20$ mmHg), does not present any change in HR even below critical PwO_2. Bradycardia, as suggested by Hughes (1973), may be regarded as an adaptive response to hypoxia, because the increased residence time of blood in the gills can help the fish to maintain the effectiveness of O_2 transfer.

According to the Starling law of the heart, bradycardia usually elicits an increase in P_{diff} allowing better cardiac filling and emptying. The absence of this effect together with the decrease in P_{DA} below its control normoxic value may be interpreted as caused by a lowered contractile strength of the myocardium resulting from a deficient O_2 supply ($PaO_2 = 2$ mmHg).

During the first 10 min of the hypoxic stress, the sturgeon adopts a continuous breathing pattern which leads to increased frequency and amplitude of opercular movements. This may limit, to some extent, the decrease in PaO_2. The hyperventilatory response to hypoxia in fish is largely driven by stimulation of gill chemoreceptors due to oxygen lack. However, the location and afferent pathways for these receptors are not well known. Burleson and Smatresk (1990) suggested that gill ventilation in channel catfish is controlled interactively by separate internally and externally oriented receptors, the latter eliciting also bradycardia. Moreover, there may be some additional action of catecholamines in extreme hypoxia. Indeed, Peyraud-Waitzenegger et al. (1979) have demonstrated that circulatory catecholamines pass the blood-brain barrier. So, as suggested by Peyraud-Waitzenegger et al. (1980), it seems possible that catecholamines are able to directly stimulate the ß-adrenoceptors of the brain centers, thus increasing the spontaneous rate of firing of the ventilatory neurons.

During the second period of the hypoxic stress (10–30 min), when PwO_2 reaches values clearly lower than critical PwO_2 for the previously determined oxyregulation, the fish presents a marked decrease of ventilation characterized by progressive reappearance of transitory apneas. It may be hypothesized that this results from a decrease in chemosensitivity and a depression of respiratory center activity due to extremely low values of PaO_2. To support a part of this hypothesis, it has been reported that chemoreceptor output declined in extreme hypoxia (Randall and Perry 1992). Then, the resumption of high ventilatory levels observed after return to normoxia allows the repayment of the oxygen debt. This hyperventilatory response could partly result from blood acidosis. Indeed, during the post-hypoxic period, PaO_2 swiftly reaches high normoxic levels which cannot explain such high ventilation rates.

The typical hypoxia-induced hyperventilation first causes a blood hypocapnia. In present conditions, this did not result in alkalosis because of the development of a metabolic acidosis by progressively amplified production of lactic acid. At the deepest hypoxia level (10 mmHg), the lack of increase in PCO_2 in spite of hypoventilation can be interpreted as a consequence of a reduction in carbon dioxide production resulting from a deficiency of O_2 supply to tissue. Moreover, a low PCO_2 level can also be linked to the increase of gas diffusing capacity predictable from high levels of catecholamines (see circulatory effects). The return to normoxia is characterized by a more massive flush of lactate. However, the pH does not fall as much as might be expected from these effect, showing the development of some adjustment to maintain the pH level as normal as possible. Indeed, the elevation of the blood lactate is accompanied by a marked increase in plasma sodium concentration suggesting a metabolic compensation of the developed acidosis. Indeed, this acidosis remained minor with regard to the large increase of the blood lactate. Another hypothesis might be a possible excretion in external medium of protons providing with lactic acid and balanced by an influx of sodium at the gill level. These exchanges could also explain the slight acidosis and the increase in plasma sodium concentration. But further investigations will be necessary to clarify this ion exchange in the branchial epithelium of the sturgeon.

Conclusions

Numerous recent studies of respiratory and cardiovascular systems in teleosts show that their ability to support oxygen changes depends largely on the responses to decreased oxygen levels including quick adjustments in cardiovascular and ventilatory activity, i.e., increased ventilatory water flow, elevated blood pressure, and bradycardia. There is also a dramatic increase in plasma levels of catecholamines, particularly if the decrease rate of aquatic oxygen tension is high (see Randall and Perry 1992 and Fritsche and Nilsson 1993). For the first time, in the present study, we demonstrated that the sturgeon responded to severe aquatic hypoxia by an increase in plasma concentration of catecholamines and cortisol. Moreover, the circulatory and ventilatory responses and the changes in acid-base status of arterial blood induced in sturgeon by deep hypoxia and, following return to normoxia, provided evidences to explain the strong resistance of this fish to oxygen depletion and the ecological success of the sturgeon in hypoxic habitats. The hypoxia-induced adjustments, reported here, were similar to those observed in teleosts submitted to identical conditions. Thus, it might be hypothesized that sturgeon survived during the evolution because it had early acquired the same adaptive processes as the most advanced fish. But, there is a need for more studies which would precisely define the ventilatory and cardiovascular control systems in the sturgeon, to propose a general model for acclimation to hypoxia.

Acknowledgments We thank Dr. Karine Pichavant-Rafini (ORPHY laboratory, EA4324) and Michel Rafini (Professor at the Language Dpt) of the Brest University, for their kindness and constant availability and their help and scientific advices. Moreover, we are extremely indebted and grateful to them for the English corrections.

Notes of the Authors These experiments were performed since 1990 to 1994. They have been investigated in the Laboratoire de Neurobiologie et Physiologie comparées, CNRS URA 1126 and the University of Bordeaux I, F-33120 Arcachon, in collaboration with the Laboratoire de Physiologie Animale, Brest University, F-29285 Brest and the IRSTEA (formerly CEMAGREF), F-33611 Cestas-Gazinet.

References

Astrup P (1956) A simple electrometric technique for the determination of carbon dioxide tension in blood and plasma, total content of carbon dioxide in plasma and bicarbonate content in 'separated' plasma at a fixed dioxide tension (40 mmHg). Scand J Clin Invest 8:33–43

Boutilier RG, Dobson G, Hoeger U, Randall DJ (1988) Acute exposure to graded levels of hypoxia in rainbow trout (*Salmo gairdneri*): metabolic and respiratory adaptations. Respir Physiol 71:69–82

Boutilier RG, Iwama GK, Heming TA, Randall DJ (1985) The apparent pK of carbon acid in rainbow trout blood plasma between 5 and 15°C. Respir Physiol 61:237–254

Burggren WW, Randall DJ (1978) Oxygen uptake and transport during hypoxic exposure in the sturgeon *Acipenser transmontanus*. Respir Physiol 34:171–183

Burleson ML, Smatresk NJ (1990) Evidence for twO oxygen-sensitive chemoreceptor loci in channel catfish, *Ictalurus punctatus*. Physiol Zool 63(1):208–221

18 Respiratory Responses to Hypoxia in the Sturgeon, *Acipenser baerii*

Dejours P (1981) Principles of comparative respiratory physiology. Elsevier, Amsterdam, New York, Oxford, p 265

Donaldson EM (1981) The pituitary-interrenal axis as an indicator of stress in fish. In: Pickering AD (ed) Stress in fish. Academic press, New York, pp 11–47

Duthie GG (1982) The respiratory metabolism of temperature-adapted flatfish at rest and during swimming activity and the use of anaerobic metabolism at moderate swimming speeds. J Exp Biol 97:259–373

Forster ME (1981) Oxygen consumption and apnea in the shortfin eel, *Anguilla australis schmidtii*. New zeal. Aust J Mar Freshwat Res 15:85–90

Fritsche R, Nilsson S (1993) Cardiovascular and ventilatory control during hypoxia. In: Rankin JC, Jensen FB (cds) Fish ecophysiology. Chapman & Hall, London, pp 180–206

Fry FEJ, Hart JS (1948) The relation of temperature to oxygen consumption in the goldfish. Biol Bull 94:66–77

Hipkins SF, Smith DG (1983) Cardiovascular events associated with spontaneous apnea in the australian short finned eel (*Anguilla australis*). J Exp Zool 227:339–348

Holeton GF, Randall DJ (1967) The effect of hypoxia upon the partial pressure of gases in the blood and water afferent and efferent to the gills of rainbow trout. J Exp Biol 46:317–327

Hughes GM (1973) Respiratory response to hypoxia in fish. Am Zool 13:475–489

Iwama GK, McGeer JC, Pawluk MP (1989) The effects of five fish anaesthetics on acid-base balance, haematocrit, blood gases, cortisol and adrenaline in rainbow trout. Can J Zool 67:2065–2073

Klyashtorin LB (1981) The ability of sturgeons (Acipenseridae) to regulate gas exchange. J Ichthyol 21:141–144

Le Moigne J, Soulier P, Peyraud-Waïtzenegger M, Peyraud C (1986) Cutaneous and gill O_2 uptake in the European eel (*Anguilla anguilla L.*) in relation to ambient PO_2, 10-400 mmHg. Respir Physiol 66:341–354

Lomholt JP, Johansen K (1979) Hypoxia acclimation in carp. How it affects O_2 uptake, ventilation, and O_2 extraction from water. Physiol Zool 52:38–49

Maruta K, Fugita K, Ito S, Nagatsu J (1984) Liquid chromatography of plasma catecholamines, with electro-chemical detection, after treatment with boric acid gel. Clin Chem 30:529–548

Matty AJ (1985) The 'adrenal' and the kidney hormones. In: Fish endocrinology. Croom Helm Pub, London and Sydney, pp 112–137

Maxime V, Peyraud-Waïtzenegger M, Claireaux G, Peyraud C (1990) Effect of rapid transfer from seawater to freshwater on respiratory variables, blood acid-base status and O_2 affinity of hemoglobin in atlantic salmon (*Salmo salar L.*) J Comp Physiol B 160:31–39

Maxime V, Pennec J-P, Peyraud C (1991) Effects of direct transfer from freshwater to seawater on respiratory and circulatory variables and acid-base status in rainbow trout. J Comp Physiol B 161(6):557–568

Mazeaud MM, Mazeaud F (1981) Adrenergic responses in fish. In: Pickering AD (ed) Stress in fish. Academic Press, New York, pp 49–75

Mazeaud MM, Mazeaud F, Donaldson EM (1977) Primary and secondary effects of stress in fish: some new data with general review. Trans Am Fish Soc 106(3):201–212

McDonald DG, Milligan CL (1992) Chemical properties of the blood. In: Hoar WS, Randall DJ, Farrell AP (eds) Fish Physiol. Vol XII, part B. Academic Press, San Diego, pp 55–133

McKenzie DJ, Taylor EW, Bronzi P, Bolis CG (1995) Aspects of cardioventilatory control in the Adriatic sturgeon (*Acipenser naccarii*). Respir Physiol 100:44–52

McKenzie DJ, Steffensen JF, Korsmeyer K, Whiteley NM, Bronzi P, Taylor EW (2007) Swimming alters response to hypoxia in the Adriatic sturgeon Acipenser naccarii. J Fish Biol 70:651–658

Metcalfe JD, Butler PJ (1989) The use of alpha-methyl-p-tyrosine to control circulating catecholamines in the dogfish *Scyliorhinus canicula*: the effects on gas exchange in normoxia and hypoxia. J Exp Biol 141:21–32

Perry SF, Reid SD (1992) The relationship between beta-adrenoceptors and adrenergic responsiveness in trout (*Oncorhynchus mykiss*) and eel (*Anguilla rostrata*) erythrocytes. J Exp Biol 167:235–250

Peyraud C (1965) Recherches sur la régulation des mouvements respiratoires chez quelques téléostéens: analyse du réflexe opto-respiratoire, Thèse de Doctorat ès Sciences, Université de Toulouse, p 258

Peyraud C, Serfaty A (1964) Le rythme respiratoire de la carpe (*Cyprinus carpio* L.) et ses relations avec le taux de l'oxygène dissous dans le biotope. Hydrobiologia 23:165–178

Peyraud-Waitzenegger M, Barthelemy L, Peyraud C (1980) Cardiovascular and ventilatory effects of catecholamines in unrestrained eels (*Anguilla anguilla* L.) J Comp Physiol 138:367–375

Peyraud-Waitzenegger M, Savina A, Laparra J, Morfin R (1979) Blood-brain barrier for epinephrine in the eel (*Anguilla anguilla* L.) Comp Biochem Physiol 23(1):35–38

Peyraud-Waïtzenegger M (1979) Simultaneous modifications of ventilation and arterial Po$_2$ by catecholamines in the eel, *Anguilla anguilla* L.: participation of alpha and beta effects. J Comp Physiol B 129:343–354

Pickering AD, Pottinger TG, Sumpter JP, Carragher JF, Le Bail PY (1991) Effects of acute and chronic stress on the levels of circulating growth hormone in the rainbow trout, *Oncorhynchus mykiss*. Gen Comp Endocrinol 83:86–93

Prosser CL (1973) Comparative animal physiology, 3rd edn. Saunders, Philadelphia, p 966

Randall DJ (1982) The control of respiration and circulation in fish during exercise and hypoxia. J Exp Biol 100:175–288

Randall DJ, Perry SF (1992) Catecholamines. In: Hoar WS, Randall DJ, Farrell AP (eds) Fish physiol. Vol XII, part B. Academic Press, San Diego, pp 255–300

Randall DJ, McKenzie DJ, Abrami G, Bondiolotti GP, Natiello F, Bolis L, Agradi E (1992) Effect of diet on responses to hypoxia in sturgeon (*Acipenser naccarii*). J Exp Biol 170:113–125

Randall DJ, Shelton G (1963) The effects of changes in environmental gas concentrations on the breathing and heart rate of a teleost fish. Comp Biochem Physiol 9:229–239

Ruer FM, Cech JJ, Doroshov SI (1987) Routine metabolism of the white sturgeon, *Acipenser transmontanus*: effect of population density and hypoxia. Aquaculture 62:45–52

Salin D (1992) La toxicité de l'ammoniaque chez l'esturgeon sibérien, *Acipenser baerii*: effets morphologiques, physiologiques, métaboliques d'une exposition à des doses sblétales et létales. Thèse No 749, Université Bordeaux I, p 134

Shelton G, Randall DJ (1962) The relation between heart beat and respiration in teleost fish. Comp Biochem Physiol 7:237–250

Tetens V, Lykkeboe G (1985) Acute exposure of rainbow trout to mild and deep hypoxia: O$_2$ affinity and O$_2$ capacitance of arterial blood. Respir Physiol 61:221–235

Thomas S, Kinkead R, Walsh PJ, WOod CM, Perry SF (1991) Desensitization of adrenaline-induced red blood cell H$^+$ extrusion *in vitro* after chronic exposure of rainbow trout to moderate environmental hypoxia. J Exp Biol 156:233–248

Truchot JP (1987) Comparative aspects of extracellular acid-base balance. In: Zoophysiology, vol 20. Springer Verlag, Berlin, Heidelberg, p 262

Ultsch GR, Jackson DC, Moalli R (1981) Metabolic oxygen conformity among lower vertebrates: the toadfish revisited. J Comp Physiol B 142:439–443

Van Raij MTM, Van den Thillart GE, Vianen GJ, Pit DS, Balm PH, Steffens AB (1996) Substrate mobilization and hormonal changes in rainbow trout (*Oncorhynchus mykiss*) and common carp (*Cyprinus carpio*) during deep hypoxia and subsequent recovery. J Comp Physiol 166:443–452

Venkatesh B, Tan CH, Lam TJ (1989) Blood steroids in the goldfish: measurement of six ovarian steroids in small volumes of serum by reverse-phase high performance liquid chromatography and radioimmunoassay. Gen Comp Endocrinol 76:397–407

Vinberg GG (1956) Rate of metabolism and food requirements of fishes. Fisheries Research Board of Canada, Translation Series no 194, p 202

Williot P, Comte S, Le Menn F (2011) Stress indicators throughout the reproduction of farmed Siberian sturgeon *Acipenser baerii* (Brandt) females. Intern Aquat Res 3:31–43

Wood CM, Shelton G (1975) Physical and adrenergic factors affecting systemic vascular resistance in the rainbow trout: a comparison with branchial vascular resistance. J Exp Biol 63:505–523

Wood CM, Shelton G (1980) Cardiovascular dynamics and adrenergic responses of the rainbow trout *in vivo*. J Exp Biol 87:247–270

Effects of Exposure to Ammonia in Water: Determination of the Sublethal and Lethal Levels in Siberian Sturgeon, *Acipenser baerii*

19

Guy Nonnotte, Dominique Salin, and Patrick Williot

Abstract

The initial aim of this work was to situate the sensitivity of the Siberian sturgeon (*Acipenser baerii* B.) to ammonia compared with other species. Ammonia toxicity in the Siberian sturgeon was analysed, and the 24 h LC50 was performed for different weight groups. This sensitivity is currently determined by normalized tests of short duration, even if their practical application is difficult. During exposure to high levels of ammonia, we also observed alterations of the branchial tissue and modifications in haematological characteristics, in the behaviour of the fishes and tetany. Thus, the interpretation of these observations remains difficult, and we propose an experimental protocol under controlled conditions of pH, oxygen partial pressure PO_2, carbon dioxide partial pressure PCO_2, temperature and ammonia concentrations in water for 72 h duration to perform additional experiments in physiology and biochemistry.

Keywords

Siberian sturgeon · *Acipenser baerii* B. · Ammonia · Toxicity · LC50 · Experimental protocol

---

G. Nonnotte (✉)
12 rue Marcel Pagnol, F-33260 La Teste de Buch, France
e-mail: guy.nonnotte@wanadoo.fr

D. Salin
109 rue Blaise Pascal, F-33160 Saint Médard en Jalles, France

P. Williot
4 rue du Pas de Madame, F-33980 Audenge, France

Introduction

Ammonia, an end product of the protein catabolism, represents 60–80% of the nitrogenous excretion of fish (Smith 1929; Luquet 1972; Vellas 1979). In intensive fish farming, excessively high doses of ammonia, issuing either from this excretion or from external pollution, can induce limited fish growth or even death. For these reasons, ammonia toxicity has been studied in numerous fish species (Table 19.1). The complete lack of data for sturgeon on this problem led us to carry out a study of ammonia toxicity on the Siberian sturgeon (*Acipenser baerii* B.) for different age groups. Thus, the intensification of sturgeon farming generates the needs to optimize all the production factors, especially the quality of water (Faure 1976; Alabaster and Lloyd 1980; Wickins 1981; Poxton and Allouse 1982). These observations will be discussed and compared to those recorded for other species. Although the initial aim of this work was to situate the sensitivity of the Siberian sturgeon (*Acipenser baerii* B.) to ammonia in comparison with other species, it was of prime importance to define the lethal and sublethal concentration to study the physiological effects of different levels of ammonia exposure.

19.1 Material and Methods

This study concerns Siberian sturgeon *(Acipenser baerii* B.), obtained from artificial reproduction and weighing between 60 mg (larvae) and 450 g juveniles). The larvae were randomly selected from one of the hatch tanks and transferred to one of the ten 3 L aquariums just before the experiment (10 larvae/aquarium). The other fish were taken from the ponds where they are usually stocked and put into the experimental structure consisting of five polyester tanks (200 L) at least 8 days before the beginning of the experiment. We used five to ten fish by tank in relation with their individual weight. During this acclimatization period, the water is renewed with sufficient flow to maintain the oxygen rate above 70% of saturation. Feeding is stopped 24 h before the beginning of test. A comparative test was carried out on trout (*Oncorhynchus mykiss*) of 34 g coming from the same fish farm. We used the same structure and the same procedure as for sturgeon. All the experiments were performed in the IRSTEA (formerly CEMAGREF) experimental hatchery at Donzacq (southwest of France) which has water with very low ammonia content and a pH and temperature which are almost constant during the whole year (Table 19.2).

We set out, using a series of static 24-h tests, to determine the lethal concentration for 50% of the fish (24 h LC50). The experimental technique followed the AFNOR NF T 90-305 (1985), Reish and Oshida (1987) recommendations. At the beginning of each 24 h test, the water was no longer renewed (static test) but was aerated by compressed air to maintain the oxygen rate above 60% of saturation. We then rapidly introduced known quantities of ammonium chloride (NH_4Cl) into four

19 Effects of Exposure to Ammonia in Water

Table 19.1 Bibliographic studies of toxicity tests

Species	Weight	Tests	Results (NH₃ mg/L)	References
Oncorhynchus mykiss		24 h LC50	0.61	Herbert and Shurben (1963)
	40 g	24 h LC50	0.41	Ball (1967)
	Eggs - 50d	24 h LC50	>3. 5	Rice and Stokes (1975)
	80 days	24 h LC50	0.068	Rice and Stokes (1975)
	Adults	24 h LC50	0.097	Rice and Stokes (1975)
	1 g	96 h LC50	0.6–0.8	Thurston et al. (1978)
	3.5 g	96 h LC50	0.5–0.6	Thurston et al. (1978)
	60 mg	96 h LC50	0.45	Thurston and Russo (1983)
	250 mg	96 h LC50	0.44	Thurston and Russo (1983)
	10 g	96 h LC50	0.62	Thurston and Russo (1983)
	250 g	96 h LC50	0.32	Thurston and Russo (1983)
	500 g	96 h LC50	0.28	Thurston and Russo (1983)
	21 g	96 h LC50	0.50	Thurston et al. (1981)
	250 g	96 h LC50	0.29	Thurston et al. (1981)
	2600 g	96 h LC50	0.16	Thurston et al. (1981)
	22 g	96 h LC50	0.4–0.6	Arthur et al. (1987)
	Larvae	96 h LC50	0.25	Bulkhalter and Kaya (1977)
Oncorhynchus kisutch	6 g	14 to 96 h LC50	0.45	Buckley (1978)
Cynoscion nebulosus	Larvae	24 h LC50	0.28	Daniels et al. (1987)
	400 mg	24 h LC50	1.38	Daniels et al. (1987)
	8.8 g	24 h LC50	1.98	Daniels et al. (1987)
Pimephales promelas	300 mg	96 h LC50	2.5	Thurston et al. (1983)
	3 g	96 h LC50	1.4	Thurston et al. (1983)
Ictalurus punctatus		24 h LC50	1.4–1.8	Tommasso et al. (1980)
		24 h LC50	2.92	Sparks (1963) in Ruffier et al. (1981)
	16 g	24 h LC50	0.7–2.1	Sheehan and Lewis (1986)
	1 g	96 h LC50	3.1	Colt and Tchobanoglous (1978)
	3–7 g	96 h LC50	1.3	Arthur et al. (1987)
	3–4 g (30 °C)	96 h LC50	3.8	Colt and Tchobanoglous (1976)
	3–4 g (22 °C)	96 h LC50	2.4	Colt and Tchobanoglous (1976)
Perca fluviatilis	14 g	24 h LC50	0.29	Ball (1967)
Rutilus rutilus	9 g	24 h LC50	0.62	Ball (1967)
Abramis brama	16 g	24 h LC50	0.58	Ball (1967))
Cyprinus carpio	300 mg	24 h LC50	1.8–2.1	Hasan and Macintosh (1986)
	300 mg	48 h LC50	1.8–1.9	Hasan and Macintosh (1986)
	300 mg	96 h LC50	1.7–1.8	Hasan and Macintosh (1986)
	30 g	48 h LC50	1.6–2.0	Dabrowska and Sikora (1986)
Acipenser baerii	60 mg–450 g	24 h LC50	1.5–2.5	This study

Table 19.2 Characteristics of water used for toxicity tests

pH	7.5 ± 0.03	T_{Amm}	9×10^{-3}
Temperature	16 ± 1 °C	HCO_3^-	4.85
Conductivity	500 µS cm^{-1}	Cl^-	0.59
Minéral fraction	89%	Na^+	0.87
Suspended matter	5 mg L^{-1}	K^+	0.06
NO_3^-	0.17	Ca^{2+}	3.75
SO_4^{2-}	0.4	Mg^{2+}	1.69
PO_4^{2-}	nd	Mn^{2+}	nd
SiO_2	0.2	Fe	nd
T.A.	4.92	Phénol	0
Σ Cations = 6.37 meq L^{-1}		Σ Anion = 6.01 meq L^{-1}	

Units: meq L^{-1} except where indicated; *nd*, no detectable

of the five tanks. We thus obtained four doses of ammonia which followed a geometrical progression and a control tank (with no ammonia). In the case of larvae, we used five to ten doses of ammonia and a control aquarium. Because of the low oxygen consumption by the larvae, no aeration was made.

During the 24 h experiment, different parameters were noted:

1. The initial and final concentration of ammonia in each tank.
2. The temperature, the pH and the oxygen content at least five times every 24 h. The last two instruments were recalibrated before each series of measurements.
3. The reactions of the fish in order to evaluate the influence and effects of ammonia on sturgeon.
4. The successive deaths: we considered a fish to be dead when there was no further opercular movement.

The concentration of ammonia in the water was determined by a spectrophotometric method with blue indophenol (AFNOR NF 90-015, 1975). We doubled all measurements and readings were carried out before 12 h. The NH_3 contents were calculated from the ammonia concentrations, the pH and the temperature using a computer program according to the formulae of Emerson et al. (1975).

The 24 h LC50 was estimated for ($NH_4^+ + NH_3$) from the totality of results obtained from the same weight range. We used a computer program which was written according to the Spearman-Karber method (Hamilton et al. 1977), the AFNOR (1985) and FAO (1987) norms (method with Log/Normal transformation). In both cases, we estimated the 95% confidence interval (see more details in Salin and Williot 1991).

For the three series of tests on sturgeons of 270 g, 1 mL blood sample in heparinized tube was taken and centrifuged (5 min at 12,000 rpm or 14,000 g) and the serum kept at 4 °C before analysis. The haematocrit was also determined after centrifugation in microtubes. The potassium and sodium contents of the serum were measured with an Eppendorf flame photometer, the chloride content using a titration method with mercuric nitrate (Schales and Schales 1941, modified). The osmotic pressure was estimated with a freezing point osmometer and the erythrocyte water content by the difference between the dry and fresh sample weights.

19.2 Results and Discussion

The observations of Siberian sturgeon during the toxicity experiments allowed us to define the following clinical signs:

1. First, the fish remain steadily at the bottom of the tanks and increase ventilation only.
2. Then, they jump, in more or less complete tetany with spasmodic and anarchic movements.
3. A loss of equilibrium then occurs, and the fish swim on their back side.
4. Finally, a very violent tetany followed by death can be observed.

19.2.1 Toxicity Tests

All results are presented as mean ± standard error of mean. Because of the low numbers of fish in the series, we used the nonparametric U test of Mann-Whitney and, for complete series, the test T of Wilcoxon (Scherrer 1984).

The results of the whole series of tests for ammonia toxicity are summarized in Table 19.3. We observed that the youngest fish (60–260 mg) are most sensitive with a 24 h LC50 of about 0.05 mmol L^{-1} NH_3 (2.5 mmol L^{-1} T_{Amm}), whereas for those weighing 450 g it is about 0.14 mmol L^{-1} NH_3 (9 mmol L^{-1} T_{Amm}). It therefore appears that the sensitivity to ammonia of Siberian sturgeon is age dependent.

This phenomenon is confirmed by the study of LC50 according to the duration of exposure (Fig. 19.1) where we can also see that the variation of the LC50 according to time is greater for the older fish than for the larvae.

Finally, it seems that the LC50 tends toward an asymptotic value. For the totality of the weight range used, there was never any deaths for doses less than 0.8 mmol $L^{-1}NH_4^+$.

(0.023 mmol L^{-1} NH_3), and there was always 100% mortality for concentrations above 12 mmol $L^{-1}NH_4^+$ (0.223 mmol L^{-1} NH_3).

The comparative test with fingerling trouts of 34 g in the same experimental conditions (tanks, quality of water, measuring and calculating techniques) allowed us to calculate a 24 h LC50 of 0.05 mmol L^{-1} NH_3 or 1.6 mmol L-1 T_{Amm} (Table 19.3). We therefore established that sturgeons of comparable weight (10–270 g) are almost twice as resistant to ammonia as trout.

The results of tests indicate that the 24 h LC50 of sturgeon (*Acipenser baerii* B.) is situated in the range of values found in other species. They are less sensitive than the trout (*Oncorhynchus mykiss*) but more sensitive than the catfish (*Ictalurus punctatus*) (Table 19.1). The tolerance level of sturgeon to ammonia is about the same as that of the carp (*Cyprinus carpio*) (Dabrowska and Sikora 1986; Hasan and Macintosh 1986). However, it is sometimes difficult to make direct comparison, because the methods of calculating the LC50 are not the same and can, in some cases, lead to uncertain results.

The difference in sensitivity to ammonia according to age (*Acipenser baerii* larvae are more sensitive than older fish) is a result already observed in other species. Thus, the eggs

Table 19.3 Experimental conditions and results of the toxicity tests

Experimental conditions

Weight	n^b	Range T_{Amm} (mmol L^{-1})	NH$_3$ (mmol L^{-1})	pH^c	Temperature^c (°C)	24h LC50^a T_{Amm} (mmol L^{-1}) Spearman Karber	AFNOR FAO	NH$_3$ (mmol L^{-1}) Spearman Karber	AFNOR FAO
Acipenser Baerii									
60 mg	8	0.78–6.2	0.02–0.12	7.7–8.1	16.5–17	3.18 (2.91–3.47)	2.61 (1.26–5.41)	0.069 (0.065–0.076)	0.062 (0.035–0.112)
65 mg	10	1.1–3.4	0.03–0.07	7.6–8.1	16.5–17	2.34 (2.21–2.48)	2.40 n.c.^{d}	0.051 (0.047–0.053)	0.048 (0.041–0.141)
260 mg	4	0.9–4.2	0.03–0.06	7.7–8.1	17–18	1.95 (1.63–2.34)	1.94 (0.63–5.97)	0.048 (0.041–0.053)	0.049 (0.018–0.153)
10 g	4	2.2–13.9	0.04–0.15	7.5–7.7	18–21	8.20 (6.31–10.66)	6.25 n.c.	0.104 (0.082–0.129)	0.083 n.c.
270 g	9	3.3–8.3	0.07–0.21	7.6–8.1	17–20	4.43 (4.27–4.60)	4.48 n.c.	0.097 (0.088–0.100)	0.101 n.c.
450 g	8	4.9–11.9	0.10–0.22	7.4–8.2	14–16	9.24 (8.34–10.22)	9.75 n.c.	0.141 (0.135–0.147)	0.130 (0.029–0.553)
Oncorhynchus mykiss									
34 g	7	0.8–2.2	0.02–0.08	7.6–8.2	16–17	1.61 (1.49–1.74)	1.51 n.c.	0.044 (0.042–0.049)	0.048 (0.012–0.153)

[a] For LC50, results and 95% confidence interval
[b] Number of concentrations used for calculation
[c] For pH and temperature, values from beginning to end of the test
[d] n.c., noncalculable

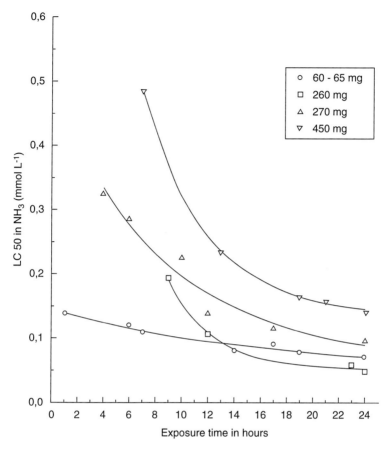

Fig. 19.1 LC50 as a function of exposure time and individual weight

are very resistant, i.e. no effect for concentrations over 0.2 mmol L^{-1} NH$_3$ (*Oncorhynchus mykiss*, Rice and Stockes 1975; *Scianeops ocelatus*, Holt and Arnold 1983; *Cynoscixm nebulosus*, Daniels et al. 1987), and larvae become more sensitive during the reabsorption of the vitelline reserve, and then the fry and adults are once again more resistant (*Ictalurus punctatus*, Colt and Tchobanoglous 1976, 1978; *Oncorhynchus mykiss*, Burkhalter and Kaya 1977; Rice and Bailey 1980; Thurston and Russo 1983).

It seems necessary to recall that the LC50 values also depend on the quality of the water used during the tests and on the execution of the latter. Thus, during our experiments, the pH and therefore the percentage of the NH$_3$ form, developed exponentially and stabilized 6–7 h after the beginning of the tests. This phenomenon is probably due to a modification in the calco-carbonic equilibrium because of the aeration (Baird et al.1979). However, with larvae of 65 mg, the 24 h LC50 value obtained when using demineralized water—where the pH remained stable—is not significantly different from that using water from the fish farm ($p = 0.05$; T test).

The 24 h LC50 values seem to depend on the ion calcium content of the water. Thus, Tomasso et al. (1980) also recorded an influence of the hardness of the water on the toxicity of ammonia to the catfish and supposed this to be due to a competition between calcium and ammonia ions. This hypothesis is confirmed by the 24 h LC50 in trout, which is usually about 0.03 mmol L^{-1} NH_3 (Ball 1967; Rice and Stokes 1975; Hillaby and Randall 1979; Haywood 1983; Thurston and Russo 1983; Meade 1985), whereas we found with the hard water of Donzacq a LC50 of 0.045 mmol L^{-1} NH_3. We can suppose that the 24 h LC50 for water with lower calcium level would be slightly different.

19.2.2 Blood Parameters

According to blood analysis, the haematocrit appears in direct proportion to the dose of ammonia, the sodium and chloride contents were stable but we recorded an increase in the potassium concentration (Fig. 19.2). We could note that the potassium levels are about 3.7 meq/L when the mortality rate is under 10%, then they increase to 6.5 meq/L when the ammonia contents cause more than 50% mortality. The water content of the erythrocytes is not modified, but the osmotic pressure of the serum of fish exposed to lethal doses (289 ± 10 mOsm, $n = 5$) is significantly higher than that of the control fish (257 ± 7 mOsm, $n = 9$) ($p = 0.028$: U test).

The haematocrit and ions levels of the control fish are very close to the results found in previous studies on *Acipenser baerii* (Ait-Fdil 1986; Bennouna 1986) or on other species of sturgeon (Magnin 1962; Potts and Rudy 1972; Natochin et al. 1975; Hunn and Christenson 1977; Lavrova et al. 1984). We notice in Fig. 19.2 an increase in the haematocrit in direct proportion to the ammonia concentration. This is not caused by an increase in volume of the erythrocytes, for their water content is not significantly modified in these experiments. This phenomenon is not always observed in other species. Indeed, as concerns the blood parameters of fish exposed to sublethal doses of ammonia, there is no modification in the number of erythrocytes in carp (Dabrowska and Wlasow 1986) nor in salmonids (Smart 1978; Buckley et al. 1979). The increase in the haematocrit of Siberian sturgeon exposed to high doses of ammonia could be due to a transfer of plasma towards the tissues. This hypothesis could explain both the evolution of haematocrit and the significant increase in osmotic pressure in *Acipenser baerii* exposed to lethal doses of ammonia. Finally, the stress caused by ammonia intoxication and the setting up of the experiment could be sufficient to explain the modification of haematocrit by a significant splenic clearance (Soivio and Oikari 1976) on *Esox lucius*. This would not however explain why this increase in haematocrit should be proportional to ammonia dose. Among the blood parameters studied, only potassium undergoes a significant modification in the presence of ammonia (U test, $p < 0.05$). Moreover, we noticed that this increase seems to appear at two stages, one for doses not causing more than 10% mortality and the other for concentrations over the 24 h LC50. It therefore seems there is a two-step phenomenon. The system regulates itself until the dose of ammonia saturates the process, then causing a strong increase in the rate of potassium in the blood.

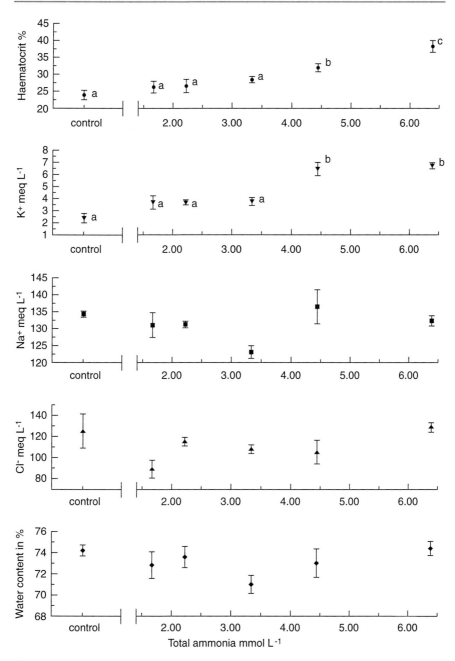

Fig. 19.2 Haematocrit, plasma ionic concentrations and water content of erythrocytes in *Acipenser baerii* in function of total ammonia concentration in water during 24 h exposure. Mean ± standard error, $n = 6$. The same letters (a, b, c) mean that there is no statistic difference ($p > 0.05$, Neuwmann-Keuls test)

19.3 Proposition of an Experimental Protocol to Study the Ammonia Toxicity for Siberian Sturgeon

By taking into account these toxicity tests and the determination of the sublethal and lethal levels of ammonia in water, a general experimental protocol was defined to perform specific studies necessary to characterize the morphological, physiological and biochemical effects of high doses of ammonia on Siberian sturgeon (see Chaps. 20–22).

Four conditions are imperative to examine ammonia toxicity or other pollutants effects:

- The batch of fishes must be homogeneous (weight, age, origin).
- Access to the internal medium of the fish (the blood) and analysis without stress must be achieved.
- The environmental water quality must be controlled and monitored during the experiments (oxygenation, PCO_2, pH, titration alkalinity, temperature).
- The intoxication level and ammonia concentrations in water must be defined.

19.3.1 Animals

Three-year-old Siberian sturgeons, *Acipenser baerii*, (1250 ± 40 g) used for the experiments reported in Chaps. 20–22, were obtained from the experimental hatchery of IRSTEA (formerly CEMAGREF). Prior to experimentation, the fish were maintained in polyester tanks with running aerated fresh water. They were fed with commercial trout pellets containing 45% protein until 1 week before the experiments.

19.3.2 Access to the Blood Compartment

The fish were anesthetized with 2-phenoxyethanol (1/200) in an oxygenated solution for approximately 10 min. The fish were cannulated via the dorsal aorta (Williot and Le Menn, see Chap. 25). The caudal artery was catheterized with Biotrol n°4 tubing, while the gills were irrigated with normoxic water containing phenoxyethanol (1/600). The catheter was inserted into a sharpened stainless-steel needle (Tuhoy n°15) in such a way that the tip of the catheter protruded just beyond the tip of the needle. The dorsal aorta was then blindly punctured just behind the anal fin with needle and catheter. When the blood went up, the catheter was pushed forward for about 10 cm and the needle was withdrawn. The catheter was fastened to the skin by silk sutures and filled with heparinized saline solution (lithium salt, Sigma reagent, at $200i.u mL^{-1}$). Following the operation (about 10 min), the fish was placed in an individual chamber and allowed to recover for at least 48 h.

During experiments, an arterial blood sample (about 1 mL) was drawn from the arterial cannula, first into two glass capillaries filled only by arterial pressure and then into a nonheparinized syringe. The haematocrit was quantified and the blood pH was measured in blood capillaries. Plasma ammonia concentration was estimated after

19 Effects of Exposure to Ammonia in Water 401

5 mn centrifugation (8000 rpm or 5223 g) with an enzymatic method (Sigma kit UV170). The measurements were performed less than 30 mn after blood sampling. Free ammonia concentration (NH_3) was calculated from total ammonia concentration in plasma and whole blood pH using the Henderson-Hasselbalch equation incorporating the pK_{Amm} value of Cameron and Heisler (1983). Each blood sample was replaced with the same volume of heparinized saline solution. Blood samples were obtained in the control conditions ($t = 0h$) and 1, 3, 6, 24, 48 and 72 h after exposure to ammonia.

19.3.3 Environmental Water

The water recirculating system used for the experiments had a total volume of 160 L. It was thermostated at $15 \pm 0.5\ °C$ and consisted of a large aerated tank from which the water was pumped ($22\ L\ mn^{-1}$) through the experimental chamber. The water was oxygenated by intensive bubbling of CO_2-free air. Water pH and PCO_2 were regulated at 8.0 ± 0.1 and $0.10\ kPa$ ($0.75\ mm\ Hg$), respectively, using a pH-CO_2-stat (Dejours et al. 1978) and manually corrected.

The titration alkalinity (see Chap. 26) was maintained at $1.90 \pm 0.05\ meq\ L^{-1}$. The device intermittently injected small quantities of pure CO_2 into the water when pH rose above a preset value. During the recovery period, a small flow of fresh water was continuously added to the system in order to maintain a low level of ammonia without changing other parameters.

The main characteristics of the water are given in Table 19.4. Water titration alkalinity and pH were controlled twice a day. The pH of a water sample was measured following equilibrium at $PCO_2 = 0.10\ kPa$ ($0.75\ mm\ Hg$), and the pH set point adjusted accordingly.

Table 19.4 Characteristics of the water from the experimental closed circuit ($n = 10$; mean $\pm$ SE)

Series	Control		Series 1	Series 2
T_{Amm} (mmol L^{-1})	0.085 ± 0.037	0.272 ± 0.112	1.851	4.523
			±0.023	±0.038
P_{NH3} (mPa)	8.8 ± 4.1	29.3 ± 2.2	173.6	402.6
			±2.9	±8.9
For all series				
Temperature (°C)	15.0 ± 0.5			
pH	8.050 [8.020–8.080]			
T.A. (meq L^{-1})	1.90 ± 0.05			
PCO_2(kPa)	0.10 (0.1% of the barometric pressure)			
PO_2(kPa)	>18.7			
Na^+ (mmol L^{-1})	2.26 ± 0.06			
Cl^- (mmol L^{-1})	1.25 ± 0.09			

T_{Amm} and P_{NH3} control values correspond with the mean $\pm$ SE at the beginning and the end of the experiments
T.A. titration alkalinity, *PCO_2* partial pressure of CO_2, *PO_2* partial pressure of O_2
1 kPa = 7.50 torr = 7.50 mmHg and 1 Torr = 133.32 Pa

19.3.4 Exposure to High External Ammonia Concentrations

Three series of experiments were performed: a control series without ammonia and two series at a sublethal concentration (1.8 mmol L^{-1} total ammonia) and around the lethal concentration for Siberian sturgeon (4.5 mmol L^{-1} total ammonia), respectively, as demonstrated by the previous toxicity tests. Ammonia in water was also measured at least twice a day by a modified phenol-hypochlorite method (AFNOR NF 90-015, 1975). Concentrations of NH_4 and NH_3 were calculated according to the table of Emerson et al. (1975).

As a conclusion, the sensitivity to ammonia of the fish, in particular Siberian sturgeon (*Acipenser baerii* B.), is currently determined by normalized tests of short duration, even if their practical application is difficult. During Siberian sturgeon exposure to high levels of ammonia, we also observed alterations of the branchial tissues and modifications in haematological characteristics, in the behaviour of the fishes and tetany. However, the interpretation of these observations remains difficult, and additional experiments must be performed following the proposed experimental protocol (see Chaps. 20–22).

Moreover, for the first time, the proposed protocol allowed to perform the physiological effects of high external ammonia concentration exposure during a long time (72 h) under controlled conditions of pH, PO_2, PCO_2, temperature and ammonia concentrations in water. The preceding experiments did not exceed 6 h (Cameron and Heisler 1983). Moreover, the results obtained in these conditions, from the control series, will bring numerous and original physiological characteristics of the Siberian sturgeon, a species which has not been much studied to date (see Chaps. 20–22 in this issue).

Acknowledgments We thank Dr. Karine Pichavant-Rafini (ORPHY laboratory, EA4324) and Michel Rafini (Professor at the Language Dpt) of the Brest University, for their kindness and constant availability, their help and their scientific advices. Moreover, we are extremely indebted and grateful to them for the English corrections.

Notes of the Authors The experiments were performed since 1990 to 1994. They have been investigated in the "Laboratoire de Neurobiologie et Physiologie comparées, CNRS URA 1126 and the University of Bordeaux I, F-33120 Arcachon," in collaboration with the IRSTEA (formerly CEMAGREF), F-33611 Cestas-Gazinet.

The present chapter was a redrawn version of a part of a PHD: Salin D (1992) La toxicité de l'ammoniaque chez l'esturgeon sibérien, *Acipenser baerii*: effets morphologiques, physiologiques, métaboliques d'une exposition à des doses sublétales et létales. Thèse N° 749, Université Bordeaux I pp134, financial grant of the IRSTEA (formerly CEMAGREF) and CNRS. Director of thesis: Truchot Jean-Paul; supervisors: Nonnotte Guy and Williot Patrick.

References

AFNOR NF T 90-015 (1975) Essais des eaux: dosage de l'azote ammoniacal 2: méthode spectrophotométrique au bleu d'indophénol, p 6

AFNOR NF T 90-305 (1985) Détermination de la toxicité aigüe d'une sustance vis à vis de la truite Arc-en-ciel, p 4

Ait-Fdil M (1986) Mise en évidence et propriétés des ATPases membranaires dans la branchie, les érythrocytes et le rein d'un Chondrostéen, l'esturgeon sibérien, *A. baerii*. Thèse N° 2157, Université Bordeaux I, p 60

Alabaster JS, Lloyd R (1980) Water quality criteria for freshwater fish: ammonia, vol 4. Butterworths, FAO London, pp 85–102

Arthur JW, West CW, Allen KN, Hedtke SF (1987) Seasonal toxicity of ammonia to five fish and nine invertebrate species. Bull Environ Contam Toxicol 38:324–331

Baird R, Bottomley J, Taitz H (1979) Ammonia toxicity and pH control in fish toxicity bioassays of treated wastewater. Water Res 13:181–184

Ball IR (1967) The relative susceptibility of some species of fresh water fish to poisons: I Ammonia. Water Res 1:767–775

Bennouna M (1986) Equilibre hydrominéral du milieu intérieur et des érythrocytes chez un Chondrostéen, l'esturgeon sibérien, A. *baerii* au cours de variations expérimentales de salinité et de température de l'environnement. Thèse N° 2148, Université Bordeaux I, p 93

Buckley JA (1978) Acute toxicity of unionized ammonia to fingerling coho salmon. Prog Fish-Cult 40(1):30–32

Buckley JA, Whitmore CM, Liming BD (1979) Effects of prolonged exposure to ammonia on the blood and the liver glycogen of coho salmon (*Oncorhynchus kisutch*). Comp Biochem Physiol 63C:297–303

Burkhalter DE, Kaya CM (1977) Effects of prolonged exposure to ammonia on fertilized eggs and sac fry of rainbow trout (*Salmo gairdneri*). Trans Am Fish Soc 106(5):470–475

Cameron JN, Heisler N (1983) Studies of ammonia in the rainbow trout: physicochemical parameters, acid-base behaviour and respiratory clearance. J Exp Biol 105:107–125

Colt J, Tchobanoglous G (1976) Evaluation of the short-term toxicity of nitrogenous compounds to channel catfish, Ictalurus punctatus. Aquaculture 8:209–224

Colt J, Tchobanoglous G (1978) Chronic exposure of channel catfish to ammonia: effects on growth and survival. Aquaculture 15:353–372

Dabrowska H, Sikora H (1986) Acute toxicity of ammonia to common carp (*Cyprinus carpio* L.) Pol Arch Hydrobiol 33(1):121–128

Dabrowska H, Wlashow T (1986) Sublethal effect of ammonia on certain biochemical and haematological indicators in common carp (*Cyprinus carpio* L.) Comp Biochem Physiol 83C(1):179–184

Daniels HV, Boyd CE, Minton RV (1987) Acute toxicity of ammonia and nitrite to spotted seatrout. Prog Fish-Cult 49:260–263

Dejours P, Armand J, Gendner JP (1978) Importance de la régulation de l'équilibre acide-base de l'eau ambiante pour l'étude des échanges respiratoires et ioniques des animaux aquatiques. C R Acad Sc Paris 287:1397–1399

Emerson KR, Russo RC, Lund RE, Thurston RV (1975) Aqueous ammonia equilibrium calculations: effect of pH and temperature. J Fish Res Board Can 32:2379–2383

Faure A (1976) Bases de la gestion de l'eau en salmoniculture intensive. La Pisciculture Française 46:11–54

Hamilton MA, Russo RC, Thurston RV (1977) Trimmed spearman-Karber method for estimating median lethal concentrations in toxicity bioassays. Environ Sci Technol 11(7):714–719

Hasan MR, Macintosh DJ (1986) Acute toxicity of ammonia to common carp fry. Aquaculture 54:97–107

Haywood GP (1983) Ammonia toxicity in teleost fishes: a review. Can Tech Rep Fish Aquat Sci 1177:35

Herbert DWM, Shurben DS (1963) A preliminary study of the effect of physical activity on the resistance of rainbow trout (*Salmo gairdnerii* Richardson) to two poisons. Annals of Applied Biology 52(2):321–326

Hillaby BA, Randall DJ (1979) Acute ammonia toxicity and ammonia excretion in rainbow trout (*Salmo gairdneri*). J Fish Res Board Can 36:621–629

Holt GJ, Arnold CR (1983) Effects of ammonia and nitrite on growth and survival of red drum eggs and larvae. Trans Am Fish Soc 112:314–318

Hun JB, Christenson LM (1977) Chemical composition of blood and bile of the shovelnose sturgeon. Prog Fish Cult 39(2):59–61

Lavrova EA, Natochin YV, Shakhamatova EL (1984) Electrolytes in the tissues of sturgeon and bony fishes in fresh and salt water. J Ichthyol 24(5):156–160

Luquet P (1972) Données sur l'alimentation des salmonidés II: Besoins alimentaires. Aliment Vie 60(5):339–347

Magnin E (1962) Recherches sur la systématique et la biologie des Acipenséridés. Ann Stn Centr Hydrobiol Appl 9:7–242

Meade JW (1985) Allowable ammonia for fish culture. Prog Fish-Cult 47(3):135–145

Natochin Y, Luk'yanenko VI, Lavrova YA, Metallov GF (1975) Cation contents of the blood serum during the marine and river periods in the life sturgeons. J Ichthyol 15(5):799–804

Potts WTW, Rudy PP (1972) Aspects of osmotic and ionic regulation in the sturgeon. J Exp Biol 56:703–715

Poxton MG, Allouse SB (1982) Water quality criteria for marine fisheries. Aquacult Engineering 1:153–191

Reish DL and Oshida PS (1987) Manual of methods in aquatic environment research, part 10: short term static bioassays. FAO Fish Tech Pap no 247, Rome, p 62

Rice SD, Bailley JE (1980) Survival, size and emergence of pink salmon, *Oncorhynchus gorbuscha*, alevins after short- and long-term exposures to ammonia. Fishery Bull 78(3):641–648

Rice SD, Stokes RM (1975) Acute toxicity of ammonia to several developmental stages of rainbow trout, Salmo Gairdneri. Fishery Bull 73:207–211

Ruffier P, Boyle W, Kleinschmidt J (1981) Short-terme acute bioassays to evaluate ammonia toxicity and effluent standards. J WPGF 53:367–377

Salin D, Williot P (1991) Acute toxicity of ammonia to Siberian sturgeon *Acipenser baerii*. In: Williot P (ed) Acipenser. CEMAGREF-DICOVA Publications, Antony, France, pp 153–167

Salin D (1992) La toxicité de l'ammoniaque chez l'esturgeon sibérien, *Acipenser baerii*: effets morphologiques, physiologiques, métaboliques d'une exposition à des doses sublétales et létales. Thèse N° 749, Université de Bordeaux I pp. 134

Schales O, Schales SS (1941) A simple and accurate method for the determination of chloride in biological fluids. J Biol Chem 140:879–884

Scherrer B (1984) Biostatistique. In: Morin G (ed). Quebec, Canada, p 850

Sheehan RJ, Lewis WM (1986) Influence of pH and ammonia salts on ammonia toxicity and water balance in young channel catfish. Trans Am Fish Soc 115:891–899

Smart GR (1978) Investigations of the toxic mechanisms of ammonia to fish gas exchange in rainbow trout exposed to acutely lethal concentrations. J Fish Biol 12:93–104

Smith HW (1929) The excretion of ammonia and urea by the gills of fish. J Biol Chem 81:727–742

Soivio A, Oikari A (1976) Haematological effects of stress on a teleost *Esox lucius* L. J Fish Biol 8:397–411

Thurston RV, Russo RC (1983) Acute toxicity of ammonia to rainbow trout. Trans Am Fish Soc 112:696–704

Thurston RV, Russo RC, Phillips GR (1983) Acute toxicity of ammonia to fathead minnows (Pimephates prometas). Trans Am Fish Soc 112:705–711

Thurston RV, Russo RC, Vinogradov GA (1981) Ammonia toxicity to fishes: effect of pH on the toxicity of the unionized ammonia species. Environ Sci & Technol 15:837–840

Thurston RV, Russo RC, Smith CE (1978) Acute toxicity of ammonia and nitrite to cutthroat trout fry. Trans Am Fish Soc 107(2):361–368

Tomasso JR, Goudie CA, Simco BA, Davis KB (1980) Effects of pH and calcium on ammonia toxicity in channel catfish. Trans Am Fish Soc 109:229–234

Vellas F (1979) L'excrétion azotée: métabolisme des composés azotés. In: Nutrition des poissons. Actes du colloque du CNERMA, Paris, CNRS Ed: 149-161

Wickins JF (1981) Water quality requirement for intensive aquaculture: a review Proc World Symp on Aquaculture. In: Heated effluents and recirculation systems, Stavanger 28–30 May 1980, I: 18-3

Consequences of High Levels of Ammonia Exposure on the Gills Epithelium and on the Haematological Characteristics of the Blood of the Siberian Sturgeon, *Acipenser baerii*

20

Guy Nonnotte, Dominique Salin, Patrick Williot, Karine Pichavant-Rafini, Michel Rafini, and Liliane Nonnotte

Abstract

Ammonia toxicity in the Siberian sturgeon, *Acipenser baerii*, a fresh water fish, was studied under controlled conditions of pH, temperature and ammonia concentrations in water.

The effects of ammonia during a 72-h exposure to lethal and sublethal doses were examined on the morphology of branchial epithelium. Some hypertrophies and necrosis were observed in proportion with the ammonia doses.

Ammonia has no effect on haematological characteristics except for the [K^+] plasmatic concentration and the erythrocyte volume for lethal levels of ammonia.

Among the numerous hypotheses to explain ammonia toxicity on fish, some refer to oxygen transport and oxygen affinity of the blood. In fact, the partial

---

G. Nonnotte (✉) • L. Nonnotte
12 rue Marcel Pagnol, F-33260 La Teste de Buch, France
e-mail: guy.nonnotte@wanadoo.fr

D. Salin
109 rue Blaise Pascal, F-33160 Saint Médard en Jalles, France

P. Williot
4 rue du Pas de Madame, F-33980 Audenge, France

K. Pichavant-Rafini
Laboratoire ORPHY EA4324, Université de Bretagne Occidentale,
6 Avenue le Gorgeu, CS 93837, 29238 Brest Cedex 3, France

M. Rafini
Département Communication, Anglais, Sciences Humaines, Université de Bretagne Occidentale, 6 Avenue le Gorgeu, CS 93837, 29238 Brest Cedex 3, France

pressure of oxygen in the blood (PaO2) decreases significantly for a lethal dose only. Moreover, the influence of ammonia on oxyphoric properties of red blood cells of Siberian sturgeon has been investigated in vivo (blood of fish exposed for 72 h to 4.5 mmol L−1 total ammonia, a lethal concentration) and in vitro (ammonia added to blood of control fish). Whatever the experimental conditions, ammonia did not modify the affinity of haemoglobin for oxygen illustrated by P50, Bohr effect, Hill number and percentage of methaemoglobin. Consequently, the toxicity of ammonia on Siberian sturgeon cannot be explained by an alteration of the gill morphology, the ionic exchanges or the oxygen-binding properties of the haemoglobin only.

Keywords

Ammonia toxicity • Gills morphology • Ionic transfer • PaO$_2$ • Oxygen affinity Siberian sturgeon • *Acipenser baerii*

Introduction

Numerous pollutants such as metals (Crespo et al. 1986), pesticides (Wendelaar Bonga 1997; Van der Oost et al. 2003; Velisek et al. 2009), ammonia (Salin and Williot 1991), nitrites (Huertas et al. 2002) and other organic substances pose serious risks to many aquatic organisms and to fish especially either in natural environment or in intensive fish farming. Accordingly, a great deal of previous research has characterized toxicity effects on gill structural changes (Mallat 1985), physiological mechanisms and behaviour (Scott and Sloman 2004) in fish exposed to contaminants. Physiological responses of fish to pollutants are dependent on the bioavailability, uptake, accumulation and disposition of contaminants within the organism and on the interactive effects of multiple contaminants (Capuzzo 1988). In this regard, physiological responses are integrators of cellular and subcellular processes and may be indicative of the overall fitness of the individual organism and of the welfare of the fish.

The physiological processes which must be targeted when faced with an exposure to pollutants are those which will contribute the most appropriately to the energy budget such as measurements of feeding, absorption, respiration, osmoregulation and excretion integrated as components of the energy balance.

The effects of ammonia exposure on the survival capacity of the Siberian sturgeon *Acipenser baerii* have been described, and the sublethal and lethal concentrations of ammonia were determined (see Chap. 19). Taking into account the determination of the sublethal and lethal levels of ammonia in water, the gill structure and the haematological characteristics of the blood of the sturgeon, *Acipenser baerii*, were examined after exposure to high levels of ammonia in the water. The modifications of gill epithelium, gill functions and the oxygen-carrying properties of the blood of the sturgeon, *Acipenser baerii*, will be considered as powerful indexes of biological effects of intoxication and detoxification processes.

20.1 Materials and Methods

All experiments were performed on three-year-old Siberian sturgeons, *Acipenser baerii*, obtained from the experimental hatchery of IRSTEA (formerly CEMAGREF). Fish were maintained in an experimental closed circuit under controlled conditions of pH, PO_2, PCO_2, temperature and ammonia concentrations in water during 72 h before the experiments (see protocol described in Chaps. 19 and 25 in this issue). Three series of experiments were carried out. A control series without ammonia and two series at a sublethal concentration (1.8 mmol L^{-1} total ammonia) and around the lethal concentration for Siberian sturgeon (4.5 mmol L^{-1} total ammonia), respectively, as demonstrated by the previous toxicity tests.

20.1.1 Histological and Ultrastructural Methods

Gill filaments were fixed in Bouin's fluid, dehydrated in ethanol and embedded in Paraplast for histological investigations. Sections of 5 μm were stained with eosin-haematoxylin.

Ultrastructural investigations were carried out after 1-h fixation in 5% glutaraldehyde in 0.1 M cacodylate buffer (pH = 7.4), washing, postfixation in 1% osmium tetroxide in 0.1 M cacodylate buffer, dehydration in ethanol and embedding either in Epon or in Araldite. The ultrathin sections were stained by the use of 0.5% alcoholic uranyl acetate and Reynold's lead citrate method.

20.1.2 Ionic and Ammonia Concentrations in Plasma

Blood samples were drawn from a heparinized Ringer-filled cannula inserted into the dorsal aorta (caudal part) and sutured in place (Salin 1992; Nonnotte et al. 1993; Williot and le Menn see Chap. 24). Fish plasma was obtained from blood collected in small centrifuge tube (400 μL) and immediately centrifuged for 5 mn at 10,000 g to separate plasma from red cells. One glass heparinized capillary (80 μL) was also filled to quantify the haematocrit. Chloride concentration [Cl^-] of fish plasma was measured with a chloride titrator Radiometer CMT10. Sodium [Na^+], potassium [K^+] and [Ca^{++}] concentrations in plasma were obtained by flame photometry (Eppendorf FCM6341) and osmotic pressure with a vapour pressure osmometer (Wescor 5500). Plasma ammonia concentration was estimated after 5 mn centrifugation (8000 rpm or 5223g) with an enzymatic method (Sigma kit UV170). The measurements were performed less than 30 mn after blood sampling. Free ammonia concentration (NH_3) was calculated from total ammonia concentration in plasma and whole blood pH using the Henderson-Hasselbalch equation incorporating the pK_{Amm} value of Cameron and Heisler (1983).

20.1.3 Oxygen Partial Pressure in Arterial Blood: PaO$_2$

Blood samples (800 µL) were drawn from the dorsal aorta cannula with a specific syringe to prevent gas exchanges and immediately injected through an oxygen electrode Radiometer E 5046, thermostated at 15 °C and calibrated with N$_2$- or air-saturated Ringer.

20.1.4 Erythrocyte Water Content

Cellular water content is used as an equivalent for cell volume. The packed cell mass was placed onto preweighed aluminium foil with a Microman Gilson (M50) positive-displacement pipette, and the cellular water content, given as litres H$_2$O/kg dry cell solid (DCS), was determined from the dry- and wet-weight samples after drying to constant weight for 24 h at 90 °C.

20.1.5 Oxygen-Carrying Properties of the Blood of the Sturgeon

Three-year-old sturgeons (about 1.5 kg live weight) were obtained from IRSTEA (formerly CEMAGREF) hatchery (St Seurin s/l'Isle, France) also and cannulated via the dorsal aorta as described previously. Sturgeons are notably tolerant to relatively high ambient ammonia levels (Salin and Williot 1991; see Chap. 19). Thus, fish were exposed for 72 h to local freshwater (15 °C; pH $\approx$ 8.0–8.1) containing 4.5 mmol L^{-1} total ammonia, which is a lethal concentration (LC 50) after 4–5 days (Chap. 19). Blood samples to be used for studies of oxygen binding were drawn from ammonia-exposed fish ($N = 5$) and from similarly treated controls ($N = 4$). Oxygen dissociation curves were constructed at a PCO$_2$ of 0.2 kPa for all samples. In addition, three samples from both ammonia-exposed and control groups were studied at additional PCO$_2$ values, 0.5, 1.0 and 2.0 kPa, in order to obtain different pH values and to evaluate the Bohr factor. The effects of adding ammonia in vitro on O$_2$ binding were studied at PCO$_2$ 0.2 kPa on blood drawn by caudal puncture from three fish that had never been exposed to waterborne ammonia. These samples were divided into two parts to which 25 µL mL^{-1} of either NaCl or NH$_4$Cl isosmotic solutions (both 320 mmol L^{-1}) were added. This resulted in a plasma ammonia concentration of about 4–5 mmol L^{-1}, which is of the same order of magnitude as in vivo mean values in sturgeon exposed for 72 h to 4.5 mmol L^{-1} ammonia in water.

To construct O$_2$-dissociation curves, blood samples (2–3 mL) were placed in Eschweiler tonometers and sequentially equilibrated at 15 °C against several (4–9) humidified gas mixtures obtained from Wösthoff pumps and containing various volume fractions of O$_2$ at the same PCO$_2$. Each equilibration step lasted 30 min. For each PO$_2$, total O$_2$ concentration was determined in duplicate using a modified Tucker chamber (Tucker 1967) on 10 µL subsamples. Dissolved oxygen was subtracted using a solubility coefficient experimentally determined on plasma from sturgeon (13.9 $\pm$ 0.2 µmol L^{-1} kPa^{-1}; mean $\pm$ SEM, $N = 7$), and the haemoglobin O$_2$ saturation

(SO$_2$) was calculated assuming that air-equilibrated samples (PO$_2$ = 20 kPa) were fully saturated. Oxygen partial pressure at half saturation (P$_{50}$) and cooperativity (slope of the Hill plot, n_H) were obtained by linear regression after Hill transformation [log SO$_2$/(1-SO$_2$) = f (log PO$_2$)]. Blood pH was measured on a number of selected samples corresponding to various PO$_2$, and the relationship between pH and SO$_2$ was found linear, allowing to interpolate values for the pH at half saturation, pH50, and for the pH difference between deoxygenated and oxygenated blood, ΔpH$_{(desoxy-oxy)}$.

In both cases, the blood samples were placed in Eschweiler tonometers which were gently shaken in a water bath at 15 °C and passed through with humidified gas mixture. To check for possible modifications of blood composition which may affect HbO$_2$-combining properties during tonometry sessions lasting up to 5 h, blood samples were routinely analysed at the beginning and at the end of each experiment for haematocrit, haemoglobin concentration (Drabkin method, Sigma kit 525A), methaemoglobin concentration (spectrophotometric, Benesch et al. 1973) and nucleoside triphosphate (NTP) concentration (Sigma kit 336UV). All of these parameters underwent significant changes over time, and the figures were then combined to calculate mean values (Tables 20.2 and 20.3). Plasma total ammonia concentration was also measured using a micromodification of the L-GLDH/NAD method (Sigma kit 170UV) and found to decrease slightly but not significantly during tonometry sessions in samples with ammonia addition in vitro (Table 20.2).

Results are given as mean ± sem. Test T of Wilcoxon and test U of Mann–Whitney were used for comparison between reference and the two experimental series (Siegel and Castellan 1988).

20.2 Results and Discussion

20.2.1 Toxicity of Ammonia: Effects of Ammonia Exposure on the Structure of the Gill Epithelium

In Siberian sturgeon, the microscopic observations of the gills of fish exposed to ammonia (Figs. 20.1, 20.2, and 20.3) reveal a modification of the epithelium of the secondary lamellae with sometimes the presence of oedema. The volume of epithelial cells is multiplied by two or three, the base of the filament is slightly turgescent, and the blood space is dilated. In sturgeon exposed to 4.5 mmol L^{-1} of total ammonia (a concentration which causes 100% mortality), the previously described phenomena are even more accentuated. In addition to the increase in cellular volume, we observed necrosis and detachment of the epithelium cells. In certain cases, the first signs of hyperplasia appeared.

These phenomena have also been recorded for carp, *Cyprinus carpio* (24 h at 30 mg L^{-1} NH$_4^+$) (Yang and Chun 1986) and for trout, *Oncorhynchus mykiss* (6 h at 34 mg L^{-1} NH$_4^+$) (Smart 1976). These alterations have classically been described in longer experiments on sublethal concentrations with, occasionally, some tissue

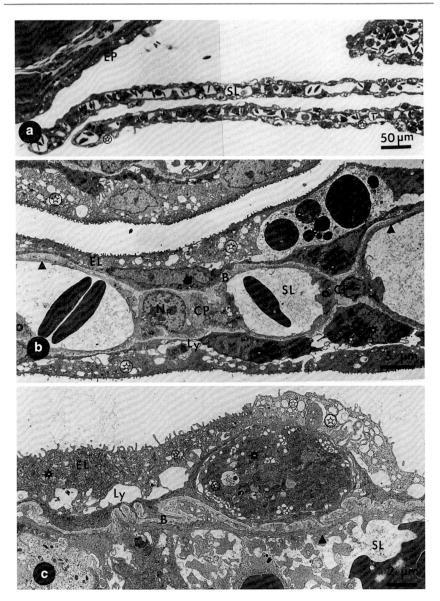

Fig. 20.1 Microscopic analysis of secondary lamellae of the Siberian sturgeon, *Acipenser baerii*, after 72-h exposure to sublethal concentration of ammonia (1.85 mmol L^{-1} T_{Am}). (**a**) A semi-thin cross section through a filament shows numerous vacuoles (☆) and necrosis (★) of the epithelium of the secondary and the primary lamellae. Meanwhile the global structure of the gills is kept, and blood sinus or pillar cells are always observable (×300). (**b**) This transmission electronic micrograph reveals that the respiratory epithelium is modified. Numerous cytoplasmic vacuoles (☆) or vesicles appear clearly as well as degenerations of cells and autophagic bodies (☆). The thin flanges of the pillar cells (▲) seem hypertrophied (◯) (×3500). (**c**) This micrograph (upper magnification) shows the vacuoles (☆) inside the cells of the epithelial external layer, numerous necrosis of cells (★) and alteration of the blood sinus wall (▲) (×7500). *B* basal lamina, *CP* pillar cell, *EL* respiratory epithelium or secondary epithelium, *EP* primary epithelium, *Ly* lymphatic space, *Np* nucleus of the pillar cell, *SL* blood sinus

20 Consequences of High Levels of Ammonia Exposure

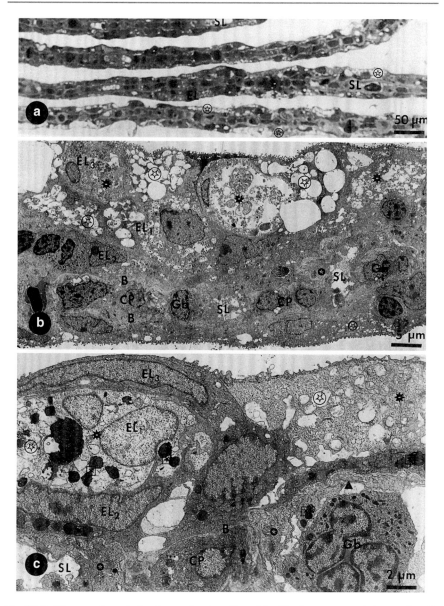

Fig. 20.2 Microscopic analysis of secondary lamellae of the Siberian sturgeon, *Acipenser baerii*, after 72-h exposure to lethal concentration of ammonia (4.5 mmol L^{-1} T_{Am}). (**a**) A semi-thin cross section through a filament shows most kinds of gills lesions: numerous vacuoles (☆) and important necrosis (✱) at the level of the respiratory epithelium, the blood sinus and pillar cells disappeared (×300). (**b**) This transmission electronic micrograph reveals an important hypertrophy of the epithelium (10–20 μm thickness), numerous cytoplasmic vacuoles (☆) and figure of degeneration (▲). The lymphatic space is not perceptible. The blood sinus is filled with leucocytes or other cellular bodies. The cytoplasmic extensions of the pillar cells are largely hypertrophied (◉) (×3000). (**c**) This micrograph (upper magnification) confirms the great number of vacuoles (☆), cellular damages (◉), the hypertrophy of the pillar cells (◉), lysosomes and electron dense bodies (✱). We can observe a hyperplasia of the epithelial external cellular layer and the appearance of a third layer of cells (EL$_3$) (×7500). *B* basal lamina, *CP* pillar cell, *EL* respiratory epithelium or secondary epithelium (EL$_1$; EL$_2$ and EL$_3$), *Gb* white red blood cells, *SL* blood sinus

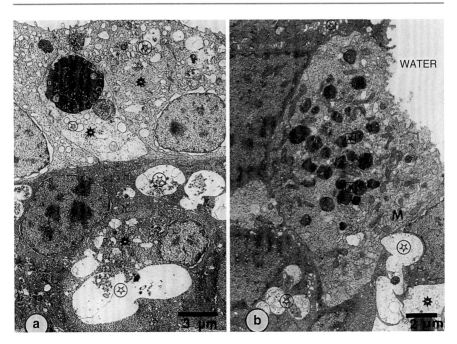

Fig. 20.3 Transmission electronic micrographs of the primary epithelium of the Siberian sturgeon, *Acipenser baerii*, after 72-h exposure to a lethal concentration of ammonia (4.5 mmol L^{-1} T_{Am}). (**a**) Exposure to ammonia provokes similar effects on primary or secondary epithelium: numerous vacuoles (☆), hypertrophy and cell damages (✶) (×6000). (**b**) On this micrograph, we can detect chloride cell damages at the level of the tubular system, disappearance of mitochondria (M) and numerous autophagic bodies (★) (×7000)

proliferation, a fusion of branchial lamellae and an increase in sensitivity to bacteria (Smart 1976; Robinette 1976; Thurston et al. 1984; Lang et al. 1987).

In Siberian sturgeon, histological observations (photonic microscopy) were confirmed by transmission electronic microscopic analysis of secondary lamellae either after an exposure of the fish to a sublethal dose of ammonia (1.85 mmol L^{-1} T_{Amm}) (Fig. 20.1) or a lethal dose of ammonia (4.5 mmol L^{-1} T_{Amm}) (Figs. 20.2 and 20.3).

Ultrastructural studies demonstrated that ammonia-induced changes occurred in the typical epithelial cells on the gill lamellae either after exposure to 1.85 mmol L^{-1} T_{Amm} (Fig. 20.1a–c) or 4.5 mmol L^{-1} T_{Amm} (Fig. 20.2a–c). Transmission electronic micrograph observations showed alterations on the primary or secondary epithelium (Fig. 20.3a). Chloride cell (ionocytes) damages were observed at the level of the tubular system, by disappearance of mitochondria and by appearance of numerous autophagic bodies (Fig. 20.3b). These changes included cytoplasmic vacuolization, mitochondrial and nuclear alterations, increased number of lysosomes and inclusions and alterations in cytoplasmic density. Most of these changes indicated cell damage or cell death. Some of the changes that have been detected in chloride cells may instead reflect decreased cellular activity.

No toxicant-specific alterations have been demonstrated after exposure to ammonia (Table 20.1) but this may only reflect the lack of ultrastructural information. It would be expected that, toxicants having different modes of toxic action, they would preferentially alter different cellular organelle, since they contain different enzyme systems. (Mallat 1985).

20.2.2 Influence of Ammonia on Haematocrit, Plasma Ionic Concentrations and Water Content of Erythrocytes

According to blood analysis, the sodium, chloride and calcium plasmatic contents (Fig. 20.4a–c) were stable. However, we recorded an increase in the potassium concentration (Fig. 20.5) for sublethal and lethal doses of ammonia. After 72-h exposure, this high concentration of $[K^+]$ in plasma was significantly different from the control for lethal concentration of ammonia only. Moreover the potassium level increased significantly both for sublethal (1.85 mmol L^{-1} total ammonia) and lethal (4.5 mmol L^{-1} total ammonia) doses of ammonia in water, although haemolysis phenomena could not be observed.

The osmotic pressure (Fig. 20.4d) increased rapidly and significantly only after lethal ammonia level exposure.

No variation was observed for the strong ion difference whatever the total ammonia concentration in water (Fig. 20.4e).

It seems clear that the Siberian sturgeon was able to perfectly regulate its ionic equilibrium 72 h (at the latest) after an exposure to sublethal level of ammonia.

On the other hand, after an exposure to a lethal level of ammonia, the phenomena, observed for a sublethal dose, were upgraded, probably due to the alterations of the gill epithelium as described before (Figs. 20.1, 20.2 and 20.3).

Table 20.1 Summary of variable effects of some toxics on the gill structure: ultrastructural studies by TEM

Toxics	Hyperplasia	Cytoplasmic vacuoles	Necrosis	Blood sinus	Lymphatic space	Chloride cells or ionocytes
Bromuride[1]	↗	↗	↗		↗	↘
Zinc[2]	↗	↗	↗	↘	↗	↗ or ↘
Mercury[3]	↗	↗	↗		↗	↘
Detergents[4]	↗	↗	↗	↘	↗	↘
pH acid[5]	↗				↘	↗
Hyperoxia[6]	↗	↗	↗	↘		↗
Ammonia[7]	↗	↗	↗	↘	↘	↘

The exposure to the toxics was short time but the concentrations were lethal

TEM transmission electronic microscopy

References: (*1*) Segers et al. (1984); (*2*) Skidmore and Towell (1972); (*3*) Tuurala and Soivio (1982); (*3*) Wobeser (1975); (*4*) Abel and Skidmore (1975); (*5*) Daye and Garside (1976); Leino and McCormick (1984); Leino et al. (1987); (*6*) Laurent and Perry (1991); (*7*) Salin (1992)

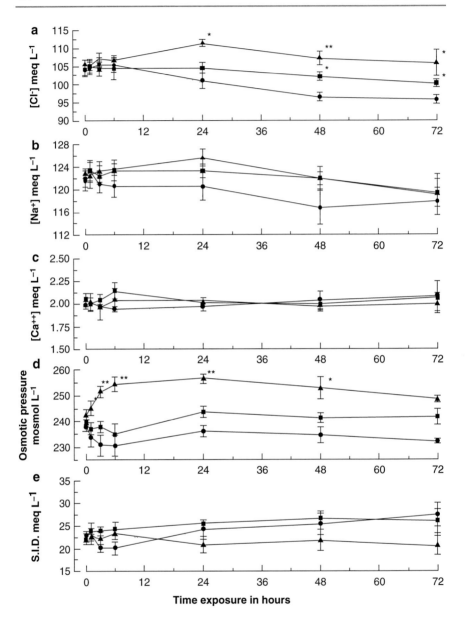

Fig. 20.4 Time course in the plasma of the concentrations of chloride (**a**), sodium (**b**) and calcium (**c**) of the osmotic pressure (**d**) and of the strong ions difference (SID) (**e**) in function of the exposure time to high levels of ammonia in water. (●) Control series; (■) 1.85 mmol L^{-1} total ammonia; (▲) 4.5 mmol L^{-1} total ammonia (Mean ± SE). $N = 6$. *$p < 0.05$; **$p < 0.01$; significant difference with the control series, Mann–Whitney test U

20 Consequences of High Levels of Ammonia Exposure

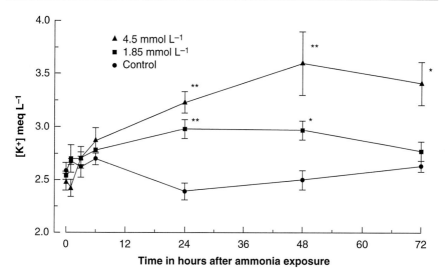

Fig. 20.5 Time course in the plasma of the concentrations of potassium [K⁺] in function of the exposure time to high levels of ammonia in water. (●) Control series; (■) 1.85 mmol L^{-1} total ammonia; (▲) 4.5 mmol L^{-1} total ammonia (Mean ± SE). $N = 6$. *$p < 0.05$; **$p < 0.01$, significant difference with the control series, Mann–Whitney test U

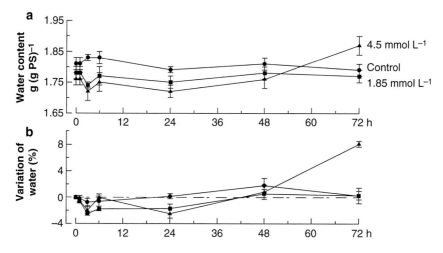

Fig. 20.6 Time course in function of the exposure time to high levels of ammonia in the water content of the erythrocytes (**a**) and of the percentage of variation of this water content (**b**). (●) Control series; (■) 1.85 mmol L^{-1} total ammonia; (▲) 4.5 mmol L^{-1} total ammonia (Mean ± SE). $N = 6$

The water content of the erythrocytes of the Siberian sturgeon is slightly modified (Fig. 20.6a) by exposure to ammonia except for a lethal dose (4.5 mmol L^{-1} total ammonia) after 72-h exposure. At this lethal dose, the water content of the red blood cells is significantly higher than that of the control fish ($p = 0016$, T test) between 48 and 72 h. This increase is 8% higher than the control value (Fig. 20.6b).

The ion levels of the control fish are very close to the results found in previous studies on *Acipenser baerii* (Ait-fdil 1986; Bennouna 1986) or on other species of sturgeon (Magnin 1962; Potts and Rudy 1972; Natochin et al. 1975; Hunn and Christenson 1977; Lavrova et al. 1984). Among the blood parameters studied, only potassium undergoes a significant modification in the presence of ammonia. Moreover, we noticed that this increase seems to appear at two stages, one for doses not causing more than 10% mortality and the other for concentrations over the 24 h LC50 (see also Chap. 19).

During exposure to ammonia in water, an accumulation of ammonia in the blood has been demonstrated (Fromm and Gillette 1968; Haywood 1983; Thurston et al. 1984). Furthermore, it has been clearly established that ammonia ion (NH_4^+) can take the place of potassium in active transports like Na^+-K^+-ATPases (Maetz and Garcia-Romeu 1964; Maetz 1973; Busacker and Chavin 1981; Claiborne et al. 1982). Moreover, Ait-fdil (1986) shows the existence of Na^+-K^+-ATPases in the gills, erythrocytes and kidneys of *Acipenser baerii*, with a greater activity in the presence of ammonia ions as compared to potassium. The substitution of K^+ by NH_4^+ would allow normal exchanges to be maintained for sodium between the blood and the tissues (Fig. 20.4). When the competition between ammonia and potassium is too strong, we can imagine that it causes an increase in the blood potassium content. This hypothesis, already put forward by Bubien and Meade (1986) for *Oncorhynchus mykiss*, needs to be verified by measuring the ammonia content in the blood and also by measuring the totality of ions in different tissues, for example, in erythrocytes (Fig. 20.7).

20.2.3 Effect of Ammonia on the Oxygen-Carrying Properties of the Blood of the Sturgeon, *Acipenser baerii*

Ammonia, a common end product of nitrogen catabolism in aquatic animals, is extremely toxic to fish, but the reasons behind this toxicity remain largely unclear. Several authors have observed a decrease of blood PO_2 or oxygen content in fish exposed to waterborne ammonia (Brockway 1950; Smart 1978) and have suggested among other possibilities that this may result from some impairment of the ability of haemoglobin to combine with oxygen. By recording spectral changes of haemoglobin during ammonia intoxication in salmon, Sousa and Meade (1977) detected a shift toward the deoxygenated form, but concluded, on the basis of essentially qualitative arguments, that this shift was caused by blood acidosis only. In fact, the possible existence of an effect of ammonia poisoning on blood O_2-binding properties has never been tested quantitatively in fish.

Among the hypotheses to explain ammonia toxicity on fish, some of them refer to oxygen transport and oxygen affinity of the blood. If ammonia intoxication causes alterations of the blood O_2-combining properties in fish, this could arise either from direct action of ammonia on the red blood cell or from more general systemic effects. The first possibility can be studied by investigating the effects of adding ammonia in vitro to normal blood on its O_2-combining properties.

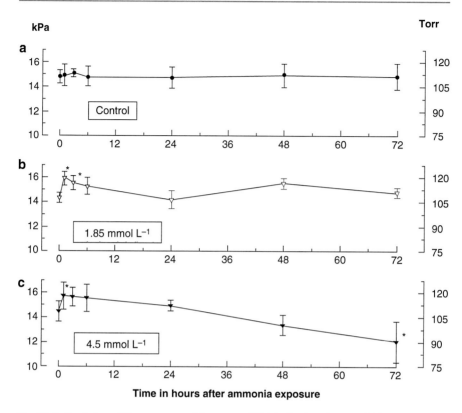

Fig. 20.7 Evolution of the oxygen partial pressure (PO$_2$) in the dorsal aorta of Siberian sturgeon after exposure to high levels of ammonia in water (Mean ± SE). $N = 6$. (**a**) Control series, (**b**) 1.85 mmol L^{-1} total ammonia, (**c**) 4.5 mmol L^{-1} total ammonia. *$p < 0.05$, significant difference with $t = 0$ h, Wilcoxon test

Comparison of oxygen binding in blood drawn from control animals and from fish exposed to waterborne ammonia has been checked for possible additional effects of ammonia intoxication. So, influence of ammonia on oxyphoric properties of red blood cells of Siberian sturgeon (*Acipenser baerii*) has been investigated in vivo (blood of fish exposed for 72 h to 4.5 mmol L^{-1} total ammonia, a lethal concentration) and in vitro (ammonia added to blood of control fish).

Data from the in vitro experimental series are summarized in Table 20.2. Although ammonia addition could be expected to induce transient red cell swelling (Motais et al. 1991), no significant difference between control and ammonia-treated samples was found for haematocrit, haemoglobin concentration and oxygen capacity. Red cell methaemoglobin and nucleoside triphosphate content were not modified upon ammonia addition. Furthermore, the presence of ammonia did not affect blood HbO$_2$-combining properties (Table 20.2). Neither the cooperativity of O$_2$ binding (as evaluated by the Hill slope, n_H) nor the oxygen affinity, i.e. the half

Table 20.2 Haematological parameters and O_2-binding properties of sturgeon blood samples ($N = 3$) before and after addition of ammonia in vitro

		Control	Ammonia addition
Haematocrit (%)		28.3 ± 1.8	29.8 ± 2.1
Haemoglobin (g L blood^{-1})		83.9 ± 4.3	81.9 ± 5.3
Methaemoglobin (%)		2.5 ± 0.1	2.7 ± 0.2
NTP/Hb$_4$ (mol mol^{-1})		1.81 ± 0.07	1.80 ± 0.12
Total ammonia (mmol L plasma^{-1})	Initial	0.13 ± 0.02	5.41 ± 1.26
	Final	0.19 ± 0.03	4.21 ± 0.79
Oxygen capacity (mL O_2 g Hb^{-1})		1.34 ± 0.07	1.39 ± 0.03
pH$_{50}$		7.862 ± 0.026	7.812 ± 0.030
ΔpH(desoxy—oxy)		0.189 ± 0.046	0.215 ± 0.014
n_H		1.48 ± 0.06	1.48 ± 0.09
P$_{50}$ at pH$_{50}$ (kPa)		1.92 ± 0.19	1.89 ± 0.14
P$_{50}$ at pH 7.8 (kPa)		2.06 ± 0.28	1.92 ± 0.21

Mean values ± SEM. Samples were added with equivalent volumes of isosmotic NaCl (control group) and NH$_4$Cl (experimental group) solutions (both 320 mmol L^{-1}). Haematocrit, haemoglobin, methaemoglobin and NTP/Hb$_4$ have been measured at the beginning and at the end of each experimental session and found to undergo no significant changes over time. n_H and P$_{50}$ were derived from O_2-binding measurements performed at 15 °C and PCO$_2$ = 0.2 kPa. P$_{50}$ values have been corrected to pH 7.8 using a Bohr factor (ΔlogP$_{50}$/ΔpH) of −0.49 (see Fig. 20.8). There are no significant differences between control and ammonia addition for any of the parameters in this Table ($P > 0.05$; Student's t-test for paired samples)

saturation O_2 tension (P$_{50}$) after correction at a common pH of 7.8, was significantly changed after ammonia addition.

Table 20.3 compares haematological data and O_2-combining properties of blood samples from control sturgeon and from animals exposed for 3 days to waterborne ammonia, resulting in plasma total ammonia concentrations of the same order of magnitude as in vitro experiments, *ca* 4 mmol L^{-1}. Values for haematocrit, haemoglobin, methaemoglobin, oxygen capacity and nucleoside triphosphate were similar to those obtained for the in vitro experimental series (Table 20.2) and revealed no significant effects of ammonia exposure. Blood oxygen affinity and cooperativity were also not significantly different in the control and ammonia-exposed groups. Figure 20.8 shows the relationship between logP$_{50}$ and pH determined on subgroups of three3 control and three ammonia-exposed sturgeons, blood pH being changed by manipulating PCO$_2$. The slope of these relationships, i.e. the Bohr factor ΔlogP$_{50}$/ΔpH, was not significantly affected by the presence of ammonia and amounted to *ca*—0.49, which indicates a relatively important Bohr effect. This also correlates with a high value of *ca* 0.15–0.21 pH units for the pH difference between deoxygenated and oxygenated blood, revealing the existence of an important Haldane effect. Thus, in this sturgeon and perhaps more generally in the ancient chondrostean group, haemoglobin properties look more like those of teleostean fish than those of elasmobranchs, which are characterized by negligible Bohr and Haldane effects (Lenfant and Johansen 1966; Butler and Metclalfe 1989; see discussion in Graham et al. 1990).

20 Consequences of High Levels of Ammonia Exposure

Table 20.3 Haematological parameters and O_2-binding properties of blood samples drawn from control sturgeon ($N = 4$) and from sturgeon exposed to waterborne ammonia (4.5 mmol L^{-1}) for 72 h ($N = 5$)

		Control	Ammonia-exposed
Haematocrit (%)		28.0 ± 3.2	27.2 ± 1.8
Haemoglobin (g L blood^{-1})		77.8 ± 6.3	77.1 ± 2.7
Methaemoglobin (%)		2.01 ± 0.10	2.40 ± 0.47
NTP/Hb$_4$ (mol mol^{-1})		1.89 ± 0.08	1.85 ± 0.06
Total ammonia (mmol L plasma^{-1})	Initial	0.11 ± 0.02	4.10 ± 0.39
	Final	0.15 ± 0.05	3.76 ± 0.41
Oxygen capacity (mL O_2 g Hb^{-1})		1.35 ± 0.05	1.34 ± 0.06
pH$_{50}$		7.899 ± 0.015	7.930 ± 0.018
ΔpH(desoxy—Oxy		0.148 ± 0.020	$0.210 \pm 0.010^*$
n_H		1.44 ± 0.07	1.62 ± 0.06
P$_{50}$ at pH$_{50}$ (kPa)		1.67 ± 0.12	1.69 ± 0.12
P$_{50}$ at pH 7.8 (kPa)		1.87 ± 0.16	1.96 ± 0.13

Mean values $\pm$ SEM. Haematocrit, haemoglobin, methaemoglobin and NTP/Hb$_4$ have been measured at the beginning and at the end of each experimental session and found to undergo no significant changes over time. n_H and P$_{50}$ were derived from O_2-binding measurements performed at 15 °C and PCO$_2$ = 0.2 kPa. P$_{50}$ values have been corrected to pH 7.8 using a Bohr factor ΔlogP$_{50}$/ΔpH) of—0.49 (see Fig. 20.8). There are no significant differences ($P > 0.05$; Student's t-test for unpaired samples) for all parameters, except ΔpH (desoxy—oxy) ($P < 0.05$)

Our P$_{50}$ values determined for the whole blood of *Acipenser baerii*, about 1.9–2.0 kPa at 15 °C and pH 7.8, reveal a distinctly higher O_2 affinity compared to those reported by Burggren and Randall (1978) for another sturgeon species, *Acipenser transmontanus* (P$_{50}$ *ca* 2.9 kPa at the same pH and temperature). However, Bohr factors are similar in both sturgeon species (−0.49 for *A. baerii* vs −0.55 for *A. transmontanus*).

Whatever the experimental conditions, ammonia did not modify the affinity of haemoglobin for oxygen illustrated by P$_{50}$, Bohr effect, Hill number and percentage of methaemoglobin.

Consequently, the toxicity of ammonia on sturgeon cannot be explained only either by alteration of the gill epithelium, their consequences on blood characteristics and ionic regulation, or by oxygen-binding properties of the haemoglobin.

Conclusions

The alterations at the gill level might have important consequences on the osmoregulatory and the respiratory capacities of the fish and might be one of the causes of the decrease in the growth rate in fish exposed to sublethal doses of ammonia for several weeks. However, it does not appear that these alterations are sufficiently significant to provoke death in the case of acute toxicity.

The acute exposure to ammonia of the Siberian sturgeon caused an important stress. Thus we could suppose that this reaction went with a catecholamine discharge (Mazeaud et al. 1977; Tomasso et al. 1981; Nikinmaa 1982; Perry and

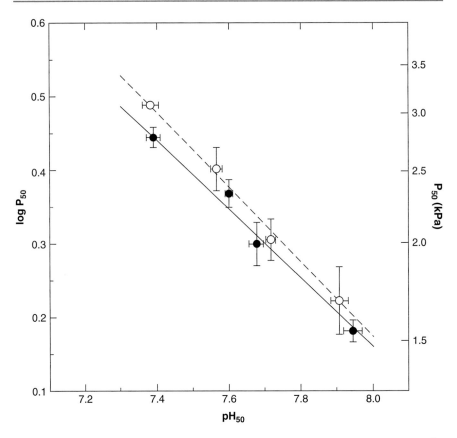

Fig. 20.8 Relationships between logP$_{50}$ and pH at 50% HbO$_2$ saturation (pH$_{50}$) for blood samples from control sturgeon (*filled circle* and *solid line*) and from sturgeon exposed to waterborne ammonia (4.5 mmol L^{-1}) for 72 h (*hollow circle* and *dashed line*). pH was modified by equilibrating each blood sample at four different PCO$_2$ values. Temperature, 15 °C. Data are reported as mean values ± SEM ($N = 3$) in each group. Equations for linear regression lines: (1) control: log $P_{50} = -0.506$ pH + 4.222; (2) ammonia exposed: log $P_{50} = -0.467$ pH + 3.896. By covariance analysis, these two regression lines are not significantly different. The equation for the joint regression line is log $P_{50} = -0488$ pH + 4.068

Wood 1989; Maxime et al. 1995; Semenkova et al. 1999). This increase of cortisol in the blood could provoke an entry of Na$^+$, Cl$^-$ and K$^+$ followed by a water influx in lethal conditions. Ammonia exposure may, at least transiently, induce swelling and an increase of the intracellular pH in red blood cells (Motais et al. 1991). This could potentially modify blood oxygen-combining properties (Soivio and Nikinma 1981).

In the conditions of the present experiments, no significant change of O$_2$-binding characteristics could be detected upon ammonia addition in vitro. Furthermore, in ammonia-exposed animals, these characteristics were found not significantly different from controls. Thus, the reported decrease of arterial O$_2$ partial pressure and content associated with ammonia intoxication in fish cannot be ascribed to any direct or indirect changes of blood HbO$_2$-combining properties. Meanwhile,

20 Consequences of High Levels of Ammonia Exposure

Table 20.4 Oxyphoric parameters (P_{50}; Hill coefficient, n_H; Bohr coefficient, β; and rate of nucleotides triphosphates (NTP)/haemoglobin in different fish species)

Espèces	P_{50} kPa	n_H (Hill)	β	NTP/Hb mmol.g^{-1}	Temp (°C)	pH	References
Oncorhynchus	3.1à 3.3			1.3	18		(1)
mykiss	3.21	2.05	−0.49	1.3	20	7.8	(2)
	2.62	1.9		1.5	15	7.9	(3)
	2.53			2.1	15		(4)
Cyprinus carpio	0.93		−0.98	1.91	20	7.9	(5)
	1.2	1.2			15	8.0	(6)
	1.2			2.18	15	8.0	(7)
Platichthys	1.15				7–10	7.9	(8)
stellatus	0.93			2.3	10		(4)
Tinca tinca	0.53	0.84	−0.64		13	7.8	(9)
		1.75					
Lota lota	1.23				15		(10)
Acipenser transmontanus	2.86		−0.55		15	7.8	(11)
Acipenser baerii	1.96	1.48	−0.53	1.83	15	7.8	(12)

References: (*1*) Soivio and Nikinma (1981); (*2*) Tetens and Lykkeboe (1981); (*3*) Tetens and Lykkeboe (1985); (*4*) Milligan and Wood (1987); (*5*) Weber and Lykkkeboe (1978); (*6*) Jensen et al. (1987); (*7*) Jensen (1990); (*8*) Wood et al. (1979); (*9*) Eddy (1973); (*10*) Cameron (1973); (*11*) Burggren and Randall (1978); (*12*) this study

this study carried out interesting original haematological data and O_2-combining properties of blood samples of sturgeon *Acipenser baerii* which can be compared with few other referent fish (Table 20.4).

Note of the Authors The experiments have been performed in the context of a PHD by Salin D (1992) untitled « La toxicité de l'ammoniaque chez l'esturgeon sibérien, *Acipenser baerii*: effets morphologiques, physiologiques, métaboliques d'une exposition à des doses sublétales et létales ». Thèse N° 749, Université Bordeaux I pp134, Director of the PhD: Truchot Jean-Paul; supervisors: Nonnotte Guy and Williot Patrick. They have been investigated in the "Laboratoire de Neurobiologie et Physiologie comparées, CNRS URA 1126 and the University of Bordeaux I, F-33120 Arcachon", in collaboration with the IRSTEA (formerly CEMAGREF), F-33611 Cestas-Gazinet, with a financial grant of the IRSTEA and CNRS.

References

Abel PD, Skidmore JF (1975) Toxic effects of an anionic detergent on the gills of rainbow trout. Water Res 9:759–765

Ait-Fdil M (1986) Mise en évidence et propriétés des ATPases membranaires dans la branchie, les érythrocytes et le rein d'un Chondrostéen, l'esturgeon sibérien, *A. baerii*. Thèse N° 2157, Université Bordeaux I, p 60

Benesch RE, Benesch R, Yung S (1973) Equations for the spectrometric analysis of haemoglobin mixtures. Anal Biochem 55:245–248

Bennouna M (1986) Equilibre hydrominéral du milieu intérieur et des érythrocytes chez un Chondrostéen, l'esturgeon sibérien, *A. baerii* au cours de variations expérimentales de salinité et de température de l'environnement. Thèse N° 2148, Université Bordeaux I, p 93

Brockway DR (1950) Metabolic products and their effects. Prog Fish Cult 12:127–129

Bubien JK, Meade TL (1986) Electrophysiological abnormalities produced by ammonium in isolated perfused brook trout, *Salvelinus fontinalis*, hearts. J Fish Biol 28:47–53

Burggren WW, Randall DJ (1978) Oxygen uptake and transport during hypoxic exposure in the sturgeon *Acipenser transmontanus*. Respir Physiol 34:171–183

Busacker GP, Chavin W (1981) Characterization of Na+/K+ATPases and Mg2+ATPases from the gill and the kidney of the goldfish (*Carassius auratus*). Comp Biochem Physiol 69B:249–256

Butler PJ, Metclalfe JD (1989) Cardiovascular and respiratory systems. In: Shuttleworth TJ (ed) Physiology of elasmobranch fishes. Berlin Springer, Verlag, pp 1–47

Cameron JN (1973) Oxygen dissociation curves and content of blood from Alaskan burbot (*Lota lota*), pike (*Esox lucius*) and grayling (*Thymallus arcticus*). Comp Biochem Physiol A 46(3):491–496

Cameron JN, Heisler N (1983) Studies of ammonia in the rainbow trout: physico-chemical parameters, acid-base behavior and respiratory clearance. J Exp Biol 105:107–125

Capuzzo JM (1988) Physiological effects of a pollutant gradient-introduction. Mar Ecol Prog Ser 46:111

Claiborne JB, Evans DH, Goldstein L (1982) Fish branchial Na+/NH4+ exchange is via basolateral Na+-K+-activated ATPases. J Exp Biol 96:431–434

Crespo S, Nonnotte G, Colin DA, Leray C, Nonnotte L, Aubree A (1986) Morphological and functional alterations induced in trout intestine by dietary cadmium and lead. J Fish Biol 28:69–80

Daye PG, Garside ET (1976) Histopathological changes in surficial tissues of brook trout, *Salvelinus fontinalis* exposed to acute and chronic level of pH. Can J Zool 54:2140–2155

Eddy FB (1973) Oxygen dissociation curves of blood of the tench, *Tinca tinca*. J Exp Biol 58:281–293

Fromm PO, Gillette JR (1968) Effects of ambient ammonia on blood ammonia and nitrogen excretion of rainbow trout. Comp Biochem Physiol 26:887–896

Graham MS, Turner JD, Wood CM (1990) Control of ventilation in the hypercapnic skate *Raja ocellata*. I. Blood and extradural fluid. Respir Physiol 80:259–277

Haywood GP (1983) Ammonia toxicity in teleost fishes: a review. Can Tech Rep Fish Aquat Sei 1177:35 p

Huertas M, Gisbert E, Rodríguez A et al (2002) Acute exposure of Siberian sturgeon (*Acipenser baerii*, Brandt) yearlings to nitrite: median-lethal concentration (LC_{50}) determination, haematological changes and nitrite accumulation in selected tissues. Aquat Toxicol 57:257–266

Hunn JB, Christenson LM (1977) Chemical composition of blood and bile of the shovelnose sturgeon. Prog Fish Cult 39(2):59–61

Jensen FB (1990) Nitrite and red cell function in carp: control factors for nitrite entry, membrane potassium, ion permeation, oxygen affinity and methaemoglobin formation. J Exp Biol 152:149–166

Jensen FB, Andersen NA, Heisler N (1987) Effects of nitrites exposure on blood respiratory properties, acid-base and electrolyte regulation in the carp, *Cyprinus carpio*. J Comp Physiol 157B:533–541

Lang T, Peters G, Hoffmann R, Meyer E (1987) Experimental investigations on the toxicity of ammonia: effects on ventilation frequency, growth, epidermal mucous cells and gill structure of rainbow trout *Salmo gairdneri*. Dis Aquat Org 3:159–165

Laurent P, Perry SF (1991) Environmental effects on fish gill morphology. Physiol Zool 64:4–65

Lavrova EA, Natochin YV, Shakhamatova EL (1984) Electrolytes in the tissues of sturgeon and bony fishes in fresh and salt water. J Ichthyol 24(5):156–160

Leino RL, McCormick JH (1984) Morphological and morphometrical changes in chloride cells of the gills of *Pimephales promelas* after chronic exposure to acid water. Cell Tis Res 236:121–128

Leino RL, McCormick JH, Jensen KM (1987) Changes in gill histology of fathead minnows and yellow perch transferred to soft water or acidified soft water with particular reference to chloride cells. Cell Tis Res 250:389–399

Lenfant C, Johansen (1966) Respiratory function in the elasmobranch *Squalus sucklei*. Respir Physiol 1:13–29

20 Consequences of High Levels of Ammonia Exposure

Maetz J (1973) Na+/NH4+, Na+/H+ exchanges and NH₃ movement across the gill of *Carassius auratus*. J Exp Biol 58:255–275

Maetz J, Garcia-Romeu F (1964) The mechanism of sodium and chloride uptake by the gill of a fresh water fish, *Carassius auratus*. J Gen Physiol 47:1209–1227

Magnin E (1962) Recherches sur la systématique et la biologie des Acipenséridés. Ann Stn Centr Hydrobiol Appl 9:7–242

Mallat J (1985) Fish gill structural changes induced by toxicants and other irritants; a statistical review. Can J Aquat Sci 42:630–648

Maxime V, Nonnotte G, Peyraud C, Williot P, Truchot JP (1995) Circulatory and respiratory effects of an hypoxic stress in the Siberian sturgeon. Respir Physiol 100:203–212

Mazeaud M, Mazeaud F, Donaldson EM (1977) Primary and secondary effects of stress in fish: some new data with a general review. Trans Am Fish Soc 106(3):201–212

Milligan CL, Wood CM (1987) Regulation of blood oxygen transport and red blood cell pHi after exhaustive activity in rainbow trout (*Salmo gairdneri*) and starry flounder (*Platichthys stellatus*). J Exp Biol 133:263–282

Motais R, Guizouarn H, Garcia-Romeu (1991) Red cell volume regulation: the pivotal role of ionic strength in controlling swelling-dependent transport systems. Biochim Biophys Acta 1075:169–180

Natochin Y, Luk'yanenko VI, Lavrova YA, Metallov GF (1975) Cation contents of the blood serum during the marine and river periods in the life sturgeons. J Ichthyol 15(5):799–804

Nikinmaa M (1982) The effects of adrenaline on the oxygen transport properties of *Salmo gairdneri* blood. Comp Biochem Physiol 71A:353–356

Nonnotte G, Maxime V, Truchot JP, Williot P, Peyraud C (1993) Respiratory responses to progressive ambient hypoxia in the sturgeon, *Acipenser baerii*. Respir Physiol 91:71–82

Perry SF, Wood VM (1989) Control and coordination of gas transfer in fishes. Can J Zool 67:2961–2970

Potts WTW, Rudy PP (1972) Aspects of osmotic and ionic regulation in the sturgeon. J Exp Biol 56:703–715

Robinette RH (1976) Effect of selected sub-lethal level of ammonia on the growth of channel catfish (*Ictalurus punctatus*). Prog Fish Cult 38(1):26–29

Salin D (1992) La toxicité de l'ammoniaque chez l'esturgeon sibérien, *Acipenser baerii*: effets morphologiques, physiologiques, métaboliques d'une exposition à des doses sblétales et létales. Thèse N° 749, Université Bordeaux I, p 134

Salin D, Williot P (1991) Acute toxicity of ammonia to siberian sturgeon *Acipenser baerii*. In: Williot P (ed) Acipenser. Cemagref Publ, Anthony, France, pp 153–167

Scott GR, Sloman KA (2004) The effects of environmental pollutants on complex fish behaviour: integrating behavioural and physiological indicators of toxicity. Aquat Toxicol 68(4):369–392

Segers JHL, Temmink JHM, Van den Berg JHL, Wegman RCC (1984) Morphological changes in the gill of carp (*Cyprinus carpio* L.) exposed to acutely toxic concentrations of methyl bromide. Water Res 18(11):1437–1441

Semenkova TB, Bayuna LV, Boev AA, Dyubin VP (1999) Effects of stress on serum cortisol levels of sturgeon in aquaculture. J Appl Ichthyol 15:270–272

Siegel S, Castellean NJ (1988) Non parametric statistics for the behavioral sciences. Mc Graw-Hill, Inc. 399p

Skidmore JF, Towell PWA (1972) Toxic effects of zinc sulphate on the gills of rainbow trout. Water Res 6:217–230

Smart G (1976) The effect of ammonia exposure in gill structure of the rainbow trout (*Salmo gairdneri*). J Fish Biol 8:471–475

Smart G (1978) Investigation of the toxic mechanisms of ammonia to fish - gas exchange in rainbow trout (*Salmo gairdneri*) exposed to acutely lethal concentrations. J Fish Biol 12:93–104

Soivio A, Nikinma M (1981) The swelling of erythrocytes in relation to the oxygen affinity of the blood of the rainbow trout, *Salmo gairdneri* R. In: Pickering AD (ed) Stress and fish. Academic Press, London, pp 103–119

Sousa RJ, Meade T (1977) The influence of ammonia on the oxygen delivery system of *Coho salmon* hemoglobin. Comp Physiol Biochem 58A:23–28

Tetens V, Lykkeboe G (1981) Blood respiratory properties of rainbow trout, *Salmo gairdneri*: response to hypoxia acclimatation and anoxic incubation of blood in vitro. J Comp Physiol 145:117–125

Tetens V, Lykkeboe G (1985) Acute exposure of rainbow trout to mild and deep hypoxia: O_2 affinity and O_2 capacitance of arterial blood. Respir Physiol 61:221–235

Thurston RV, Russo RC, Luedtke RJ, Smith CE, Meyn EL, Chakoumakos C, Wang KC, Brown CJD (1984) Chronic toxicity of ammonia to rainbow trout. Trans Am Fish Soc 113:56–73

Tomasso JR, Davis KD, Simco BA (1981) Plasma corticosteroid dynamics in channel catfish (*Ictalurus punctatus*) exposed to ammonia and nitrite. Can J Fish Aquat Sci 38:1106–1112

Tucker VA (1967) Method for oxygen content and dissociation curves on microliter blood samples. J Appl Physiol 23:410–414

Tuurala H, Soivio A (1982) Structural and circulatory changes in the secondary lamellae of *Salmo gairdneri* gills after sub-lethal exposures to dehydroabietic acid and zinc. Aquat Toxicol 2:21–29

Van der Oost R, Beyer J, Vermeulen NPE (2003) Fish bioaccumulation and biomarkers in environmental risk assessment: a review. Environ Toxicol and Pharmacol 13(2):57–149

Velisek J, Svobodova Z, Machova J (2009) Effects of bifenthrin on some haematological, biochemical and histopathological parameters of common carp (*Cyprinus carpio* L.) Fish Physiol and Biochem 35(4):583–590

Weber RE, Lykkkeboe G (1978) Respiratory adaptations in carp blood: influence of hypoxia, red cell organic phosphates, divalent cations and CO_2 on haemoglobin-oxygen affinity. J Comp Physiol 128:127–137

Wendelaar Bonga SE (1997) The stress response in fish. Physiological Review 77(3):591–625

Wobeser G (1975) Acute toxicity of methyl mercury chloride and mercuric chloride for rainbow trout (*Salmo gairdneri*) fry and fingerlings. J Fish Res Board Can 32:2005–2013

Wood CM, Mc Mahon DR, Donald M (1979) Respiratory gas exchange in the resting starry flounder, *Platichthys stellatus*: a comparison with other teleosts. J Exp Biol 78:167–179

Yang HC, Chun SK (1986) Histopathological study of acute toxicity of ammonia on common car (Cyprinus carpio). Bull Koraen Fish Soc 19(3):249–256

Acid-Base Balance and Ammonia Loading in the Siberian Sturgeon *Acipenser baerii*, Exposed to High Concentrations of Ammonia

21

Guy Nonnotte, Dominique Salin, Patrick Williot, Karine Pichavant-Rafini, Michel Rafini, and Liliane Nonnotte

Abstract

The ammonia concentration was studied in the blood of the Siberian sturgeon, *Acipenser baerii*, with and without high concentration in water. In control conditions, the ammonia excretion was mainly due to NH_3 diffusion according to the partial pressure gradient. So the distribution of the total ammonia followed the pH gradient between water and blood. After exposure to high concentration of ammonia, the blood concentration increased quickly during the early hours and a metabolic alkalosis appeared. P_{NH3} was stabilized after less than 24-h exposure, while the total ammonia concentration increased continuously as blood pH returned to initial values. If we take into account the real pH in the branchial boundary layer and not in the water of the circuit to calculate P_{NH3}, we could conclude that the evolution of the blood ammonia concentration could be explained by NH_3 diffusion along its partial pressure and that the ammonia exchanges by simple diffusion were dominant. However, whatever the ammonia

---

G. Nonnotte (✉) · L. Nonnotte
12 rue Marcel Pagnol, F-33260 La Teste de Buch, France
e-mail: guy.nonnotte@wanadoo.fr

D. Salin
109 rue Blaise Pascal, F-33160 Saint Médard en Jalles, France

P. Williot
4 rue du Pas de Madame, F-33980 Audenge, France

K. Pichavant-Rafini
Laboratoire ORPHY EA4324, Université de Bretagne Occidentale,
6 Avenue le Gorgeu, CS 93837, 29238 Brest Cedex 3, France

M. Rafini
Département Communication, Anglais, Sciences Humaines, Université de Bretagne Occidentale, 6 Avenue le Gorgeu, CS 93837, 29238 Brest Cedex 3, France

concentration in water, the fish quickly reached a P_{NH3} equilibrium with the water and seemed incapable of preventing the invasion of the blood by ammonia. That also explained the initial metabolic alkalosis. But these results did not allow to attribute the ammonia toxicity to NH_3.

Keywords

Siberian sturgeon • *Acipenser baerii* • Acid-base balance • Ammonia toxicity • Nonionic diffusion

Introduction

Ammonia is the main end products of nitrogen catabolism in bony fish. It is very soluble and acts as a weak base. It is well established that the excretion of ammonia is accomplished almost entirely by the gill, but there is little agreement on the nature of the excretion mechanisms: if simple diffusive NH_3 loss has often been considered as the predominant mode, ammonium diffusion and exchange of ammonium for sodium or protons have been proposed to account for ammonia movements in various situations, particularly in conditions of high external ammonia concentrations.

In these conditions, if ammonia crosses the gill epithelium in nonionized NH_3 form, then it would be protonated in the blood to form ammonium ion, and an alkalosis would result. In addition, in the steady state, if NH_3 diffusion is the dominant mode of ammonia transfer, ammonia distribution should approach a balance with equality of NH_3 partial pressure outside and inside. Previous investigations have instead reported a P_{NH3} unbalance in high external ammonia experiments which was taken as evidence for sodium ammonium or proton ammonium exchange mechanisms in addition to NH_3 diffusion. However a steady state may not have been attained in those short-term experiments. We have attempted to clarify this problem using the Siberian sturgeon as an experimental animal. The effects of high ammonia concentrations in water and on blood ammonia and acid-base balance were followed during 4 days of exposure with the aim to reach a steady state.

21.1 Materials and Methods

The experimental protocol was described in details in Chap. 19.

Sturgeons were cannulated (see Williot and Le Menn, Chap. 24) in dorsal aorta and were placed in a closed recirculating device thermostated at 15 °C. The water was oxygenated by intensive bubbling of CO_2 free air. The pH was regulated at 8.0 and the PCO_2 at 0.1 kPa using a pH-CO_2 stat system. The titration alkalinity was manually corrected at 1.9 mEq/L (see Chap. 25).

Three series of experiments were performed: one control series without ammonia, one series at 1.85 mmol L^{-1} T_{amm} (total ammonia) which is a sublethal concentration, and another series at 4.5 mmol L^{-1} T_{amm}, near the lethal concentration. Blood samples were drawn from the caudal artery catheter prior to and after 1, 3, 6, 24, 48, and 72 h of ammonia exposure. The blood pH was measured with a thermostated microelectrode; PCO_2 and HCO_3^- contents were estimated using Astrup method (1956) (see Chap. 25). Plasma ammonia was quantified by an enzymatic method. For blood and water, P_{NH_3} was calculated from total ammonia concentration and pH using the pK and the solubility coefficient of Cameron and Heisler (1983).

21.2 Results and Discussion

High external ammonia concentrations provoke a rapid and significant increase in blood pH during the first hours of exposure with a maximum of 6 h in both series (Fig. 21.1). Then the initial pH recovered within 48–72 h. Recovery took place more rapidly in the first series as an effect of ammonia concentration. This initial alkalosis was probably the result of an NH_3 entry and its subsequent conversion to ammonium ion.

When we represented the time course of the acid-base parameters on a pH-bicarbonate diagram (Davenport 1974), no significant modification in the acid-base balance was observed for the control situation (Fig. 21.2).

The pH control values are quite similar to the data given by Burggren and Randall (1978) on *Acipenser transmontanus*, but there is, to our knowledge, no report on the blood ammonia content for this family. However, our values are in the range of those of other teleost fish (Heisler 1984; Claiborne and Evans 1988; Cameron and Heisler 1983; Maetz 1972).

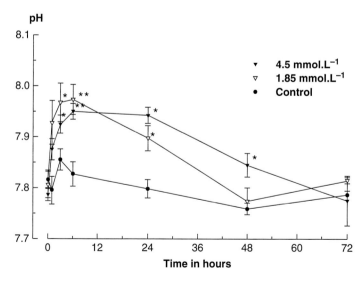

Fig. 21.1 Time course of the blood pH vs. the exposure time at sublethal and lethal concentrations of total ammonia (mean ± SE). *$p < 0.05$; **$p < 0.01$, significant difference with control series, test U of Mann–Whitney

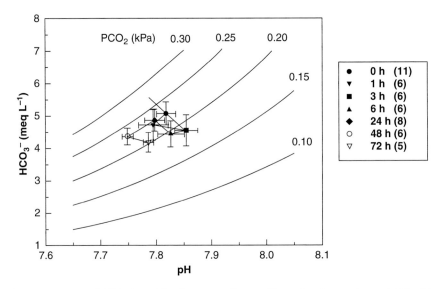

Fig. 21.2 Diagram [HCO$_3^-$] vs. pH or Davenport diagram illustrating changes in arterial blood in control conditions. The straight line at 0 h represents the buffer line. Mean ± SE, ($n=11$). No significant difference appears during the experiment (p ˃ 0.05, T test of Wilcoxon)

In the two series with high (sublethal and lethal) external ammonia concentrations, these diagrams underlined a transient increase of the blood pH (Fig. 21.3). These diagrams [HCO$_3^-$] vs. pH show that this alkalosis was metabolic. The recovery was achieved by both a decrease of PCO$_2$ (respiratory alkalosis due to a hyperventilation reaction) and a decrease of [HCO$_3^-$] (metabolic acidosis). In the first series which corresponds to the sublethal concentration, pH, PCO$_2$, and [HCO$_3^-$] recovered their initial values after 72 h. But for the second series near the lethal concentration, only the pH recovered after 72 h.

To summarize, an exposure to high concentration of ammonia induced a metabolic alkalosis which was completely recovered in the case of a sublethal concentration but not with a lethal concentration of ammonia. As this alkalosis is in accordance with the theory of nonionic diffusion of ammonia, could NH3 diffusion account solely for ammonia transfer in such experiments?

Ammonia is very soluble in water and acts as a weak base according to the equation:

$$NH_3 + H_3O^+ \leftrightarrow NH_4^+ + H_2O \quad \rightarrow \quad K_{Amm} = [NH_3][H_3O^+]/[NH_4^+]$$

Since the pK of this equilibrium is approximately 9.5 (Emerson et al. 1975; Cameron and Heisler 1983) and the pH of fresh water or fish are near neutrality, less than 4% will be present as free NH$_3$ molecules and the remainder as NH$_4^+$ ions. It is also well established that the excretion of ammonia is accomplished almost entirely by the gill, the contribution of the kidney accounting only for less than 5% (Smith 1929; Kormanik and Cameron 1981; Evans 1982; Medale et al. 1991). In spite of the importance of this phenomenon, recent reviews have pointed out that

21 Acid-Base Balance and Ammonia Loading in the Siberian Sturgeon

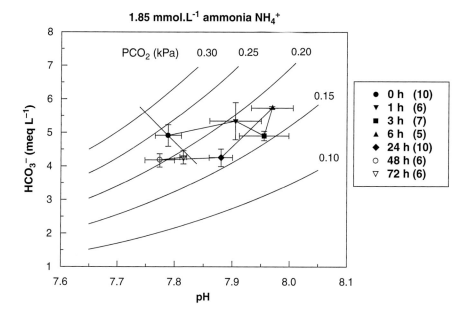

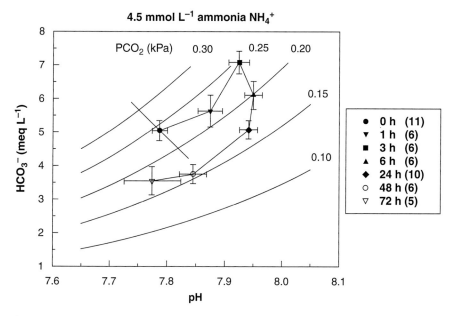

Fig. 21.3 Diagram [HCO_3^-] vs. pH or Davenport diagram illustrating the time course of acid-base parameters in arterial blood of Siberian sturgeon submitted to either a sublethal concentration of ammonia in water (**a**) (1.85 mmol L^{-1} total ammonia) or a lethal concentration (**b**) (4.5 mmol L^{-1} total ammonia). Mean ± SE, ($n = 6$ *for each series*). No significant differences appear during the experiments ($p > 0.05$, T test of Wilcoxon)

there is very little agreement on the nature of the mechanisms by which ammonia is excreted by the teleost fish (Kormanik and Cameron 1981; Evans and Cameron 1986; Randall and Wright 1989). Simple diffusive loss of NH_3 can account for an important part of ammonia transfer in the teleosts in fresh water without ammonia (Evans and Cameron 1986; Wright and Wood 1985; Maetz 1973; Cameron and Heisler 1983), but the possibility of an ion exchange between Na^+ or H^+ and NH_4^+ is also proposed and could represent up to 30% of total ammonia excretion (Maetz 1973; Girard and Payan 1980; Wright et al. 1989; Evans et al. 1989).

When the concentrations of ammonia increase in the water (during pollution or in intensive fish culture ponds), ammonia acts as a toxic (Brocway 1950; Haywood 1983; Thurston et al. 1978, 1981, 1983; Colt and Armstrong 1979; Meade 1985). Even if the real mechanism of toxicity for fish is not really known, some possibilities have been pointed out: an action on the ability of oxygen uptake through the gill epithelium (Smart 1976, 1978), an increase in the energy expenditure (Buckley et al. 1979; Walker and Schenker 1970), and an action on the central nervous system (Levi et al. 1974; Smart 1978; Arillo et al. 1981). Whatever the real causes of the toxicity are, the problem of the entry of ammonia in the fish received little attention compared with the determination of its toxic level.

To try to answer the questions posed by the entry of ammonia in the fish, the evolution of total ammonia (T_{amm}) and P_{NH3} in the plasma of sturgeon during high exposure was analyzed, and the values obtained for these two parameters were compared between water and plasma (Fig. 21.4). In control conditions, plasma total ammonia was distributed along the pH gradient. Just after exposure to high external concentration, plasma total ammonia increased rapidly. In series 1 (1.85 mmol L^{-1} T_{amm}), the plasma [T_{Amm}] reached the water value after 24–48 h and then continued to slightly increase. In series 2 (4.5 mmol L^{-1} T_{amm}), the plasma (T_{Amm}) remained below that of water, but after 72 h, the difference was not significant. This study suggests that this unusual result could be explained by the duration of the experiment. Indeed, in this work, the movements of ammonia were followed during 72 h instead of only less than 6 h in other studies. If we consider only the first 6 h in our experiment, the plasma total ammonia was clearly below that of water, but the system was not yet in a steady state.

If the changes in P_{NH3} plasma of sturgeon (Fig. 21.4) are analyzed in place of total ammonia values, P_{NH3} in water and plasma are not different in control conditions, and a rapid increase of P_{NH3} during the first hours of the two ammonia exposure series is detected. But in this case, the system reached a steady state after 6–24 h with no additional increase in spite of an inward gradient, and the plasma P_{NH3} never reached the water level. From 24 to 72 h, the plasma content was around 45% of the water value. This is in contradiction with the nonionic diffusion theory according to which P_{NH3} should be near the balance on both sides of a biological barrier. In fact, if the entry of ammonia followed the P_{NH3} gradient, a steady state should have been attained without any difference in P_{NH3}.

Several published studies on fish have reported similar unbalances during exposure to high concentrations of ammonia. This apparent contradiction to nonionic diffusion could be explained by a comment on the calculation of P_{NH3} in water. It has recently been demonstrated that, because of the CO_2 excretion through the gill epithelium, the water is acidified by CO_2 hydration into HCO_3^- catalyzed by the carbonic anhydrase

21 Acid-Base Balance and Ammonia Loading in the Siberian Sturgeon

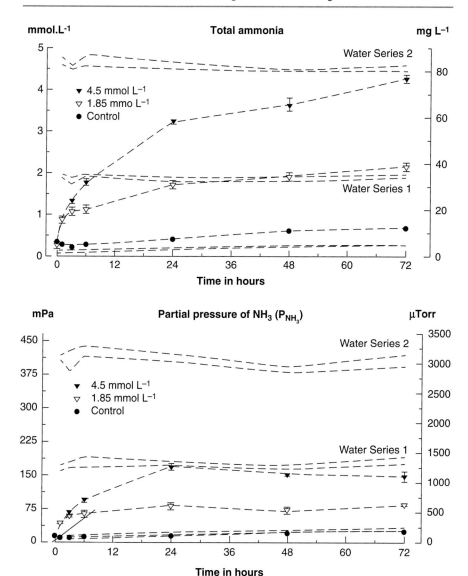

Fig. 21.4 Time course of the total ammonia (T_{amm}) (*above*) and of the NH$_3$ partial pressure (P_{NH_3}) (*below*) in the plasma of the Siberian sturgeon as a function of the exposure time to ammonia. Mean ± SE. The *dashed lines* indicate the T_{amm} and P_{NH_3} in water. Series 1, 1.85 mmol L^{-1} T_{amm}, and series 2, 4.5 mmol L^{-1} T_{amm}. Mean ± SE. Water pH and PCO$_2$ were regulated at 8.0 ± 0.1 and 0.10 kPa, respectively

in the gill apical epithelium or in the mucus (Wright et al. 1986, 1989; Conley and Mallat 1988; Playle and Wood 1989). Thus, the pH of water near the gill epithelium is more acid than in the bulk water. If water pH is more acid, then more ammonia should be converted to ammonium ions, and water P_{NH_3} would be lower than the one calculated with bulk water pH. Table 21.1 gives the estimated pH difference (Δ pH) between

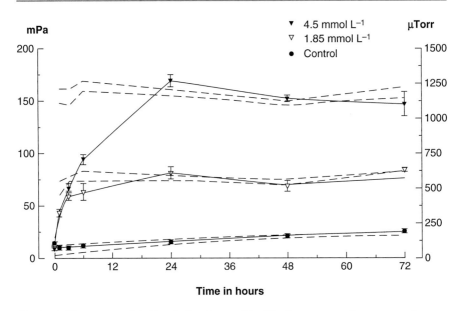

Fig. 21.5 Time course of the P_{NH3} in the plasma of the Siberian sturgeon after an ammonia exposure in water (1.85 mmol L^{-1} and 4.5 mmol L^{-1} T_{amm}) and of a theoretical P_{NH3} in water in the boundary layer near the gill epithelium. The theoretical values of P_{NH3} in boundary layer of water were calculated from a theoretical pH obtained with the values of the initial pH in bulk water and the evaluated ΔpH (see Table 21.1). As seen on this figure, this equality was reached by a reduction of only 0.4 unit of water pH and reached during the period of 24–72 h. The *dashed lines* indicate the T_{amm} and theoretical P_{NH3} in water

bulk and gill water for which P_{NH3} would be equal on both sides of the gill epithelium after 24–72 h of exposure to 1.85 mmol L^{-1} and 4.5 mmol L^{-1} T_{amm}, and it appears that a decrease of 0.3–0.4 pH unit is sufficient to obtain the equality of P_{NH3} between plasma and water in these conditions of ammonia exposure.

This theoretical decrease in water pH near the gill epithelium in the range of values was found on other species with water pH similar to our experiments (Wright et al. 1989). It is possible to recalculate a theoretical value of P_{NH3} in water using this Δ pH and the total ammonia concentration. The results are given in Fig. 21.5. As seen on this figure, this equality was reached by a reduction of only 0.4 unit of water pH. P_{NH3} was calculated from a theoretical pH obtained with the values of the initial pH in bulk water and the estimated ΔpH (see Table 21.1).

Table 21.1 Theoretical values of the pH differences (ΔpH) between the bulk water and the water near the gill epithelium (mean ± SE), P_{NH3} in water and plasma being equal between 24 and 72-h ammonia exposure

Time		24 h	48 h	72 h	Mean
Témoin		0.081 ± 0.048	0.082 ± 0.030	0.071 ± 0.016	0.079 ± 0.022
	n	8	6	5	19
1.85 mmol L^{-1}		0.344 ± 0.028	0.392 ± 0.031	0.338 ± 0.017	0.355 ± 0.016
	n	10	6	6	22
4.5 mmol L^{-1}		0.390 ± 0.015	0.402 ± 0.012	0.460 ± 0.033	0.410 ± 0.012
	n	10	6	5	21

Conclusions

This study suggests that even in conditions of high external concentrations, ammonia distribution between water and fish could be explained on the basis of predominant nonionic diffusion. This phenomenon has two consequences: firstly, whatever the concentration of ammonia in water, the fish reaches a balance of P_{NH3} with the environmental water quickly and seems unable to prevent the entry of ammonia, and secondly the entry of NH_3 in the sturgeon, as a function of partial pressure gradient, induces a metabolic alkalosis which results from the protonization of NH_3 into NH_4^+ in the blood.

If we take into account the real pH in the branchial boundary layer and not in the water of the circuit to calculate P_{NH3}, we could conclude that the evolution of blood ammonia concentration could be explained by NH_3 diffusion along its partial pressure and that the ammonia exchanges by the simple diffusion were dominant. However, whatever the ammonia concentration in water, the fish reached quickly to a P_{NH3} equilibrium with the water and seemed incapable of preventing the invasion of the blood by ammonia. That also explained the initial metabolic alkalosis. But these results did not allow to attribute the ammonia toxicity to NH_3.

In spite of this conclusion, the problem of the toxicity of ammonia remains. It cannot be due to NH_3 only but also certainly to a high level of T_{amm} (NH_4^+ and un-ionic ammonia NH_3) in the blood.

Note of the Authors The experiments have been performed in the context of a PHD: Salin D (1992) La toxicité de l'ammoniaque chez l'esturgeon sibérien, *Acipenser baerii*: effets morphologiques, physiologiques, métaboliques d'une exposition à des doses sublétales et létales. Thèse N° 749, Université Bordeaux I pp134, Director of the PhD, Truchot Jean-Paul; supervisors, Nonnotte Guy and Williot Patrick. They have been investigated in the "Laboratoire de Neurobiologie et Physiologie comparées, CNRS URA 1126 and the University of Bordeaux I, F-33120 Arcachon," in collaboration with the IRSTEA (formerly CEMAGREF), F-33611 Cestas-Gazinet with a financial grant of the IRSTEA and CNRS.

References

Arillo A, Margiocco C, Melodia F, Mensi P, Schenone G (1981) Ammonia toxicity mechanism in fish: studies on rainbow trout (*Salmo gairdneri* R.) Ecotoxico Environ Safety 5:316–328

Astrup P (1956) A simple electrometric technique for the determination of carbon dioxide tension in blood and plasma, total content of carbon dioxide in plasma and bicarbonate content in 'separated' plasma at a fixed dioxide tension (40 mmHg). Scand J Clin Invest 8:33–43

Brocway DR (1950) Metabolic products and their effects. Prog Fish Cult 12(3):127–129

Buckley JA, Whitmore CM, Liming BD (1979) Effects of prolonged exposure to ammonia on the blood and the liver glycogen of coho salmon (*Oncorhynchus kisutch*). Comp Biochem Physiol 63C:297–303

Burggren WW, Randall DJ (1978) Oxygen uptake and transport during hypoxic exposure in the sturgeon *Acipenser transmontanus*. Respir Physiol 34:171–183

Cameron JN, Heisler N (1983) Studies of ammonia in the rainbow trout: physicochemical parameters, acid-base behaviour and respiratory clearance. J Exp Biol 105:107–121

Claiborne JB, Evans DH (1988) Ammonia and acid-base balance during high ammonia exposure in a marine teleost (*Myoxocephalus octodecimspinosus*). J Exp Biol 96:431–434

Colt J, Armstrong D (1979) Nitrogen toxicity to fish, crustaceans and molluscs. Department of civil Engineering. University of California Davis, California, p 30

Conley DM, Mallat J (1988) Histochemical localisation of Na+/K+−ATPase and carbonic anhydrase activity in the gills of 17 fish species. Can J Zool 66:2398–2405

Davenport HW (1974) The ABC of acid-base chemistry. Univ Press, Chicago, p 158

Emerson KR, Russo RC, Lund RE, Thurston RV (1975) Aqueous ammonia equilibrium calculations: effect of pH and temperature. J Fish Res Board Can 32:2379–2383

Evans DH (1982) Mechanism of acid extrusion by two marine fishes: the teleost *Opsanus beta* and the elasmobranch *Squalus acanthias*. J Exp Biol 97:289–299

Evans DH, More KJ, Robbins SL (1989) Modes of ammonia transport across the gill epithelium of the marine teleost *Opsanus beta*. J Exp Biol 144:339–356

Evans DH, Cameron JN (1986) Gill ammonia transport. J Exp Biol 239:17–23

Girard JO, Payan P (1980) Ion exchange through respiratory and chloride cells in freshwater and seawater adapted teleosts. Am J Phys 238:R260–R268

Haywood GP (1983) Ammonia toxicity in teleost fishes: a review. Can Tech Rep Fish Aquat Sei 1177:35

Heisler N (1984) Acid-base regulation in fishes. In: Hoard WS, Randall DJ (eds) Fish physiology, vol X. Academic Press, New York, pp 315–401

Kormanik GA, Cameron JN (1981) Ammonia excretion in animals that breathe water: a review. Mar biol Lett 2:11–23

Levi G, Morisi G, Coletti A, Catanzaro R (1974) Free amino acid in fish brain: normal levels and changes upon exposure to high ammonia concentrations in vivo and upon incubation of brain slices. Comp Biochem Physiol 49A:623–636

Maetz J (1972) Branchial sodium exchange and ammonia excretion in the goldfish *Carassius auratus*. Effects of ammonia-loading and temperature changes. J Exp Biol 56:601–620

Maetz J (1973) $Na^+/NH4^+$, Na^+/H^+ exchanges and NH_3 movement across the gill of *Carassius auratus*. J Exp Biol 58:215–275

Meade JW (1985) Allowable ammonia for fish culture. Prog Fish Cult 47(3):135–145

Medale F, Blanc D, Kaushik SJ (1991) Studies on the nutrition of Siberian sturgeon *Acipenser baerii*. II utilization of dietary non-protein energy by sturgeon. Aquaculture 93:143–154

Playle RC, Wood C (1989) Water chemistry changes in the gill micro-environment of rainbow trout. J Comp Physiol 159B:527–537

Randall D, Wright PA (1989) The interaction between carbon dioxide and ammonia excretion and water pH in fish. Can J Zool 67:2936–2942

Smart GR (1976) The effect of ammonia exposure on gill structure of the rainbow trout (*Salmo gairdneri*). J Fish Biol 8:471–475

Smart GR (1978) Investigations of the toxic mechanisms of ammonia to fish gas exchange in rainbow trout exposed to acutely lethal concentrations. J Fish Biol 12:93–104

Smith HW (1929) The excretion of ammonia and urea by the gills of fish. J Biol Chem 81:727–742

Thurston RV, Russo RC, Smith CE (1978) Acute toxicity of ammonia and nitrite to cutthroat trout fry. Trans Am Fish Soc 107(2):361–368

Thurston RV, Russo RC, Vinogradov GA (1981) Ammonia toxicity to fishes: effect of pH on the toxicity of the unionized ammonia species. Environ Sci Technol 15:837–840

Thurston RV, Russo RC, Phillips GR (1983) Acute toxicity of ammonia to fathead minnows (*Pimephales promelas*). Trans Am Fish Soc 112:705–711

Walker CO, Schenker S (1970) Pathogenesis of hepatic encephalopathy with special reference to the role of ammonia. Am J Chim Nutri 23(5):619–632

Wright PA, Wood C (1985) An analysis of branchial ammonia excretion in the freshwater rainbow trout: effects of environmental pH change and sodium uptake blockade. J Exp Biol 114:329–353

Wright PA, Heming T, Randall DJ (1986) Downstream pH changes in water flowing over the gills of rainbow trout. J Exp Biol 126:499–512

Wright PA, Randall DJ, Perry SF (1989) Fish gill water boundary layer: a site of linkage between carbon dioxide and ammonia excretion. J Comp Physiol 158B:627–635

Effects of Exposure to Ammonia on Plasma, Brain and Muscle Concentrations of Amino Acids and Adenyl Nucleotides in the Siberian Sturgeon, *Acipenser baerii*

22

Guy Nonnotte, Dominique Salin, Patrick Williot, Karine Pichavant-Rafini, Michel Rafini, and Liliane Nonnotte

Abstract

Neurological disorders appear extremely rapidly for elevated doses of ammonia in Siberian sturgeon baerii. To explain this phenomenon, the effects of ammonia during a 72-h exposure in control conditions and for a lethal and a sublethal doses were examined on the concentration of free amino-acids and adenyl nucleotides in the Siberian sturgeon plasma, muscle and brain.

Results showed an important role of the glutamine-glutamate-aspartate group in the ammonia tissue detoxification process and a drop of ATP in the brain.

Moreover the hypothesis of a detoxification by an increase of urea concentration in plasma was verified, but no significant difference appeared between the three series.

The important role of the synthesis of the aspartate seems specific to Siberian sturgeon.

G. Nonnotte (✉) · L. Nonnotte
12 rue Marcel Pagnol, F-33260 La Teste de Buch, France
e-mail: guy.nonnotte@wanadoo.fr

D. Salin
109 rue Blaise Pascal, F-33160 Saint Médard en Jalles, France

P. Williot
4 rue du Pas de Madame, F-33980 Audenge, France

K. Pichavant-Rafini
Laboratoire ORPHY EA4324, Université de Bretagne Occidentale, 6 Avenue le Gorgeu, CS 93837, 29238 Brest Cedex 3, France

M. Rafini
Département Communication, Anglais, Sciences Humaines, Université de Bretagne Occidentale, 6 Avenue le Gorgeu, CS 93837, 29238 Brest Cedex 3, France

© Springer International Publishing AG, part of Springer Nature 2018
P. Williot et al. (eds.), *The Siberian Sturgeon (Acipenser baerii*, Brandt, 1869)
Volume 1 - Biology, https://doi.org/10.1007/978-3-319-61664-3_22

The implications of this amino-acid in the synaptic transmission and the raise of the concentrations of the other amino-acids may take to secondary processes which represent another form of toxicity. As in mammals, the origin of the ammonia toxicity seems to be metabolic, in particular at the nervous system level for a great part.

Keywords

Siberian sturgeon • *Acipenser baerii* • Ammonia toxicity • Brain • NTP • Free amino acids • Detoxification • Urea

Introduction

The observations of the Siberian sturgeon, *Acipenser baerii*, during the toxicity experiments (Chap. 19) allowed us to define the following clinical signs:

1. The fish remain steadily at the bottom of the tanks and increase ventilation only.
2. They jump, in more or less complete tetany with spasmodic and anarchic movements.
3. A loss of equilibrium then occurs and the fish swim on their back side.
4. A very violent tetany followed by death can be observed.

The secretion of mucus in fish submitted to sublethal doses of ammonia was increased significantly in particular at the gill level.

These reactions—loss of equilibrium, spasmodic and erratic movements, hyperventilation—vary both with time and the dose of ammonia, on the one hand, and with the individual fish, on the other hand. They have often been described in many species of fish submitted to acute ammonia exposures (Smart 1976, 1978; Colt and Tchobanoglous 1976; Thurston et al. 1978, 1981, 1983; Buckley et al. 1979; Hillaby and Randall 1979; Ruffier et al. 1981; Haywood 1983) and may be compared with the phenomenon reported by Hillaby and Randall (1979) in trout 5 min after an injection of 300 pmol of $NH_4Cl/100$ g of fish.

The increase of urea concentrations in plasma, considered as a possible process of detoxification, associated to the increase of ammonia concentration in plasma, could explain these symptoms. To verify this hypothesis of detoxification of NH_3, the urea plasmatic concentrations in Siberian sturgeon were determined in control conditions and when the fish were submitted to sublethal or lethal concentrations of total ammonia in water during 72 h.

Modifications of biochemical reactions at the central nervous system level, namely, a decrease in ATP levels, amino-acid concentrations, and an increase of glutamic acid, a neurotransmitter, can also explain the previous observed reactions: the hyperventilation (Chap. 21), the convulsions and the erratic movements.

In this context, adenylate energy charge, adenyl nucleotides concentrations in the muscle and the brain of Siberian sturgeon as well as free amino acids concentrations in plasma, brain and muscle were quantified after exposure to high ammonia concentration in water.

22.1 Materials and Methods

The experimental protocol was described in details in Chap. 19.

Sturgeons were cannulated (see Williot and Le Menn, Chap. 24, Nonnotte et al. 1993, Salin 1992) in dorsal aorta and were placed in a closed recirculating device thermostatted at 15 °C. The water was oxygenated by intensive bubbling of CO_2-free air. The pH was regulated at 8.0 and the PCO_2 at 0.1 kPa using a pH-CO_2 stat system. The titration alkalinity was manually corrected at 1.9 mEq L^{-1} (see Chap. 25).

Three series of experiments were performed: one control series without ammonia, one series at 1.85 mmol L^{-1} total ammonia which is a sublethal concentration and another series at 4.5 mmol L^{-1} total ammonia, near the lethal concentration. Blood samples were drawn from the caudal artery catheter prior to and after 1, 3, 6, 24, 48 and 72 h of ammonia exposure.

Fish plasma was obtained from blood collected in small centrifuge tube (450 µL) and immediately centrifuged for 5 mn at 8000 g to separate plasma from red cells. Plasma ammonia concentration was estimated with an enzymatic method (Sigma kit UV170).

20 µL of plasma were used to quantify the urea concentration by a colorimetric method (Sigma kit 535). Immediately after sampling, 400 µL of plasma were treated to determine the free amino-acid concentration. At 4 °C, the plasma sample was mixed (2v/1v) with sulfosalicylic acid buffered by lithium citrate at pH 2.2 and centrifuged for 5 mn at 5000 g. The supernatant was quickly frozen and stocked at −20 °C.

The brain and muscles were quickly sampled after decapitation and treated at 4 °C. The entire manipulation did not exceed 5 mn. Samples were homogenised in cold water (400 µL) with a Potter mill and an ultrasonic homogeniser. 20 µL of homogenate were used to quantify the protein concentration by Bradford method (Biorad kit). 350 µL was mixed (2v/1v) at 4 °C with sulfosalicylic acid buffered by lithium citrate at pH 2.2 and centrifuged (temp. 4 °C) for 5 mn at 5000 g for brain samples or mixed with deionised water for muscle 700 µL per 100 mg initial tissue. The supernatant was quickly frozen and stocked at −20 °C.

Free amino acids concentrations were obtained with an auto analyser Beckman Gold 600 by colorimetric method with ninhydrin reaction.

The plasmatic NTP (ATP, GTP, UTP) concentration was measured by the phosphoglycerate kinase (kit Sigma 3366UV).

Adenyl nucleotide (ATP, ADP, AMP) concentrations in brain and muscle tissues were determined after N_2 liquid freezing of tissues by a luciferin-luciferase method (Sigma kit FLE-50).

The adenylate energy charge was calculated by AEC = (ATP + 1/2ADP)/ (ATP + ADP + AMP).

22.2 Results and Discussion

The urea concentrations in the plasma of the sturgeon are similar to those observed in other ammonotelic fish (Vellas 1979). No significant difference (Fig. 22.1) appeared between either of the series ($p > 0.05$, Kruskall–Wallis H-test) or the different samples in a single series ($p > 0.05$, U test).

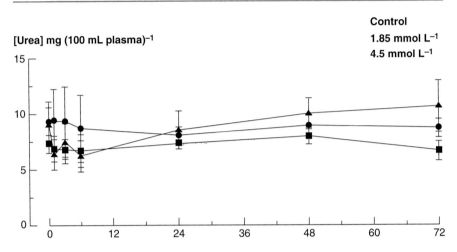

Fig. 22.1 Time course of the urea plasmatic concentration in Siberian sturgeon in control conditions and submitted to sublethal (1.85 mmol L^{-1}) or lethal (4.5 mmol L^{-1}) concentrations of total ammonia in water during 72 h. Mean ± SE; $n = 10$

The absence of variation had already been noted in *Periophthalmus cantonensis* submitted to high levels of ammonia (Iwata 1988) and can be explained by the urea production pathway from the purines and not by the fixation of ammonia in the ornithine cycle as in mammals (Brown and Cohen 1960). Moreover, urea represents less than 10% of the nitrogen excretion in *Acipenser baerii* as in fresh water teleosts (Smith 1929; Vellas and Serfaty 1975; Vellas 1979; Medale et al. 1991). Olson and Fromm (1971) reached the same conclusion for trout exposed to various levels of ammonia during short periods as in our experiments.

Modifications of biochemical reactions at the central nervous system level, in particular a decrease in ATP levels, amino-acid concentrations and an increase of glutamic acid, a neurotransmitter, (Walker and Schenker 1970; Levi et al. 1974) can also explain the observed hyperventilation, the convulsions and the erratic movements (Schwartz 1958 in Mutch and Banister 1983; Smart, 1976; Bubien and Meade 1986).

In control conditions, the main process of ammonia production in fish is the transamination of amino acids into glutamate. The glutamate is also a major amino donor to other amino acids in subsequent transamination reactions catalysed by the glutamate dehydrogenase (GDH), enzyme located in the liver (Pequin and Serfaty 1963; Walton and Conwey 1977), the organic acid α-ketoglutarate and NH$_3$ being the derived products according to Fig. 22.2.

Other reactions may be involved in ammonia production such as the conversion of AMP into IMP (Cooper and Plum 1987) or the following reaction catalysed by glutaminase (Fig. 22.3).

These reactions are reversible and allow to decrease ammonia concentrations in the plasma of fish submitted to high levels of ammonia in water concomitantly to a decrease of ATP and an increase in glutamine which can explain the observed neurological disorders (Raabe 1988; Cooper and Plum 1987).

The adenylate energy charge [ATP + 1/2ADP]/[ATP + ADP + AMP] and adenyl nucleotide concentrations in the muscle and the brain of the Siberian sturgeon were

22 Effects of Exposure to Ammonia on Plasma, Brain and Muscle Concentrations of AA

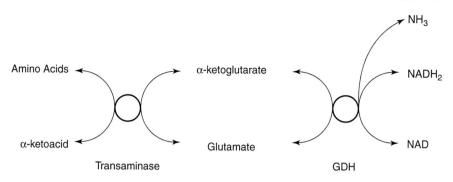

Fig. 22.2 Ammonia production in fish by the transamination of amino acids into glutamate

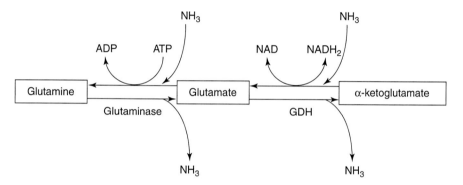

Fig. 22.3 Ammonia production by reaction catalysed by glutaminase

determined in control situation and after 72-h exposure to ammonia in water at lethal concentration of 4.5 mmol L^{-1} in water.

As clearly observed in Table 22.1, a major decrease of ATP and of the total adenyl nucleotide concentrations occurs in the brain of the intoxicated fishes, whereas these variations were not significant in their muscles.

Similar decreases of ATP and of the total adenyl nucleotides (Σ AN) were described in trout (Arillo et al. 1981), in goldfish (Schenone et al. 1982) and in mammals in hepatic encephalopathy.

The stability of the adenylate energy charge combined with the decrease of the concentration in amino acids in the brain (see later) allowed to explain the decrease of ATP: it is due either to a higher demand of energy and/or to a lesser production of ATP.

Therefore, different hypotheses can explain the decrease of ATP concentrations.

- First, an increase of the Na$^+$-K$^+$-ATPase activity by substitution of potassium ions by NH$_4^+$
- Secondly, the synthesis of glutamine, a reaction catalysed by the glutamine synthetase, which consumes much ATP (Walton and Conwey 1977)
- Thirdly, an increase of the neuronal activity and of the active mechanisms of ionic transfers

Table 22.1 Adenylate energy charge and adenyl nucleotides concentration in the muscle and the brain of the Siberian sturgeon in control situation and after 72-h exposure ammonia in water (4.5 mmol L^{-1})

	Muscle		Brain	
	Control	Intoxicated	Control	Intoxicated
AEC	0.867 ± 0.013	0.857 ± 0.021	0.878 ± 0.012	0.856 ± 0.016
ATP	2.5 ± 0.27	1.76 ± 0.49	1.45 ± 0.23	0.81 ± 0.9*
ADP	0.40 ± 0.07	0.33 ± 0.09	0.36 ± 0.09	0.21 ± 0.01
AMP	0.22 ± 0.08	0.12 ± 0.03	0.04 ± 0.02	0.04 ± 0.01
ΣAN	2.91 ± 0.25	1.82 ± 0.44	1.86 ± 0.28	1.06 ± 0.09*

AEC (adenylate energy charge) = [ATP + 1/2ADP]/[ATP + ADP + AMP]; ΣAN: sum of the adenyl nucleotides

Mean ± SE; $n = 6$; *$p < 0.02$, U test-Mann–Whitney

Concerning the second hypothesis, the free amino-acid concentrations in the plasma, the muscle and the brain of the sturgeon (Table 22.2 and Fig. 22.5) were monitored before and after exposure to ammonia. Table 22.2 shows that lysine represents 28% of the total essential free amino acids (AA) measured in plasma in control conditions without ammonia, leucine (19.5%), valine (15.5%) and threonine (9.5%). These four essential amino acids represent 72.5% of the determined essential amino acids. The lesser free amino acids are the phosphoserine and the cysteine. The glutamate and the glutamine represent 0.3–4.0% of the total free amino acids, respectively.

Meanwhile, this distribution may be variable not only between different species but also for the same species. Leucine, lysine, valine and threonine are often the more important AA. Serine, proline and taurine concentrations may also reach very high levels (Ogata 1985; Ogata and Aral 1985; Ogata and Murai 1988). Finally, the glutamine concentration seems higher in carp than in sturgeon.

Figure 22.4 shows the time course of the free amino acids in Siberian sturgeon plasma during 72 h in control conditions and after exposure to ammonia in water.

In control conditions the sum of the AA decreases by about 15%. The concentration of lysine, ornithine, arginine and serine showed the most important decrease, while valine, isoleucine and leucine increased. These variabilities may be due to the starvation diet enforced upon the fish for 8 days and during the experimentation.

After an exposure to 1.85 mmol L^{-1} ammonia during 72 h, this drop increased from 15% to 75%. Specifically, for glutamine and glutamate, this decrease was important between 48 h and 72 h. It was in accordance with the interaction of a detoxification process. It may explain the evolution of the ammonia concentration in the blood as previously described in Chap. 21.

So, glutamine seems to be used as a buffer for temporary stocking excess of ammonia before its excretion. This process can limit the physiological disturbances provoked by the sublethal concentration of ammonia in water.

The time course of the free amino acids in the plasma of sturgeons exposed to lethal concentration of ammonia in water (4.5 mmol L^{-1} ammonia) was more complex. After a decrease of 30% of the sum of the free amino acids in the plasma between 0 h and 24 h, a great increase (+70%) was observed between 24 h and 72 h.

22 Effects of Exposure to Ammonia on Plasma, Brain and Muscle Concentrations of AA

Table 22.2 Free amino-acid concentrations in plasma, white muscle and brain of Siberian sturgeon in control conditions without ammonia (Mean ± SE)

Name	Abb.	Plasma µmol L^{-1}			Muscle µmol L^{-1}			Cerveau µmol L^{-1}		
		m	SE	n	m	SE	n	m	SE	n
Essential amino acids										
Arginine	Arg	99.74	±7.44	12	0.37	±0.20	3	1.29	±0.12	5
Histidine	His	47.35	±3.59	11	1.62	±0.40	4	3.46	±0.31	5
Isoleucine	Ile	166.66	±6.78	12	2.68	±0.52	4	0.73	±0.13	5
Leucine	Leu	313.52	±15.24	12	4.19	±0.91	4	1.61	±0.26	5
Lysine	Lys	464.35	±38.34	12	1.85	±0.39	4	4.01	±0.44	5
Méthionine	Met	57.18	±4.39	12	1.40	±0.26	4	0.92	±0.15	5
Phenylalanine	Phe	66.12	±2.63	12	1.38	±0.18	4	0.43	±0.18	5
Threonine	Thr	152.41	±11.22	12	3.26	±0.57	4	2.91	±0.31	4
Tryptophan	Trp		Nd		1.40	±0.82	4		Nd	
Valine	Val	252.84	±9.30	12	4.12	±0.83	4	1.74	±0.22	5
Non-essential amino acids										
Alanine	Ala	116.52	±12.60	12	19.27	±2.87	4	4.85	±0.45	5
Asparagine	Asn		Nd			Nd			Nd	
Aspartic acid	Asp	6.24	±0.52	12	0.88	±0.09	4	6.88	±0.23	5
Cysteine	Cys	4.13	±0.67	11		Nd		0.22	±0.11	4
Glutamic acid	Glu	6.81	±0.64	12	1.08	±0.19	4	28.10	±3.47	5
Glutamine	Gln	96.69	±5.30	12	4.60	±1.05	4	45.36	±2.33	5
Glycine	Gly	17.08	±1.67	12	5.04	±1.10	4	8.02	±0.26	5
Proline	Pro	37.13	±5.19	12	2.39	±0.28	4		Nd	
Serine	Ser	117.72	±11.2	12	5.01	±1.21	4	5.29	±0.89	5
Tyrosine	Tyr	47.34	±3.31	12	1.17	±0.15	4	0.34	±0.15	3
Amino acids, monoamines and other molecules										
ß-alanine	ßAla		Nd		35.09	±7.19	4		Nd	
Γ-Aminobutyrate	GABA		Nd			Nd		12.32	±1.82	5
Ammoniac	NH$_4^+$	347.25	±46.78	12	36.41	±9.03	4	13.56	±2.53	5
Cystathionine		6.53	±2.01	10	0.32	±0.24	3	0.64	±0.10	4
Ornithine	Orn	114.43	±12.77	12	0.38	±0.13	4	0.80	±0.07	5
Phosphoserine	Pho	0.06	±0.01	9	0.00	±0.00	4	0.01	±0.00	5
Taurine	Tau	37.32	±3.13	12	26.93	±6.28	4	44.95	±4.93	5

Units: µmol L^{-1} in plasma and µmol g protein^{-1} in muscle and brain. *Abb* abbreviations, *nd* non-detected

Specifically, the concentrations of taurine, alanine, glutamine and glutamate rose sharply at 72 h (+225%, +300%, +52% and 95%, respectively).

In contrast, the concentrations of ornithine, threonine and serine decreased in the same proportion as in the other series, certainly due to the starvation diet of the fish.

The increase of alanine was also interesting. It might explain a great disturbance of the hepatic metabolism in particular the glycogenogenesis (Arillo et al. 1981; Begum 1987).

The increases of glutamine-glutamate concentrations confirm the hypothesis of the ammonia trapping role of both these molecules after ammonia exposure. But in

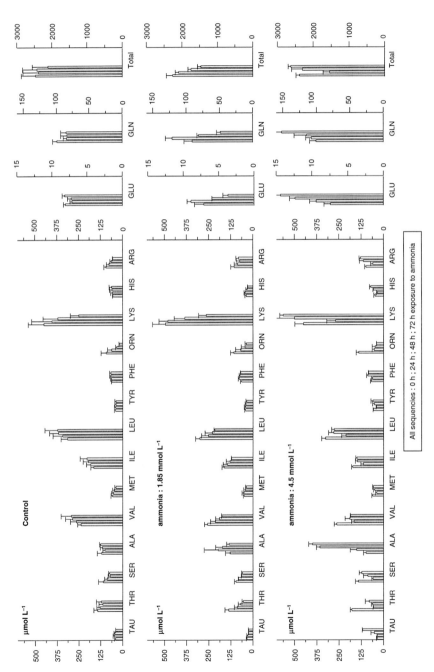

Fig. 22.4 Time course of the free amino acids in the plasma of the Siberian sturgeon in control condition without ammonia in water and submitted to sublethal (1.85 mmol L^{-1}) or lethal (4.5 mmol L^{-1}) concentrations of total ammonia in water during 72 h. Mean ± SE; $n = 6$

lethal conditions the process was insufficient since the concentration of ammonia in the blood increased (Chap. 21).

These studies demonstrated the fundamental role of glutamine and suggested that the general increase of free amino acids in the plasma might be due to transamination reactions from the glutamate when the synthesis of glutamine and glutamate were saturated. That is certainly the case in lethal exposure, whereas, at sublethal conditions, the increase of ammonia concentration in the blood is stopped and a regulatory mechanism can occur.

For the muscle (Table 22.2), 70% of the determined amino acids in control conditions are represented by ß-alanine (27%), taurine (21%), alanine (14%), glycine (4%) and serine (4%). Glutamine, glutamate and aspartate represent 3.6, 0.9 and 0.7% of the total, respectively. These concentrations in free amino acids in the Siberian sturgeon are similar to those reported in other species (Dabrowska and Wlashow 1986; Iwata 1988; Yokoyama and Nakazoe 1991). The differences are probably due to the specific nutrition and not to the differences in species (Dabrowska 1984).

After an exposure to 1.85 mmol L^{-1} of ammonia during 72 h (Fig. 22.5), the concentration of glycine may be greatly modified (+180%) as it was also described in carp (Dabrowska and Wlashow 1986) and in mudskipper (Iwata 1988). The glutamate increases (+70%) like the aspartate (+60%) while the glutamine decreases (−15%). As the total of the essential amino-acid concentrations decreases (−50%), the total concentration of amino acids remains identical. The ammonia concentration of the muscle tissue is unchanged as in control conditions.

An exposure to 4.5 mmol L^{-1} ammonia during 72 h (Fig. 22.5) provokes an important increase of glutamate (+80%), glutamine (+30%) and aspartate (+60%), and the ammonia concentration of the muscle increases (+40%).

The most pertinent of these results are the great increases of glutamate, glutamine and aspartate. This peculiarity has already been described in carp (Dabrowska and Wlashow 1986) and in mudskipper (Iwata 1988). For the glutamate-glutamine complex, its increase may be due to the process of detoxification by catching of NH_3, formation of glutamate and then conversion into glutamine. The formation of glutamine in the muscle is proportional to the ammonia concentration (Iwata 1988) and followed by an exit into the blood. This time course of the muscular glutamine corroborates the one observed in the plasma (Fig. 22.4).

In the brain (Table 22.2), the most concentrated free amino acids are the glutamine (24%), the glutamate (15%), the GABA (6.5%) and the taurine (24%). So the neurotransmitter AA concentrations in the brain are about ten times more elevated than in the muscle, which confirms the predominant role of these AA in the synaptic transmission at the central nervous system level. The other AA represent only 2% of the total AA concentration. For example, the ß-Alanine concentration was 35 μmol g protein^{-1} in the muscle and not detected in the brain tissue.

The free amino-acid concentrations in the brain tissue of the Siberian sturgeon are similar to that of the trout (Arillo et al. 1981), of the carp (Dabrowska and Wlashow 1986) and of goldfish (Schenone et al. 1982; Levi et al. 1974). In these species, the taurine, the glutamine, the glutamate and the GABA are the most concentrated AA.

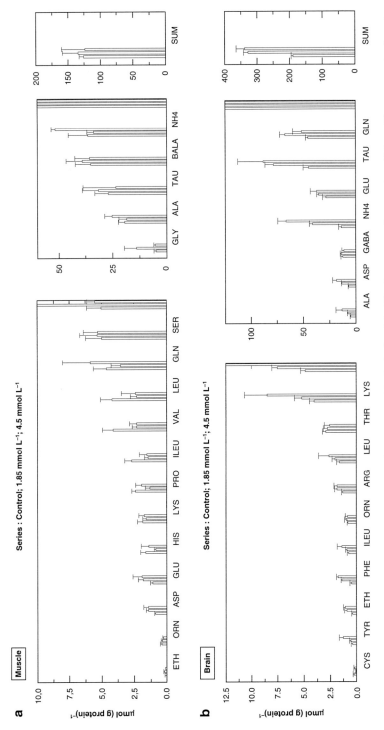

Fig. 22.5 Concentration of the free amino acids in the white muscle and in the brain of the Siberian sturgeon in control condition without ammoniac in water and submitted to sublethal (1.85 mmol L^{-1}) or lethal (4.5 mmol L^{-1}) concentrations of total ammonia in water during 72 h. Mean ± SE; $n = 6$

The ammonia concentrations in the brain of the sturgeon, in control conditions, are lower than in the muscle, as in other fish. Conversely, as in mammals, the high concentrations in glutamine and glutamate suggest a better protection of this tissue if the ammonia concentration in the internal medium increases (Levi et al. 1974; Schenone et al. 1982; Cooper and Plum 1987). It might also be explained by a higher activity of the glutamine synthetase (GS) in the brain than in the other tissues (Walton and Conwey 1977).

After an exposure to 1.85 mmol L^{-1} of ammonia during 72 h (Fig. 22.5), the amino acids pool, specifically the concentrations of the amino acids neurotransmitters such as aspartate (+100%), glutamate and GABA (+20%), increase largely. The glutamine concentration rises also (+50%) as does the ammonia concentration (+200%).

An exposure to 4.5 mmol L^{-1} of ammonia during 72 h (Fig. 22.5) amplified these modifications (glutamate +130% and aspartate +170%).

The evolution of the concentration of aspartate in Siberian sturgeon is in agreement with the one reported for goldfish (Levi et al. 1974) and common carp (Dabrowska and Wlashow 1986). The rise is certainly due, as in the muscle, to transamination processes from the glutamate.

Finally, the increases of aspartate and glutamate associated with the decrease of ATP concentration must provoke neurologic troubles corresponding to the ones described in Chap. 19.

Conclusions

The Siberian sturgeon exposed to high levels of ammonia in water seems to use the formation of the complex glutamate-glutamine preferably to urea synthesis. These reactions are important to temporarily reduce the ammonia concentration in the tissues, by transforming NH3 into another less toxic chemical form.

A specificity in Siberian sturgeon is the important role of the synthesis of aspartate.

The implications of these amino acids in the synaptic transmission and the raise of the concentrations of the other amino acids may take to secondary processes which represent another form of toxicity. As in mammals, the origin of the ammonia toxicity seems to be metabolic, in particular at the nervous system level for a great part.

Acknowledgments We are indebted to F. Parrot-Rouleau (Biochemistry Laboratory, CHR Bordeaux, France) for determination of AA and P. Raymond (INRA, Villenave d'Ornon, France) for determination of nucleotides concentrations.

Notes of the Authors The experiments were performed since 1990 to 1994. They have been investigated in the "Laboratoire de Neurobiologie et Physiologie comparées, CNRS URA 1126 and the University of Bordeaux I, F-33120 Arcachon", in collaboration with the IRSTEA (formerly CEMAGREF), F-33611 Cestas-Gazinet. The present chapter was a redrawn version of a part of a PHD: Salin D (1992) La toxicité de l'ammoniaque chez l'esturgeon sibérien, *Acipenser baerii*: effets morphologiques, physiologiques, métaboliques d'une exposition à des doses sblétales et létales. Thèse N° 749, Université Bordeaux I pp134, financial grant of the IRSTEA (formerly CEMAGREF) and CNRS. Director of thesis, Truchot Jean-Paul; supervisors, Nonnotte Guy and Williot Patrick.

References

Arillo A, Margiocco C, Melodia F, Mensi P, Schenone G (1981) Ammonia toxicity mechanism in fish: studies on rainbow trout (*Salmo gairdneri* R.) Ecotoxicol Environ Saf 5:316–328

Begum SJ (1987) Biochemical adaptive responses in glucose metabolism of fish (*Tilapia mossambica*) during ammonia toxicity. Curr Sci 56(14):705–708

Brown GW, Cohen PP (1960) Activities of urea enzymes in various higher and lower vertebrates. Biochem J 75:82–91

Bubien JK, Meade TL (1986) Electrophysiological abnormalities produced by ammonium in isolated perfused brook trout, *Salvelinus fontinalis*, hearts. J Fish Biol 28:47–53

Buckley JA, Whitmore CM, Liming BD (1979) Effects of prolonged exposure to ammonia on the blood and the liver glycogen of coho salmon (*Oncorhynchus kisutch*). Comp Biochem Physiol 63C:297–303

Colt J, Tchobanoglous G (1976) Evaluation of the short-term toxicity of nitrogenous compounds to channel catfish, *Ictalurus punctatus*. Aquaculture 8:209–224

Cooper AJL, Plum F (1987) Biochemistry and physiology of brain ammonia. Physiol Rev 67(2):440–517

Dabrowska H (1984) Effect of dietary protein on free amino acid content in rainbow trout (*Salmo gairdneri* R.) muscles. Comp Biochem Physiol 77A:553–556

Dabrowska H, Wlashow T (1986) Sub-lethal effect of ammonia on certain biochemical and haematological indicators in common carp (*Cyprinus carpio* L.) Comp Biochem Physiol 83C:179–184

Haywood GP (1983) Ammonia toxicity in teleost fishes: a review. Can Tech Rep Fish Aquat Sci 1177:35

Hillaby BA, Randall DJ (1979) Acute ammonia toxicity and ammonia excretion in rainbow trout (Salmo Gairdneri). J Fish Res Board Can 36:621–629

Iwata K (1988) Nitrogen metabolism in the mudskipper *Periophthalmus cantonensis*: changes in free amino-acids and related compounds in various tissues under conditions of ammonia loading with special reference to its high ammonia tolerance. Comp Biochem Physiol 91A:499–508

Levi G, Morisi G, Coletti A, Catanzaro R (1974) Free amino acid in fish brain: normal levels and changes upon exposure to high ammonia concentrations in vivo and upon incubation of brain slices. Comp Biochem Physiol 49A:623–636

Medale F, Blanc D, Kaushik SJ (1991) Studies on the nutrition of Siberian sturgeon *Acipenser baerii* II utilization of dietary non-protein energy by sturgeon. Aquaculture 93:143–154

Mutch BJC, Banister EW (1983) Ammonia metabolism in exercise and fatigue: a review. Med Sci Sports Exerc 15(1):41–50

Nonnotte G, Maxime V, Truchot JP, Williot P, Peyraud C (1993) Respiratory responses to progressive ambient hypoxia in the sturgeon, *Acipenser baerii*. Respir Physiol 91:71–82

Ogata H (1985) Post-feeding changes in distribution of free amino-acids and ammonia in plasma and erythrocytes of carp. Bull Jap Soc Sci Fish 51:1705–1711

Ogata H, Aral S (1985) Comparison of free amino-acids contents in plasma, whole blood and erythrocytes of carp, coho salmon, rainbow trout and channel catfish. Bull Jap Soc Sci Fish 51:1181–1186

Ogata H, Murai T (1988) Changes in ammonia and amino-acids levels in the erythrocytes and plasma of the carp, *Cyprinus carpio*, during passage through the gills. J Fish Biol 33:471–479

Olson KR, Fromm PO (1971) Excretion of urea by two teleosts exposed to different concentrations of ambient ammonia. Comp Biochem Physiol 40A:999–1007

Pequin L, Serfaty A (1963) L'excretion ammoniacale chez un téléostéen dulcicole la carpe (*Cyprinus carpio*). Comp Biochem Physiol 10:315–324

Raabe W (1988) Neuronal effects of ammonia. In: Soeters PB, Wilson JHP, Meijer AJ, Homs E (eds) Advances in ammonia metabolism and hepatic encephalopathy. Elsevier Sci Pub, Netherlands, pp 349–355

Ruffier P, Boyle W, Kleinschmidt J (1981) Short-term acute bioassays to evaluate ammonia toxicity and effluent standards. J WPGF 53(3):367–377

Salin D (1992) La toxicité de l'ammoniaque chez l'esturgeon sibérien, *Acipenser baerii*: effets morphologiques, physiologiques, métaboliques d'une exposition à des doses sublétales et létales. Thèse N° 749, Université Bordeaux I, p 134

Schenone G, Arillo A, Margiocco C, Melodia F, Mensi P (1982) Biochemical bases for environmental adaptation in goldfish (*Carassius auratus* L.): resistance to ammonia. Ecotoxicol Environ safety 6:479–488

Smart GR (1976) The effect of ammonia exposure on gill structure of the rainbow trout (*Salmo gairdneri*). J Fish Biol 8:471–475

Smart GR (1978) Investigations of the toxic mechanisms of ammonia to fish gas exchange in rainbow trout exposed to acutely lethal concentrations. J Fish Biol 12:93–104

Smith HW (1929) The excretion of ammonia and urea by the gills of fish. J Biol Chem 81:727–742

Thurston RV, Russo RC, Smith CE (1978) Acute toxicity of ammonia and nitrite to cutthroat trout fry. Trans Am Fish Soc 107(2):361–368

Thurston RV, Russo RC, Vinogradov GA (1981) Ammonia toxicity to fishes: effect of pH on the toxicity of the unionized ammonia species. Environ Sci Technol 15:837–840

Thurston RV, Russo RC, Phillips GR (1983) Acute toxicity of ammonia to fathead minnows (*Pimephales promelas*). Trans Am Fish Soc 112:705–711

Vellas F (1979) L'excrétion azotée: métabolisme des composés azotés. In: Nutrition des poissons. Actes du colloque du CNERMA. CNRS Ed, Paris, pp 149–161

Vellas F, Serfaty A (1975) Le metabolism ammoniacal chez un téléostéen dulcicole la carpe (*Cyprinus carpio*). Cahiers Lab Montereau 2:5–14

Walker CO, Schenker S (1970) Pathogenesis of hepatic encephalopathy with special reference to the role of ammonia. Am J Chin Nutr 23(5):619–632

Walton MJ, Conwey CB (1977) Aspects of ammoniagenesis in rainbow trout, *Salmo gairdneri*. Comp Biochem Physiol 57B:143–149

Yokoyama M, Nakazoe J (1991) Effects of dietary protein levels on free amino acid and glutathione contents in the tissue of rainbow trout. Comp Biochem Physiol 99A:203–206

The Importance of Water Quality in Siberian Sturgeon Farming: Nitrite Toxicity

23

Enric Gisbert

Abstract

The maintenance of good water quality is of primary importance in aquaculture systems. Although nitrites usually occur in low concentrations, their presence in aquaculture systems at elevated levels is a potential problem due to its well-documented toxicity. Nitrites are not a problem in flow-through systems, but nitrite may become a serious problem in recirculating systems where water is reused. Nitrites are formed by nitrification, the process in which nitrifying bacteria oxidises ammonia into nitrite and then into nitrate. Nitrites in the ambient water can be actively taken up across the gill epithelium and get accumulated in the internal medium resulting in many physiological disturbances (oxidation of haemoglobin, ionic unbalance, liver damage, cardiovascular problems) that may result in the animal's death. The median-lethal concentration of nitrite in juvenile (172 g) Siberian sturgeon after 72 h of exposure was 130 mg/L in water with high chloride content (130.5 mg/L). Levels of Cl$^-$ in water are especially important in freshwater species to prevent/reduce nitrite toxicity, since nitrite is a competitive inhibitor of Cl$^-$ uptake and vice versa. In any case, in aquaculture systems nitrites levels are recommended to be below than 1.0 mg NO$_2$/L. Siberian sturgeon exposed to toxic levels of nitrite showed several signs of behavioural distress characterised by an increase of ventilatory activity, erratic and torpid swimming, loss of equilibrium and overturning swimming. In this chapter, the effects of nitrite intoxication, as well as the recovery of animals exposed to an acute episode of nitrite intoxication are presented and discussed.

Keywords

Nitrite • Toxicity • Osmoregulation • Water quality

E. Gisbert
IRTA-Sant Carles de la Ràpita, Unitat de Cultius Experimentals, Crta. Poble Nou km 5.5, 43540 Sant Carles de la Ràpita, Tarragona, Spain
e-mail: enric.gisbert@irta.cat

© Springer International Publishing AG, part of Springer Nature 2018
P. Williot et al. (eds.), *The Siberian Sturgeon (Acipenser baerii*, Brandt, 1869)
Volume 1 - Biology, https://doi.org/10.1007/978-3-319-61664-3_23

Introduction

Fish perform all their vital functions in water. As they are totally dependent upon water for respiratory purposes, feeding and growth, excretion of wastes, maintaining a salt balance and reproducing, understanding the physical and chemical qualities of water is critical to successful aquaculture. Thus, the success or failure of an aquaculture operation is based in water quality, and consequently, the maintenance of *good water quality* is of primary importance for both survival and optimum growth performance in aquaculture systems, regardless of the use of well and surface waters. Most water quality problems experienced in different production systems (e.g. ponds, open-flow tanks, RAS etc.) are associated with low dissolved oxygen levels and high *fish waste* metabolite concentrations in the water. In this sense, if all the oxygen demands are met, the second factor that becomes limiting is the accumulation of nitrogenous compounds in the water. The major source of nitrogen (up to 90%) in aquaculture systems is from fish feed and is produced through the normal metabolic processes of fish. Thus, the metabolites of concern for maintaining the proper water quality include total ammonia nitrogen (TAN), unionised ammonia (NH_3), nitrite (NO_2), nitrate (NO_3) (to a lesser extent), dissolved carbon dioxide (CO_2), suspended solids and non-biodegradable organic matter. In this sense, the levels of metabolites in water that can have an adverse effect on growth are generally an order of magnitude lower than those tolerated by fish for survival. However, among the above-mentioned compounds, this chapter is mainly focused on the effects of nitrites on Siberian sturgeon condition and health, since other parameters affecting water quality are considered in other parts of this book.

23.1 Nitrite in Aquatic Environments

Nitrite is a natural component of the nitrogen cycle in aquatic ecosystems (Fig. 23.1). Although it usually occurs in low concentrations, its presence in the environment at elevated levels is a potential problem due to its well-documented toxicity to animals. Nitrite is formed by *nitrification*, which is defined as the biological process in which bacteria use reduced nitrogen compounds, such as ammonia, as food for nitrifying bacteria. Ammonia is the major end product of protein catabolism excreted by fish, but it is also produced by the decomposition of urea, fish faeces, and uneaten feed remaining in the bottom of the tank. In brief, nitrification is a two-step process, where ammonia is first oxidised to nitrite and then nitrite is oxidised to nitrate. In aquaculture facilities, ammonia is oxidised by *Nitrosomonas* spp. bacteria, though in the aquatic environment other bacteria are likely more important. Nitrite is then further oxidised to nitrate by another group of nitrite-nitrate conversion bacteria *Nitrobacter* spp. This issue is of special relevance in fish culture conditions since nitrate is relatively non-toxic to fish and may safely accumulate in the tank until it is flushed out by replacement water or converted to gaseous nitrogen (N_2) by anaerobic heterotrophs and lost to the atmosphere in a process known as denitrification (Lawson 1995).

23 Nitrite Toxicity in the Siberian Sturgeon

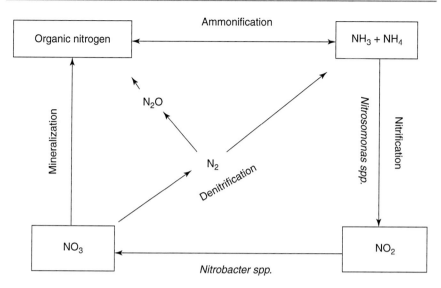

Fig. 23.1 Schematic view of the nitrogen cycle occurring in the aquatic ecosystems as well as in aquaculture systems in which ammonia is oxidised by chemoautotrophic bacteria *Nitrosomonas* spp. and *Nitrobacter* spp. (modified from Lawson 1995)

The two steps in this process are normally carried out sequentially, since the first step has a higher kinetic reaction rate than the second step, and the overall kinetics is usually controlled by ammonia oxidation, and as a result there is no appreciable amount of nitrite accumulation as nitrite is converted into nitrate fairly rapidly. This means that in systems where oxygen is not limiting, the conversion of ammonia to nitrite is the rate-limiting step in the total process. The following reactions show the basic chemical conversions occurring during ammonia oxidation by *Nitrosomonas* spp. (1) and *Nitrobacter* spp. (2):

1. $NH_4^+ + 1.5O_2 \rightarrow NO_2^- + 2H^+ + H_2O + 84 \text{kcal/molammonia}$,
2. $NO_2^- + 0.5O_2 \rightarrow NO_3^- + 17.8 \text{kcal/molnitrite}$.

Nitrification requires oxygen and proceeds most efficiently when the level of oxygen dissolved is near to saturation. Based on the above-mentioned equations, between 4.0 and 4.6 kg of oxygen is required to completely oxidise 1 kg of ammonia into nitrates. This process is also more efficient at pH values between 7.0 and 8.0 and at temperatures between 27 °C and 35 °C, which may be a problem in farming sturgeon species that are reared in cold-temperate water temperatures. However, this range of pH values may be affected by the history and condition under which the bacteria were cultured, and recent studies have shown that the optimum range of pH for nitrification can vary from 7.0 to 9.0 (Chen et al. 2006). As *Nitrobacter* spp. are less tolerant of low temperatures than *Nitrosomonas* spp. nitrite can accumulate at low temperatures in aquaculture systems. However, water generally does not have enough residence time in flow-through systems for nitrite to become a problem, but nitrite may become a serious nightmare in recirculating systems where the water is continuously reused.

Since the conversion of ammonia is a biological process, time is required for the bacterial population to develop sufficient biomass in biological filters in order to remove the toxic nitrogen load (Fig. 23.2). The bacteria that oxidise ammonia must develop first and produce the nitrite before the bacteria that use nitrite for food can grow. As mentioned before, nitrifying bacteria do not grow well below a pH of 7 and will cease to provide nitrification if system pH falls into the acidic range (Wheaton 1993). This is not a problem in fish ponds or open-flow systems, but this could be a problem in recirculation systems (RAS) if water pH is not properly balanced. Thus, if pH rises to near 8 or above, even tiny amounts of ammonia in the water will be toxic to the fish, but if pH falls below 7, the bacteria will quit and ammonia will soar to levels dangerous even at the relatively safe low pH. Carbonate buffering provided by sodium bicarbonate represents a relatively easy and affordable solution to maintain pH in the ideal range of between 7 and 8 in water recirculating systems. In addition, it is important to have in mind that excessively high ammonia and/or nitrite concentrations are not only toxic to the fish but also to nitrifying bacteria. The major inhibitory agents are un-ionised ammonia (NH_3) and nitrous acid (HNO_3). In particular, NH_3 inhibits *Nitrosomonas* spp. at 10–150 mg/L and *Nitrobacter* spp. between 0.1 and 1.0 mg/L, whereas HNO_3 inhibits both groups of bacteria at 0.22–218 mg/L. In this sense, nitrite oxidation is more sensitive to environmental stresses in biological filters than ammonia oxidation, and consequently, nitrite accumulation

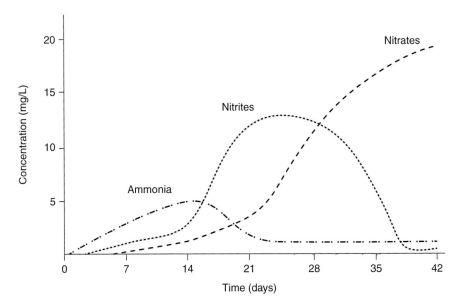

Fig. 23.2 Nitrification curve in recirculating systems after setting up a biological filter and showing the time lapse needed for nitrifying bacteria to grow out in the biological filters in order to convert ammonia into nitrites and nitrites into nitrates (modified from Wheaton 1993)

may result in systems with rapidly changing environmental conditions (Lawson 1995; Pillay and Kutty 2005).

23.2 Nitrite Toxicity

Aquatic animals are at higher risk of *nitrite intoxication* than terrestrial animals, since nitrites in the ambient water can be actively taken up across the gill epithelium by means of chloride cells towards the internal medium. In this sense, freshwater fish are hyperosmotic to their environment and require an active uptake of ions across the gills to compensate for ions lost with the urine and via passive efflux across the gills. As Jensen (2003) and Tomasso and Grossel (2005) revised, the mechanisms of ion uptake in the gills of freshwater fish are based on the fact that a H^+-ATPase situated in the apical membrane of epithelial cells extrudes protons and creates the driving force for Na^+ entry via sodium channels. The protons originate from hydration of CO to H^+ and HCO_3^-, catalysed by carbonic anhydrase inside the epithelial cells, and the formed HCO_3^- serves as counter ion for Cl^- uptake via an apical Cl^-/HCO_3^- exchange mechanism. In this way, the H^+-ATPase is thought to energise Cl^- uptake from dilute freshwater by creating a favourable gradient for the apical exit of HCO_3^-. However, this system of ion transport through gills lacks selectivity, since nitrites (NO_2^-) have affinity for the branchial Cl^- uptake mechanism, so whenever nitrites are present in the ambient water, a part of the Cl^- uptake will be shifted to NO_2^- uptake and, as a result, nitrites accumulate in the organism. Nitrite concentrations in the blood plasma may be more than 60 times higher than the concentrations in the surrounding medium (Fontenot and Isely 1999). However, some freshwater fishes (Fundulidae and Anguillidae) do not concentrate nitrites in their plasma, regardless of environmental nitrite levels and show a high degree of resistance to nitrite intoxication. This fact might be due to different rates of Cl^- uptake by the gill which may explain the observed differences in NO_2^- uptake and consequent toxicity among freshwater fishes (Tomasso and Grossel 2005).

Nitrite also accumulates in other tissues such as gills, liver, brain and muscle (Margiocco et al. 1983). In the only study conducted in Siberian sturgeon exposed to environmental nitrite, Huertas et al. (2002) found that nitrite levels in the gills and plasma of fish exposed to 130, 180 and 275 mg NO_2/L were lower than the concentration of the toxicant in the environment, which seemed to indicate that Siberian sturgeon can be able to regulate plasmatic nitrite levels and keep them lower than those in the environment. In most of the species studied so far, nitrite is actively concentrated in body fluids and some tissues like that in the liver and brain, reaching higher concentrations than those registered in the water (see reviews in Jensen 2003 and Kroupova et al. 2005).

Blood appears to be the primary target of nitrite action. From the blood plasma, nitrite diffuses into red blood cells, where it oxidises the iron in haemoglobin from a ferrous to a ferric state. The affected haemoglobin, so-called methaemoglobin (MetHb) or ferrihaemoglobin, has a brownish colour and is not capable of performing its main function, the transport of oxygen in the circulatory system (Tomasso

1994). The reaction kinetics for the oxidation of haemoglobin (with haem iron as Fe^{2+}) by nitrite into methaemoglobin (with haem iron as Fe^{3+}) is characterised by an initial lag period followed by an autocatalytic phase, and the stoichiometry for the overall reaction is

$$4Hb(Fe^{2+})O_2 + 4NO_2^- + 4H^+ \rightarrow 4Hb(Fe^{3+}) + 4NO_3^- + O_2 + 2H_2O$$

A visible symptom of high methaemoglobin levels is the brown colour of blood and gills. Fish with high percentages of the haemoglobin in the methaemoglobin form are not capable of delivering adequate oxygen levels to their tissues, and consequently they suffer from a functional hypoxia even though environmental oxygen concentrations are high and above the safe levels for the species. Methaemoglobinaemia was observed in Siberian sturgeon juveniles exposed to different nitrite levels ranging from 25 to 275 mg/L (Huertas et al. 2002), but nitrite intoxication did not affect the total haemoglobin nor haematocrit levels in Siberian sturgeon as it is shown in Fig. 23.3. However, several studies, including the above-mentioned one in Siberian sturgeon (Tomasso 1994; Doblander and Lackner 1997; Huertas et al. 2002; Kroupova et al. 2005 among others), have revealed that nitrite-induced methaemoglobinaemia and the subsequent functional anaemia may not be

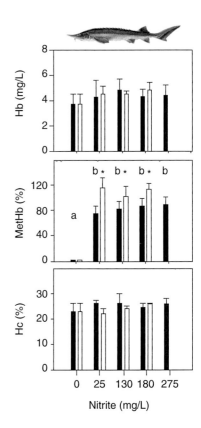

Fig. 23.3 Haemoglobin (Hb) concentration (mg/L) and percentage of methaemoglobin (MetHb) and haematocrit (Hc) of Siberian sturgeon yearlings exposed to different nitrite concentrations (modified from Huertas et al. 2002). *Black* bars correspond to torpid/moribund animals and white bars to surviving fish at the end of the 72-h nitrite exposure from different levels of nitrites. Values are expressed as mean ± standard deviation of the mean. Means not sharing a common letter are significantly different ($P < 0.05$), whereas the asterisk denotes statistically significant differences between moribund and surviving fish exposed to the same concentration of nitrites ($P < 0.05$). Data modified from Huertas et al. (2002)

the primary cause of death, since surviving specimens exposed to different levels of environmental nitrite had higher levels of MetHb than moribund animals. In particular, Huertas et al. (2002) showed that torpid/moribund and surviving fish exposed to (275 mg NO$_2$/L) had MetHb values similar to those of animals exposed to 25 mg NO$_2$/L of (66.9 ± 8.6% in torpid and 84.8 ± 7.1% in surviving fish vs. 56.0 ± 8.6% in moribund and 86.7 ± 11.6% in surviving fish). In this context, MetHb formation is countered by metHb reductase systems, primarily the NADH-dependent methaemoglobin reductase. At any given nitrite concentration a quasi-steady MetHb level is reached, reflecting a balance between Hb oxidation and reduction. However, as the load of nitrites gradually increases during nitrite exposure, MetHb concentrations are forced upwards, and levels of 70–85% of the total Hb are not uncommon at late stages of nitrite exposure, as it was found in Siberian sturgeon (Fig. 23.4). The amount of MetHb necessary to kill, to reduce the growth of or to prevent the normal behaviour of fish varies with species and with environmental conditions. As a rough rule of thumb, MetHb concentrations in excess of 50% are considered threatening to fish (Kroupova et al. 2005). Although MetHb formation has traditionally been viewed as a key physiological disturbance, mortality in fish during nitrite

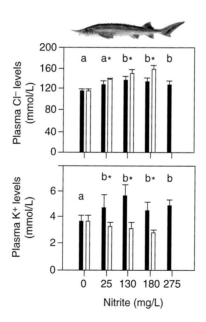

Fig. 23.4 Plasma chloride (Cl$^-$) and potassium (K$^+$) concentrations in Siberian sturgeon yearlings exposed to different nitrite concentrations. No data is shown regarding plasma Ca^{2+} (1.3 ± 0.05 mM) and Na$^+$ (119.1 ± 3.1 mM) levels, since there were no differences between torpid/moribund animals and surviving fish exposed to nitrites. *Black* bars correspond to torpid/moribund animals and white bars to surviving fish at the end of the 72-h nitrite exposure from different levels of nitrites. Values are expressed as mean ± standard deviation of the mean. Means not sharing a common letter are significantly different ($P < 0.05$), whereas the asterisk denotes statistically significant differences between moribund and surviving fish exposed to the same concentration of nitrites ($P < 0.05$). Data modified from Huertas et al. (2002)

exposure can be associated with both low and high MetHb levels, showing that other effects are important. Thus, the toxicity of nitrite results from a combination of effects rather than from any single effect in particular.

As Jensen (2003) reviewed, fish can accommodate relatively high MetHb levels at rest, but the decrease in blood O_2 content will limit the scope for activity. Thus, high MetHb levels impair swimming performance. The decline in arterial O_2 content also eventually creates problems with ensuring a sufficient O_2 delivery to the tissues in resting fish. Tissue hypoxia and anaerobic energy production becomes reflected in elevated plasma lactate concentrations and may result in histopathological changes in organs such as the liver, central metabolic organ of the organism with a predominant role in the intermediary metabolism, important functions in lipid storage and detoxification processes. Hepatocyte degeneration as a consequence of liver hypoxia, damage of lysosomal and microsomal membranes in the liver and the formation of DNA-damaging nitroso derivates have been reported to occur in the liver of nitrite-exposed fish (Jensen 2003).

Metahaemoglobinaemia is not the only physiological disorder caused by the accumulation of nitrites in the water. As Siberian sturgeon is hyperosmotic in relation to its environment, it requires an active uptake of ions across the gills in order to osmoregulate and compensate their passive lost across the gill epithelia and urine. As a consequence, osmoregulatory mechanisms are particularly active, and in particular chloride cells. Thus, in order to maintain chloride homeostasis, the active branchial Cl^- uptake needs to balance the passive branchial Cl^- efflux and the Cl^- lost via the urine; however, nitrite is actively taken up across the gills in competition with Cl^-. When this nitrogenous compound is present in the environment, an ionic imbalance resulting from a significant increase in plasma chloride (hyperchloraemia) and potassium (hyperkalaemia) levels is observed in Siberian sturgeon (Fig. 23.4). However, the exposure to environmental nitrites did not affect plasma Ca^{2+} levels (Huertas et al. 2002). According to Gisbert et al. (2004), the nitrite-induced hyperchloraemia may be due to the hypertrophy and hyperfunctionality of chloride cells (Fig. 23.5), as shown by the high percentage of hypertrophied chloride cells (86.1% in nitrite-exposed fish vs. 10% in fish not exposed to the toxicant), as well as an increase in gill Na^+-K^+ ATPase activity, which was almost doubled in fish exposed to the toxicant with regard to fish kept in clean water. In addition, a concomitant loss of K^+ and Cl^- from skeletal musculature and their subsequent accumulation in plasma could also be responsible for the high levels of plasmatic K^+ and Cl^- found in Siberian sturgeon exposed to nitrites, which seemed to be supported by the elevated levels of accumulated nitrite in the muscular tissue. The above-mentioned K^+ efflux from skeletal musculature also resulted in a severe extracellular hyperkalaemia, which has been described as a common symptom of nitrite toxicity episodes in several fish species. The rise in extracellular K^+ levels is unfavourable for the heart and other excitable tissues by causing depolarisation that could potentially lead to heart failure and nerve malfunction. Also, the decline in intracellular K^+ content is critical by its possible influence on muscular metabolism

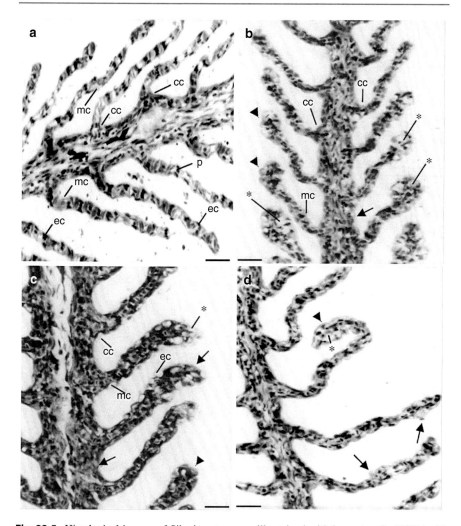

Fig. 23.5 Histological images of Siberian sturgeon gills stained with haematoxylin-VOF (**a, b**) and haematoxylin-eosin (**c, d**) dyes. (**a**) Control fish. Note the organisation of gill lamellae in epithelial (*ec*), chloride (*cc*), mucous (*mc*), and pillar cells (*pc*). (**b**) Gill of a nitrite-exposed fish after 24 h of recovery in nitrite-free conditions. Note the clubbing at the tips of secondary lamellae (*arrowhead*), the presence of aneurysms (*asterisk*), the hyperplasia of the primary lamellar epithelium (*arrow*), the hypertrophy of chloride (*cc*) and the mucous cells (*mc*). (**c**) Gill of a nitrite-exposed fish after 96 h of recovery. Note the hyperplasia of epithelia of the primary and secondary lamellae (*arrow*), the prevalence of secondary lamellar clubbing (*arrowhead*), epithelial lifting (*asterisk*) and the hypertrophy of epithelial (*ec*), chloride (*cc*) and mucous cells (*mc*). (**d**) Gill of nitrite-exposed fish after 10 days of recovery. Note the low incidence of epithelial lifting (*arrow*), aneurysms (*asterisk*) and secondary lamellar clubbing (*arrowhead*). Scale bars: 20 µm. Data modified from Gisbert et al. (2004)

and function (Jensen 2003). In addition, even though it has not been evaluated in Siberian sturgeon, the available literature indicates that nitrite toxicity may potentially affect some cardiovascular functions. In this sense, a rapid and persistent increase in the heart rate has been observed in rainbow trout exposed nitrite, which suggested a nitrite-induced vasodilatation possibly induced by the nitric oxide generated from the nitrite. In addition, the heart rate variability (i.e. variability in the time elapsed between consecutive heartbeats) also decreased in nitrite-exposed trout, and this reduced heart rate variability may be due to a critical change in the automatic control of the heart, and it seems to reflect the physiological deterioration of this vital organ (Jensen 2003).

Finally, it is also important to mention that changes in swimming behaviour may reflect how a fish is sensing and responding to its environment (Martins et al. 2012), and consequently, nitrite intoxication in Siberian sturgeon can be visually detected by fish displaying abnormal behaviour. In this sense, Siberian sturgeon yearlings (172 g in body weight) exposed to toxic levels of environmental nitrite (130–275 mg/L) showed several signs of behavioural distress characterised by increase of ventilatory activity, erratic and torpid swimming, loss of equilibrium and overturning swimming (Huertas et al. 2002; Gisbert et al. 2004). Considering the profound consequences ventilatory activity has on the homoeostasis of a fish's internal environment such as control of cellular oxygen status, oxidative stress and internal pH regulation, high ventilatory activity can be both a passive consequence of water quality and a driver promoting various physiological dysfunctions in the organism (Martins et al. 2012). In addition, the lost of equilibrium and overturning swimming style may be indicators of severe neural damage caused by nitrite intoxication. Consequently, sturgeon farmers and researchers should have a thorough knowledge of fish swimming activity under normal rearing conditions, noting any short-term changes in swimming, such as reduced or elevated swimming speed or increased manoeuvre complexity, as an acute indicator of potentially detrimental health and welfare conditions (Martins et al. 2012).

The lethal effects of nitrite toxicity in Siberian sturgeon result from a combination of disrupting effects on multiple physiological functions, including ion regulatory, respiratory, cardiovascular and excretory processes rather than from any single effect in particular. In this species, the median-lethal concentration of nitrite after 72 h of exposure (72-h LC_{50}) was 130 mg/L in water with high chloride content (130.5 mg/L) (Huertas et al. 2002). In that study, 25% of fish exposed to 25 mg NO_2/L died as a consequence of nitrite intoxication, while 66% and 10% survived the 72-h exposure to 180 and 275 mg NO_2/L, respectively. The median-lethal concentration of nitrite for Siberian sturgeon was much higher than values reported for another sturgeon species like the shortnose sturgeon *A. brevirostrum* (Fontenot et al. 1998) and other freshwater species like rainbow trout, catfish tilapia and European eel (Huertas et al. 2002; Jensen 2003). However, results from different species cannot be directly compared due to different fish size and water quality. In this sense, the levels of Cl^- are especially important in freshwater species to prevent/reduce nitrite toxicity, since several studies have revealed that nitrite is a competitive inhibitor of Cl^- uptake and vice versa, and this would explain why an increase of water Cl^- concentration protects fish from both nitrite uptake in the gills and nitrite toxicity. Indeed, adding Cl^- to the water is the single most important method to protect freshwater fish against nitrite contamination (Jensen 2003; Matsche et al. 2012). In this sense, calcium chloride is more effective than sodium chloride in reducing nitrite toxicity (Fontenot et al. 1998). In addition to water quality (e.g. pH, temperature, cation, anion and oxygen concentrations), nitrite toxicity is also dependent on the length of exposure to the toxicant, fish size and age and individual fish susceptibility (Kroupova et al. 2005). In any case and regardless of data on lethal toxicity values for this nitrogenous compound, in aquaculture systems nitrites levels are recommended to below than 1.0 mg NO_2/L (Pillay and Kutty 2005). Regarding nitrates, which are the end product of the nitrification process, these nitrogenous metabolites are lesser toxic than nitrites and do not generally represent any problem for the fish farmer. In recirculating systems, $NO3^-$ levels are controlled by daily water exchanges, but in some systems with low water flow rates, this parameter has become increasingly important, and concentration levels should be lower than 10 mg NO_3/L (Pillay and Kutty 2005). According to Hamlin (2006), the 96-h LC_{50} for Siberian sturgeon of 7–700 g in body weight ranged between 397 and 1028 mg NO_3/L nitrate-N. Surprisingly, the former author found that young Siberian sturgeon were far more tolerant to elevated nitrate levels than their adult counterparts indicating an increased susceptibility to nitrate with increasing size. In addition, different nitrate toxicity susceptibilities were reported in Siberian sturgeon, which have been attributed to different cohort variability, with certain cohorts being more sensitive to elevated nitrate than others (Hamlin 2006).

23.3 Recovery from Nitrite Exposure

Episodes of nitrite contamination are often transient in aquaculture systems, making it important to obtain information on *nitrite elimination* from body compartments and the reversibility of the above-mentioned nitrite-induced physiological effects. Siberian sturgeon exposed to toxic levels of environmental nitrite have the capacity to recover their normal physiological condition (Gisbert et al. 2004). The loss of nitrite to the environment occurs both across the gills and via the urine, and the recovery in MetHb results from the methaemoglobin reductase activity, which may be slightly up-regulated by nitrite exposure (Jensen 2003). In this sense, several authors described that when fish returned to nitrite-free water following an episode of nitrite exposure, blood MetHb levels decreased at rates that were slightly lower than the preceding increase rate. However, MetHb levels in nitrite-exposed Siberian sturgeon (130 mg NO_2/L for 18 h) decreased much more slowly during the recovery time than the preceding increase during the intoxication episode, since maximum MetHb levels were observed at 18 h post-nitrite exposure, while fish did not recover normal levels of functional haemoglobin (0% MetHb) until 96 h after the transfer of fish into nitrite-free conditions (Fig. 23.6; Gisbert et al. 2004).

The recovery of the normal balance of plasma electrolyte in Siberian sturgeon yearlings exposed to an acute nitrite intoxication episode (130 mg NO_2/L for 18 h)

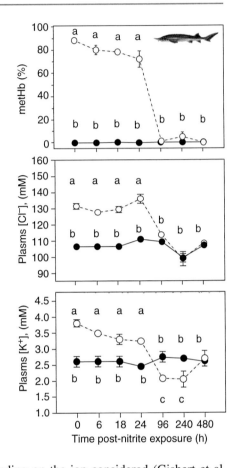

Fig. 23.6 Recovery dynamics of MetHb, plasma chloride (Cl⁻) and potassium (K⁺) levels in Siberian sturgeon yearlings exposed to 130 mg NO₂/L during 18 h. No data is shown regarding plasma Ca^{2+} (1.2 ± 0.02 mM) and Na^+ (115.4 ± 1.1 mM) levels, since there were no differences between nitrite-exposed fish and those kept in nitrite-free water during the 18 h recovery period. *White circles* correspond to nitrite-exposed fish, and *black circles* correspond to control fish. Means not sharing a common letter are significantly different ($P < 0.05$). Values are expressed as mean ± standard deviation of the mean. Data modified from Gisbert et al. (2004)

followed two different dynamics depending on the ion considered (Gisbert et al. 2004). Normal plasma Cl⁻ levels were observed between 24 and 96 h post-nitrite exposition and paralleled the decrease in nitrite levels in all measured body compartments. In the absence of environmental nitrite, the progressive recovery of the normal functionality of hypertrophied branchial chloride cells, coupled with the decrease of the efflux of Cl⁻ from the skeletal musculature, might reduce the input of this anion into the plasma. In addition, hyperchloraemia and the excess of plasma Cl⁻ accumulated during the nitrite intoxication episode might be eliminated via passive diffusion through the gill epithelium as a consequence of the large concentration gradient between plasma and water, since plasma Cl⁻ values (131.5 ± 1.6 mM) were ca. 60 times higher than those measured in the water (2.2 mM). In contrast, the dynamics of plasma K⁺ recovery from hyperkalaemia were different from those observed for chloride (Fig. 23.6). Just before the complete recovery of normal plasma K⁺ levels, which was not observed until 20 days in nitrite-free conditions, two different phases were detected with regard to plasma K⁺ values. During the first 24 h post-nitrite exposition, fish still showed a severe hyperkalaemia, while from 96 h to 10 days post-exposition, plasma K⁺ levels progressively decreased and were even lower than those measured in control fish. The higher Na^+-K^+ ATPase activities measured in the kidney of nitrite-exposed fish during the recovery period seemed to indicate that this organ played an important role in the elimination of the excess of plasmatic K^+ via the urine. However, this did not seem to be the only mechanism used by Siberian sturgeon to compensate for the K^+ unbalance. As Knudsen and Jensen (1997) reported that the skeletal musculature might recover its normal K^+ levels at the expense of the K^+ contained in the extracellular compartment, thus reducing plasma K^+ levels, whereas the overall plasma K^+ deficit that appeared to develop during this period of recovery originated; it might be rectified via dietary K^+ intake (Jensen 2003).

23.4 Concluding Remarks

Considering the potential toxicity of nitrogen waste compounds affecting fish health and welfare, water quality parameters should be monitored to serve as guide for managing aquaculture systems in order to detect potential conditions that can adversely affect the growth and physiological disturbances of fish. In cases where nitrite toxicity problems are encountered, the above-mentioned information on the effects of nitrite toxicity in Siberian sturgeon, as well as the recovery mechanisms of this species after an acute episode of nitrite intoxication, can help in the diagnosis and treatment. Thus, careful monitoring and data collection will remain useless unless it influences decisions regarding water management. This becomes more important as cost to implement various aquaculture management schemes (i.e. aeration, water exchange, inputs) increases. Most of the water quality problems, including nitrite accumulation and toxicity, can be solved with adequate water exchange. Thus, if large quantities of water suitable for aquaculture were available, monitoring would not be as critical and high production levels can be targeted, whereas if water is limited, the risk of encountering water quality and disease problems increases as one goes for more intensive culture.

References

Chen S, Ling J, Blancheton J-P (2006) Nitrification kinetics of biofilm as affected by water quality factors. Aquac Eng 34:179–197

Doblander C, Lackner R (1997) Oxidation of nitrite to nitrate in isolated erythrocytes: a possible mechanism for adaptation to environmental nitrite. Can J Fish Aquat Sci 54:157–161

Fontenot QC, Isely JJ, Tomasso J (1998) Acute toxicity of ammonia and nitrite to shortnose sturgeon fingerlings. Prog Fish Cult 60:315–318

Fontenot QC, Isely JJ, Tomasso JR (1999) Characterization and inhibition of nitrite uptake in shortnose sturgeon fingerlings. J Aquat Anim Health 11(1):76–80

Gisbert E, Rodríguez A, Cardona L (2004) Recovery of Siberian sturgeon yearlings after an acute exposure to environmental nitrite: changes in the plasmatic ionic balance, Na^+–K^+ ATPase activity, and gill histology. Aquaculture 239:141–154

Hamlin HJ (2006) Nitrate toxicity in Siberian sturgeon (*Acipenser baerii*). Aquaculture 253:688–693

Huertas M, Gisbert E, Rodríguez A et al (2002) Acute exposure of Siberian sturgeon (*Acipenser baerii*, Brandt) yearlings to nitrite: median-lethal concentration (LC_{50}) determination, haematological changes and nitrite accumulation in selected tissues. Aquat Toxicol 57:257–266

Jensen FB (2003) Nitrite disrupts multiple physiological functions in aquatic animals. Comp Biochem Physiol Part A 135:9–24

Knudsen PK, Jensen FB (1997) Recovery from nitrite-induced methaemoglobinaemia and potassium balance disturbances in carp. Fish Physiol Biochem 16:1–10

Kroupova H, Machova J, Svobodova Z (2005) Nitrite influence on fish: a review. Vet Med–Czech 50:461–471

Lawson TN (1995) Fundamentals of aquaculture engineering. Chapman & Hall, New York

Margiocco C, Arillo A, Mensi P et al (1983) Nitrite bioaccumulation in *Salmo gairdneri* Rich. And hematological consequences. Aquat Toxicol 3:261–270

Martins CIM, Galhardo L, Noble C et al (2012) Behavioural indicators of welfare in farmed fish. Fish Physiol Biochem 38:17–41

Matsche MA, Markin E, Donaldson E et al (2012) Effect of chloride on nitrite-induced methaemoglobinemia in Atlantic sturgeon, *Acipenser oxyrinchus oxyrinchus* (Mitchill). J Fish Dis 35:873–885

Pillay TVR, Kutty MN (2005) Aquaculture: principles and practices, 2nd edn. Wiley-Blackwell, Oxford, UK

Tomasso JR (1994) Toxicity of nitrogenous wastes to aquaculture animals. Rev Fish Sci 2:291–314

Tomasso JR Jr, Grossel R (2005) Physiological basis for large differences in resistance to nitrite among freshwater and freshwater-acclimated euryhaline fishes. Environ Sci Technol 39:98–102

Wheaton FW (1993) Aquaculture engineering. Krieger Publishing Company, Malabar, Florida

Part IV

Specific Methods

Cannulation in the Cultured Siberian Sturgeon, *Acipenser baerii* Brandt

24

Patrick Williot and Françoise Le Menn

Abstract

The chapter deals with a detailed description on how to equip a sturgeon with an intra-aortic cannula. In a first section, the advantages of such a procedure are recalled. Later on, a brief overview cannula in sturgeon is given. Further, the list of material and the chronology of process are provided. An overall effectiveness is pointed out as well as its safety aspect.

Keywords

Acipenser baerii • Cannulation • Methodology • Dorsal aorta

Introduction

Cannulation of animals is a needed method when aiming at administrating a long-term treatment or searching how some blood parameters are changing throughout a given period of time, usually for 24 h or some days at the very most. Working with aquatic animals, the objective is of the utmost importance as the method avoids any multiple handling possibly accompanied with deleterious and unexpected consequences on the physiology of the animal.

In the late 1980s and early 1990s, a time where sturgeon investigations were being developed in some western countries, namely, France, new methodologies were promoted, one of them being the cannulation which was achieved by the two authors of the present chapter. The objectives were to document: (a) the stress due to ammonia intoxication (Salin 1992), (b) the stress due to a hypoxia (Nonnotte et al. 1993, Maxime et al. 1995), (c) the impact of spawning procedures on both stress indicators and reproduction potential during the final steps of reproduction

P. Williot (✉)
4, Rue du pas de madame, 33980 Audenge, France
e-mail: williot.patrick@neuf.fr

F. Le Menn
26, Rue Gustave Flaubert, 33600 Pessac, France

© Springer International Publishing AG, part of Springer Nature 2018
P. Williot et al. (eds.), The Siberian Sturgeon (*Acipenser baerii*, Brandt, 1869)
Volume 1 - Biology, https://doi.org/10.1007/978-3-319-61664-3_24

465

(Williot 1997; Williot et al. 2011), (d) the effects of spawning procedures on steroid profiles during the final steps of reproduction (Williot 1997; Chap. 17) and (e) the reference content for all blood parameters (Williot 1997; Williot et al. 2011).

The present cannulation methodology was kindly transmitted to UC Davis (USA) and applied rapidly by Belanger et al. (2001) and further by Lankford et al. (2003).

In the meantime, on the one hand, a similar technique was applied to explore the effects of hypercapnia on blood of the White sturgeon (*Acipenser transmontanus*) (Crocker and Cech 1998), and on the other hand a cannulation of the dorsal aorta through the roof of the mouth was implanted in Adriatic sturgeon (*Acipenser naccarii*) (Di Marco et al. 1999).

The objective of the present chapter is to give a detailed description of the process set up by the authors allowing one to apply the methodology in further physiological investigations on sturgeon.

24.1 Needed Material

24.1.1 Fish and Holding Tanks

For fish weighing in the range of 7–9 kg (Williot et al. 2011), 2 m diameter tanks (Fig. 24.1) were used to hold one spawner per tank. In fact, the procedure can be applied to a wider range of fish weight. Water depth was maintained around maximum 50 cm to allow checking the fish and blood sampling very easily. Water supply was oxygen-saturated via an aerated-enriched column placed vertically on the periphery of the tank. The outlet is in central position and represented by a grid.

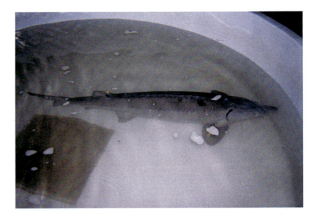

Fig. 24.1 A 2 m diameter tank for holding the fish

24.1.2 Cannulation

A V-shaped table equipped with grown moss and supplying water through a small pipe to be inserted in the mouth of the fish is needed (Fig. 24.2). A disinfectant (usually an iodine solution called betadine) is used to wash the skin area as well as the tools to be inserted (syringe and cannula).

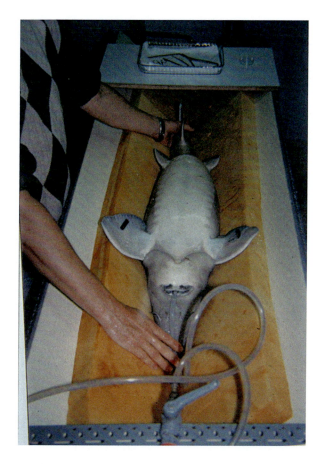

Fig. 24.2 A V-shaped table with fish with a small pipe in its mouth allows the fish breathing

Fig. 24.3 A complete Tuohy syringe

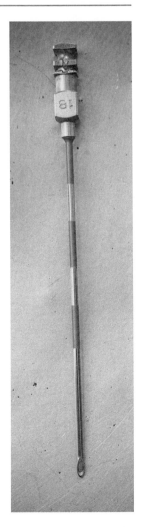

The Tuohy syringe (or trocar) (No 15 or 17) (Figs. 24.3, 24.4, 24.5 and 24.6) allows positioning the cannula into the vasculature. The cannula is inserted into the Tuohy syringe. A cannula (n°3404; internal–external diameters are 0.76–1.22 mm) is composed of a 2.5-m-long polyethylene capillary (Biotrol, France) (Fig. 24.7).

Fig. 24.4 The two parts of a Tuohy syringe

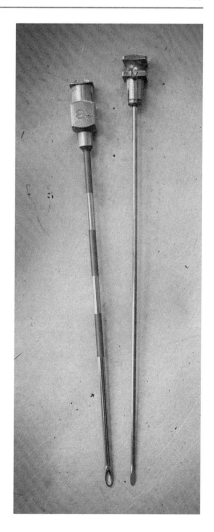

Scalpel, needle, current syringe, scissors and suture material complete the needed surgical equipment. The cannula is implanted behind the anal fin, as shown Fig. 24.8.

The needed remaining materials are (a) a physiological solution (Table 24.1), (b) a lighter to close up the external extremity of the cannula, (c) a piece of Styrofoam (as a float) (Fig. 24.9) and (d) an elastic band.

Fig. 24.5 A magnified view of the lower part of the Tuohy syringe

Fig. 24.6 Another magnified view of the lower part of the Tuohy syringe. It gives a good indication on the angle that has to be respected in inserting the Tuohy syringe into the fish

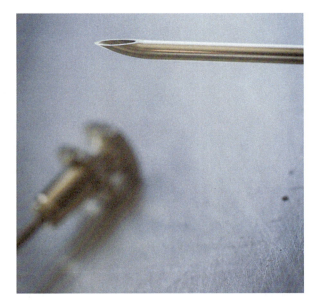

Fig. 24.7 A still packed cannula

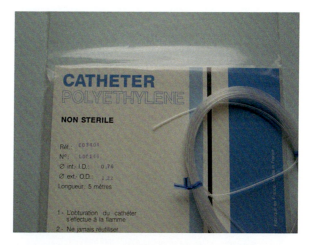

Fig. 24.8 Ventral view of a cannulated fish. *1. Upper white arrow*, coming out of the cannula; *2. lower red arrows*, three subcutaneous bridges

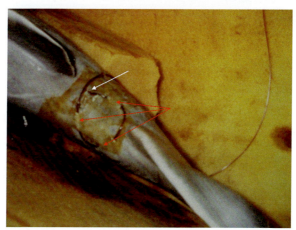

Table 24.1 Physiological solutions used to fill up the cannula

Components/characteristics	Solution used by Salin (1992)	Siberian sturgeon (SIS) medium Williot et al. (1991); Williot (1997)
NaCl	130 mmol/L	128 mmol/L
KCl	2.71 mmol/L	2.7 mmol/L
$CaCl_2$	1.94 mmol/L	1.5 mmol/L
$MgCl_2$	0.9 mmol/L	0.84 mmol/L
$NaHCO_3$	5 mmol/L	
Glucose	1 mmol/L	
Heparin-Li	200 ui/mL	
Na_2SO_4		0.7 mmol/L
Hepes		20 mmol/L
pH		7.55
NaOH (1 M)		16–20 mL
pH		8-8.1
Osmotic pressure		260–280 mosmol/L

Fig. 24.9 View of a cannulated fish with the cannula (*white arrow*) and Styrofoam float (*black arrow*) allowing easy, secure and non-stressful blood sample

24.2 Chronology of the Process

(a) Anaesthetise the fish (for about 5mn in a bath containing clove oil (emulsified in ethanol; 1:10, clove oil/ethanol) added to water at a concentration of 40 ppm) (Williot et al. 2011).
(b) Set up the fish ventral side-up on V-shaped table with a continuous supply of oxygenated water in the mouth (Fig. 24.2) according to the method described by Doroshov et al. (1983).
(c) Prepare one of the ends of the cannula with a bevelled in order to suit the end of the Tuohy syringe and smooth things over. Drive it into the Tuohy syringe.
(d) Fill up the cannula with the physiological solution.
(e) Disinfect the small area behind the genital papilla.
(f) Place the Tuohy syringe with a correct angle in the vertical symmetry plan of the fish, and then sink it very carefully into the musculature up to the blood which is seen entering the capillary when the aorta is reached.
(g) Sink the capillary for about 10 cm into the vasculature and then take off the syringe.
(h) Push the physiological solution inside the cannula until blood is no more visible in the cannula with a current syringe.
(i) Close the cannula with a flame with a lighter or with a cap.
(j) Perform the three subcutaneous bridges with the scalpel and then the needle.
(k) Thread the cannula through the three subcutaneous bridges.
(l) Make three cross-stitches in between the bridges.
(m) Fix the end of cannula onto the Styrofoam piece with an elastic band.
(n) The equipped fish is ready to be put back into its tank.

This operation lasted 20–25 min for experienced people.

24.3 Effectiveness

The present methodology of sturgeon's cannulation proved to be extremely useful, especially in studying the effects of stressors. The most delicate step is to place the cannula properly into the caudal vasculature. In the majority of the cases, the cannula was in function for up to 3 days. In one experiment, despite the care in attaching the cannula, two females out of 16 lost their cannula, i.e. 12% (Williot et al. 2011). Despite the cannulas, ovulated eggs were currently collected, and fish were again blood sampled afterwards. Further, no one fish died which means that the cannula methodology is safety.

It is worth mentioning that this procedure could be applied for longer time (i.e. 1 week) without neither haemorrhage nor infection. Finally, the method could be also applied to the dorsal vein.

References

Belanger JM, Son JH, Laugero KD et al (2001) Effects of short-term management stress and ACTH injections on plasma cortisol levels in cultured white sturgeon, *Acipenser transmontanus*. Aquaculture 203:165–176

Crocker CE, Cech JJ Jr (1998) Effects of hypercapnia on blood-gas and acid-base status in the white sturgeon, *Acipenser transmontanus*. J Comp Physiol B 168:50–60

Di Marco P, McKenzie DJ, Mandish A et al (1999) Influence of sampling conditions on blood chemistry values of Adriatic sturgeon *Acipenser naccarii* (Bonaparte, 1836). J Appl Ichthyol 15:73–77

Doroshov SI, Clark WH, Lutes PB et al (1983) Artificial propagation of the white sturgeon, *Acipenser transmontanus* Richardson. Aquaculture 32:93–104

Lankford SE, Adams TE, Cech JJ Jr (2003) Time of day and water temperature modify the physiological stress response in green sturgeon, *Acipenser medirostris*. Comp Biochem Physiol A Mol Integr Physiol 135:291–302

Maxime V, Nonnotte G, Peyraud C (1995) Circulatory and respiratory effects of a hypoxic stress in the Siberian sturgeon. Respir Physiol 100:203–212

Nonnotte G, Maxime V, Truchot JP (1993) Respiratory responses to progressive ambient hypoxia in the sturgeon, *Acipenser baerii*. Respir Physiol 91:71–82

Salin D (1992) La toxicité de l'ammoniaque chez l'esturgeon sibérien, *Acipenser baerii*: effets morphologiques, physiologiques et métaboliques d'une exposition à des doses sublétales et létales. Thèse n° 749, Université de Bordeaux 1, p 134 + annexes

Williot P, Brun R, Rouault T et al (1991) Management of female breeders of the Siberian sturgeon, *Acipenser baeri* Brandt: first results. In: Williot P (ed) ACIPENSER. Cemagref Publ, Antony, France, pp 365–379

Williot P (1997) Reproduction de l'esturgeon sibérien (*Acipenser baerii* Brandt) en élevage: gestion des génitrices, compétence à la maturation in vitro de follicules ovariens et caractéristiques plasmatiques durant l'induction de la ponte. Thèse n°1822, Université Bordeaux I, p 227

Williot P, Comte S, Le Menn F (2011) Stress indicators throughout the reproduction of farmed Siberian sturgeon, *Acipenser baerii* (Brandt) females. IAR 3:31–43

Some Basic Methods in Respiratory Physiology Studies Applied in the Siberian Sturgeon

25

Guy Nonnotte, Patrick Williot, Karine Pichavant-Rafini, Michel Rafini, and Liliane Nonnotte

Abstract

Numerous studies have shown that extracellular acid-base status in fish is greatly sensitive to small variations of temperature, PCO_2, and bicarbonate system in water. Whereas control of temperature and oxygenation of the environmental water is a usual practice in physiological studies on fish and even though water is directly in contact with the gills, the main site of acid-base and ionic regulations, the environmental acid-base status (pH, alkalinity, PCO_2) has rarely been paid much attention.

The purpose of this chapter is to recall and to describe a methodological approach to manage the water acid-base system which must be perfectly controlled to study numerous physiological regulation mechanisms in fishes.

Moreover, a particular attention was dedicated to different methods used to determine the extracellular acid-base balance in fish. At present, the use of automatic clinical blood analyzers allows to obtain either blood PO_2, PCO_2, and pH simultaneously or the total CO_2 in plasma if using a CO_2 analyzer. But the use of these devices remains problematic at a temperature lower than 37 °C. The aim of this review is also to recall that the best results are obtained using the Astrup interpolation method (1956) which was applied in the Siberian sturgeon research,

---

G. Nonnotte (✉) • L. Nonnotte
12 rue Marcel Pagnol, F-33260 La Teste de Buch, France
e-mail: guy.nonnotte@wanadoo.fr

P. Williot
4 rue du Pas de Madame, F-33980 Audenge, France

K. Pichavant-Rafini
Laboratoire ORPHY EA4324, Université de Bretagne Occidentale,
6 Avenue le Gorgeu, CS 93837, 29238 Brest Cedex 3, France

M. Rafini
Département Communication, Anglais, Sciences Humaines, Université de Bretagne Occidentale, 6 Avenue le Gorgeu, CS 93837, 29238 Brest Cedex 3, France

© Springer International Publishing AG, part of Springer Nature 2018
P. Williot et al. (eds.), *The Siberian Sturgeon (Acipenser baerii*, Brandt, 1869)
Volume 1 - Biology, https://doi.org/10.1007/978-3-319-61664-3_25

respectively, by Salin (La toxicité de l'ammoniaque chez l'esturgeon sibérien, *Acipenser baerii:* effets morphologiques, physiologiques, métaboliques d'une exposition à des doses sblétales et létales. Thèse No 749, Université Bordeaux I, p 134, 1992), Nonnotte et al. (Respir Physiol 91:71–82, 1993), and Maxime et al. (Respir Physiol 100:203–212, 1995).

Keywords
Methodology • Acid-base • Water alkalinity • pH • Bicarbonate • Carbon dioxide partial pressure • *Acipenser baerii*

Abbreviations

a_H	Ions H^+ activity
αCO_2	Carbon dioxide solubility
αO_2	Oxygen solubility
ß	Buffer value of the blood = $\Delta HCO3^-]/\Delta pH$
C_a	Concentration of the added strong acid
CO_3^{--}	Carbonate ion
γ_H	Ions H^+ activity coefficient
H_2CO_3	Carbonic acid
HCO_3^-	Bicarbonate ion
K_i	Equilibrium constant
N_a	Acid normality (0.1 N)
PCO_2	Carbon dioxide partial pressure in kPa
pH	$-\log a_H$ or $a_H = 10^{-pH}$
PO_2	Oxygen partial pressure in kPa
PwO_2	Oxygen partial pressure in water in kPa
pH_b	Blood pH
$PwCO_2$	Carbon dioxide partial pressure in water in kPa
t	Temperature in °C
TA	Alkalinity in mEq L^{-1}
V_a	Added acid volume in µL
V_{eq}	Acid volume added at the equivalent point
V_0	Water sample volume in mL or µL

Introduction

The carbonate ($CO_2/H_2CO_3/HCO_3^-/CO_3^{--}$) chemical system is a universal buffer in environmental water such as in body fluids.

The extracellular compartment of the fish is an aqueous solution which contains buffer chemical substances of which the most important are the bicarbonate and protein anions. Moreover, bicarbonates represent the major form of alkalinity in natural waters, and the ambient water must also be considered as an aqueous solution, the acid-base state of which may be modified by physical processes and/or biological exchanges.

Water alkalinity (TA) is a chemical measurement of water's ability to neutralize acids. Alkalinity is also a measure of water's buffering capacity or its ability to resist changes in pH upon the addition of acids or bases. The partitioning of CO_2 from the atmosphere and the weathering of carbonate minerals in rocks and soil represent its major sources. Other salts of weak acids, such as borate, silicates, ammonia, phosphates, and organic bases from natural organic matter, may be present in small amounts.

In both cases, the status of the CO_2–H_2CO_3–HCO_3^-–CO_3^{--} chemical system is pointed out as the environmental water and the body fluids of the fish may contain these similar buffer substances.

Measurements of Siberian sturgeon respiratory metabolism, acid-base status, and some physiological consequences of a hypoxic or ammonia exposure stress were extensively studied for the first time by Salin and Williot (1991), Salin (1992), Nonnotte et al. (1993), and Maxime et al. (1995). During these experiments, at low water PCO_2 (0.10 kPa) and temperature (15.0 ± 0.5 °C), the acid-base state of the freshwater in the experimental device could have been altered by respiratory and ionic exchanges between the fish and the water and by exchanges of carbon dioxide (CO_2) with the atmosphere. Water pH, partial pressure of carbon dioxide ($PwCO_2$), and the concentrations of $[HCO_3^-]$ and $[CO_3^{--}]$ had to be kept as constant as possible during experimentation to prevent physiological disturbances.

Although the experimental processes were briefly reported in Chaps. 18 and 20, it was essential to specify and detail the experimental methods and procedures used.

First, the aim of the present chapter is thus to present a simplified method to determine alkalinity (TA) which might be used not only in laboratory experiments but most generally in water quality control.

Secondly, the description of the extracellular acid-base balance of the fish requires knowledge of the extracellular pH, the partial pressure of carbon dioxide of the blood ($PCO_{2,b}$), the plasmatic bicarbonate (HCO_3^-), and carbonate (CO_3^{--}) concentrations which were not directly measurable. The aim of the present section is also to describe an interpolation method used to determine the acid-base status of Siberian sturgeon in varied environmental conditions.

25.1 Carbonate Buffers in Physiology

A classic buffer is a combination of a weak acid and its conjugate salt. According to the Brönsted-Lowry concept, an acid (AH) is a substance which may dissociate into a base A^- and a proton H^+. A base is a substance which may accept a proton according to the reaction

$$AH \leftrightarrow A^- + H^+ \tag{25.1}$$

The AH–A⁻ system is referred to as an acid-conjugate base system. The acid dissociates more or less according to its strength, the temperature, and the salinity of the solution. Its equilibrium constant (dissociation constant) K is defined by the eq,

$$\left[A^-\right]\left[H^+\right]/\left[AH\right]=K \tag{25.2}$$

The higher the constant K is, the stronger the acid AH is.

There exist many acid-conjugate base systems such as $H_2PO_4^-$–HPO_4^{--}, H_2CO_3–HCO_3^-, and HCO_3^-–CO_3^{--}, each with their own equilibrium constant.

These systems constitute buffers since addition of H^+ leads to the bonding up of some H^+ with A^- to form the undissociated acid AH (Eq. 25.1). In a solution of a given pH containing these systems, all the acids and bases are in equilibrium with the same concentration of $[H^+]$.

The major reactions involved in the carbonate system are

$$CO_2 + H_2O \leftrightarrow H_2CO_3 \tag{25.3}$$

$$H_2CO_3 \leftrightarrow HCO_3^- + H^+ \tag{25.4}$$

$$HCO_3^- \leftrightarrow CO_3^{--} + H^+ \tag{25.5}$$

$$\left[H^+\right] = K_1\left[H_2CO_3\right]/\left[HCO_3^-\right] = K_2\left[HCO_3^-\right]/\left[CO_3^{--}\right] \tag{25.6}$$

$$pH = pK_1 + \log\left(\left[HCO_3^-\right]/\left[H_2CO_3\right]\right) = pK_2 + \log\left(\left[CO_3^{--}\right]/\left[HCO_3^-\right]\right) \tag{25.7}$$

The respective amounts of each acid and its conjugate base depend on the value of the equilibrium constants, K_1 for the H_2CO_3–HCO_3^- system and K_2 for the HCO_3^-–CO_3^{--} system, respectively.

Since K_2 of the system HCO_3^-–CO_3^{--} is low (high pK), this buffer only comes into play at high pH, and only then does a significant amount of carbonate exist (Fig. 25.1).

25.2 Water as an Acid-Base System: A Methodological Approach to Determine the Water Alkalinity

Although numerous studies have shown that extracellular acid-base state is greatly sensitive to small variation of water PCO₂, less attention has been devoted to water acid-base balance, whereas control of temperature and water oxygenation is a common practice and easier to achieve in physiological studies on fish.

In the experiments reported by Salin (1992), Nonnotte et al. (1993), and Maxime et al. (1995), the water volume in which the animal is restrained is continuously bubbled with CO_2-free air, thereby allowing a constant PO_2, a CO_2 washout, and an increased

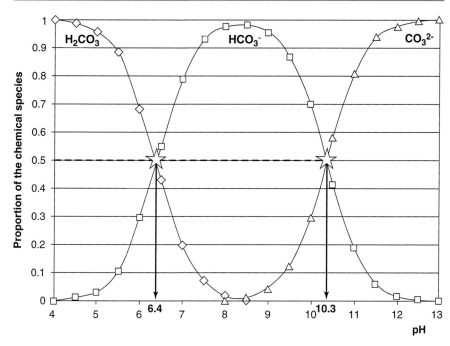

Fig. 25.1 Diagram of distribution of $CO_2/H_2CO_3/HCO_3^-/CO_3^{--}$ chemical system as a function of the pH of the water solution. From Eq. 25.6 and this diagram, $[H_2CO_3] = [HCO_3^-]$ for a pH = 6.4 and $pK_1 = 6.4$; $[HCO_3^-] = [CO_3^-]$ for a pH = 10.3 and $pK_2 = 10.3$
In water, pH < 8.2, TA was equivalent to $[HCO_3^-]$ and $[CO_3^{--}]$ is negligible. At pH < 4.5, the HCO_3^- ions disappeared

pH. As a given pH set point is reached, the signal of a pH electrode triggers the opening of a valve, injecting a small flow of CO_2 gas into the bubbling air. Thanks to this CO_2 input, the pH decreases to a set point where the CO_2 flow stops (Dejours et al. 1978). The pH set point corresponds to a given PCO_2 predetermined by equilibrium of the water against known gas mixtures and measuring pH with the same electrode.

Indeed, according to Eqs. 25.6 and 25.7, as long as the water temperature, salinity, and alkalinity (TA) are constant, a given pH must correspond to a given value of water PCO_2. So water alkalinity has to be controlled frequently and must be adjusted to a constant value.

In freshwater, since bicarbonates represent the major form of alkalinity for pH < 8.2, TA was given essentially by.

$$TA \approx \left[HCO_3^-\right] \text{in mEq L}^{-1} \qquad (25.8)$$

Culberson et al. (1970) proposed a very fast method, but in our experiments, TA water reference value was determined by a micro-modified Gran method (Culberson et al. 1970; Gran 1952; Gran et al. 1981; Truchot 1987) which is the most accurate method for determining alkalinity.

25.2.1 Principles, Measurements, and Calculations

During the titration of a basis by a strong acid, the equivalence point of the chemical reaction is the point at which chemically equivalent quantities of acid and base have been mixed. In other words, the moles of acid are equivalent to the moles of base. When neutralization (equivalent point) is reached, the initial alkalinity (TA) of the sample is equal to the added strong acid so.

$$TA \circ V_0 = V_{eq} \circ C_a \quad (25.9)$$

with V_0, volume of the sample; C_a, concentration of the added strong acid; and V_{eq}, the added acid volume at the equivalent point.

After the addition of an acid volume V_a, the quantity of remaining alkalinity at a given time (TA$_t$) in V_0 is equal to TA$_t(V_0 + V_a)$ with TA$_t <$ TA.

From (Eq. 25.9), if $V_a \cdot C_a$ is equal to the quantity of added acid, we can write.

$$TA_t(V_0 + V_a) = TA \circ V_0 - V_a \circ C_a \quad (25.10)$$

$$TA_t(V_0 + V_a) = C_a(V_{eq} - V_a) \quad (25.11)$$

In the Gran titration method (Gran 1952), there are two possibilities to determine V_{eq}:

1. Before the equivalent point ($pK_a = 6.4$; $V_a < V_{eq}$) where HCO_3^- and formed H_2CO_3 remain (Fig. 25.1).
2. After the equivalent point ($V_a > V_{eq}$; pH < 3.5) where all the carbonate system has been changed in carbonic acid. In these conditions, [H+] has not been neutralized and is proportional to V_a (Fig. 25.2).

In the proposed micromethod, the second possibility was retained.

Fig. 25.2 Linear regression of the function $a_H(V_0 + V_a) = N_a \cdot V_a \cdot \gamma_H -$ TA$\cdot V_0 \cdot \gamma_H \cdot y = a_H(V_0 + V_a)$ in µL; $x = V_a$ in µL; the slope is given by $N_a \cdot \gamma_H = 0.0827$; TA$\cdot V_0 \gamma_H$ was equal to 5.567. So TA was equal to 2.235 mEq L^{-1}. V_0, water sample volume (3000 µL); V_a, added acid volume in µL; N_a, acid normality (0.1 N); TA, alkalinity (mEq L^{-1})

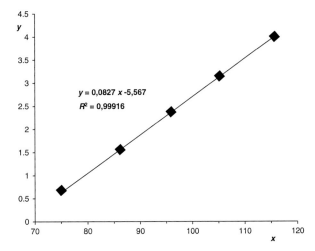

25 Some Basic Methods in Respiratory Physiology Studies

In this case the remaining alkalinity

$$TA_t \approx -\left[H^+\right] = a_h / \gamma_H \tag{25.12}$$

with a_H, ions H^+ activity and pH $= -\log a_H$ or $a_H = 10^{-pH}$, and γ_H, ions H^+ activity coefficient.

When combining Eqs. (25.10–25.12) and since C_a is equal to the acid normality, N_a (HCl being a monoacid), alkalinity, pH, sample volume, and added acid volumes are linked by the following equations:

$$a_H \left(V_0 + V_a\right) = N_a \, V_a \, \gamma_H - TA \, V_0 \tag{25.13}$$

$$10^{-pH} \left(V_0 + V_a\right) = N_a \, V_a \, \gamma_H - TA \, V_0 \, \gamma_H \tag{25.14}$$

V_0, water sample volume (3000 µL)
V_a, added acid volume in µL
N_a, acid normality (0.1 N)
TA, alkalinity (mEq L^{-1})

For $3.75 < pH < 4.00$ and $V_a > V_{eq}$, $10^{-pH} (V_0 + V_a)$ (see Eq. 25.14) represents the quantity of [H+] in the solution. Since [H+] is not neutralized, this quantity is directly proportional to V_a, and function Eq. 25.14 corresponds to a linear function $10^{-pH} (V_0 + V_a) = f$ (V_a). The slope $(N_a \cdot \gamma_H)$ of the straight line allows the determination of γ_H and $-TA \cdot V_0 \, \gamma_H$; the intercept of the best-fit regression line allows to calculate V_{eq} and TA (Fig. 25.2).

The regression coefficient of the curve is generally close to one, and the reproducibility is better than 0.1%. The function may be fit with a least squares procedure to find slope and intercept.

25.2.2 Procedure

Table 25.1 and Fig. 25.2 give an example of determination of TA in an experimental circuit.

Water sample was titrated with a volume (V_0) of HCl 0.1 N until pH reached a value lower than 4.0. At this pH, [HCO$_3^-$] is negligible since pK_1 for the H$_2$CO$_3$– HCO$_3^-$ is near 6–6.5:

1. A pH meter is used throughout the analysis to determine PH. The pH meter is standardized using buffers of 4.0–7.0
2. 3 mL water of the experimental circuit is measured, sampled with a glass syringe and added to an adapted beaker.
3. The pH probe and a small stir bar are inserted.

Table 25.1 Acid volume added $(V_a$, HCl 0.1 N) to titrate 3 mL water (V_0) $(2.89 < pH < 3.657)$

pH	3.657	3.306	3.117	2.995	2.890
V_a (µL)	75	85.5	95.8	105.2	115.7

The concentration of bicarbonate ions becomes negligible

482 G. Nonnotte et al.

4. The initial pH is determined.
5. The sample is titrated to approximatively pH 3.5 with HCl 0.1 N, and the volume of titrant (HCl) used is recorded.
6. The titration is continued in near 10 µL increments (V_a) with a microburette, and the pH and the titrant volume added after each addition are recorded. The sample is continuously mixed and pH reading is allowed to stabilize before pH recording. After 1 min equilibrium, the new pH value is noted.
7. The sample is titrated to a final pH of approximately 2.8. Several points between pH 3.5 and 2.8 must have been obtained.

25.2.3 Validity of this Micromethod

For the determination of the TA by this modified Gran method, the titration of the water sample by HCl was made after the equivalent point (see Fig. 25.1). In this case, all the carbonate system was changed in H_2CO_3.

For $3.75 < pH < 4.00$ and $V_a > V_{eq}$, $a_H (V_0 + V_a)$ (see Eq. 25.13) represents the quantity of [H+] in the solution. Since [H+] is not neutralized, this quantity is directly proportional to V_a, and function (Eq. 25.14) must be used and is represented by a straight line, the intercept giving $V_a = V_{eq}$.

If the solution is opened to the atmosphere, the sample might lose CO_2 during the titration. In our method, this problem does not exist since $2.8 < pH < 3.5$ and the V_{eq} value obtained by this method must be used.

In the experiment reported by Salin (1992), pH and TA in water samples, collected with a glass syringe twice a day in the experimental circuit, were immediately measured and appropriate corrections made. The TA value was adjusted at 1.9 mEq/L by adding appropriate amounts of HCl or NaOH. Following equilibration at $PCO_2 = 0.10$ kPa at 15 °C, the pH of a circuit water sample was measured and the pH set point of the pH-stat adjusted accordingly.

After control, the values were fixed as such: pH = 8.050; TA = 1.90 mEq L^{-1}; and $PCO_2 = 0.1$ kPa (Salin 1992).

Similar measurements can also be performed to determine the acid-base transfer between the fish and the ambient water by confinement or recirculation techniques (Heisler 1984; Truchot 1987, 1992; Gaumet et al. 1994).

25.3 Acid-Base State of Blood in the Siberian Sturgeon

In a given body fluid, there exist many acid-conjugate base systems such as H_2CO_3/HCO_3^- and HCO_3^-/CO_3^{--}.

These systems constitute buffers (see above) and must be described by the Eqs. 25.3–25.7.

A description of the acid-base state in the blood requires knowledge of at least the pH, the PCO_2, and the bicarbonate (plus carbonate if necessary) concentrations. The easiest way to calculate the bicarbonate concentration in the arterial blood is the Henderson-Hasselbalch equation.

25 Some Basic Methods in Respiratory Physiology Studies

Since pH = −log[H⁺] and according to Eqs. 25.2 and 25.4, the relation between H^+, HCO_3^-, and H_2CO_3 may be written (Eq. 25.7) as such:

$$pH = pK_1 + \log\left(\left[HCO_3^-\right]/\left[H_2CO_3\right]\right) = pK_2 + \log\left(\left[CO_3^{--}\right]/\left[HCO_3^-\right]\right)$$

Since the concentration of H_2CO_3 is directly proportional to PCO_2,

$$\left[H_2CO_3\right] = \alpha CO_2 \circ PCO_2 \left(\text{Henry's law}\right) \tag{25.15}$$

αCO_2 is the solubility coefficient of CO_2 in plasma.
The Henderson–Hasselbalch equation is thereby given by:

$$pH = pK_1 + \log\left(\left[HCO_3^-\right]/\alpha CO_2 \circ PCO_2\right) = pK_2 + \log\left(\left[CO_3^{--}\right]/\left[HCO_3^-\right]\right) \tag{25.16}$$

However, the pH measured in an erythrocyte-containing blood is the plasma pH, and this equation applied to blood concerns in fact the plasma.

For Siberian sturgeon, the values retained for pK_1 and pK_2 were, respectively, $pK_1 = 6.7811 - 0.0893$ pH and $pK_2 = 9.80$ at 15 °C, and the solubility coefficient αCO_2 in plasma was 0.0535 mmol L⁻¹ mmHg⁻¹ (from Boutilier et al. 1985).

At present, the assessment of extracellular acid-base balance and metabolic state is routinely performed with blood gas analyzers in water breathers (fish) and air breathers (mammals) during physiological experiments.

Nevertheless, it's important to consider Tables 25.2 and 25.3. These tables give the oxygen and carbon dioxide solubility in water or air and the usual values of

Table 25.2 The solubility coefficient α of O_2 and CO_2 in freshwater (FW), seawater (SW), and air as a function of temperature

| Temp. (°C) | FW | | SW35g/L | | Air |
| | αO_2 | αCO_2 | αO_2 | αCO_2 | α gaz |
	(μmol L⁻¹ mmHg⁻¹)				
5	2.52	84.17	2.03	70.79	57.68
10	2.23	70.57	1.83	60.00	56.66
15	2.01	60.23	1.67	51.58	55.68
20	1.82	51.89	1.54	44.74	54.73

Table 25.3 Extracellular acid-base variables in a water breather (Siberian sturgeon) and an air breather (rat)

	Temp. (°C)	pH	PCO₂ (kPa)	HCO₃⁻ (mmol L⁻¹)
Siberian sturgeon Salin (1992)	15	7.82	0.24	5.07
Rat Brun-Pascaud et al. (1982)	37	7.47	4.60	25.5

Temp, temperature; PCO₂ (kPa), carbon dioxide partial pressure in plasma; HCO₃⁻ (mmol L⁻¹), plasmatic bicarbonate concentration

extracellular acid–base variables in a water breather (Siberian sturgeon) and an air breather (rat).

Although the pH values (simulated at 20 °C) were close, great differences appeared in PCO_2 and $[HCO_3^-]$.

This derives from the oxygen concentration in water (0.239 mmol L^{-1}) being lower than in air (8.40 mmol L^{-1}) and the respiratory equivalent in water breathers (5.6 L of water ventilated to extract 1mmolO$_2$) being higher than in air breathers (0.9 L of air ventilated to extract 1 mmolO$_2$) (see Table 25.3). Consequently, PCO_2 is higher in rat than in sturgeon, and to maintain a close pH, the $[HCO_3^-]$ increased in rat concurrently.

Blood gas analyzers report a wide range of results, but the only parameters directly measured are the oxygen (PO_2) and carbon dioxide (PCO_2) partial pressures and the blood pH. The bicarbonate concentrations are obtained from the pH (pH electrode), PCO_2 being measured with a carbon dioxide electrode maintained at 37 °C and calculated by the Siggaard-Andersen nomogram derived from a series of in vitro experiments relating pH, PCO_2, and bicarbonates.

The carbon dioxide electrode (Severinghaus and Bradley 1958) is a modified pH electrode in contact with sodium bicarbonate solution and separated from the blood specimen by a rubber or Teflon semipermeable membrane. Carbon dioxide, but not hydrogen ions, diffuses from the blood sample across the membrane into the sodium bicarbonate solution, producing hydrogen ions and a change in pH. Hydrogen ions are produced in proportion to PCO_2 and are measured by the pH-sensitive glass electrode. As with the pH electrode, the Severinghaus electrode must be maintained at 37 °C and be calibrated with gases of known PCO_2, and the integrity of the membrane is essential. Because diffusion of CO_2 into the electrolyte solution is required, the response time is slow, i.e., 2–3 min even at 37 °C.

The solubility of all gases, the pH, and the constants K_1 and K_2 depend on the temperature, and below 37 °C, the response of the CO_2 electrode may be excessively slow, and large measurement errors may remain. This Severinghaus method seems to be inaccurate at low blood PCO_2 in water breathers (Truchot 1987) and in case of hypothermy in mammals, for example (Leon et al. 2012).

Because of the difficulty of obtaining accurate measurements directly in aquatic animals, blood PCO_2 is often calculated with the Henderson-Hasselbalch equation, from the total carbon dioxide concentration measurement and pH determinations using CO_2 analyzer (Cameron 1971). But such procedures can also be performed at temperatures (37 °C) higher than that of the animal to improve the sensitivity of the sensors, and the total CO_2 measured in this way may include not only bicarbonate but also carbonate and protein-linked carbamate (Truchot 1987).

Thus, in aquatic animals, good results are usually obtained with the Astrup (1956), an interpolation technique for acid-base measurement, based on pH, the use of the Davenport diagram to determine the base (approximately HCO_3^-) deficit or excess as an expression of metabolic acidosis or alkalosis and the arterial PCO_2 as an expression of respiratory acidosis or alkalosis. This method was based on pH measurements performed on blood aliquots equilibrated at two known PCO_2 lower and higher than the expected value at the temperature as poïkilotherm animals.

25.3.1 Determination of Blood pH (pH$_b$), [HCO$^{3-}{}_b$], and PCO$_{2,b}$ in Siberian Sturgeon

This method is based on pH measurements performed on a blood aliquot (50 µL) immediately after sampling (pH$_b$) and on two aliquots of the same blood (2 × 50 µL) equilibrated at PCO$_{2,1}$ = 0.1 kPa and PCO$_{2,2}$ = 0.5 kPa, bracketing the expected value (Fig. 25.3).

Complete equilibration of the samples at 15 °C and low PCO$_2$ values should always be carefully checked, since it may take a long time (30 min).

The blood pH$_1$ and pH$_2$ were measured after complete equilibration at PCO$_{2,1}$ and PCO$_{2,2}$. After interpolation on a pH-log PCO$_2$ diagram (Fig. 25.1), PCO$_{2,b}$ was calculated by.

$$\log PCO_{2,b} = \log PCO_{2,1} + \left(\log PCO_{2,2} - \log PCO_{2,1} \right)\left(pH_b - pH_1 \right)/\left(pH_2 - pH_1 \right)$$

[HCO$_3^-$]$_1$ and [HCO$_3^-$]$_2$ had to be determined using Henderson-Hasselbalch equation.

The sum of bicarbonates and carbonates in the blood was obtained by

$$\left[HCO_3^- + CO_3^{2-} \right]_b = \alpha CO_2 \circ PCO_{2,b} \, K_1 \left(1 + 2K_2/a_H \right)/a_H \text{ with } a_H = 10^{-pH} a_H,$$

ions H$^+$ activity and $\left[CO_2 \text{ total} \right] = \left[HCO_3^- + CO_3^{2-} \right]_b + \alpha CO_2 \circ PCO_{2,b}$.

The buffer value of the blood (ß) was given by

$$\beta = \left[HCO_3^- \right]_1 - \left[HCO_3^- \right]_2)/\left(pH_1 - pH_2 \right) \text{ and } \beta = \alpha CO_2$$
$$\left[10\left(pH_1 + \log PCO_{2,1} - pK_1 \right) - 10\left(pH_2 + \log PCO_{2,2} - pK_2 \right) \right]/\left[pH_1 - pH_2 \right]$$

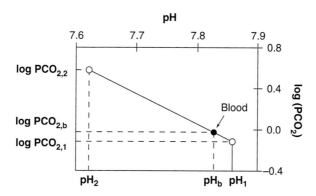

Fig. 25.3 Example of determination of blood PCO$_2$ (PCO$_{2b}$) in Siberian sturgeon by the Astrup method (PCO$_{2,1}$ = 0.1 kPa, PCO$_{2,2}$ = 0.5 kPa). pH$_b$, pH$_1$, and pH$_2$ were measured at 15 °C, using a Radiometer microelectrode G 298A calibrated by Radiometer precision buffer solution types S1500 and S1510 and connected to a Radiometer PHM72 pH meter

αCO_2 and pK_1 had to be calculated as a function of the experimental temperature (t) with the following formula given by Boutilier et al. (1985):

$\alpha CO_2 = 0.10064 - 5.4431 \times 10^{-3}\,t + 2.1776 \times 10^{-4}\,t^2 - 4.9731 \times 10^{-6}\,t^3 + 4.5288 \times 10^{-8}\,t^4$

with t in °C,
and
$pK_1 = 6.4755t - 0.0187 + (1.1704 - 0.1672\,pH) \log t + 0.1073\,pH - 0.7511$.

For Siberian sturgeon at 15 °C, Salin (1992) retained $pK_1 = 6.7811 - 0.0893\,pH$, $pK_2 = 9.80$, and $\alpha CO_2 = 0.0535$ mmol L^{-1} mmHg.

There are various ways to represent the relation between pH, HCO_3^-, and PCO_2 graphically.

Fig. 25.4 is the [HCO_3^-] vs. pH diagram introduced by Van Slyke (Peters and Van Slyke 1931) and used by Davenport (1969). To a given pH and [HCO_3^-] corresponds a given PCO_2. When blood PCO_2 is kept constant, a low pH corresponds to a low [HCO_3^-] and iso-PCO_2 lines or CO_2 isobars can be drawn for a given temperature (15 °C in these experiments) and for a given blood. On the [HCO3$^-$] vs. pH diagram, many properties of the blood and physiological phenomena may be represented (see legends Fig. 25.4). The relationship between [HCO_3^-] and pH is linear, and the slope of this strait line,

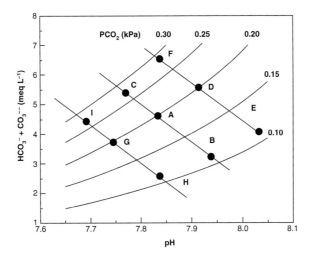

Fig. 25.4 [HCO_3^- + CO3^{--}] vs. pH diagram (Davenport 1969). A, initial experimental point; A → B, uncompensated respiratory alkalosis; A → B → H, respiratory alkalosis compensated by a metabolic acidosis; A → C, uncompensated respiratory acidosis; A → C → F, respiratory acidosis compensated by a metabolic alkalosis; A → D, uncompensated metabolic alkalosis; A → D → F, metabolic alkalosis compensated by a respiratory acidosis; A → G, uncompensated metabolic acidosis; A → G → H, metabolic acidosis compensated by a respiratory alkalosis; A → E, metabolic alkalosis and respiratory alkalosis; A → I. metabolic acidosis and respiratory acidosis. For a complete presentation of the principle and use of [HCO_3^-] vs. pH diagram, the book of Davenport (1969) should be consulted. Davenport deals with human blood, but its general considerations are valid for any blood (Dejours 1981)

β, corresponds to the buffer value of the blood, $\beta = \Delta HCO_3^-]/\Delta pH$ expressed in mmol L^{-1} pH unit^{-1}.

Each point is characterized by its pH, $[HCO_3^-]$, and PCO_2. The straight line is the buffer line which represents the blood titration curve by CO_2. During "pure" respiratory alkalosis or acidosis, the representative point of the concomitant changes of $[HCO_3^-]$ and pH moves along the buffer line. During "pure" metabolic alkalosis or acidosis, the representative points of the acid-base state move along an iso-PCO_2 line.

Many environmental factors may be involved in metabolic processes, for example, the production of lactic acid during exercise or hypoxia (see Chap. 18). But studies quantifying acid-base transfers associated with extracellular and/or intracellular acid-base changes are very complex and may thus be required to obtain a better understanding of the processes (Stewart 1978; Truchot 1987). General methods used to quantify inter-compartmental transfer of acid-base relevant substances in fish have been reviewed in detail by Heisler (1984).

Conclusions

The experimental processes described in this chapter are useful not only for scientists but also for fish farmers. For example, the efficiency of water recirculating systems in aquaculture depends on the components used in its design. Typically, each system will include units with capabilities for pH control and maintenance of appropriate alkalinity and PCO_2 consequently.

Many reviews analyzed the acid-base status of fish in various environmental conditions, but the acid-base parameters of the water were often ignored and not taken into account to discuss the results. The quality of the water (pH, alkalinity, and PCO_2) seriously influences the results and conclusions as demonstrated by Shaughnessy et al. (2015) in *Acipenser transmontanus*, Truchot et al. (1980) in *Scyliorhinus canicula*, and Claiborne and Heisler (1984) in *Cyprinus carpio*, for example (see also review Truchot 1987).

Moreover, to attain a most exact understanding of acid-base balance in Siberian sturgeon, it would often be appropriate to plot changes of pH, $[HCO_3^-]$ as functions of PCO_2, and to use a representation such as the $[HCO_3^-]$ vs. pH diagram. Such considerations are rarely encountered mainly because of the limitations of measurement techniques and the complexity of the system.

In acid-base physiological studies in sturgeon or other fishes, the methodology of blood sampling is also paramount, whatever the experimental protocols and/or the pH–PCO_2 method of determination. The cannulation of the dorsal aorta must be largely preferred (Salin 1992; Williot 1997; Williot et al. 2011; Williot and Le Menn Chap. 24) than caudal puncture sampling to prevent stress and pollution with venous blood (Shaughnessy et al. 2015).

In sturgeons, various vascular cannulations were used to study the blood parameters in Adriatic sturgeon (Randall et al. 1992; Di Marco et al. 1999), white sturgeon (McEnroe and Cech 1985; Crocker et al. 2000), green sturgeon (Belanger et al. 2001), and Siberian sturgeon *Acipenser baerii* (Salin 1992; Nonnotte et al. 1993; Maxime et al. 1995; Williot 1997, Williot et al. 2011; Williot and Le Menn Chap. 24) in response to different treatments and environmental conditions. Caution must be exercised to ensure that the method of anesthesia will not interfere with subsequent analysis. Recovery of the fish from anesthesia even with the oxygen supply maintained requires 48–72 h before normal physiological status can be observed (Heisler 1978; Claiborne and Heisler 1984). Taking this observation into account during the experimental procedure is essential for the unbiased determination of normal blood parameters.

But whatever the methods which were used to chronically implant blood catheters, repeated blood sampling was achieved with the absence of a stress response during sampling of the experimental fish which may also bias the results.

Acknowledgments We wish to gratefully acknowledge the efficient help of Dr. Nonnotte Philippe, Research Engineer Geochemistry/TI-MS at the Geosciences Department of the Brest University (France), IUEM, UMR 6538, and to Christophe Nonnotte (S/A Flight Tests Aircraft Manager BSEMD) Airbus Industry, Toulouse (France), in drawing the figures.

Notes of the Authors Respiration is the process by which animals take in oxygen necessary for cellular metabolism and release carbon dioxide that accumulates in their body as a result of the expenditure of energy. Techniques used for measuring oxygen consumption were described in Chaps. 18 and 52. This chapter concerns carbon dioxide release and acid-base status for the sturgeon *Acipenser baerii* only.

References

Astrup P (1956) A simple electrometric technique for the determination of carbon dioxide tension in blood and plasma, total content of carbon dioxide in plasma and bicarbonate content in 'separated' plasma at a fixed dioxide tension (40 mm hg). Scand J Clin Invest 8:33–43

Belanger JM, Son JH, Laugero KD, Moberg GP, Doroshov SI, Lankford SE, Cech JJ (2001) Effects of short-term management stress and ACTH injections on plasma cortisol levels in cultured white sturgeon, *Acipenser transmontanus*. Aquaculture 203:165–176

Boutilier RG, Iwama GK, Heming TA, Randall DJ (1985) The apparent pK of carbon acid in rainbow trout blood plasma between 5 and 15°C. Respir Physiol 61:237–254

Brun-Pascaud M, Gaudebout C, Blayo MC, Pocidalo JJ (1982) Arterial blood gases and acid-base status in awake rats. Respir Physiol 48:45–57

Cameron JN (1971) Rapid method for determination of total carbon dioxide in small blood samples. J Appl Physiol 31:632–634

Claiborne JB, Heisler N (1984) Acid-base regulation and ion transfers in the carp (*Cyprinus carpio*): pH compensation during and after exposure to environmental hypercapnia. J Exp Biol 108:25–43

Crocker CE, Farrell AP, Gamperl AK, Cech JJ (2000) Cardio-respiratory responses of white sturgeon to environmental hypercapnia. Am J Phys 279:R617–R628

Culberson C, Pytkowicz RM, Hawley JE (1970) Seawater alkalinity determination by the pH method. J Mar Res 28:15–21

Davenport HW (1969) The ABC of acid-base chemistry. Univ Press, Chicago, p 119

Dejours P, Armand J, Gendner JP (1978) Importance de la regulation de l'équilibre acide-base de l'eau ambiante pour l'étude des échanges respiratoires et ioniques des animaux aquatiques. C R Acad Sci Paris 287:1397–1399

Dejours P (1981) Principles of comparative respiratory physiology. Elsevier, Amsterdam, New York, Oxford, p 265

Di Marco P, McKenzie DJ, Mandich A, Bronzi P, Cataldi E, Cataudella S (1999) Influence of sampling conditions on blood chemistry values of Adriatic sturgeon *Acipenser naccarii*. J Appl Ichtyol 15:73–77

Gaumet F, Boeuf G, Truchot JP, Nonnotte G (1994) Effects of environmental water salinity on blood acid-base status in juvenile turbot (*Scophthalmus maximus* L.) Comp Biochem Physiol 109A:985–994

Gran G (1952) Determination of the equivalence point in potentiometric precipitation titrages part II. Analyst 77:661–671

Gran G, Johansson A, Johansson S (1981) Automatic titrage by stepwise addition of equal volumes of titrant in potentiometric precipitation titrages part VII. Analyst 106:1109–1118

Heisler N (1978) Bicarbonate exchange between body compartments after changes of temperature in the larger spotted dogfish (*Scyliorhinus stellaris*). Respir Physiol 33:145–160

Heisler N (1984) Acid-base regulation in fishes. In: Ws H, Randall DJ (eds) Fish physiology, vol 10. Pt A. Academic Press, London New York, pp 315–401

Leon K, Pichavant-Rafini K, Quemener E, Sébert P, Egreteau PY, Ollivier H, Carré JL, L'Her E (2012) Oxygen blood transport during experimental sepsis: effect of hypothermia. Crit Care Med 40:912–918

Maxime V, Nonnotte G, Peyraud C, Williot P, Truchot JP (1995) Circulatory and respiratory effects of an hypoxic stress in the Siberian sturgeon. Respir Physiol 100:203–212

McEnroe M, Cech JJ (1985) Osmoregulation in juvenile and adult white sturgeon, *Acipenser transmontanus*. Environ Biol Fish 14:23–30

Nonnotte G, Maxime V, Truchot JP, Williot P, Peyraud C (1993) Respiratory responses to progressive ambient hypoxia in the sturgeon, *Acipenser baerii*. Respir Physiol 91:71–82

Peters JP and Van Slyke DD (1931) Quantitative clinical chemistry. Volume 1. Interpretation. Chapter XII: Hemoglobin and oxygen. Chapter XVIII: Carbonic acid and acid-base balance. Baltimore, Md., The Williams and Wilkins Co

Randall DJ, McKenzie DJ, Abrami G, Bondiolotte GP, Natiello F, Bronzi P, Bolis L, Agradi E (1992) Effects of diet on responses to hypoxia in sturgeon (*Acipenser naccarii*). J Exp Biol 170:113–125

Salin D (1992) La toxicité de l'ammoniaque chez l'esturgeon sibérien, *Acipenser baerii*: effets morphologiques, physiologiques, métaboliques d'une exposition à des doses subtétales et létales. Thèse No 749, Université Bordeaux I, p 134

Salin D, Williot P (1991) Acute toxicity of ammonia to siberian sturgeon *Acipenser baerii*. In: Williot P (ed) Acipenser. Cemagref Publ, Anthony France, pp 153–167

Shaughnessy CA, Baker DW, Brauner CJ, Morgan JD, Bystriansky JS (2015) Interaction of osmoregulatory and acid-base compensation in white sturgeon (*Acipenser transmontanus*) during exposure to aquatic hypercarbia and elevated salinity. J Exp Biol 218:2712–2719

Severinghaus JW, Bradley AF (1958) Electrodes for blood PO_2 and PCO_2 determinations. J Appl Physiol 13:515–520

Stewart PA (1978) Independent and dependent variables of acid-base control. Respir Physiol 33:9–26

Truchot JP, Toulmond A, Dejours P (1980) Blood acid-base balance as a function of water oxygenation: a study at two different ambient CO2 levels in the dogfish, *Scyliorhinus canicula*. Respir Physiol 41:13–28

Truchot JP (1987) Comparative aspects of extracellular Acid-Base balance. Zoophysiology, vol. 20. Springer Verlag, Berlin, Heidelberg, p 262

Truchot JP (1992) Acid-base changes on transfer between sea and fresh water in the Chinese crab, *Eriocheir sinensis*. Respir Physiol 87:419–427

Williot P (1997) Reproduction de l'esturgeon sibérien (*Acipenser baerii* Brandt) en élevage: gestion des génitrices, compétence à la maturation in vitro de follicules ovariens et caractéristiques plasmatiques durant l'induction de la ponte. Thèse no 1822, Université Bordeaux I, p 227

Williot P, Comte S, Le Menn F (2011) Stress indicators throughout the reproduction of farmed Siberian sturgeon *Acipenser baerii* (Brandt) females. Inern Aquat Res 3:31–43

Conclusions of the Volume 1: Recent Progress in Biology of *Acipenser baerii*, an Exciting Subject for Further Academic Research and Development of Aquaculture

The Siberian sturgeon book included two volumes. The first one was dedicated to the biology, ecology and physiology of the Siberian sturgeon *Acipenser baerii* and would be interesting not only for all scientists but also for all those working with this species. An equivalent importance was given to synthesis of updated already published data and new results never published.

This volume presents knowledges in biology and ecology, anatomy, genetics, ontogeny, behaviour and sensory research, nutrition, biology and physiology of the reproduction and, finally, ecophysiology.

Fish represent the largest and most diverse group of vertebrates. Their evolutionary position relative to other vertebrates and their ability to adapt to a wide variety of environments make them ideal for studying the evolutions of vertebrates. A number of their characteristics make them excellent experimental models for all studies in biology.

The sturgeons are a group of fish that are members of a phylogenetically ancient group, diverging from modern teleosts approximatively 210 million years ago (Grande and Bemis 1996). They are representative of the infraclass Chondrostei, order Acipenseriformes, i.e. semi-ossified fish.

The Siberian sturgeons inhabit the northward flowing rivers of Siberia from the Ob in the west to the Kolyma River in the east. Two different forms have been identified inhabiting the Siberian rivers. One feeds in the estuaries and river deltas and spawns at considerable distances upstream; the other lives in the Lake Baikal system (Chaps. 1 and 2). The range of the Siberian sturgeon is exceptionally broad, encompassing from latitudes of 73–74°N at the mouth of the Lena and Ob Bay to 48–49°N in the Chernyi Irtysh and Selenga rivers and longitudinally over 97° (Dryagin 1948, Votinov et al. 1975). The Siberian sturgeon cannot be considered as anadromous or even as semi-anadromous for numerous reasons: (a) the location and time of their spawning, (b) their capacity to adapt to weak salinities (lower than 10‰), and (c) the entire life cycle of the Siberian sturgeon that is bound to freshwater (Chaps. 1 and 2). Thus, the migrations of Siberian sturgeon, including the Ob and Baikal populations, whose migrations are extensive, can be considered as

© Springer International Publishing AG, part of Springer Nature 2018
P. Williot et al. (eds.), The Siberian Sturgeon (*Acipenser baerii*, Brandt, 1869)
Volume 1 - Biology, https://doi.org/10.1007/978-3-319-61664-3

potamodromous (Bemis and Kynard 1997; Ruban 2005). The Siberian sturgeon exhibits a complex of specific adaptations determining the species great ecological plasticity, allowing this species to inhabit a wide and diverse habitat range.

After an approach on ecological characteristics of Siberian sturgeon, an anatomic description of the Siberian sturgeon *A. baerii* was useful (Chap. 3) for all those working and yet not very familiar with this semi-ossified fish.

A large part of this issue concerns the ontogeny in this species. Recent studies bringing new, accurate information through ontogeny of the axial skeleton in chondrostei (Chap. 4) are presented. The axial skeleton is involved in body support and is subjected to strong mechanical constraints during swimming. Therefore, a poor vertebral mineralization will favour axial anomalies to appear in the region where the vertebrae support the strongest constraints. Identifying the region of the vertebral axis that is predominantly affected by anomalies would allow to (a) understand where the strongest constraints are applied and (b) perform investigations on the vertebral structure in this particular region of healthy specimens in order to highlight adapted features resisting to these constraints. It remains to be established the context which may perturbed this phenomenon whether it is involved in the abnormalities in shape, i.e. bend fish, observed in all farmed sturgeon stock. The potential involved diet quality remains to study (Leray et al. 1985; Williot et al. 2004). In this context, the actual synthesis presented in Chap. 10 brings a valuable exercise in the field of nutritional requirements which has been poorly investigated as shown by the lack of recent research.

The swimming characteristics of Siberian sturgeon are carried in Chap. 12. *Acipenser baerii* presents a few resemblances in swimming behaviour, capacity and functional morphology of other sturgeon species and teleosts, and its swimming characteristics are closely bond up with development of locomotor organ (i.e. axial skeleton, Chap. 4) and ontogenetic behaviour (Chap. 9). However, it is distinguishable from other fish by vertical swimming, drift behaviour and benthic behaviour which appear successively.

The illustration of embryogenesis of Siberian sturgeon is reported in Chap. 8. These data may serve as reference guides for further academic studies on the early development in Siberian sturgeon and to hatchery managers.

Field data and laboratory studies on the behaviour and development of Siberian sturgeon at early life intervals have been associated (Chap. 9). The available informations are discussed under two different points of views, the biological/ecological approach and the aquaculture one.

Two different approaches, concerning the sex determination, are presented in Chap. 5 and in Chap. 6. Despite the diversity of molecular methodological approaches that have been used, no sex marker has been identified in sturgeon today.

The recent investigation on sex determination that has been made possible by DNA methylation and by genetic analysis seems strongly to suggest the fundamental question of a genetic control of the sex. That has been discussed in both chapters and could provide new informations for future research. This is a very attractive question because at present in fish farm, it seems that all cohorts present an equilibrate sex ratio (Williot and Brun 1998) which strongly suggest a genetic control of the sex. Indeed, most of sturgeons farm aim at producing caviar.

Conclusions of the Volume 1

The analysis of transposable elements (TEs) represents a somewhat recent approach dealing which the phylogeny (Chap. 7). In this paper, a first global analysis of transposable element diversity focused on the Siberian sturgeon. This is the first study. The authors show that many families of TEs are represented in the germ cell-containing gonad transcriptome of *Acipenser baerii*. These results suggest that the genome of this chondrostei contains many families of active TEs which might contribute to its evolution. The data obtained in this study demonstrate the probable high TE diversity which has been maintain in fish, coelacanth and amphibians but strongly reduced in birds and mammals (Volff et al. 2003; Chalopin et al. 2015). The sequencing of multiple sturgeon species and characterization of their TE repertoires will interest both scientists and fish farmers with the development of molecular strategy to improve and ensure the caviar production.

The structure and the function of both olfactory and gustatory systems are reviewed in Chap. 10. The recent data about morphology of olfactory organ and taste buds and their cell composition, innervation and development in sturgeon ontogeny are brought. Special emphasis to the functional characteristics of olfactory and gustatory systems and their role in sturgeon behaviour related to feeding, spawning and migration is reported.

For the first time in Siberian sturgeon, an extremely detailed description of the central nervous system (CNS) is provided with Chap. 13, in particular the chemical neuroanatomy of the hypothalamo-hypophyseal system. Through these anatomical studies of the diencephalon, the authors analysed in details the preoptic-hypothalamic-hypophyseal system of Siberian sturgeon. This system is located at the base of the brain and has a neurosecretory role exerted by hypophysiotropic neurons most of them located in the preoptic and hypothalamic periventricular region. The hypothalamus of Siberian sturgeon is characterized by the importance of a very large ventricle and the presence of a majority of cerebrospinal fluid-contacting (CSF-C) cell types that are also observed in teleosts. Most of the neuropeptides and neurohormones found in tetrapods are present in sturgeons.

Oogenesis (Chap. 14) of Siberian sturgeon *Acipenser baerii* is studied using light and electron microscopy. Five stages have been identified and correlated with physiological state of the ovarian follicle constituted by the oocyte surrounded by its cellular (theca and granulosa) and a-cellular (zona radiata) layers: Stages I and II before vitellogenesis, Stages III and IV during vitellogenesis and Stage V during maturational processes. This very great interest work is the reference in the field of sturgeon. It has been carried out with regard to reproduction, but its use is potentially also applied to a caviar production system.

The sperm and spermatozoa characteristics in the Siberian sturgeon are presented in Chap. 15. Siberian sturgeon usually produces high volume of semen with relatively low sperm and protein concentration, which is partially explained by the atypical testicular morphology (see Chap. 3) where spermatozoa are mixed with urine during passage through the kidneys to the Wolffian ducts.

The concentrations of spermatozoa, proteins, ions and, therefore, osmolality are lower than in teleosts. Moreover, the structure of spermatozoa as well as fertilization process is unique among fishes.

The main characteristics of sturgeon spermatozoa are an elongated head with an acrosome containing acrosomal proteins. The flagellum is equipped with a fin for more efficient movement. During penetration into the egg micropyle, the acrosome undergoes acrosomal reactions, which include formation of fertilization filament and opening of posterolateral projections. The fertilization filament activates the egg and causes the formation of a perivitelline space, while the posterolateral projections serve as an anchor against release from the micropyle. The acrosomal reaction has been recognized to be important for fertilization and development.

Hormonal profiles of steroid hormones (E2, T, 11-KT and 17, 20βP) are reported for the first time in Siberian sturgeon (Chap. 16). Moreover, they were determined on non-stressed females of Siberian sturgeon from before the hormonal injection up to post-ovulation time. Not surprisingly, E2 remained stable at a very low level. The other steroids showed a peak post-injection and before ovulation, which of 11-KT being later than that of T. Later on, the levels declined. Findings are discussed with regard to the literature especially regarding their biological role.

In an updated synthesis on gonadal steroids in sturgeon, especially in the Siberian sturgeon, alike in teleosts, E2 increases during the vitellogenesis (Chap. 16). Synthesis, plasmatic levels and biological activity of T, 11KT and diverse C21 and C19 are reported. The 17, 20, 21P seems to be a good candidate as mediator of gonadotropin to induce oocyte maturation. But this hypothesis remains to be verified in vivo.

The last part of this issue concerns the ecophysiology of *Acipenser baerii*.

In front of a decrease of the oxygen level in water (Chap. 18), *Acipenser baerii* is able to maintain a standard O_2 uptake during progressive hypoxia down to a critical ambient PO_2 (Pc). Below Pc, anaerobic metabolism is initiated, as indicated by (a) lactate release to the blood, (b) metabolic acidosis and (c) repayment of an O_2 debt after fish returns to normoxia. These findings are in contrast with those reported by Burggren and Randall (1978) for similarly sized specimens of another sturgeon species, *Acipenser transmontanus*. These authors found a steady decrease not only of oxygen uptake but also of gill ventilatory flow rate during progressive hypoxia and concluded that *Acipenser transmontanus* behaves as an oxyconformer. A more recent study by McKenzie et al. (2007) in the Adriatic sturgeon, *Acipenser naccarii*, proposes another interpretation. The Adriatic sturgeon behaves as an oxyconformer when exposed to progressive hypoxia under static conditions but as an oxyregulator when it was allowed to swim at a low sustained speed. In this case, the critical partial pressure of oxygen in water was evaluated to a result which is close near to the one obtained in Siberian sturgeon (Chap. 18).

The Siberian sturgeon responded to severe aquatic hypoxia by an increase in plasma concentration of catecholamines and cortisol. Moreover, the circulatory and ventilatory responses and the changes in acid-base status of arterial blood induced in sturgeon by deep hypoxia and, following return to normoxia, provided evidences to explain the strong resistance of this fish to oxygen depletion. But, there is a need for more studies which would precisely define the ventilatory and cardiovascular control systems in the Siberian sturgeon. Intensive farming and using of "recirculating aquaculture system" (RAS) required the analysis of the underlying eco-dependent physiological mechanisms at every levels of organization. Numerous pollutants such as

Conclusions of the Volume 1

ammonia, nitrites and other organic substances pose serious risks to fish. Accordingly, a great deal of previous research has characterized toxicity effects on physiological mechanisms, gill structural changes and behaviour in fish exposed to contaminants (Mallat 1985). Physiological responses of fish to pollutants are dependent on the bioavailability, uptake, accumulation of contaminants within the organism and on the interactive effects of multiple contaminants. In this regard, physiological responses are integrators of cellular and subcellular processes and may be indicative of the overall fitness of the individual organism and of the welfare of the fish.

For example, the effects of ammonia exposure on the survival capacity of the Siberian sturgeon have been described in Chaps. 19–22. Sublethal and lethal concentrations of ammonia were determined in Chap. 19. The results of tests indicate that the tolerance level of the Siberian sturgeon to ammonia is about the same as that of the carp; they are less sensitive than the trout but more sensitive than the catfish. However, it is sometimes difficult to make direct comparison, because the methods of calculating the LC50 are not the same and can, in some cases, lead to uncertain results. Moreover, the fish gill functions and the gill structural changes induced by toxicants will be considered as powerful indexes of biological effects of intoxication and detoxification processes (Chap. 20).

For the first time in this species, two important points appear that concern ammonia toxicity. The evolution of blood ammonia concentrations (Chap. 21) can be explained by NH_3 diffusion along its partial pressure if the real pH in the branchial boundary layer is taken into account. Concerning the toxicity of ammonia, results in Chap. 22 show an important role of the glutamine-glutamate-aspartate group of amino acids in tissue ammonia detoxification process and a drop of ATP in the brain. These results allow to bring out some new hypothesis on the causes of ammonia toxicity for fish, and further investigations are necessary to specify the role of the glutamate. *Acipenser baerii*, an ammonotelic fish, seems an excellent candidate to study the toxic effects of the ammonia on the pool of neurotransmitter amino acids and adenylic nucleotides glutamate not only for fish but indirectly also in human pathology (hepatic encephalopathy).

Nitrite (NO_2) is typically an intermediate product when ammonium is transformed into nitrate by microscopic organisms and is also a matter of great concern for intensive aquaculture (Hamlin et al. 2008). The high density of fish is associated with a large production of waste products, including ammonia excreted by the fish, with the potential accumulation of ammonia and nitrite to toxic levels. The tolerance limits of nitrites for the Siberian sturgeon have been extensively studied in Chap. 23. In cases where nitrite toxicity problems are encountered, the information on the effects of nitrite toxicity in Siberian sturgeon, as well as the recovery mechanisms of this species after an acute episode of nitrite intoxication, can help in the diagnosis and treatment.

Considering the potential toxicity of nitrogen waste compounds affecting fish health and welfare, not only water quality parameters should be monitored to serve as guide for managing aquaculture systems but also the optimization of the nutritional requirements and diet composition of *Acipenser baerii* (Chap. 10) to limit the toxic effects, the conditions that can adversely affect the growth and physiological disturbances that can be avoided.

Thus, if large quantities of water suitable for aquaculture were available, monitoring would not be as critical, and high production levels can be targeted, whereas if water is limited, the risk of encountering water quality and disease problems increases as one goes for more intensive culture.

The survival capacity and moreover the welfare of a living organism, i.e. the fish (Chap. 44, second volume), depends on their aptitude to fulfil both conditions: (1) firstly, if they increase their metabolic resources as much as possible to face up to natural environmental constraints and (2), secondly, minimize their routine energetic needs and those linked to the adaptation to the environment.

In these conditions, if the difference between the energetic available resources and the real need increases, all the physiological functions are optimized. The growth and welfare of the fish are ensured. On the contrary any exposure to stress situation decreases the metabolic capacity of the animal which can reach its physiological limits and die.

Sturgeon species have been object of numerous research over the past 20 years (outside the USSR). Two sturgeon species, one being euryhaline, *Acipenser transmontanus* (white sturgeon), and the other a freshwater species, *Acipenser baerii* (Siberian sturgeon), have been the most frequently studied species until 2005 (Jarić and Gessner 2011).

By these 25 chapters, *Acipenser baerii* appears an undeniably useful model for academic research not only for other sturgeon species and teleost studies but also for aquaculture development being given its economical interest for production of caviar. That will be confirmed through the 25 chapters constituting the volume 2.

Audenge, France	P. Williot
La Teste de Buch, France	G. Nonnotte
Krasnodar, Russia	M. Chebanov

References

Bemis WE, Kynard B (1997) Sturgeons rivers: an introduction to acipenceriform biogeography and life story. Environ Biol Fish 48(1/4):167–183

Burggren WW, Randall DJ (1978) Oxygen uptake and transport during hypoxic exposure in the sturgeon *Acipenser transmontanus*. Respir Physiol 34:171–183

Chalopin D, Naville M, Plard F et al (2015) Comparative analysis of transposable elements highlights mobilome diversity and evolution in vertebrates. Genome Biol Evol 7:567–580

Dryagin PA (1948) O nekotorykh morfologicheskikh i biologicheskikh otlichiyakh osetra, obitayushchego v rekakh Yakutii ot sibirskogo osetra *Acipenser baerii* Brandt (On some morphological and biological distinctions between sturgeon occurring in the rivers of Yakutiya and the Siberian sturgeon, *Acipenser baerii* Brandt). Zool Zhurnal 27:525–534

Grande L, Bemis W (1996) Interrelationships of Acipenseriformes, with comments on "Chondrostei". In: Stiassny MLJ, Parenti LR, Jonson GD (eds) Interrelationsips of fishes. Academic Press, New York, pp 85–115

Hamlin HJ, Michaels JT, Beaulaton CM, Graham WF, Dutt W, Steinbach P, Losordo TM, Schrader KK, Kl M (2008) Comparing denitrification rates and carbon sources in commercial scale upflow denitrification biological filters in aquaculture. Aquac Eng 38:79–92

Jarić I, Gessner J (2011) Analysis of publications on sturgeon research between 1996 and 2010. Scientometrics 90(2):715–735

Leray C, Nonnotte G, Roubaud P, Léger C (1985) Incidence of (n-3) essential fatty acid deficiency on trout reproduction processes. Reprod Nutr Dev 25(3):567–581

Mallat J (1985) Fish gill structural changes induced by toxicants and other irritants; a statistical review. Can J Aquat Sci 42:630–648

McKenzie DJ, Steffensen JF, Korsmeyer K, Whiteley NM, Bronzi P, Taylor EW (2007) Swimming alters response to hypoxia in the Adriatic sturgeon *Acipenser naccarii*. J Fish Biol 70:651–658

Ruban GI (2005) The Siberian sturgeon *Acipenser baerii* Brandt. Species structure and ecology. World Sturgeon Conservation Society. Special Publication No 1, Norderstedt, p 203

Volff JN, Bouneau L, Ozouf-Costaz C et al (2003) Diversity of retrotransposable elements in compact pufferfish genomes. Trends Genet 19:674–678

Votinov NP, Zlokazov VN, Kasyanov VP, Setsko RI (1975) Sostoyanie zapasov osetra v vodoemakh Sibiri i meropriyatiya po ikh uvelicheniyu (Status of sturgeon reserves in the rivers of Siberia and measures aimed to increase these reserves). Sredneuralskoe Knizhnoe lzdatelstvo, Sverdlovsk, p 94

Williot P, Brun R (1998) Ovarian development and cycles in cultured Siberian sturgeon, *Acipenser baeri*. Aquat Living Resour 11(2):111–118

Williot P, Rouault T, Rochard E, et al. (2004) French attempts to protect and restore *Acipenser sturio* in the Gironde: status and perspectives, the research point of view. In: Gessner J & Ritterhoff J (eds). Bundesamt für Naturschutz, 101, pp 83–99